Universitext

W0230351

For further volumes:
www.springer.com/series/223

Steffen Roch · Pedro A. Santos ·
Bernd Silbermann

Non-commutative Gelfand Theories

A Tool-kit for Operator Theorists and Numerical Analysts

 Springer

Steffen Roch
Technische Universität Darmstadt
Fachbereich Mathematik
Schloßgartenstraße 7
64289 Darmstadt
Germany
roch@mathematik.tu-darmstadt.de

Prof. Bernd Silbermann
Technische Universität Chemnitz
Fakultät für Mathematik
Reichenhainer Straße 39
09126 Chemnitz
Germany
bernd.silbermann@mathematik.tu-chemnitz.de

Pedro A. Santos
Universidade Técnica de Lisboa
Instituto Superior Técnico
Departamento de Matemática
Av. Rovisco Pais, 1
1049-001 Lisboa
Portugal
pedro.santos@ist.utl.pt

ISBN 978-0-85729-182-0   e-ISBN 978-0-85729-183-7
DOI 10.1007/978-0-85729-183-7
Springer London Dordrecht Heidelberg New York

British Library Cataloguing in Publication Data
A catalogue record for this book is available from the British Library

Mathematics Subject Classification (2010): 45E05, 45E10, 46H10, 46N20, 46N40, 47A53, 47B35, 65R20

Dedicated to the memory of Israel Gohberg (1928–2009).

Preface

The central notion in this book is that of a local principle. Local principles provide an abstract frame for the natural and extremely useful idea of localization, i.e. to divide a global problem into a family of local problems. The local principles the reader will encounter in this text are formulated in the language of Banach algebras and can be characterized as non-commutative Gelfand theories. They now form an integral part of the theory of Banach and C^*-algebras, and they provide an indispensable tool to study concrete problems in operator theory and numerical analysis.

More than thirty years ago, Douglas derived the Gohberg-Krupnik symbol calculus for singular integral operators via a combination of a local principle (which now bears his name) and Halmos' two projections theorem. Around the same time Kozak proved the equivalence between the stability of an operator sequence and the invertibility of a related element in a certain Banach algebra which he then studied by Simonenko's local principle. Since that time there have appeared dozens of papers where the idea of localization has been used, further developed, and applied in several contexts. As the outcome of this development, we now have a powerful, rich and beautiful theory of algebraic localization, the principles and results of which are widely scattered in the literature. The lack of a general context, and the use of different notation from paper to paper make it difficult for the researcher and the graduate student to familiarize themselves with the theory behind local principles and to make use of these results to study their own problems.

It is this defect that the present book seeks to solve. It started as a much simpler task: an updated re-edition of an out-of-print report [168], back in 1998. The changing objectives and professional obligations of the authors kept on increasing the scope and delaying the work. After more than ten years, we are finally able to present it. We think that the delay has been worth it, and the reader has a readable and useful text in his hands.

It is our intention that this work be a basic but complete introduction to local principles, formulated in the language of Banach algebras, that allows the reader to get a general view of the area and enables him to read more specialized works. Many results that appeared in periodicals or reports, and can be hard to find, are presented, streamlined and contextualized here, and the relations with other results are made clear. Some results which were in complete form available only in the Russian literature, like the local principle by Simonenko, are included. And finally, a few existing gaps in the theory are filled in with full proofs, which appear here for the first time.

The text starts with a chapter on the relevant notions for local principles of Banach algebra theory. As such, the first part can serve as a textbook for a one semester graduate course on Banach algebras with emphasis on local principles. Exercises and examples are given throughout the text. We focus on applications to singular integral operators and convolution type operators on weighted Lebesgue spaces. The choice of applications is the result both of our particular interests as researchers and of the genetic inheritance of the text, which was born as a report on algebras of convolution type operators.

Most figures in the book were produced with the help of *Mathematica*[1]. A couple of figures were produced with *Adobe Illustrator*[2]. The authors acknowledge the research center CMA and its successor CEAF (Portugal) for travel and meeting support during the writing of the book.

We would like to thank our colleagues Marko Lindner and Helena Mascarenhas for their stimulating and helpful discussions during the work on this text. We are specially grateful to Alexei Karlovich, who carefully read the manuscript and gave many valuable suggestions for its improvement.

Our sincere thanks goes to Springer and Karen Borthwick for including the book in the Universitext series and for pleasant and helpful co-operation.

Chemnitz, Darmstadt, Lisboa   *Steffen Roch*
December 2009   *Pedro A. Santos*
                                                                                   Bernd Silbermann

---

[1] Mathematica is a registered trademark of Wolfram Research, Inc.
[2] Adobe Illustrator is a trademark of Adobe, Inc.

Contents

Part II Case Studies

Introduction

This is a text on tools which can help to solve invertibility problems in Banach algebras. The number of such problems is much larger than one might guess at first glance. Of course, the most obvious invertibility problem one has in mind is the question of whether a given operator is invertible in an algebra of operators, or whether a given function is invertible in an algebra of functions. A classical example, solved by Wiener, is the question of whether the inverse of a non-vanishing function in the algebra of functions with absolutely convergent Fourier series belongs to the same algebra again.

But there are many more problems in analysis, operator theory, or numerical analysis which turn out to be equivalent to invertibility problems in suitably associated Banach algebras. For example, think of the question of whether a given bounded linear operator on a Banach space possesses the Fredholm property, i.e., whether its kernel and its cokernel are linear spaces of finite dimension. One of the equivalent characterizations of this notion states that an operator has the Fredholm property if and only if its image in the Calkin algebra is invertible. Hence, Fredholmness is indeed an invertibility problem.

For technical reasons, this invertibility problem is often studied in a suitable subalgebra of the Calkin algebra. An example which will be treated in detail in this text is the smallest closed subalgebra of the Banach algebra of all bounded linear operators on $L^2(\mathbb{T})$ which contains all singular integral operators with continuous coefficients. Here, $\mathbb{T}$ denotes the complex unit circle, and $L^2(\mathbb{T})$ is the related Lebesgue space with respect to the normalized Lebesgue measure. The Calkin image of this subalgebra is a commutative C^*-algebra, and hence subject to the Gelfand-Naimark theorem. In particular, this algebra proves to be isometrically isomorphic to the algebra $C(\mathbb{T}) \times C(\mathbb{T})$, where $C(\mathbb{T})$ is the algebra of all complex-valued continuous functions on $\mathbb{T}$. In that sense, we know all about this algebra. However, if the coefficients are merely piecewise continuous, this subalgebra of the Calkin algebra is no longer commutative, and the classical Gelfand theory fails.

A different collection of examples stems from a problem in numerical analysis. To solve an operator equation $Au = v$ numerically, one chooses a sequence of operators A_n which act on finite-dimensional spaces, and which converges strongly to A, and one replaces the equation $Au = v$ by the sequence of finite linear systems $A_n u_n = v_n$ with suitable approximations v_n of the right-hand side v. A crucial question is whether the sequence (A_n) is stable, i.e., whether the operators A_n are invertible for large n and whether the norms of their inverses are uniformly bounded.

The stability of a sequence of operators is again equivalent to an invertibility problem. For simplicity, assume that each A_n is an $n \times n$ matrix. We consider the direct product $\mathscr{E}$ of the sequence $(\mathbb{C}^{n \times n})_{n \geq 1}$ of algebras, i.e., the set of all bounded sequences $(A_n)_{n \geq 1}$ of matrices $A_n \in \mathbb{C}^{n \times n}$. Provided with pointwise defined operations and the supremum norm, this set becomes a Banach algebra. Further, we write $\mathscr{G}$ for the restricted product of the sequence $(\mathbb{C}^{n \times n})_{n \geq 1}$, i.e., for the collection of all sequences $(G_n)_{n \geq 1} \in \mathscr{E}$ such that $\|G_n\| \to 0$ as $n \to \infty$. The set $\mathscr{G}$ is a closed two-sided ideal of $\mathscr{E}$. Now a simple Neumann series argument shows that a sequence $(A_n) \in \mathscr{E}$ is indeed stable if and only if its coset $(A_n) + \mathscr{G}$ is invertible in the quotient algebra $\mathscr{E}/\mathscr{G}$. Hence, stability is also an invertibility problem.

Commutative Banach algebras are subject to the Gelfand theory, one of the most beautiful pieces of functional analysis. The essence of this theory is given by the following observation: To each unital commutative Banach algebra $\mathscr{A}$, there is associated a compact Hausdorff space $M_{\mathscr{A}}$ such that $\mathscr{A}$ can be represented (up to elements in the radical) as an algebra $\widehat{\mathscr{A}}$ of continuous functions on $M_{\mathscr{A}}$. More precisely: there is a continuous homomorphism $\widehat{} : \mathscr{A} \to \widehat{\mathscr{A}}$ (called the Gelfand transform) which has the radical of $\mathscr{A}$ as its kernel, and which owns the following property: an element a is invertible in $\mathscr{A}$ if and only if its Gelfand transform $\widehat{a}$ does not vanish on $M_{\mathscr{A}}$.

To state the latter fact in a different way note that, for each $x \in M_{\mathscr{A}}$, the point evaluation $a \mapsto \widehat{a}(x)$ defines a homomorphism from $\mathscr{A}$ onto $\mathbb{C}$, and one can show that every non-trivial homomorphism from $\mathscr{A}$ onto $\mathbb{C}$ is of this form. Thus, an element a of a unital commutative Banach algebra $\mathscr{A}$ is invertible if and only if $\varphi(a) \neq 0$ for every non-trivial homomorphism $\varphi : \mathscr{A} \to \mathbb{C}$.

For general (non-commutative) unital Banach algebras $\mathscr{A}$, non-trivial homomorphisms from $\mathscr{A}$ onto $\mathbb{C}$ need not exist. Think of the Banach algebra $\mathbb{C}^{n \times n}$ with $n > 1$, which does not possess non-trivial ideals. To derive a theory for non-commutative Banach algebras which can serve as a substitute for the classical Gelfand theory it is therefore necessary to allow for more general homomorphisms on $\mathscr{A}$ rather than homomorphisms into $\mathbb{C}$. We shall see that such generalizations of the Gelfand theory indeed exist, provided the underlying algebra is not too far from a commutative algebra. In particular, we shall consider two classes of Banach algebras which satisfy this assumption: algebras which possess a rich center, and algebras which fulfill a standard polynomial identity. The center of an algebra consists of all elements which commute with each other element of the algebra. Thus, commutative algebras are algebras which coincide with their center, and algebras with a large center can thus be considered as close to commutative algebras. On the other hand, polynomial identities serve as a substitute for the simplest polynomial identity $ab = ba$, which characterizes the commutative algebras. In that sense also, algebras with polynomial identity are close to commutative algebras.

Algebras with a large center occur in many places. So it is no wonder that concepts for their study have been worked out since the nineteen-sixties. These concepts were called local principles, because the underlying ideas resemble the method of *localization* or *freezing of coefficients* widely used in the theory of partial differential equations. Local principles can indeed be considered as non-commutative Gelfand theories in the sense that they associate to a given Banach algebra $\mathscr{A}$ with a non-trivial central subalgebra $\mathscr{C}$, a family of Banach algebras $\mathscr{A}_\tau$ with continuous homomorphisms $W_\tau : \mathscr{A} \to \mathscr{A}_\tau$ – labeled by the maximal ideals of $\mathscr{C}$ – such that an element $a \in \mathscr{A}$ is invertible in $\mathscr{A}$ if and only if its "shadow" $W_\tau(a)$ is invertible in $\mathscr{A}_\tau$ for every τ. Of course, one should expect that the invertibility of the elements $W_\tau(a)$ is easier to verify than the invertibility of a itself, in which case the local principle provides an effective tool to study invertibility.

The book starts with a concise exposition about Banach algebra theory centered around the notions of invertibility and spectrum. In Chapter 2, we study several local principles, namely the local principles by Allan-Douglas, Simonenko, and Gohberg-Krupnik. In their original form, they appeared about 40 years ago, and the relationship between them was not fully understood. The latter changed in the last years, mainly thanks to the introduction of new technical ingredients like norm-preserving localization, local inclusion theorems, and theorems of Weierstrass type.

Chapter 2 is concluded by a discussion of Krupnik's generalization of Gelfand theory to Banach algebras that fulfill a standard polynomial identity, the so-called PI-algebras. The latter proved to be extremely useful to study Banach algebras generated by idempotents (with some relations between them), which is the subject of Chapter 3. Our goal is to present this material, which until now has been spread over many publications, in a systematic way. These first three chapters form the first part of the book.

The second part of this text deals with case studies where local principles are applied to various particular Banach algebras generated by bounded operators of a special type or generated by approximation sequences of special operator classes. For instance, we shall consider algebras generated by one-dimensional singular integral operators with piecewise continuous coefficients on composed curves acting on L^p-spaces with Khvedelidze weights, and algebras generated by Wiener-Hopf and Hankel operators with piecewise continuous generating functions. The local principles will be employed to derive criteria for the Fredholm property of operators in these algebras. However we will not deal with index computation since it is not a matter of local theories but is of a global nature.

Among the concrete examples of algebras of approximation sequences we shall be concerned with in Chapter 6 are algebras of the finite sections method and of spline Galerkin methods for one- and two-dimensional singular integral and Wiener-Hopf operators. A peculiarity of these algebras is that their center is trivial in many cases. It is therefore a further goal of Chapter 6 to introduce some tools, the so-called lifting theorems, which allow one to overcome these difficulties by passing to a suitable quotient algebra which then has a nice center.

The authors have tried to make this book as easy to read as possible, giving special attention to coherence of notation throughout the book. Usually the font of a symbol will give an immediate clue to the type of mathematical object it represents. For example, sets (like those in the complex plane) are usually represented by the same font as the one used for the real line or the complex plane ($\mathbb{R}$, $\mathbb{C}$), while general curves in the complex plane are represented by Γ. Algebras and ideals are usually represented in a calligraphic font, as in $\mathscr{A}$, $\mathscr{B}$, $\mathscr{C}$, etc. Lower case letters a, b, c can either represent elements of an abstract algebra (with e the identity), or functions. In the case of functions we reserved f and g for continuous functions, u and v for elements of Lebesgue spaces, but $i - n$ we left for indexes. The imaginary unit is represented by $\mathbf{i}$. Upper case roman letters A, B, C etc, usually represent operators, whereas H and W (specified by additional parameters) are used to designate homomorphisms.

Part I
Non-commutative Gelfand Theories

Chapter 1
Banach algebras

Banach algebras provide a framework for many of the local principles. This chapter summarizes the material from Banach algebra theory which is needed in order to understand the following chapters. A reader who is acquainted with the basics of Banach algebras can certainly skip this chapter. On the other hand, we have tried to present the text in a (nearly) self-contained manner. Thus, all basic results are provided with a proof, whereas we have to refer to the literature for the proofs of some results which are mentioned only as an aside. This concerns in particular some results on C^*-algebras, which form certainly the most important subclass of Banach algebras, but which will not play a distinguished role in this text.

We have also included and systematized in the present chapter some results, for example on matrix algebras and inverse-closedness, that are hard to find in text-books.

1.1 Basic definitions

In this section, we collect some basic notions and facts concerning Banach and C^*-algebras, their ideals and homomorphisms. For more comprehensive and detailed introductions to this topic, see the references in the notes and comments section, at the end of the chapter.

1.1.1 Algebras

Let $\mathbb{F}$ be a field with zero element 0 and identity 1. In what follows, only the fields $\mathbb{Q}$ of the rational numbers, $\mathbb{R}$ of the real numbers, and $\mathbb{C}$ of the complex numbers will occur. We will also use the notation $\mathbb{K}$ to refer to one of the fields $\mathbb{R}$ and $\mathbb{C}$.

An *algebra* over the field $\mathbb{F}$ is a linear space $\mathscr{A}$ over $\mathbb{F}$ together with an additional bilinear operation

S. Roch et al., *Non-commutative Gelfand Theories*, Universitext, DOI 10.1007/978-0-85729-183-7_1, © Springer-Verlag London Limited 2011

$$\mathscr{A} \times \mathscr{A} \to \mathscr{A}, \qquad (a, b) \mapsto ab,$$

called multiplication, which satisfies the associativity law

$$(ab)c = a(bc) \qquad \text{for all } a, b, c \in \mathscr{A}.$$

We also refer to algebras over $\mathbb{R}$ and $\mathbb{C}$ as *real* and *complex algebras*.

The algebra $\mathscr{A}$ is *commutative* or *Abelian* if $ab = ba$ for all $a, b \in \mathscr{A}$. An element $e \in \mathscr{A}$ is called a *unit element* or an *identity* if $ae = ea = a$ for all $a \in \mathscr{A}$. The unit element is unique if it exists. Algebras which possess a unit element are called *unital*. In a unital algebra, we define $a^0 := e$.

A non-empty subset $\mathscr{B}$ of an algebra $\mathscr{A}$ which forms an algebra with respect to the inherited operations is called a *subalgebra* of $\mathscr{A}$. If $\mathscr{A}$ is unital and if the unit element of $\mathscr{A}$ belongs to $\mathscr{B}$, then $\mathscr{B}$ is called a *unital subalgebra* of $\mathscr{A}$. Notice that a subalgebra can be a unital algebra without being a unital subalgebra.

Example 1.1.1. The zero-dimensional linear space $\{0\}$ forms an algebra with respect to the multiplication $0 \cdot 0 = 0$. This algebra has an identity element which coincides with the zero element, and it is the only algebra with this property. When speaking henceforth on unital algebras we will always mean an algebra with at least two elements. $\qquad\square$

Example 1.1.2. The field $\mathbb{F}$ itself can be considered as a unital algebra with respect to its natural operations. $\qquad\square$

Example 1.1.3. Given an algebra $\mathscr{A}$, the set $\mathscr{A}^{n \times n}$ of all $n \times n$ matrices with entries in $\mathscr{A}$ becomes an algebra with respect to the standard matrix operations. If $\mathscr{A}$ has an identity element e, then the diagonal matrix $\text{diag}(e, e, \dots, e)$ serves as the identity element of $\mathscr{A}^{n \times n}$.

For $n > 1$, the set of all matrices $(a_{ij}) \in \mathscr{A}^{n \times n}$ with $a_{ij} = 0$ unless $i = j = 1$ forms a non-unital subalgebra of $\mathscr{A}^{n \times n}$ which also can be considered as a unital algebra with identity element $\text{diag}(e, 0, \dots, 0)$. $\qquad\square$

Example 1.1.4. Given a non-commutative algebra $\mathscr{A}$, the set

$$\text{Cen } \mathscr{A} := \{c \in \mathscr{A} : ca = ac \text{ for any } a \in \mathscr{A}\}$$

forms a subalgebra of $\mathscr{A}$. It is called the *center* of $\mathscr{A}$. $\qquad\square$

A linear subspace $\mathscr{J}$ of an algebra $\mathscr{A}$ is called a *left ideal* if

$$aj \in \mathscr{J} \quad \text{for all } a \in \mathscr{A} \text{ and } j \in \mathscr{J}.$$

Analogously, right ideals are defined. An *ideal* of an algebra is a subspace which is both a left and a right ideal. Ideals are subalgebras. The algebra $\mathscr{A}$ itself and the zero ideal $\{0\}$ are the *trivial ideals* of $\mathscr{A}$. An algebra without non-trivial ideals is said to be *simple*. An ideal $\mathscr{J}$ of $\mathscr{A}$ is called *proper* if $\mathscr{J} \neq \mathscr{A}$. If the algebra $\mathscr{A}$ is unital, then no proper left-sided ideal of $\mathscr{A}$ contains the identity element. In particular, proper ideals are never unital.

Given an algebra $\mathscr{A}$ and an ideal $\mathscr{J}$ of $\mathscr{A}$, the quotient $\mathscr{A}/\mathscr{J}$ is defined as the set of all cosets $a + \mathscr{J}$ of elements $a \in \mathscr{A}$. There is a natural linear structure on $\mathscr{A}/\mathscr{J}$ which makes the quotient into a linear space. Moreover, provided with the multiplication $(a + \mathscr{J})(b + \mathscr{J}) := ab + \mathscr{J}$, the quotient becomes an algebra, the *quotient algebra of $\mathscr{A}$ by $\mathscr{J}$*. If the algebra $\mathscr{A}$ has an identity e, then $e + \mathscr{J}$ is the identity element of $\mathscr{A}/\mathscr{J}$.

Let $\mathscr{A}$ and $\mathscr{B}$ be algebras over the same field $\mathbb{F}$. A *homomorphism* from $\mathscr{A}$ into $\mathscr{B}$ is a mapping W which reflects the algebraic structure, i.e., W is $\mathbb{F}$-linear and multiplicative. The latter means that

$$W(ab) = W(a)W(b) \quad \text{for all } a, b \in \mathscr{A}.$$

In the case that $\mathscr{A}$ and $\mathscr{B}$ are unital algebras with unit elements e and f, respectively, a homomorphism $W : \mathscr{A} \to \mathscr{B}$ is said to be *unital* if $W(e) = f$. The inverse of a bijective homomorphism is a homomorphism again. Bijective homomorphisms are called *isomorphisms*, and algebras with an isomorphism between them are called *algebraically isomorphic*. Algebraic isomorphy is an equivalence relation in the category of algebras.

There is a close connection between ideals and homomorphisms: The kernel $\operatorname{Ker} W = \{a \in \mathscr{A} : W(a) = 0\}$ of a homomorphism $W : \mathscr{A} \to \mathscr{B}$ is an ideal of $\mathscr{A}$ and, conversely, every ideal $\mathscr{J}$ of an algebra $\mathscr{A}$ is the kernel of the associated canonical homomorphism

$$\mathscr{A} \to \mathscr{A}/\mathscr{J}, \quad a \mapsto a + \mathscr{J}.$$

Let $W : \mathscr{A} \to \mathscr{B}$ be a homomorphism and $\mathscr{C}$ a subalgebra of $\mathscr{A}$. The restriction of W onto $\mathscr{C}$ is a homomorphism from $\mathscr{C}$ into $\mathscr{B}$ which we denote by $W|_{\mathscr{C}}$. Further, if $\mathscr{J}$ is an ideal of $\mathscr{A}$ which lies in the kernel of W, then

$$\mathscr{A}/\mathscr{J} \to \mathscr{B}, \quad a + \mathscr{J} \mapsto W(a)$$

defines the *quotient homomorphism* of W by $\mathscr{J}$. Thus, every homomorphism $W : \mathscr{A} \to \mathscr{B}$ factors into $W_2 \circ W_1$ where W_1 is the canonical (surjective) homomorphism from $\mathscr{A}$ onto $\mathscr{A}/\mathscr{J}$ with $\mathscr{J} = \operatorname{Ker} W$, and W_2 is the (injective) quotient homomorphism of W by $\mathscr{J}$.

Let $\mathscr{A}$ be an algebra over $\mathbb{K}$. A mapping $a \mapsto a^*$ of $\mathscr{A}$ into itself is an *involution* if, for all $a, b \in \mathscr{A}$ and all $\lambda, \mu \in \mathbb{K}$,

$$(a^*)^* = a, \quad (\lambda a + \mu b)^* = \overline{\lambda} a^* + \overline{\mu} b^* \quad \text{and} \quad (ab)^* = b^* a^*$$

where the bar stands for complex conjugation (which becomes the identity map for real algebras). The element a^* is also called the *adjoint of a*. Algebras with involution are also called *-*algebras*.

Involutions are surjective. Hence, if $\mathscr{A}$ has an identity e, then $e^* = e$. An element a of a *-algebra is called *self-adjoint* or *Hermitian* if $a^* = a$ and *normal* if $aa^* = a^*a$. A non-empty subset $\mathscr{M}$ of a *-algebra is called *symmetric* if $m \in \mathscr{M}$ implies $m^* \in \mathscr{M}$. Symmetric ideals are also called *-ideals. A homomorphism $W : \mathscr{A} \to \mathscr{B}$ between *-algebras over the same field $\mathbb{K}$ is *symmetric* or a *-*homomorphism* if $W(a^*) = W(a)^*$ for every element $a \in \mathscr{A}$. If $\mathscr{J}$ is a *-ideal of the *-algebra $\mathscr{A}$ then

$$(a + \mathscr{J})^* := a^* + \mathscr{J}$$

defines an involution on the quotient algebra $\mathscr{A} / \mathscr{J}$. This gives a correspondence between symmetric ideals and kernels of symmetric homomorphisms.

A bijective *-homomorphism is called a *-*isomorphism*, and the *-algebras $\mathscr{A}$ and $\mathscr{B}$ are called *-*isomorphic* if there is a *-isomorphism between them.

We will write $\mathscr{A} \cong \mathscr{B}$ in order to indicate that there is an algebraic (continuous, symmetric) isomorphism between the algebras (Banach algebras, C^*-algebras, respectively) $\mathscr{A}$ and $\mathscr{B}$.

1.1.2 Banach and C^*-algebras

An algebra $\mathscr{A}$ over $\mathbb{K}$ is *normed* if it is a normed linear space over $\mathbb{K}$ and if

$$\|ab\| \leq \|a\| \|b\| \quad \text{for all } a, b \in \mathscr{A}.$$

This condition implies the continuity of the multiplication:

$$\|xy - ab\| = \|xy - xb + xb - ab\| = \|x(y - b) + (x - a)b\| \leq \|x\| \|y - b\| + \|x - a\| \|b\|$$

for arbitrary elements $a, b, x, y \in \mathscr{A}$.

If $\mathscr{A}$ is unital then one requires in addition that the identity element has norm 1. If the underlying linear space of a normed algebra is a Banach space, then the algebra is called a *Banach algebra* over $\mathbb{K}$.

The natural substructures of Banach algebras are their *closed* subalgebras and their *closed* ideals, and the natural morphisms between Banach algebras are the *continuous* homomorphisms. If there is a continuous isomorphism between Banach algebras $\mathscr{A}$ and $\mathscr{B}$ over the same field $\mathbb{K}$, then $\mathscr{A}$ and $\mathscr{B}$ are called *topologically isomorphic*. By Banach's theorem ([160, Theorem III.11]), the inverse of a continuous isomorphism is continuous again. Thus, topological isomorphy is an equivalence relation in the category of Banach algebras. The algebras $\mathscr{A}$ and $\mathscr{B}$ are called *isometrically isomorphic* if there is an isometric isomorphism between them.

Closed subalgebras of Banach algebras are often defined in terms of their generators. We agree upon the following convention. Given a unital Banach algebra $\mathscr{A}$

and elements $a_1, \ldots, a_n \in \mathscr{A}$, we let $\mathrm{alg}\{a_1, \ldots, a_n\}$ stand for the smallest closed subalgebra of $\mathscr{A}$ which contains the elements $a_1, \ldots, a_n$ and the identity. The elements $a_1, \ldots, a_n$ are called the *generators* of $\mathrm{alg}\{a_1, \ldots, a_n\}$ and we say then that the algebra $\mathscr{A}$ is *generated by* the elements $a_1, \ldots, a_n \in \mathscr{A}$. The number of generators of an algebra can be infinite.

If $\mathscr{A}$ is a Banach algebra and $\mathscr{J}$ a closed ideal of $\mathscr{A}$, then the quotient algebra $\mathscr{A}/\mathscr{J}$ becomes a Banach algebra on defining the norm by $\|a + \mathscr{J}\| := \inf_{j \in \mathscr{J}} \|a + j\|$, and the canonical homomorphism from $\mathscr{A}$ onto $\mathscr{A}/\mathscr{J}$ is a contraction (i.e., its norm is not greater than one). Thus, the duality between ideals and kernels of homomorphisms discussed above implies a duality between closed ideals and kernels of continuous homomorphisms.

A Banach algebra $\mathscr{B}$ with an involution * is called a *Banach *-algebra* if the involution acts as an isometry, and it is called a *C^*-algebra* if

$$\|a^* a\| = \|a\|^2 \quad \text{for } a \in \mathscr{A}. \tag{1.1}$$

C^*-algebras are Banach *-algebras: the C^*-axiom (1.1) implies $\|a\|^2 \le \|a^*\|\,\|a\|$, whence $\|a\| \le \|a^*\|$, and changing the roles of a and a^* gives $\|a\| = \|a^*\|$.

The following basic facts indicate why it is much more convenient to work in C^*-algebras than in general Banach algebras.

Theorem 1.1.5.
  (i) *Closed ideals of C^*-algebras are symmetric.*
  (ii) *Quotients of C^*-algebras by their closed ideals are C^*-algebras.*
  (iii) **-Homomorphisms between C^*-algebras are contractions.*
  (iv) *Injective *-homomorphisms between C^*-algebras are isometries.*

Assertion (iii) is a result on *automatic continuity* since the continuity of the *-homomorphism is not a priori required.

Parts of the following isomorphism theorems are well known from general ring theory. Their important new aspects are that all the algebras are C^*-algebras again, i.e., the range Im W of W is closed, $\mathscr{K}$ is an ideal of $\mathscr{A}$, and $\mathscr{B} + \mathscr{J}$ is closed in Theorems 1.1.6, 1.1.7 and 1.1.8, respectively. Note that the order of the isomorphism theorems is not unique in the literature.

Theorem 1.1.6 (First isomorphism theorem). *Let $\mathscr{A}$ and $\mathscr{B}$ be C^*-algebras and* $\mathrm{W} : \mathscr{A} \to \mathscr{B}$ *a *-homomorphism. Then* Im W *is a C^*-subalgebra of $\mathscr{B}$, and there is a natural *-isomorphism*

$$\mathscr{A}/\mathrm{Ker}\,\mathrm{W} \cong \mathrm{Im}\,\mathrm{W}$$

which is given by the quotient homomorphism of W *by* Ker W.

Theorem 1.1.7 (Second isomorphism theorem). *Let $\mathscr{J}$ be a closed ideal of a C^*-algebra $\mathscr{A}$, and $\mathscr{K}$ a closed ideal of $\mathscr{J}$. Then $\mathscr{K}$ is a closed ideal of $\mathscr{A}$, and there is a natural *-isomorphism*

$$(\mathscr{A}/\mathscr{K})/(\mathscr{J}/\mathscr{K}) \cong \mathscr{A}/\mathscr{J}.$$

Theorem 1.1.8 (Third isomorphism theorem). *Let $\mathscr{A}$ be a C^*-algebra, $\mathscr{B}$ a C^*-sub-algebra of $\mathscr{A}$, and $\mathscr{J}$ a closed ideal of $\mathscr{A}$. Then $\mathscr{B} + \mathscr{J}$ (= the algebraic sum) is a C^*-subalgebra of $\mathscr{A}$, and there is a natural *-isomorphism*

$$(\mathscr{B} + \mathscr{J})/\mathscr{J} \cong \mathscr{B}/(\mathscr{B} \cap \mathscr{J}).$$

We mention a few examples which illustrate the richness of the class of Banach algebras. The verification of the aforementioned properties is left as an exercise. Some of these examples will be considered in detail later on.

Example 1.1.9. Let $\mathscr{A}$ be a normed algebra. Then each of the following expressions defines a norm on the matrix algebra $\mathscr{A}^{n \times n}$:

$$\|[a_{ij}]\|_l := \max_{1 \le i \le n} \sum_{j=1}^{n} |a_{ij}| \quad \text{and} \quad \|[a_{ij}]\|_c := \max_{1 \le j \le n} \sum_{i=1}^{n} |a_{ij}|$$

and these norms are equivalent. If $\mathscr{A}$ is a Banach algebra, then $\mathscr{A}^{n \times n}$ is a Banach algebra with respect to each of the given norms. Besides those already mentioned, there is a multitude of other norms on $\mathscr{A}^{n \times n}$ with the same properties (as one knows from linear algebra). In contrast to this, if $\mathscr{A}$ is a C^*-algebra, then there is a unique norm on $\mathscr{A}^{n \times n}$ which makes $\mathscr{A}^{n \times n}$ into a C^*-algebra with respect to the involution $(a_{ij})^* := (a_{ji}^*)$ (again as known from linear algebra where the spectral norm plays this role).   □

Example 1.1.10. Consider a compact Hausdorff space X (a compact subset of the plane, for example). The set of all continuous functions $f : X \to \mathbb{C}$ becomes a commutative unital complex Banach algebra with respect to pointwise defined operations and the maximum norm, that is,

$$(f+g)(s) := f(s) + g(s), \quad (\lambda f)(s) := \lambda f(s), \quad (fg)(s) := f(s)g(s),$$

and $\|f\| := \max\{|f(x)| : x \in X\}$. This algebra will be denoted by $C(X)$. If $X = [a, b]$ is a subinterval of the real line, we will sometimes write $C[a, b]$ in place of $C([a, b])$. Provided with the involution $(f^*)(x) := \overline{f(x)}$ (with the bar referring to complex conjugation), $C(X)$ becomes a C^*-algebra.

For each closed subset A of X, the set $\{f \in C(X) : f(x) = 0 \text{ for } x \in A\}$ is a closed ideal of $C(X)$.   □

Example 1.1.11. Let X be not a compact but a locally compact Hausdorff space and consider the space $L^\infty(X)$ of all essentially bounded measurable functions on X, with the norm

$$\|f\|_\infty := \operatorname*{esssup}_{x \in X} |f(x)|.$$

Then $C_0(X)$, the closure in $L^\infty(X)$ of the set of all continuous functions $f : X \to \mathbb{C}$ with compact support, becomes a non-unital commutative Banach algebra when provided again with pointwise defined operations and the supremum norm. $\square$

Example 1.1.12. Let $\mathbb{T}$ denote the complex unit circle $\{\xi \in \mathbb{C} : |\xi| = 1\}$. Consider the set $\mathscr{W}$ of all functions $f : \mathbb{T} \to \mathbb{C}$ with absolutely convergent power series representation, i.e.,

$$f(\xi) = \sum_{n \in \mathbb{Z}} f_n \xi^n \quad \text{with} \quad \sum_{n \in \mathbb{Z}} |f_n| < \infty.$$

Being absolutely convergent, the series converges uniformly. Thus, by the Weierstrass' criterion of uniform convergence, f is a continuous function. We define pointwise operations in $\mathscr{W}$ and the norm by

$$\|f\|_{\mathscr{W}} = \left\| \sum_{n \in \mathbb{Z}} f_n \xi^n \right\|_{\mathscr{W}} := \sum_{n \in \mathbb{Z}} |f_n|.$$

Then $\mathscr{W}$ becomes a unital Banach algebra, the so-called *Wiener algebra*. $\square$

Example 1.1.13. Let X be a Banach space over $\mathbb{K}$ and $\mathscr{L}(X)$ the Banach space of the bounded linear operators on X with norm

$$\|T\|_{\mathscr{L}} := \sup_{\|x\| \leq 1} \|Tx\|.$$

With multiplication defined as the composition of operators, $\mathscr{L}(X)$ becomes a unital Banach algebra over $\mathbb{K}$. This algebra is non-commutative if $\dim X > 1$. In the case $X = H$ is a Hilbert space, the involution $A \mapsto A^*$, the Hilbert space adjoint, makes $\mathscr{L}(H)$ into a C^*-algebra. $\square$

1.1.3 Unitization

Let $\mathscr{A}$ be a non-unital algebra over a field $\mathbb{F}$. There are several ways to embed $\mathscr{A}$ into a unital algebra. The most obvious one is to consider the product $\widetilde{\mathscr{A}} := \mathscr{A} \times \mathbb{F}$ which becomes a unital algebra with identity $(0, 1)$ on defining the operations

$$(a, \lambda) + (b, \mu) := (a + b, \lambda + \mu), \quad \alpha(a, \lambda) := (\alpha a, \alpha \lambda),$$

$$(a, \lambda)(b, \mu) := (ab + \mu a + \lambda b, \lambda \mu).$$

The set $\widehat{\mathscr{A}}$ of all pairs $(a, 0)$ with $a \in \mathscr{A}$ forms a subalgebra of $\widetilde{\mathscr{A}}$, and the embedding $a \mapsto (a, 0)$ establishes an isomorphism between $\mathscr{A}$ and $\widehat{\mathscr{A}}$. Moreover, $\mathscr{A} \cong \widehat{\mathscr{A}}$

is an ideal of $\widetilde{\mathscr{A}}$, and $\widetilde{\mathscr{A}}/\widehat{\mathscr{A}} \cong \mathbb{F}$. Thus, $\widetilde{\mathscr{A}}$ is the smallest unital extension of $\mathscr{A}$, its so-called *minimal unitization*.

In case $\mathscr{A}$ is a normed algebra over the field $\mathbb{K}$, there is a natural norm on $\widetilde{\mathscr{A}}$ given by

$$\|(a, \lambda)\| := \|a\| + |\lambda|. \tag{1.2}$$

With this choice, the embedding of $\mathscr{A}$ into its minimal unitization becomes an isometry, and $\mathscr{A} \cong \widehat{\mathscr{A}}$ is a closed ideal of $\widetilde{\mathscr{A}}$. Since $\widetilde{\mathscr{A}}$ and $\widehat{\mathscr{A}}$ differ by the one-dimensional linear space $\mathbb{K}$ only, the minimal unitization of a Banach algebra is a Banach algebra again. Of course, there are other norms on $\widetilde{\mathscr{A}}$ besides (1.2) which lead to the same results.

If $\mathscr{A}$ is a *-algebra, then $(a, \lambda)^* := (a^*, \bar{\lambda})$ defines an involution on $\widetilde{\mathscr{A}}$. But it turns out that (1.2) does not define a C^*-norm on $\widetilde{\mathscr{A}}$. Instead, one considers

$$\|(a, \lambda)\|_* := \sup\{\|ab + \lambda b\| : b \in \mathscr{A}, \|b\| = 1\}. \tag{1.3}$$

Theorem 1.1.14. *If $\mathscr{A}$ is a non-unital C^*-algebra, then (1.3) defines a C^*-norm on the minimal unitization of $\mathscr{A}$.*

Proof. We will check only the two main details. First we show that $\|(a, \lambda)\|_* = 0$ implies that $a = 0$ and $\lambda = 0$. If $\lambda = 0$, then $\|(a, 0)\|_* = \|a\|$, and the implication holds. If $\lambda \neq 0$, then one can arrange by scalar multiplication that $\lambda = -1$. If $\|(a, -1)\|_* = 0$, then the estimate

$$\|ab - b\| = \|ab + (-1)b\| \leq \|(a, -1)\|_*$$

implies that $ab = b$ for each $b \in \mathscr{A}$ with norm 1 and, hence, for each b. Taking adjoints one finds $ba^* = b$ for each $b \in \mathscr{A}$. In particular, $a = aa^* = a^*$ and $ab = ba = b$. Thus, a is the identity element of $\mathscr{A}$, a contradiction.

The second point we will check is the C^*-axiom for the norm (1.3). From

$$\begin{aligned}
\|(a, \lambda)\|_*^2 &= \sup\{\|ab + \lambda b\|^2 : b \in \mathscr{A}, \|b\| = 1\} \\
&= \sup\{\|b^*(a^*ab + \lambda a^*b + \bar{\lambda}ab + \bar{\lambda}\lambda b)\| : b \in \mathscr{A}, \|b\| = 1\} \\
&\leq \|(a, \lambda)^*(a, \lambda)\|_* \\
&\leq \|(a, \lambda)^*\|_* \|(a, \lambda)\|_*
\end{aligned}$$

we get $\|(a, \lambda)\|_* \leq \|(a, \lambda)^*\|_*$. Replacing (a, λ) by $(a, \lambda)^*$ gives the reverse inequality, and the estimate

$$\|(a, \lambda)\|_*^2 \leq \|(a, \lambda)^*(a, \lambda)\|_* \leq \|(a, \lambda)\|_*^2$$

finally yields the C^*-axiom. ∎

We will provide the minimal unitization of a C^*-algebra *always* with the norm (1.3), and we will omit the subscript $_*$ in what follows.

Formally, the minimal unitization construction can be applied to every algebra, being unital or not. To avoid trivialities (and for serious mathematical reasons, see the proof of the previous theorem), we will apply this construction only to non-unital algebras.

1.1.4 Matrix algebras

Here we collect some further material on matrix algebras (see Examples 1.1.3 and 1.1.9). The first result characterizes the ideals of a matrix algebra, and the second one provides necessary and sufficient conditions for an algebra to be a matrix algebra. Both results are formulated first in a purely algebraic setting and then for Banach algebras.

Proposition 1.1.15. *Let $\mathscr{A}$ be an algebra with identity, and let $\mathscr{J}$ be an ideal of $\mathscr{A}^{n \times n}$. Then there is an ideal $\mathscr{G}$ of $\mathscr{A}$ such that $\mathscr{J} = \mathscr{G}^{n \times n}$.*

Proof. Let $\mathscr{G}$ denote the set of all elements $g_{11} \in \mathscr{A}$ for which there exist elements $g_{ij} \in \mathscr{A}$ such that

$$\begin{bmatrix} g_{11} \cdots g_{1n} \\ \vdots \quad \vdots \\ g_{n1} \cdots g_{nn} \end{bmatrix} \in \mathscr{J}.$$

It is easy to see that $\mathscr{G}$ is an ideal of $\mathscr{A}$. To prove the inclusion $\mathscr{J} \subseteq \mathscr{G}^{n \times n}$, let $g = [g_{ij}]_{i,j=1}^{n} \in \mathscr{J}$. Write e_{lm} for the lm^{th} matrix unit in $\mathscr{A}^{n \times n}$, i.e., e_{lm} is the matrix which has the identity element e at its jk^{th} entry whereas all other entries are zero. Then

$$e_{1i} g e_{j1} = \begin{bmatrix} g_{ij} \, 0 \ldots 0 \\ 0 \; 0 \ldots 0 \\ \vdots \; \vdots \; \ddots \; \vdots \\ 0 \; 0 \ldots 0 \end{bmatrix}$$

whence $\mathscr{J} \subseteq \mathscr{G}^{n \times n}$. The proof of the reverse inclusion is left as an exercise. ∎

Corollary 1.1.16. *Let $\mathscr{A}$ be a Banach algebra with identity, and let $\mathscr{J}$ be a closed ideal of $\mathscr{A}^{n \times n}$. Then there is a closed ideal $\mathscr{G}$ of $\mathscr{A}$ such that $\mathscr{J} = \mathscr{G}^{n \times n}$.*

Indeed, one easily checks that the ideal $\mathscr{G}$ defined in the previous proof is closed if $\mathscr{J}$ is closed.

Theorem 1.1.17. *Let $\mathscr{A}$ be a unital algebra over a field $\mathbb{F}$. If $\mathscr{A}$ contains a unital subalgebra $\mathscr{A}_0$ which is isomorphic to $\mathbb{F}^{n \times n}$, then there exists a unital subalgebra $\mathscr{D}$ of $\mathscr{A}$ such that $\mathscr{A}$ is isomorphic to $\mathscr{D}^{n \times n}$. Moreover, the centers of $\mathscr{A}$ and $\mathscr{D}$ coincide.*

Proof. Let $E_{jk} \in \mathbb{F}^{n \times n}$ denote the matrix the jk^{th} entry of which is $1 \in \mathbb{F}$ whereas all other entries are $0 \in \mathbb{F}$. Let μ stand for the isomorphism between $\mathscr{A}_0$ and $\mathbb{F}^{n \times n}$, and let $e_{jk} \in \mathscr{A}_0$ be the uniquely determined elements such that $\mu(e_{jk}) = E_{jk}$. Note that $e_{jk}e_{im} = \delta_{ki}e_{jm}$. For $a \in \mathscr{A}$ and $j, k = 1, \ldots, n$, set

$$w_{jk}(a) := \sum_{i=1}^{n} e_{ij}ae_{ki}$$

and

$$w(a) := [w_{jk}(a)]_{j,k=1}^{n} \in \mathscr{A}^{n \times n}.$$

We will prove that w is the desired isomorphism. Clearly w is linear, and it is easy to check that $w(a)w(b) = w(ab)$. Thus, w is an algebra homomorphism. Further, the identity

$$\sum_{m}\sum_{l} e_{m1}w_{ml}(a)e_{1l} = \sum_{m}\sum_{l}\sum_{i} e_{m1}e_{im}ae_{li}e_{1l} = \sum_{m}\sum_{l} e_{mm}ze_{ll} = a \qquad (1.4)$$

shows that the kernel of w is trivial. Let $\mathscr{D} := w_{11}(\mathscr{A})$. The set $\mathscr{D}$ is a linear space, and since

$$w_{11}(a)w_{11}(b) = \sum_{i}\sum_{j} e_{i1}ae_{1i}e_{j1}be_{1j} = \sum_{i} e_{i1}ae_{11}be_{1i} = w_{11}(ae_{11}b)$$

is also an algebra. We will show next that $w_{jk}(\mathscr{A}) = \mathscr{D}$ for all $1 \leq j, k \leq n$. Define elements $p_{j'j} := \sum_{i=1}^{n} e_{l(i,j'),l(i,j)}$ where

$$l(i, j) = \begin{cases} j' & \text{if} \quad i = j, \\ j & \text{if} \quad i = j', \\ i & \text{if} \quad i \neq j, j'. \end{cases}$$

Note that $p_{j'j}^2 = e$. Now let $a \in \mathscr{A}$ and $j, k \in \{1, \ldots, n\}$. Then

$$w_{jk}(a) = \sum_{i=1}^{n} e_{ij}ae_{ki} = \sum_{i=1}^{n} e_{ij}p_{j1}^2 ap_{1k}^2 e_{ki} = \sum_{i=1}^{n} e_{i1}a'e_{1i} = w_{11}(a')$$

with $a' := p_{j1}ap_{1k}$. Thus, w maps $\mathscr{A}$ into $\mathscr{D}^{n \times n}$. To see that w maps $\mathscr{A}$ onto $\mathscr{D}^{n \times n}$, let $[f_{jk}]_{j,k=1}^{n} \in \mathscr{D}^{n \times n}$ and put $f := \sum_{i}\sum_{m} e_{im}f_{im}$. Then

$$w_{jk}(f) = \sum_{l}\sum_{i}\sum_{m} e_{lj}e_{im}f_{im}e_{kl}$$

$$= \sum_{l}\sum_{i}\sum_{m} f_{im}e_{lj}e_{im}e_{kl} \qquad (\text{since } f_{im} \in \mathscr{D})$$

$$= \sum_{l} f_{jk}e_{ll} = f_{jk}.$$

Hence, w is an algebra isomorphism between $\mathscr{A}$ and $\mathscr{D}^{n \times n}$.

It remains to show that the centers of $\mathscr{A}$ and $\mathscr{D}$ coincide. First note that $de_{jk} = e_{jk}d$ for all $d \in \mathscr{D}$ and $j, k \in \{1, \ldots, n\}$. For $c \in \text{Cen } \mathscr{A}$, we have

$$w_{11}(c) = \sum_{i=1}^{n} e_{i1} c e_{1i} = \sum_{i=1}^{n} e_{ii} c = c.$$

Hence, $c \in \mathscr{D}$ and, consequently, $c \in \mathrm{Cen}\,\mathscr{D}$. Conversely, let $c \in \mathrm{Cen}\,\mathscr{D}$. Then $\mathrm{diag}(c,\ldots,c)$ is in the center of $\mathscr{D}^{n \times n}$, whence $w^{-1}(\mathrm{diag}(c,\ldots,c)) \in \mathrm{Cen}\,\mathscr{A}$. But

$$w^{-1}(\mathrm{diag}(c,\ldots,c)) = \sum_{i}\sum_{m} e_{im} c_{im} = \sum_{i} e_{ii} c = c,$$

so we conclude that $c \in \mathrm{Cen}\,\mathscr{A}$, and the proof is finished.   ∎

Now let $\mathscr{A}$ be a Banach algebra and endow $\mathscr{D}^{n \times n}$ with a matrix norm, say

$$\|[d_{jk}]\| := \max_{j} \sum_{k} \|d_{jk}\|. \tag{1.5}$$

Corollary 1.1.18. *Let $\mathscr{A}$ be a unital Banach algebra over $\mathbb{C}$, which contains a unital subalgebra $\mathscr{A}_0$ isomorphic to $\mathbb{C}^{n \times n}$. Then the algebra $\mathscr{D}$ defined by the previous theorem is closed, and w is a continuous isomorphism between $\mathscr{A}$ and $\mathscr{D}^{n \times n}$.*

Proof. Put $c = \max_{j,k} \|e_{jk}\|$. Then

$$\|w(a)\| = \max_{j} \sum_{k} \|w_{jk}(a)\| \le \max_{j} \sum_{k}\sum_{i} c^2 \|a\| = n^2 c^2 \|a\|,$$

and

$$\|a\| = \left\| \sum_{j}\sum_{k} e_{j1} w_{jk}(a) e_{1k} \right\| \le nc^2 \|w(a)\|$$

by (1.4).   ∎

1.1.5 A flip elimination technique

We conclude this introductory section by an elementary flip elimination technique which is useful for studying particular algebras and will be important, for instance, in Section 5.7.

An element p of an algebra is called an *idempotent* if $p^2 = p$. If $\mathscr{A}$ is an algebra with identity e, an element $j \in \mathscr{A}$ is called a *flip* if $j^2 = e$. We will see that certain algebras which contain a flip can be identified with 2×2 matrix algebras without flip. Again we start with a purely algebraic setting.

Let $\mathscr{A}$ be an algebra with identity e which contains an idempotent p and a flip j such that $jpj = e - p$. The mapping $M : \mathscr{A} \to \mathscr{A}^{2 \times 2}$,

$$a \mapsto \begin{bmatrix} pap & pa(e-p) \\ (e-p)ap & (e-p)a(e-p) \end{bmatrix}$$

and the associated mapping $L : \mathscr{A} \to \mathscr{A}^{2\times2}$,

$$L(a) := \begin{bmatrix} e & 0 \\ 0 & j \end{bmatrix} M(a) \begin{bmatrix} e & 0 \\ 0 & j \end{bmatrix} = \begin{bmatrix} pap & pajp \\ pjap & pjajp \end{bmatrix}.$$

are injective homomorphisms. Now assume that $\mathscr{A}$ is generated by the flip j and by (the elements of) an algebra $\mathscr{B}$ with p in its center and with $j\mathscr{B}j \subseteq \mathscr{B}$. Then every element $a \in \mathscr{A}$ can be written as a sum $a_1 + a_2 j$ with $a_1, a_2 \in \mathscr{B}$, and this decomposition is unique. Indeed, let a_1, a_2 be elements of $\mathscr{B}$ such that $a_1 + a_2 j = 0$. Multiplying this equality from both sides by p and taking into account that p belongs to the center of $\mathscr{B}$, we find that $pa_1 p = 0$. Similarly, $(e-p)a_1(e-p) = 0$, whence

$$a_1 = pa_1 p + (e-p)a_1(e-p) = 0.$$

But then a_2 must be zero, too. Now let $a = a_1 + a_2 j$ with $a_1, a_2 \in \mathscr{B}$. Then

$$L(a) = \begin{bmatrix} pa_1 p + pa_2 jp & pa_1 jp + pa_2 p \\ pja_1 p + pja_2 jp & pja_1 jp + pja_2 p \end{bmatrix}$$

$$= \begin{bmatrix} pa_1 p & pa_2 p \\ pja_2 jp & pja_1 jp \end{bmatrix}$$

$$= \begin{bmatrix} pa_1 p & pa_2 p \\ p\tilde{a}_2 p & p\tilde{a}_1 p \end{bmatrix}$$

where we wrote $\tilde{a}_i$ for $ja_i j \in \mathscr{B}$. Hence, L maps $\mathscr{A}$ into $[p\mathscr{B}p]^{2\times2}$. To show that L is onto, let a_1, a_2, a_3, a_4 be elements of $\mathscr{B}$ and set

$$a := pa_1 p + pa_2 j(e-p) + (e-p)ja_3 p + (e-p)ja_4 j(e-p).$$

Then $a \in \mathscr{A}$, and it easy to see that

$$L(a) = \begin{bmatrix} pa_1 p & pa_2 p \\ pa_3 p & pa_4 p \end{bmatrix}.$$

So we just proved the following result.

Proposition 1.1.19. *Let $\mathscr{B}$ be an algebra with identity e which contains an idempotent p in its center. Let $\mathscr{A}$ be the algebra generated by $\mathscr{B}$ and a flip j with the properties $j\mathscr{B}j \subseteq \mathscr{B}$ and $jpj = e - p$. Then every element $a \in \mathscr{A}$ can be written uniquely as the sum $a_1 + a_2 j$ with $a_1, a_2 \in \mathscr{B}$, and the mapping*

$$L : \mathscr{A} \to [p\mathscr{B}p]^{2\times2}, \quad a \mapsto \begin{bmatrix} pa_1 p & pa_2 p \\ p\tilde{a}_2 p & p\tilde{a}_1 p \end{bmatrix}$$

with $\tilde{a} := jaj$ is an isomorphism.

The result for the Banach case is exactly the same.

Corollary 1.1.20. *Let $\mathscr{B}$ be a Banach algebra with identity e the center of which contains an idempotent p. Let $\mathscr{A}$ be the Banach algebra generated by $\mathscr{B}$ and a flip j with the properties $j\mathscr{B}j \subseteq \mathscr{B}$ and $jpj = e - p$. Then every element $a \in \mathscr{A}$ can be written uniquely as the sum $a_1 + a_2 j$ with $a_1, a_2 \in \mathscr{B}$, and the mapping*

$$L : \mathscr{A} \to [p\mathscr{B}p]^{2 \times 2}, \quad a \mapsto \begin{bmatrix} pa_1p & pa_2p \\ p\tilde{a}_2p & p\tilde{a}_1p \end{bmatrix}$$

with $\tilde{a} := jaj$ is a continuous isomorphism.

The difference between the formulations of Proposition 1.1.19 and its corollary is the meaning of the word *generated*. In the proposition it means *algebraically generated*, whereas in the corollary it stands for *algebraically generated and completed*.

Proof. The assertions follow either as in the algebraic setting or are straightforward. Let us verify, for example, that every element $a \in \mathscr{A}$ can be written as the sum $a_1 + a_2 j$ with $a_1, a_2 \in \mathscr{B}$. Let $a \in \mathscr{A}$, and let $a_1^{(n)} + a_2^{(n)} j$ with $a_1^{(n)}, a_2^{(n)} \in \mathscr{B}$ be elements which converge to a. Then the elements

$$p\left(a_1^{(n)} + a_2^{(n)} j\right) p + (e - p)\left(a_1^{(n)} + a_2^{(n)} j\right)(e - p) = pa_1^{(n)}p + (e - p)a_1^{(n)}(e - p)$$

converge to $pap + (e - p)a(e - p)$ which implies that $pap + (e - p)a(e - p) \in \mathscr{B}$. Similarly, the elements

$$p\left(a_1^{(n)} + a_2^{(n)} j\right) jp + (e - p)\left(a_1^{(n)} + a_2^{(n)} j\right) j(e - p) = pa_2^{(n)}p + (e - p)a_2^{(n)}(e - p)$$

converge to $pajp + (e - p)aj(e - p)$ which implies that $pajp + (e - p)aj(e - p) \in \mathscr{B}$. Consequently,

$$\begin{aligned} a &= pap + (e - p)a(e - p) + pa(e - p) + (e - p)ap \\ &= [pap + (e - p)a(e - p)] + [pajp - (e - p)aj(e - p)]j, \end{aligned}$$

where the expressions in brackets belong to $\mathscr{B}$. ∎

Note that if $\mathscr{A}$ and $\mathscr{B}$ are C^*-algebras and if p and j are self-adjoint elements, then L is actually an isometric *-isomorphism.

1.1.6 Exercises

Exercise 1.1.1. Check all the details in the construction of a quotient algebra. Consider both the pure algebraic setting and the setting of Banach algebras.

Exercise 1.1.2. Let $\mathscr{A}$ be a unital normed algebra and $a \in \mathscr{A}$. Show that the subalgebra $\mathrm{alg}\{a\}$ of $\mathscr{A}$ is the closure of the set of polynomials with complex coefficients in a and that $\mathrm{alg}\{a\}$ is a commutative subalgebra of $\mathscr{A}$.

Exercise 1.1.3. Two norms $\|\cdot\|_1$ and $\|\cdot\|_2$ on a normed algebra $\mathscr{A}$ are called equivalent if there is a positive constant C such that $C^{-1}\|a\|_1 \leq \|a\|_2 \leq C\|a\|_1$ for each $a \in \mathscr{A}$.

a) Show that equivalence of norms is an equivalence relation.
b) Show that the property $\|e\| = 1$ in the definition of a normed unital algebra is not essential in the following sense: Every normed algebra with $\|e\| \neq 1$ possesses an equivalent norm $\|\cdot\|_1$ which makes it to a normed algebra with $\|e\|_1 = 1$.

Exercise 1.1.4. Let $\mathscr{J}$ be a closed ideal of a unital normed algebra $\mathscr{A}$. Show that the canonical homomorphism from $\mathscr{A}$ onto the quotient algebra $\mathscr{A}/\mathscr{J}$ is continuous with norm 1.

Exercise 1.1.5. Prove all statements of Examples 1.1.9,1.1.10, 1.1.11, 1.1.12 and 1.1.13. For the Wiener algebra, one can consult [171].

Exercise 1.1.6. Give an example of a non-closed ideal of the Banach algebra $C[0,1]$.

Exercise 1.1.7. Consider the Wiener algebra $\mathscr{W}$ in Example 1.1.12.

a) Given functions f and g in $\mathscr{W}$ with power series representations $\sum_{n \in \mathbb{Z}} f_n \xi^n$ and $\sum_{n \in \mathbb{Z}} g_n \xi^n$, respectively, find the representation of the product fg as a power series.
b) Consider the Banach space $l^1(\mathbb{Z})$ of all sequences $x : \mathbb{Z} \to \mathbb{C}$ with $\|x\| := \sum_{n \in \mathbb{Z}} |x_n| < \infty$. Define a product in $l^1(\mathbb{Z})$ which makes this space a Banach algebra which is isomorphic to the Wiener algebra.

Exercise 1.1.8. Prove that the Lebesgue space $L^1(\mathbb{R})$ with the usual norm and with the *convolution product* defined by

$$(f * g)(t) := \int_{-\infty}^{+\infty} f(t-s)g(s)\,ds$$

is a commutative Banach algebra. Does it have a unit?

Exercise 1.1.9. Check the details of the unitization construction.

Exercise 1.1.10. Check the details in the proofs of Proposition 1.1.15 and Theorem 1.1.17.

Exercise 1.1.11. Let $p_1, \ldots, p_n$ be elements of a unital Banach algebra $\mathscr{A}$ such that $p_i \notin \{0, e\}$, $p_i^2 = p_i$ and $\sum_{i=1}^{n} p_i = e$. Endow $\mathscr{A}^{n \times n}$ with the matrix norm (1.5) and define the mapping

$$M : \mathscr{A} \to \mathscr{A}^{n \times n}, \quad a \mapsto [p_i a p_j]_{i,j=1}^{n}.$$

Set $P := \mathrm{diag}(p_i)$. Prove that M is a Banach algebra isomorphism between $\mathscr{A}$ and $P\mathscr{A}^{n \times n}P$.

1.2 Invertibility and spectrum

1.2.1 Invertibility

Let $\mathscr{A}$ be an algebra with identity e over a field $\mathbb{F}$. An element $a \in \mathscr{A}$ is *left* (*right*, resp. *two-sided*) *invertible* if there exists an element $b \in \mathscr{A}$ such that $ba = e$ ($ab = e$, resp. $ab = ba = e$). Two-sided invertible elements are called invertible. It is easy to see that the two-sided inverse of an element $a \in \mathscr{A}$ is uniquely determined if it exists. We denote it by a^{-1}. Obviously,

$$(a^{-1})^{-1} = a \quad \text{and} \quad (ab)^{-1} = b^{-1}a^{-1}$$

for all invertible elements $a, b \in \mathscr{A}$. Thus, the set $\mathscr{G}_{\mathscr{A}}$ of all invertible elements of $\mathscr{A}$ forms a group with respect to multiplication.

Let $\mathscr{A}_1, \mathscr{A}_2$ be unital algebras over the same field. A homomorphism $\mathrm{smb} : \mathscr{A}_1 \to \mathscr{A}_2$ is called a *symbol mapping* if it has the property that $a \in \mathscr{A}_1$ is invertible if and only if $\mathrm{smb}(a) \in \mathscr{A}_2$ is invertible.

The following observation is often useful.

Lemma 1.2.1. *Let $\mathscr{A}$ be an algebra with identity $e \neq 0$, and let $a, b \in \mathscr{A}$. If $e - ab$ is invertible, then $e - ba$ is invertible.*

Indeed, one can easily check that $(e - ba)^{-1} = e - b(ab - e)^{-1}a$.

Theorem 1.2.2 (Neumann series). *Let $\mathscr{A}$ be a unital Banach algebra over $\mathbb{K}$.*

(i) *If $u \in \mathscr{A}$ and $\|u\| < 1$, then $e - u$ is invertible and $(e - u)^{-1} = \sum_{n=0}^{\infty} u^n$.*
(ii) *If $a \in \mathscr{A}$ is invertible and $\|w\| < \|a^{-1}\|^{-1}$, then $a - w$ is also invertible. Moreover,*

$$\|(a - w)^{-1}\| \leq \frac{\|a^{-1}\|}{1 - \|a^{-1}\|\,\|w\|} \tag{1.6}$$

and

$$\|(a - w)^{-1} - a^{-1}\| \leq \frac{\|a^{-1}\|^2\|w\|}{1 - \|a^{-1}\|\,\|w\|}. \tag{1.7}$$

Proof. Due to

$$\left\|\sum_{n=0}^{\infty} u^n\right\| \leq \sum_{n=0}^{\infty} \|u^n\| \leq \sum_{n=0}^{\infty} \|u\|^n = \frac{1}{1 - \|u\|},$$

the Neumann series $\sum_{n=0}^{\infty} u^n$ converges absolutely in the norm of $\mathscr{A}$, and its sum is just the inverse of $e - u$:

$$(e - u)(e + u + u^2 + \ldots) = (e + u + u^2 + \ldots) - (u + u^2 + \ldots) = e.$$

Now let $a \in \mathscr{A}$ be an arbitrary invertible element, and let $w \in \mathscr{A}$ with $\|w\| < \|a^{-1}\|^{-1}$. Then $\|a^{-1}w\| \leq \|a^{-1}\|\|w\| < 1$, whence it follows that the element $e - a^{-1}w$ is invertible and that its inverse is $\sum_{n=0}^{\infty}(a^{-1}w)^n$. But then $a - w = a(e - a^{-1}w)$ is also invertible, and its inverse is $\sum_{n=0}^{\infty}(a^{-1}w)^n a^{-1}$. Moreover,

$$\|(a-w)^{-1}\| \leq \|a^{-1}\| \sum_{n=0}^{\infty} (\|a^{-1}\|\|w\|)^n \leq \frac{\|a^{-1}\|}{1 - \|a^{-1}\|\|w\|}$$

and

$$\|(a-w)^{-1} - a^{-1}\| \leq \|a^{-1}\|\|(a-w)^{-1}\|\|a - (a-w)\|$$

which proves the second assertion.   ∎

Corollary 1.2.3. *The group $\mathcal{G}_{\mathscr{A}}$ of the invertible elements of a unital Banach algebra is open, and the mapping $x \mapsto x^{-1}$ is a homeomorphism of $\mathcal{G}_{\mathscr{A}}$ onto itself.*

Proof. If $a \in \mathscr{A}$ is invertible then, by the previous theorem, $\mathcal{G}_{\mathscr{A}}$ contains the open neighborhood $\{w \in \mathscr{A} : \|a - w\| < \|a^{-1}\|^{-1}\}$ of a. Thus, $\mathcal{G}_{\mathscr{A}}$ is open.

For the second assertion, let (x_n) be a sequence in $\mathcal{G}_{\mathscr{A}}$ such that $x_n \to x \in \mathcal{G}_{\mathscr{A}}$. We have to show that then $x_n^{-1} \to x^{-1}$. This follows immediately from (1.7) by choosing $a := x$ and $w := x - x_n$.   ∎

Example 1.2.4. Theorem 1.2.2 is important from several points of view. For instance, it immediately implies an iterative method to calculate the solution of equations of the form $x - Ax = y$ where y is a given element of a Banach space X and $A \in \mathscr{L}(X)$ is an operator with $\|A\| < 1$. For $y \in X$, define

$$
\begin{aligned}
x_0 &:= y \\
x_1 &:= y + Ax_0 = y + Ay \\
x_2 &:= y + Ax_1 = y + Ay + A^2 y \\
&\ \vdots \\
x_n &:= y + Ax_{n-1} = S_n y \quad \text{with} \quad S_n = \sum_{k=0}^{n} A^k.
\end{aligned}
$$

If $\|A\| < 1$ then $\sum_{k=0}^{\infty} A^k$ converges, $x_n \to x = (I-A)^{-1} y$, and the error of the nth approximation can be estimated by

$$\|x - x_n\| \leq \|(I-A)^{-1} - S_n\|\|y\| \leq \left(\sum_{k=n+1}^{\infty} \|A\|^k \right) \|y\| = \frac{\|A\|^{n+1}}{1 - \|A\|}\|y\|.$$

                                                                                    □

1.2.2 Spectrum

Let $\mathscr{A}$ be a unital algebra over $\mathbb{F}$ and let $a \in \mathscr{A}$. The *resolvent set* of a is the set

$$\rho_{\mathscr{A}}(a) := \{\lambda \in \mathbb{F} : \lambda e - a \in \mathcal{G}_{\mathscr{A}}\}.$$

Its complement $\sigma_{\mathscr{A}}(a) := \mathbb{F} \setminus \rho_{\mathscr{A}}(a)$ is called the *spectrum* of a in $\mathscr{A}$. When the underlying algebra is understood we will denote the resolvent set and the spectrum of an element a simply by $\rho(a)$ and $\sigma(a)$. For $\lambda \in \rho(a)$, the element $R_\lambda(a) := (\lambda e - a)^{-1}$ is called the *resolvent* of a at λ, and the function

$$\rho(a) \to \mathscr{A}, \quad \lambda \mapsto (\lambda e - a)^{-1}$$

is called the *resolvent function of a*.

The following is the spectral mapping theorem for polynomials.

Proposition 1.2.5. *Let $\mathscr{A}$ be a complex algebra with identity e and let p be a polynomial with complex coefficients of positive degree. Then, for all $a \in \mathscr{A}$ and $b \in \mathscr{G}_{\mathscr{A}}$,*

$$\sigma(p(a)) = p(\sigma(a)) \quad \text{and} \quad \sigma(b^{-1}) = (\sigma(b))^{-1}.$$

Proof. For every $\lambda \in \mathbb{C}$, there is a polynomial q such that

$$p(a) - p(\lambda)e = (a - \lambda e)q(a).$$

Thus, if $\lambda \in \sigma(a)$, then $p(\lambda) \in \sigma(p(a))$, whence the inclusion $p(\sigma(a)) \subseteq \sigma(p(a))$. For the reverse inclusion, let $\lambda \in \sigma(p(a))$, and let

$$p(x) - \lambda = \alpha(x - \mu_1) \ldots (x - \mu_n)$$

be the factorization of the polynomial $p - \lambda$ into linear factors. Then

$$p(a) - \lambda e = \alpha(a - \mu_1 e) \ldots (a - \mu_n e),$$

and at least one of the factors $a - \mu_i e$ is not invertible. Thus, $\mu_i \in \sigma(a)$ for some i, which implies that $\lambda = p(\mu_i) \in p(\sigma(a))$.

The second assertion is evident. ∎

Proposition 1.2.6. *Let $\mathscr{A}$ be an algebra over $\mathbb{F}$ with identity e, and let $a, b \in \mathscr{A}$. Then*

$$\sigma(ab) \setminus \{0\} = \sigma(ba) \setminus \{0\}. \tag{1.8}$$

The proof of the proposition above is just a modification of the proof in Lemma 1.2.1 and is left as an exercise.

The resolvent $R_\lambda(a)$ can also be considered as a function of the element a.

Proposition 1.2.7. *In a unital Banach algebra over $\mathbb{K}$, the resolvent $R_\lambda(a)$ depends continuously both on λ and on a.*

Proof. The function $\rho(a) \times \mathscr{A} \to \mathscr{A}$, $(\lambda, a) \mapsto \lambda e - a$ is clearly continuous in the product topology of $\rho(a) \times \mathscr{A}$. By Corollary 1.2.3, the inversion is also a continuous

function, and since the composition of continuous functions is continuous again, the
result follows.   ∎

Example 1.2.8. For $n \in \mathbb{N}$, consider the algebra $\mathbb{K}^{n \times n}$ of the $n \times n$ matrices with
entries in $\mathbb{K}$ as an algebra over $\mathbb{K}$. The spectrum of an element of $\mathbb{K}^{n \times n}$ is just the
set of its eigenvalues in $\mathbb{K}$. Thus, if $\mathbb{K} = \mathbb{R}$, it can happen that the spectrum of an
element is empty, as the matrix

$$\begin{bmatrix} 0 & -1 \\ 1 & 0 \end{bmatrix}$$

without real eigenvalues shows.   □

Example 1.2.9. Let X be a compact Hausdorff space and consider the complex
algebra $C(X)$. A function $f \in C(X)$ is invertible if and only if it has no zero on X.
Thus, the spectrum $\sigma(f)$ coincides with the set of values of f on X; i.e., $\sigma(f) =$
$f(X)$.   □

The main properties of the spectrum of an element of a complex Banach algebra
are summarized in the following. Whereas the boundedness and the closedness of
the spectrum also hold for real Banach algebras, the existence of points in the spectrum is a peculiarity of complex algebras (recall Example 1.2.8). The proof makes
use of some concepts from complex analysis which can be found in [1], for instance.

Theorem 1.2.10. *Let $\mathscr{A}$ be a complex Banach algebra with identity e, and let $a \in$
$\mathscr{A}$. Then:*

  (i) *the resolvent set of a is open in $\mathbb{C}$, and the mapping $\zeta \mapsto (a - \zeta e)^{-1}$ from the
       resolvent set of a into $\mathscr{A}$ is analytic;*
 (ii) *the spectrum of a is a non-empty and compact subset of the complex plane.*

Proof. (i) Let ζ_0 be in the resolvent set of a, and set $a_0 := (a - \zeta_0 e)^{-1}$. For every
$\zeta \in \mathbb{C}$, one has $a - \zeta e = a - \zeta_0 e - (\zeta - \zeta_0)e$. So we get from Theorem 1.2.2 that
if $\|(\zeta - \zeta_0)e\| = |\zeta - \zeta_0| < \|a_0\|^{-1}$, then ζ is in the resolvent set of a and the
inverse of $a - \zeta e$ is given by $\sum_{n=0}^{\infty}(\zeta - \zeta_0)^n a_0^{n+1}$. This series converges absolutely
on $\{\zeta \in \mathbb{C} : |\zeta - \zeta_0| < \|a_0\|^{-1}\}$, and every function $\zeta \mapsto (\zeta - \zeta_0)^n a_0^{n+1}$ is analytic
on this disk.

(ii) The spectrum of a is a closed subset of $\mathbb{C}$ by assertion (i). Moreover, if $|\lambda| > \|a\|$
then, by Theorem 1.2.2 again, the element $e - \lambda^{-1}a$ is invertible. Hence, the element
$a - \lambda e = -\lambda(e - \lambda^{-1}a)$ is invertible, and its inverse is

$$(a - \lambda e)^{-1} = \sum_{n=0}^{\infty} \lambda^{-n-1} a^n \tag{1.9}$$

where the series converges absolutely. Thus, $|\lambda| \leq \|a\|$ for all $\lambda \in \sigma(a)$, and $\sigma(a)$
is a compact subset of $\mathbb{C}$.

Assume that $\sigma(a)$ is empty. Then $\lambda \mapsto (a - \lambda e)^{-1}$ is an analytic function on $\mathbb{C}$ by part (i). Moreover, this function is uniformly bounded. Indeed, from (1.9) we obtain

$$\|(a - \lambda e)^{-1}\| \leq \sum_{n=0}^{\infty} |\lambda|^{-n-1} \|a\|^n = (|\lambda| - \|a\|)^{-1}$$

for all $|\lambda| > \|a\|^{-1}$. Thus, if $|\lambda| > 2\|a\|$, then $\|(a - \lambda e)^{-1}\| \leq \|a\|^{-1}$. Further, the resolvent function is obviously bounded on the closed disk $\{\lambda \in \mathbb{C} : |\lambda| \leq 2\|a\|\}$. Thus, by Liouville's theorem, the function $\lambda \mapsto (a - \lambda e)^{-1}$ is constant. Since $\lim_{\lambda \to \infty} (a - \lambda e)^{-1} = 0$, the constant value of this function is 0, which is impossible. ∎

The following is an important consequence of the non-emptiness of the spectrum.

Corollary 1.2.11 (Gelfand–Mazur theorem). *Let $\mathscr{A}$ be both a complex Banach algebra with identity $e \neq 0$ and a skew field. Then $\mathscr{A}$ is isomorphic to the field $\mathbb{C}$.*

Proof. Let $a \in \mathscr{A}$. By the preceding proposition, there is a complex number λ such that $a - \lambda e$ is not invertible. The only non-invertible element of a skew field is 0. Hence, $a = \lambda e$ and $\mathscr{A} = \mathbb{C}$. ∎

1.2.3 Spectral radius

Let $\mathscr{A}$ be an algebra over $\mathbb{K}$ with identity e. If the spectrum of an element $a \in \mathscr{A}$ is not empty, then the (possibly infinite) number

$$r_{\mathscr{A}}(a) := \sup\{|\lambda| : \lambda \in \sigma_{\mathscr{A}}(a)\}$$

is called the *spectral radius* of a. If the underlying algebra is evident, we will also write $r(a)$ for the spectral radius of a. Thus, $r(a)$ is the radius of the smallest closed disk in the complex plane with center at 0 that contains the spectrum of the element a. Notice that, for a real or complex Banach algebra $\mathscr{A}$ with identity, it is an immediate consequence of Theorem 1.2.2 that

$$r(a) \leq \|a\| \quad \text{for all } a \in \mathscr{A}. \tag{1.10}$$

Unless stated otherwise, *algebra* will mean *complex algebra* hereafter.

Theorem 1.2.12. *Let $\mathscr{A}$ be a Banach algebra with identity e. For every $a \in \mathscr{A}$, the limit $\lim_{n \to \infty} \|a^n\|^{1/n}$ exists, and it coincides with the spectral radius $r(a)$ of a.*

Proof. Let $\lambda \in \sigma(a)$. Then, by Proposition 1.2.5, $\lambda^n \in \sigma(a^n)$. The estimate (1.10) gives $|\lambda^n| \leq \|a^n\|$ and, consequently, $|\lambda| \leq \|a^n\|^{1/n}$ for all positive integers n. Thus, $r(a) \leq \|a^n\|^{1/n}$ for all positive integers n, and this implies

$$r(a) \leq \inf \|a^n\|^{1/n}.$$

Further, recall from assertion (i) of Theorem 1.2.10 that the function $\lambda \mapsto (a - \lambda e)^{-1}$ is analytic in $\{\lambda \in \mathbb{C} : |\lambda| > r(a)\}$. On the other hand, if $|\lambda|$ is large enough, then $(a - \lambda e)^{-1} = \sum_{n \geq 0} \lambda^{-n-1} a^n$, and the series converges absolutely. By Hadamard's formula, the radius of convergence of this series is just $\limsup \|a^n\|^{1/n}$. Since the series represents the Laurent series of $(a - \lambda e)^{-1}$, and it cannot be analytic outside its convergence radius we must have

$$\limsup \|a^n\|^{1/n} \leq r(a).$$

The estimates obtained imply both the existence of the limit and the formula for the spectral radius.   ∎

The formula for the spectral radius is remarkable: It connects purely algebraic quantities (the spectral radius of a) with metric quantities (the norms of a^n).

If two elements commute, one has the following, the proof of which is left as an exercise.

Proposition 1.2.13. *Let $\mathscr{A}$ be a Banach algebra with identity e and let $a, b \in \mathscr{A}$. If $ab = ba$, then*

$$r(ab) \leq r(a)r(b) \quad and \quad r(a+b) \leq r(a) + r(b).$$

1.2.4 Continuity of the spectrum

The trivial example of the numbers $1/n$ shows that the limit of a sequence of invertible elements is not necessarily invertible again.

Proposition 1.2.14. *Let $\mathscr{A}$ be a Banach algebra with identity e, and let $a_n, a \in \mathscr{A}$ with $a_n \to a$.*

  (i) *If a is invertible, then the a_n are invertible for all sufficiently large n, and $a_n^{-1} \to a^{-1}$.*

  (ii) *If the a_n are invertible, and if $\sup \|a_n^{-1}\| < \infty$, then a is invertible, and $a_n^{-1} \to a^{-1}$.*

Proof. (i) The invertibility of the a_n is a consequence of Theorem 1.2.2, which also yields the uniform boundedness of the norms of the a_n^{-1}. Thus,

$$\|a_n^{-1} - a^{-1}\| \leq \|a_n^{-1}\| \|a - a_n\| \|a^{-1}\| \to 0.$$

(ii) One has $\|e - a_n^{-1} a\| \leq \|a_n^{-1}\| \|a_n - a\| \to 0$. Hence, $a_n^{-1} a = e - (e - a_n^{-1} a)$ is invertible for all sufficiently large n, whence the left invertibility of a follows. The right invertibility can be shown analogously. Thus, a is invertible. The convergence of a_n^{-1} to a^{-1} can be checked as in part (i). ∎

The conditions in assertion (ii) can be considerably weakened in the commutative context.

Proposition 1.2.15. *Let $\mathscr{A}$, a, a_n be as in Proposition 1.2.14. If the a_n are invertible, if $a_n a = a a_n$ for all n, and if $\sup r(a_n^{-1}) < \infty$, then a is invertible, and $a_n^{-1} \to a^{-1}$.*

Proof. We have

$$r(e - a_n^{-1}a) = r(a_n^{-1}(a_n - a)) \le r(a_n^{-1}) r(a_n - a) \le r(a_n^{-1}) \|a_n - a\| \to 0$$

by Proposition 1.2.13 and by (1.10). Hence, for all sufficiently large n, $1 \notin \sigma(e - a_n^{-1}a)$ so that $a_n^{-1}a = e - (e - a_n^{-1}a)$ is invertible, whence the invertibility of a. The convergence of a_n^{-1} to a^{-1} is a consequence of part (i) of Proposition 1.2.14. ∎

Next we consider continuity properties of the spectrum of an element a as a function of a. Let $(M_n)_{n \in \mathbb{Z}^+}$ be a sequence of subsets of the complex plane $\mathbb{C}$. The *partial limiting set* or *limes superior* $\limsup_{n \to \infty} M_n$ consists of all points $m \in \mathbb{C}$ which are a partial limit of a sequence (m_n) with $m_n \in M_n$. Limiting sets are closed.

Proposition 1.2.16. *Let $\mathscr{A}$ be a Banach algebra with identity e, and let $a_n, a \in \mathscr{A}$ with $a_n \to a$. Then*

$$\limsup_{n \to \infty} \sigma(a_n) \subseteq \sigma(a).$$

Proof. Let $\lambda \in \limsup \sigma(a_n)$, and let (λ_{n_k}) with $\lambda_{n_k} \in \sigma(a_{n_k})$ be a sequence which converges to λ. Then $a_{n_k} - \lambda_{n_k} e$ converges to $a - \lambda e$. Assume $a - \lambda e$ to be invertible. Then $a_{n_k} - \lambda_{n_k} e$ is invertible for all sufficiently large k due to Proposition 1.2.14 (i), which is a contradiction. ∎

Proposition 1.2.17. *Let $\mathscr{A}$ be a Banach algebra with identity e, let $a \in \mathscr{A}$, and let V be a neighborhood of $0 \in \mathbb{C}$. Then there is a number $\delta > 0$ such that*

$$\sigma(b) \subseteq \sigma(a) + V \quad \text{for all} \quad b \in \mathscr{A} \text{ with } \|b - a\| < \delta.$$

A mapping ϕ of a topological space S into the set of the subsets of a topological space T is *upper semi-continuous* if, for each $s_0 \in S$ and each neighborhood U of $\phi(s_0)$, there is a neighborhood V of s_0 such that

$$\phi(s) \subseteq U \quad \text{for all } s \in V.$$

In that sense, the mapping $a \mapsto \sigma(a)$ of a Banach algebra $\mathscr{A}$ into the compact subsets of $\mathbb{C}$ is upper semi-continuous. The proof of Proposition 1.2.17 follows easily from the next result by specifying $S = \{a \in \mathscr{A} : \|a\| \le r\}$ and $T = \{z \in \mathbb{C} : |z| \le r\}$ with some sufficiently large r.

Lemma 1.2.18. *Let ϕ be a mapping of a metric space S into the set of the closed subsets of a compact metric space T. Then ϕ is upper semi-continuous if and only if $\limsup_{n \to \infty} \phi(s_n) \subseteq \phi(s)$ for all sequences $s_n \to s$.*

Proof. Let ϕ be upper semi-continuous, and let U be a neighborhood of $\phi(s)$. Then, to every sequence $s_n \to s$, $\phi(s_n) \subseteq U$ for all sufficiently large n, hence, $\limsup \phi(s_n) \subseteq U$. Thus, $\limsup \phi(s_n)$ lies in the intersection of all open neighborhoods of $\phi(s)$, and this intersection coincides with $\phi(s)$ because $\phi(s)$ is closed.

Conversely, let $\limsup \phi(s_n) \subseteq \phi(s)$ for every sequence $s_n \to s$, but assume ϕ to be not upper semi-continuous. Then there exists a neighborhood U of $\phi(s)$ as well as a sequence (m_{n_k}) of points $m_{n_k} \in \phi(s_{n_k}) \setminus U$. The sequence (m_{n_k}) possesses a convergent subsequence (because T is compact). The limit of this subsequence does not belong to U and thus, not to $\phi(s)$. This contradicts the hypothesis. ∎

The upper semi-continuity of the mapping $a \mapsto \sigma(a)$ is – without further conditions – the sharpest continuity property that can be obtained. This is illustrated by the following example of a sequence of quasinilpotent elements (i.e., of elements with spectrum $\{0\}$) which converges to a non-quasinilpotent element. Non-quasinilpotent elements which can be obtained in this manner are called *limpotent*. This example goes back to Kakutani and is taken, as it stands, from Rickart [162]. The notion *limpotent element* was proposed by Gohberg and Krupnik [72].

Example 1.2.19 (Kakutani). Let H be a separable Hilbert space and $(e_m)_{m=1}^{\infty}$ be an orthonormal basis of H. Consider the sequence of scalars

$$\alpha_m = \exp(-k), \quad \text{for } m = 2^k(2l+1),$$

where $k, l = 0, 1, 2, \ldots$, and define the weighted shift operator A on H by

$$A e_m = \alpha_m e_{m+1}, \quad m = 1, 2, \ldots.$$

The norm of A is just $\sup_m |\alpha_m|$. Observe also that

$$A^n e_m = \alpha_m \alpha_{m+1} \ldots \alpha_{m+n-1} e_{m+n}$$

and hence $\|A^n\| = \sup_m (\alpha_m \alpha_{m+1} \ldots \alpha_{m+n-1})$. By the definition of the α_m,

$$\alpha_1 \alpha_2 \ldots \alpha_{2^t-1} = \Pi_{j=1}^{t-1} \exp(-j 2^{t-j-1}),$$

and therefore

$$(\alpha_1 \alpha_2 \ldots \alpha_{2^t-1})^{1/2^{t-1}} > \left(\Pi_{j=1}^{t-1} \exp\left(-(j/2^{j+1})\right) \right)^2.$$

If we set $\beta := \sum_{j=1}^{\infty} j/2^{j+1}$, then

$$\exp(-2\beta) \leq \lim_{n \to \infty} \|A^n\|^{1/n} = r(A).$$

In particular, A is not quasinilpotent. Now define operators A_k by

$$A_k e_m := \begin{cases} 0 & \text{if} \quad m = 2^k(2l+1), \\ \alpha_m e_{m+1} & \text{if} \quad m \neq 2^k(2l+1). \end{cases}$$

The operators A_k are *nilpotent* (that is, a certain power of A_k is zero, namely $A_k^{2k+1} = 0$). On the other hand,

$$(A - A_k)e_m = \begin{cases} e^{-k}e_{m+1} & \text{if} \quad m = 2^k(2l+1), \\ 0 & \text{if} \quad m \neq 2^k(2l+1). \end{cases}$$

Therefore, $\|A - A_k\| = e^{-k}$, so that $\lim A_k = A$. $\square$

The next result shows that commutativity forces the spectrum to be continuous.

Proposition 1.2.20. *Let $\mathscr{A}$ be a Banach algebra with identity e, let $a \in \mathscr{A}$, and let V be a neighborhood of $0 \in \mathbb{C}$. Then there is a $\delta > 0$ such that*

$$\sigma(b) \subseteq \sigma(a) + V \qquad \text{and} \qquad \sigma(a) \subseteq \sigma(b) + V$$

for all b with $\|b - a\| < \delta$ and $ab = ba$.

Proof. The first inclusion follows from Lemma 1.2.18. For a proof of the second one, we can assume without loss that V is a disk with center at zero and with radius $\varepsilon > 0$.

Suppose the assertion to be false. Then there is a sequence (b_n) such that $\lim b_n = a$ and $ab_n = b_n a$, but $\sigma(a) \not\subseteq \sigma(b_n) + V$ for every n. For $n \in \mathbb{N}$, choose $x_n \in \sigma(a) \setminus (\sigma(b_n) + V)$, and let x_0 be a partial limit of the sequence (x_n) (which exists due to the compactness of $\sigma(a)$). Then $|x_0 - x| \geq \varepsilon$ for every $x \in \sigma(b_n)$ and every n or, equivalently, dist $(\sigma(b_n - x_0 e), 0) \geq \varepsilon$. Hence, $r((b_n - x_0 e)^{-1}) \leq \varepsilon^{-1}$ (cf. Exercise 1.2.6), whence via Proposition 1.2.15 the invertibility of $a - x_0 e = \lim(b_n - x_0 e)$ follows. This contradicts the assumption $x_0 \in \sigma(a)$. ∎

Observe that the commutativity of b_n with a has only been used to obtain a contradiction via Proposition 1.2.15. Thus, if it happens that the boundedness of the spectral radii $(r(b_n^{-1} - xe))$ implies the boundedness of $(\|b_n^{-1} - xe\|)$, then we again have a contradiction, and continuity follows. In particular we see that the spectrum is a continuous function when restricted to the set of the normal elements of a C^*-algebra, since in this case $r(b) = \|b\|$ (see Proposition 1.2.36 below). This result is due to Newburgh [127], who also proved the following refinement of Proposition 1.2.17.

Proposition 1.2.21. *Let $\mathscr{A}$ be a Banach algebra with identity $e \neq 0$, let $a \in \mathscr{A}$, and let $a_n \in \mathscr{A}$ be elements with $\lim a_n = a$. Then, if n is large enough, $\sigma(a_n)$ has points in the neighborhood of each connected component of $\sigma(a)$.*

Corollary 1.2.22. *Limpotent elements of Banach algebras have connected spectra.*

1.2.5 Subalgebras and invertibility

All algebras considered in this section are supposed to be unital. Let $\mathscr{A}$ be an algebra and $\mathscr{B}$ be a subalgebra of $\mathscr{A}$ containing the identity. If $b \in \mathscr{B}$, what is the relation between $\sigma_{\mathscr{B}}(b)$ and $\sigma_{\mathscr{A}}(b)$? Evidently, if b is not invertible in $\mathscr{A}$, then it cannot be invertible in $\mathscr{B}$. Thus, $\sigma_{\mathscr{A}}(a) \subseteq \sigma_{\mathscr{B}}(a)$. But if $b \in \mathscr{B}$ is invertible in $\mathscr{A}$, then it might happen that $b^{-1} \notin \mathscr{B}$ and thus, the spectra of b in $\mathscr{A}$ and $\mathscr{B}$ can be different. This fact is illustrated by the following example.

Example 1.2.23 (The disk algebra). Consider the unit circle $\mathbb{T} = \{z \in \mathbb{C} : |z| = 1\}$ and the open unit disk $\mathbb{D} = \{z \in \mathbb{C} : |z| < 1\}$, and let $\mathscr{A} := C(\mathbb{T})$. The *disk algebra* $\mathbb{A}$ is the closure in $\mathscr{A}$ of the algebra of all polynomials p in $z \in \mathbb{T}$ (the reason for the name will become clear from Exercise 2.1.2).

The spectrum of the polynomial $p(z) := z$, when considered as an element of the algebra $\mathscr{A}$ of all continuous functions, is $\mathbb{T}$ by Example 1.2.9. We are going to determine the spectrum of p, now considered as an element of the disk algebra $\mathbb{A}$. Since $\|p\|_\infty = 1$, $\sigma_{\mathbb{A}}(p)$ is contained in the closed unit disk. We claim that it coincides with that disk. Let $|\lambda| < 1$ and suppose that $\lambda \notin \sigma_{\mathbb{A}}(p)$. Then there is an $f \in \mathbb{A}$ such that $(p - \lambda)f = 1$. Let (p_n) be a sequence of polynomials which converges to f uniformly on $\mathbb{T}$. Then, for any $\varepsilon > 0$, there exists an $n_0 \in \mathbb{N}$ such that

$$\sup\{|p_n(z) - p_m(z)| : z \in \mathbb{T}\} = \|p_n - p_m\|_\infty < \varepsilon$$

for $n, m > n_0$. By the maximum modulus principle,

$$\sup\{|p_n(z) - p_m(z)| : z \in \mathbb{D}\} < \varepsilon$$

for $n, m > n_0$. Consequently, the limit function $g := \lim p_n$ is analytic on $\mathbb{D}$ and continuous on its closure. Moreover, $g_{|\mathbb{T}} = f$. By the same argument, since $p_n(p - \lambda) \to 1$ uniformly on $\mathbb{T}$, one has $p_n(p - \lambda) \to 1$ uniformly on $\mathbb{D}$. In particular, $g(z)(z - \lambda) = 1$ for each $z \in \mathbb{D}$. Choosing $z := \lambda$, we arrive at a contradiction. Thus, $\mathbb{D} \subset \sigma_{\mathbb{A}}(p)$, whence $\sigma_{\mathbb{A}}(p) = \mathbb{T} \cup \mathbb{D}$. In particular, $\sigma_{\mathbb{A}}(p) \neq \sigma_{\mathscr{A}}(p)$. □

The following notion will help to clarify the situation.

Definition 1.2.24. Let $\mathscr{A}$ be a normed algebra. An element $z \in \mathscr{A}$ is said to be a *left (right) topological divisor of zero* of $\mathscr{A}$ if there exists a sequence (z_n) in $\mathscr{A}$ such that

(i) $\|z_n\| = 1$ for all $n \in \mathbb{N}$;
(ii) $\lim_{n \to \infty} z z_n = 0$ ($\lim_{n \to \infty} z_n z = 0$).

An element which is both a left and right topological divisor of zero is simply called a *topological divisor of zero*.

Example 1.2.25. Let z denote the identical mapping of the interval $[0, 1]$. The function z is a topological divisor of zero of the algebra $C[0, 1]$. For, one easily checks that the functions defined by

$$z_n(t) := \begin{cases} 1 - nt & \text{if } 0 \leq t < 1/n, \\ 0 & \text{if } 1/n \leq t \leq 1 \end{cases}$$

fulfill the conditions of the definition. □

Proposition 1.2.26. *Left (right) topological divisors of zero of a unital normed algebra cannot be invertible.*

Proof. Let $\mathscr{B}$ be a unital normed algebra and suppose $z \in \mathscr{B}$ to be a left topological divisor of zero. Then there exists a sequence (z_n) in $\mathscr{B}$ such that $\|z_n\| = 1$ and $zz_n \to 0$. If z would be invertible, then $z^{-1}zz_n = z_n \to 0$. But this is impossible because $\|z_n\| = 1$ for all $n \in \mathbb{N}$. The reasoning for a right topological divisor of zero is the same. ∎

Every topological divisor of zero of a subalgebra of a given normed algebra is also a topological divisor of zero of the algebra itself. This is because the definition of a topological divisor of zero only involves norm properties (which do not change when passing to a subalgebra).

Corollary 1.2.27. *If z is a left (right) topological divisor of zero in a subalgebra of a unital normed algebra $\mathscr{A}$ then z cannot be invertible in $\mathscr{A}$.*

Thus, topological divisors of zero are "fundamentally" non-invertible. There is no larger normed algebra in which they might become invertible.

We establish one more relation between $\mathscr{G}_{\mathscr{A}}$ and the set of topological divisors of zero of an algebra $\mathscr{A}$.

Proposition 1.2.28. *Let $\mathscr{A}$ be a Banach algebra. Then every element in the boundary of $\mathscr{G}_{\mathscr{A}}$ is a topological divisor of zero of $\mathscr{A}$.*

Proof. Let z be in the boundary of $\mathscr{G}_{\mathscr{A}}$. We will show that z is a right topological divisor of zero. In the same way, one can check that z is also a left topological divisor of zero.

Since $\mathscr{G}_{\mathscr{A}}$ is open, z is not invertible and there is a sequence (x_n) in $\mathscr{G}_{\mathscr{A}}$ which converges to z. Set $z_n := x_n^{-1}/\|x_n^{-1}\|$ for $n \in \mathbb{N}$. We check that the sequence (z_n) meets the conditions of Definition 1.2.24. Evidently, $\|z_n\| = 1$. Further,

$$\|z_n z\| = \frac{\|x_n^{-1}z\|}{\|x_n^{-1}\|} = \frac{\|(x_n^{-1}z - e) + e\|}{\|x_n^{-1}\|} \leq \frac{\|x_n^{-1}z - e\|}{\|x_n^{-1}\|} + \frac{1}{\|x_n^{-1}\|}$$

$$\leq \frac{\|x_n^{-1}\|\,\|z - x_n\|}{\|x_n^{-1}\|} + \frac{1}{\|x_n^{-1}\|} = \|z - x_n\| + \frac{1}{\|x_n^{-1}\|}.$$

The first term on the right-hand side tends to zero. For the second one notice that, since z is not invertible, $x_n^{-1}z$ cannot be invertible. Thus, $\|e - x_n^{-1}z\| \geq 1$ by Neumann series. This implies that

$$1 \leq \|e - x_n^{-1}z\| = \|x_n^{-1}(x_n - z)\| \leq \|x_n^{-1}\| \, \|x_n - z\|,$$

whence $\|x_n^{-1}\|^{-1} \leq \|x_n - z\|$, too. The conclusion is that $z_n z \to 0$, that is, z is a right topological divisor of zero. ∎

Proposition 1.2.29. *Let $\mathscr{B}$ be a closed unital subalgebra of a unital Banach algebra $\mathscr{A}$ and let $b \in \mathscr{B}$. Then*

$$\sigma_{\mathscr{A}}(b) \subseteq \sigma_{\mathscr{B}}(b), \quad \partial \sigma_{\mathscr{B}}(b) \subseteq \partial \sigma_{\mathscr{A}}(b), \quad and \quad r_{\mathscr{B}}(b) = r_{\mathscr{A}}(b),$$

where ∂M refers to the boundary of the set $M \subset \mathbb{C}$.

Proof. The first inclusion is obvious and has been already mentioned. For the second inclusion, let $\mu \in \partial \sigma_{\mathscr{B}}(b)$. Then $\mu e - b$ lies in the boundary of $\mathscr{G}_{\mathscr{B}}$ and is, thus, a topological divisor of zero in $\mathscr{B}$ by Proposition 1.2.28. By Corollary 1.2.27, $\mu e - b$ is not invertible in $\mathscr{A}$, i.e., $\mu \in \sigma_{\mathscr{A}}(b)$. In fact one even has $\mu \in \partial \sigma_{\mathscr{A}}(b)$, because $\mu \in \partial \sigma_{\mathscr{B}}(b)$ is the limit of a sequence in $\rho_{\mathscr{B}}(b)$, and this sequence is also contained in $\rho_{\mathscr{A}}(b)$. Finally, the equality of the spectral radii is an immediate consequence of the formula for the spectral radius stated in Theorem 1.2.12. ∎

This proposition shows that, when one goes from an algebra $\mathscr{B}$ to a larger algebra, the spectrum of an element can only be reduced at the cost of interior points, not losing any points in its boundary. Conversely, when passing to a subalgebra, the spectrum of an element can only increase by suppressing "holes", not by increasing its boundary. To be precise, by a *hole* in $\sigma_{\mathscr{A}}(a)$ we mean a bounded, connected component of the resolvent set $\rho_{\mathscr{A}}(a)$.

Theorem 1.2.30. *Let $\mathscr{A}$ be a unital Banach algebra, $\mathscr{B}$ a closed unital subalgebra of $\mathscr{A}$ and $b \in \mathscr{B}$.*

(i) *Let B be a hole of $\sigma_{\mathscr{A}}(b)$. Then either $B \subset \sigma_{\mathscr{B}}(b)$ or $B \cap \sigma_{\mathscr{B}}(b) = \emptyset$.*
(ii) *If $\rho_{\mathscr{A}}(b)$ is connected, then $\sigma_{\mathscr{A}}(b) = \sigma_{\mathscr{B}}(b)$.*

Proof. Given a hole B, define $B_1 := B \setminus \sigma_{\mathscr{B}}(b)$ and $B_2 := B \cap \sigma_{\mathscr{B}}(b)$. Then $B_1 \cap B_2 = \emptyset$ and $B_1 \cup B_2 = B$.

Clearly, B_1 is open. Since $\partial \sigma_{\mathscr{B}}(b) \subseteq \sigma_{\mathscr{A}}(b)$ and $B \cap \sigma_{\mathscr{A}}(b) = \emptyset$, one has $B_2 = B \cap \mathrm{int}\, \sigma_{\mathscr{B}}(b)$ which implies that B_2 is open, too. Since B is connected, either B_1 or B_2 must be empty. This proves assertion (i).

For a proof of assertion (ii), we first show that $\sigma_{\mathscr{B}}(b) \setminus \sigma_{\mathscr{A}}(b)$ is open. Suppose there is a point

$$\lambda \in \partial(\sigma_{\mathscr{B}}(b) \setminus \sigma_{\mathscr{A}}(b)) \cap (\sigma_{\mathscr{B}}(b) \setminus \sigma_{\mathscr{A}}(b)).$$

Then λ is the limit of a sequence (λ_n) in the complement of $\sigma_{\mathscr{B}}(b) \setminus \sigma_{\mathscr{A}}(b)$, i.e., in $\rho_{\mathscr{B}}(b) \cup \sigma_{\mathscr{A}}(b)$. There cannot be infinitely many of the λ_n in $\sigma_{\mathscr{A}}(b)$ since otherwise $\lambda \in \sigma_{\mathscr{A}}(b)$ due to the closedness of spectra. Thus, λ is the limit of a sequence in $\rho_{\mathscr{B}}(b)$. But then

$$\lambda \in \partial \rho_{\mathscr{B}}(b) = \partial \sigma_{\mathscr{B}}(b) \subseteq \partial \sigma_{\mathscr{A}}(b) \subseteq \sigma_{\mathscr{A}}(b)$$

by Proposition 1.2.29, which is again impossible. Thus, $\sigma_{\mathscr{B}}(b) \setminus \sigma_{\mathscr{A}}(b)$ is open, and the connected set $\rho_{\mathscr{A}}(a)$ is the union of its disjoint open subsets $\rho_{\mathscr{B}}(b)$ and $\sigma_{\mathscr{B}}(b) \setminus \sigma_{\mathscr{A}}(b)$. Since $\rho_{\mathscr{B}}(b)$ is non-empty by Theorem 1.2.2, the set $\sigma_{\mathscr{B}}(b) \setminus \sigma_{\mathscr{A}}(b)$ must be empty. ∎

In applications one often meets situations where one is interested in the invertibility of an element b in an algebra $\mathscr{A}$, but where technical limitations force one and allow one to study the invertibility of b in a suitable subalgebra of $\mathscr{A}$ only. This causes no problem if the subalgebra is inverse-closed in the following sense.

Definition 1.2.31. Let $\mathscr{A}$ be an algebra with unit e and $\mathscr{B}$ a unital subalgebra of $\mathscr{A}$. We say that $\mathscr{B}$ is *inverse-closed* in $\mathscr{A}$ if every element of $\mathscr{B}$, which is invertible in $\mathscr{A}$, is also invertible in $\mathscr{B}$.

There are large classes of algebras for which the inverse-closedness can be granted. For example, this happens for closed symmetric subalgebras of C^*-algebras, as we shall see below. But in general it might be a difficult task to verify the inverse-closedness of a subalgebra of a given algebra. The following assertions provide some simple, but often effective, criteria for deciding the inverse-closedness of a given algebra.

Recall that a subset of the complex plane is called *thin* if it does not contain inner points (with respect to $\mathbb{C}$).

Corollary 1.2.32. *Let $\mathscr{A}$ be a unital Banach algebra and $\mathscr{B}$ a unital closed subalgebra of $\mathscr{A}$. If $\mathscr{B}$ contains a dense subalgebra $\mathscr{B}_0$ the elements of which have thin spectra in $\mathscr{B}$ then $\mathscr{B}$ is inverse-closed in $\mathscr{A}$.*

Proof. Let $c \in \mathscr{B}_0$. Since $\sigma_{\mathscr{B}}(c)$ is thin, we have $\sigma_{\mathscr{B}}(c) = \sigma_{\mathscr{A}}(c)$ by Proposition 1.2.29, whence it follows that the elements in $\mathscr{B}_0$ are invertible in $\mathscr{B}$ if and only if they are invertible in $\mathscr{A}$. Now let $c \in \mathscr{B}$ be invertible in $\mathscr{A}$. Approximate c by a sequence $(c_n) \subseteq \mathscr{B}_0$. For large n, the c_n are invertible in $\mathscr{A}$, too, and it is easy to check that $c_n^{-1} \to c^{-1}$ as $n \to \infty$. But $c_n^{-1} \in \mathscr{B}$, and since $\mathscr{B}$ is a closed subalgebra we conclude that $c^{-1} \in \mathscr{B}$. ∎

Lemma 1.2.33. *Let $\mathscr{B}$ be a unital subalgebra of a unital algebra $\mathscr{A}$, and let $\mathscr{J} \subset \mathscr{B}$ be an ideal of $\mathscr{A}$. Then $\mathscr{J}$ is an ideal of $\mathscr{B}$, and if the quotient algebra $\mathscr{B}/\mathscr{J}$ is inverse-closed in $\mathscr{A}/\mathscr{J}$, then $\mathscr{B}$ is inverse-closed in $\mathscr{A}$.*

Proof. We only prove the second assertion. Let $b \in \mathscr{B}$ be invertible in $\mathscr{A}$. Then the coset $b + \mathscr{J}$ is invertible in $\mathscr{A}/\mathscr{J}$. Since $\mathscr{B}/\mathscr{J}$ is inverse-closed in $\mathscr{A}/\mathscr{J}$ by assumption, $b + \mathscr{J}$ is already invertible in $\mathscr{B}/\mathscr{J}$. Thus, there are elements $b' \in \mathscr{B}$ and $j \in \mathscr{J}$ such that $bb' = e + j$. Multiplying by b^{-1} we obtain $b' = b^{-1} + b^{-1}j$, whence $b^{-1} = b' - b^{-1}j \in \mathscr{B}$. ∎

Note that the converse of Lemma 1.2.33 does not hold. For a counter-example see Exercise 1.4.9.

As far as we know, it is still unknown if the inverse-closedness of $\mathscr{B}$ in $\mathscr{A}$ implies the inverse-closedness of the matrix algebra $\mathscr{B}^{n \times n}$ in $\mathscr{A}^{n \times n}$ for $n > 1$. But in the simple but important special case when $\mathscr{B}$ is commutative, the answer is affirmative. The proof is based on the following lemma, which is taken from [108].

Lemma 1.2.34. *Let $\mathscr{A}$ be an algebra with identity e, and let $A = [a_{ij}] \in \mathscr{A}^{n \times n}$ be a matrix whose entries commute. Then A is invertible in $\mathscr{A}^{n \times n}$ if and only it the determinant $\det A$ (defined in the common way) is invertible in $\mathscr{A}$.*

Proof. We denote the identity matrix $\mathrm{diag}(e, \dots, e)$ in $\mathscr{A}^{n \times n}$ by e_n. Let $\mathrm{adj} A$ denote the transpose of the matrix of cofactors from A. Using the commutativity of the entries of A one can easily check that $\mathrm{adj} A A = A \, \mathrm{adj} A = \det A \, e_n$. Hence, if $\det A$ is invertible in $\mathscr{A}$ then A is invertible in $\mathscr{A}^{n \times n}$.

Let now A be invertible in $\mathscr{A}^{n \times n}$ and $A^{-1} = [c_{ij}]$. We will show that the entries c_{ij} commute and that they also commute with the entries of A. Let r be any of the entries of A. Then

$$[rc_{ij}] = rA^{-1} = A^{-1}A(rA^{-1}) = (A^{-1}r)AA^{-1} = A^{-1}r = [c_{ij}r],$$

hence, r commutes with the entries of A^{-1}. Repeating this argument, now with r taken from the c_{ij}, and using the fact just proved we get that the entries of A^{-1} also commute. Thus, the identities $AA^{-1} = A^{-1}A = e_n$ imply that

$$\det A \det(A^{-1}) = \det(A^{-1}) \det A = e,$$

hence, $\det A$ is invertible in $\mathscr{A}$.   ∎

Proposition 1.2.35. *Let $\mathscr{B}$ be a commutative unital subalgebra of a unital algebra $\mathscr{A}$. If $\mathscr{B}$ is inverse-closed in $\mathscr{A}$, then $\mathscr{B}^{n \times n}$ is inverse-closed in $\mathscr{A}^{n \times n}$.*

Proof. If $A \in \mathscr{B}^{n \times n}$ is invertible in $\mathscr{A}^{n \times n}$ then, by Lemma 1.2.34, $\det A \in \mathscr{B}$ is invertible in $\mathscr{A}$. It follows that $\det A$ is invertible in $\mathscr{B}$ due to the inverse-closedness property. Using the lemma again, we conclude that A is invertible in $\mathscr{B}^{n \times n}$.   ∎

1.2.6 The spectrum of elements of C^*-algebras

Let $\mathscr{A}$ be an involutive algebra with identity e. Remember that an element $a \in \mathscr{A}$ is called

- *normal* if $aa^* = a^*a$,
- *self-adjoint* or *Hermitian* if $a = a^*$,
- *unitary* if $a^*a = aa^* = e$,
- *an isometry* if $a^*a = e$,
- *a partial isometry* if $aa^*a = a$.

Proposition 1.2.36. *Let $\mathscr{A}$ be a C^*-algebra with identity $e \neq 0$.*

(i) *If a is normal, then $r(a) = \|a\|$.*
(ii) *If a is isometric, then $r(a) = 1$.*
(iii) *If a is unitary, then $\sigma(a) \subseteq \mathbb{T}$.*
(iv) *If a is self-adjoint, then $\sigma(a) \subseteq \mathbb{R}$.*

Proof. (i) We have $\left\|a^{2^n}\right\|^2 = \left\|(a^*)^{2^n} a^{2^n}\right\| = \left\|(a^*a)^{2^n}\right\|$. The self-adjointness of a^*a yields

$$\left\|(a^*a)^{2^n}\right\| = \left\|(a^*a)^{2^{n-1}}\right\|^2 = \ldots = \|a^*a\|^{2^n} = \|a\|^{2^{n+1}}.$$

Thus, $\left\|a^{2^n}\right\| = \|a\|^{2^n}$, and by Theorem 1.2.12,

$$r(a) = \lim \left\|a^{2^n}\right\|^{2^{-n}} = \lim \|a\|^{2^n 2^{-n}} = \|a\|.$$

(ii) It is

$$\|a^n\|^2 = \|(a^*)^n a^n\| = \|(a^*)^{n-1} a^* a a^{n-1}\| = \|(a^*)^{n-1} a^{n-1}\|$$
$$= \ldots = \|a^*a\| = \|e\| = 1.$$

Consequently, $r(a) = \lim \|a^n\|^{1/n} = 1$.

(iii) The spectrum of a is contained in the closed unit disk $\{\lambda \in \mathbb{C} : |\lambda| \leq 1\}$ by assertion (ii). Since a^{-1} is also unitary, the spectrum of a^{-1} is contained in the closed unit disk, too. The equality $\sigma(a^{-1}) = \sigma(a)^{-1}$ implies that $\sigma(a) \subseteq \mathbb{T}$.

(iv) Self-adjoint elements are normal. Hence, $r(a) = \|a\|$ by assertion (i). It remains to show that $a - \lambda e$ is invertible for all $\lambda \in \mathbb{C} \setminus \mathbb{R}$ with $|\lambda| \leq \|a\|$. Choose a real number μ with $1/\mu > \|a\|$. Then $e + i\mu a$ is invertible. Define

$$u := (e - i\mu a)(e + i\mu a)^{-1}.$$

The element u is unitary, and because of $|(1 - i\mu\lambda)(1 + i\mu\lambda)^{-1}| \neq 1$, the element $(1 - i\mu\lambda)(1 + i\mu\lambda)^{-1} e - u$ is invertible by assertion (iii). Now we have

$$(1 - i\mu\lambda)(1 + i\mu\lambda)^{-1} e - u =$$
$$= (1 - i\mu\lambda)(1 + i\mu\lambda)^{-1} e - (e - i\mu a)(e + i\mu a)^{-1}$$
$$= (1 + i\mu\lambda)^{-1}[(1 - i\mu\lambda)(e + i\mu a) - (1 + i\mu\lambda)(e - i\mu a)](e + i\mu a)^{-1}$$
$$= (1 + i\mu\lambda)^{-1}[2\mu i(a - \lambda e)](e + i\mu a)^{-1},$$

hence, $a - \lambda e$ is invertible. ∎

An element of a C^*-algebra with identity is called *positive* if it is self-adjoint and if its spectrum is in $[0, \infty[$.

Proposition 1.2.37. *An element of a C^*-algebra $\mathscr{A}$ with identity is positive if and only if it is of the form b^*b for some $b \in \mathscr{A}$.*

Finally, we mention two useful results concerning invertibility in C^*-algebras. The proofs are left as exercises (see Exercises 1.2.24 and 1.2.25, where also some hints are given).

Theorem 1.2.38 (Inverse-closedness). *Let $\mathscr{A}$ be a C^*-algebra with identity e, and let $\mathscr{B}$ be a C^*-subalgebra of $\mathscr{A}$ containing e. Then $\mathscr{B}$ is inverse-closed in $\mathscr{A}$.*

Theorem 1.2.39 (Semi-simplicity). *Let $\mathscr{A}$ be a unital C^*-algebra and $r \in \mathscr{A}$ be an element having the property that, whenever an element $a \in \mathscr{A}$ is invertible, then $a + r$ is also invertible. Then $r = 0$.*

We will see in Proposition 1.3.6 that this property is indeed equivalent to the semi-simplicity of $\mathscr{A}$, which will be defined in Section 1.3.

1.2.7 Exercises

Exercise 1.2.1. Consider a unital algebra $\mathscr{A}$, and let $a \in \mathscr{A}$. Prove that if a is nilpotent (i.e., if $a^n = 0$ for some $n \in \mathbb{N}$), then $\sigma(a) = \{0\}$.

Exercise 1.2.2. Give an example of a C^*-algebra $\mathscr{A}$ with identity and of an element $a \in \mathscr{A}$ such that $\sigma(a^*a) \neq \sigma(aa^*)$. Prove that this cannot happen if $\mathscr{A} = \mathbb{C}^{n \times n}$. Show that A^*A and AA^* are unitarily equivalent for each matrix $A \in \mathbb{C}^{n \times n}$, that is, there is a unitary matrix C such that $C^*A^*AC = AA^*$.

Exercise 1.2.3. Prove Proposition 1.2.6.

Exercise 1.2.4. Let $\mathscr{A}$ be a unital algebra and $a \in \mathscr{A}$. Show the *resolvent identity*

$$R_\lambda(a) - R_\mu(a) = (\mu - \lambda)R_\lambda(a)R_\mu(a)$$

for $\mu, \lambda \in \rho(a)$.

Exercise 1.2.5. Show that if the product of two commuting elements of a unital algebra is invertible then both elements are invertible.

Exercise 1.2.6. Let $\mathscr{A}$ be a Banach algebra with identity e and let $a \in \mathscr{A}$ be invertible. Show that $\|a^{-1}\| \geq r(a^{-1}) = \text{dist}(\sigma(a), 0)^{-1}$.

Exercise 1.2.7. Consider the shift operator

$$V : l^2(\mathbb{Z}^+) \to l^2(\mathbb{Z}^+), \quad (x_0, x_1, x_2, \ldots) \mapsto (0, x_0, x_1, x_2, \ldots).$$

Show that V is a bounded operator and calculate its norm. Prove that V is left invertible but not right invertible, and determine its spectrum.

Exercise 1.2.8. Let $\mathscr{A}$ be a unital Banach algebra, and let (a_n) be a sequence in $\mathscr{A}$ with limit $a \in \mathscr{A}$. Prove that if $\alpha_n \in \sigma(a_n)$ and $\alpha_n \to \alpha$, then $\alpha \in \sigma(a)$.

Exercise 1.2.9. Let $\mathscr{A}$ be a unital Banach algebra, and let a, b be commuting elements of $\mathscr{A}$ (i.e., $ab = ba$). Show that

$$r(ab) \leq r(a)r(b) \quad \text{and} \quad r(a+b) \leq r(a) + r(b).$$

Exercise 1.2.10. Let $\mathscr{A}$ be a unital Banach algebra. Prove that:

a) $r(a^2) = r^2(a)$ for each $a \in \mathscr{A}$;
b) the following statements are equivalent:

    i. there exists $c > 0$ such that $c\|a\|^2 \leq \|a^2\|$ for all $a \in \mathscr{A}$;
    ii. there exists $d > 0$ such that $d\|a\| \leq r(a)$ for all $a \in \mathscr{A}$;

c) $\|a\|^2 = \|a^2\|$ for all $a \in \mathscr{A}$ if and only if $\|a\| = r(a)$ for all $a \in \mathscr{A}$.

Exercise 1.2.11. Let $\mathscr{A}$ be a unital Banach algebra. Define the *exponential* of $a \in \mathscr{A}$ by

$$\exp(a) := \sum_{n=0}^{\infty} \frac{a^n}{n!}. \tag{1.11}$$

Show that:

a) the series converges absolutely for every $a \in \mathscr{A}$;
b) $\|\exp(a)\| \leq \exp(\|a\|)$;
c) if $ab = ba$, then $\exp(a+b) = \exp(a)\exp(b)$;
d) $\exp(a)$ is invertible for all $a \in \mathscr{A}$, and $(\exp(a))^{-1} = \exp(-a)$;
e) if H is a Hilbert space and if (A_n) is a sequence in $\mathscr{L}(H)$ which converges strongly to an operator $A \in \mathscr{L}(H)$, then the sequence $(\exp(A_n))$ converges strongly to $\exp(A)$. (A sequence (A_n) of operators converges strongly to A if $\|A_n x - Ax\| \to 0$ for each $x \in H$.)

Exercise 1.2.12. Let $\mathscr{B}$ be a closed unital subalgebra of a Banach algebra $\mathscr{A}$, and let $b \in \mathscr{B}$ be an element for which $\sigma_{\mathscr{B}}(b)$ has no interior points. Show that then $\sigma_{\mathscr{B}}(b) = \sigma_{\mathscr{A}}(b)$.

Exercise 1.2.13. Let $\mathscr{B}$ be a finite-dimensional unital subalgebra of an algebra $\mathscr{A}$. Prove that $\mathscr{B}$ is inverse-closed in $\mathscr{A}$.

Exercise 1.2.14. Prove that the center of a unital algebra $\mathscr{A}$ is inverse-closed in $\mathscr{A}$. More generally, show that, for each subset B of $\mathscr{A}$, its commutator $\{c \in \mathscr{A} : cb = bc \text{ for all } b \in B\}$ is an inverse-closed subalgebra of $\mathscr{A}$.

Exercise 1.2.15. Let $\mathscr{B}$ be a Banach algebra with identity e, $\mathscr{A}$ a closed and inverse-closed subalgebra of $\mathscr{B}$ with $e \in \mathscr{A}$, and $W : \mathscr{A} \to \mathscr{B}$ a bounded unital homomorphism. Show that

$$\mathscr{C} := \{a \in \mathscr{A} : W(a) \in \mathscr{A}\}$$

contains the identity element and that $\mathscr{C}$ is a closed and inverse-closed subalgebra of $\mathscr{A}$.

Exercise 1.2.16. Zorn's lemma implies that every commutative subalgebra of a Banach algebra $\mathscr{A}$ with identity is contained in a maximal commutative (= Abelian) subalgebra of $\mathscr{A}$, i.e., in a commutative subalgebra which is not properly contained in a larger commutative subalgebra. Show that maximal Abelian subalgebras (usually abbreviated to *masas*) are inverse-closed.

Exercise 1.2.17. Let $\mathscr{A}$ be an algebra over $\mathbb{F}$ with identity e, and let $p \neq e$ be a non-zero idempotent in $\mathscr{A}$.

a) Show that $\mathrm{alg}\{p\}$ consists of all linear combinations $\alpha p + \beta q$ with $\alpha, \beta \in \mathbb{F}$ where $q := e - p$.
b) Determine the spectrum of $\alpha p + \beta q$ in $\mathrm{alg}\{p\}$.
c) Determine $\exp(a)$ for each element $a \in \mathrm{alg}\{p\}$. (The exponential is defined by (1.11).)

Exercise 1.2.18. Let $\mathscr{A}$ be an algebra over $\mathbb{F}$ with identity e, and let j be an element of $\mathscr{A}$ with $j^2 = e$. Describe the algebra $\mathrm{alg}\{j\}$, and derive an invertibility criterion for the elements of this algebra. (Hint: consider the element $(e + j)/2$ and use Exercise 1.2.17.)

Exercise 1.2.19. Let $\mathscr{A}$ be an algebra with identity e, and let $p \in \mathscr{A}$ an idempotent with complementary idempotent $q := e - p$.

a) Prove that $e + qap$ is invertible for every element $a \in \mathscr{A}$.
b) Let $a \in \mathscr{A}$. Prove that $ap + q$ is invertible if and only if $pap + q$ is invertible.
c) Let $a, b \in \mathscr{A}$ be invertible elements with $ab = ba$. Prove that $ap + bq$ is invertible if and only if $pa + qb$ is invertible.

Exercise 1.2.20. Let $\mathscr{A}$ be a unital algebra, $\mathscr{B}$ an inverse-closed subalgebra of $\mathscr{A}$ and $p \in \mathscr{B}$ an idempotent. Show that:

a) $p\mathscr{A}p := \{pap : a \in \mathscr{A}\}$ is a unital algebra with identity p;
b) $p\mathscr{B}p$ is inverse-closed in $p\mathscr{A}p$.

The algebra $p\mathscr{A}p$ is called a *corner* of $\mathscr{A}$.

Exercise 1.2.21. Let $\mathscr{A}$ be an algebra with identity e, $\mathscr{J}$ an ideal of $\mathscr{A}$ and p an idempotent in $\mathscr{A}$. Show that:

a) the set $\mathscr{B}$ of all elements $pap + j + (e - p)$ with $a \in \mathscr{A}$ and $j \in \mathscr{J}$ is a unital subalgebra of $\mathscr{A}$;
b) $\mathscr{B}$ is inverse-closed in $\mathscr{A}$.

Exercise 1.2.22. Let $\mathscr{A}$ be an algebra with identity element e, $p \in \mathscr{A}$ an idempotent, $q := e - p$, and let a be an invertible element of $\mathscr{A}$. Show that pap is invertible in $p\mathscr{A}p$ if and only if $qa^{-1}q$ is invertible in $q\mathscr{A}q$. Verify *Kozak's identity*

$$(pap)^{-1} = pa^{-1}p - pa^{-1}q(qa^{-1}q)^{-1}qa^{-1}p.$$

Exercise 1.2.23. Prove the following properties of inverse-closed algebras.

a) The intersection of a family of inverse-closed subalgebras of an algebra is inverse-closed again.
b) Let $\mathscr{A}$ be a unital Banach algebra and $\mathscr{B}$ be a unital closed subalgebra of $\mathscr{A}$ with the following property: there is a dense subset $\mathscr{B}_0 \subset \mathscr{B}$ such that every element in $\mathscr{B}_0$ which is invertible in $\mathscr{A}$ possesses an inverse in $\mathscr{B}$. Then $\mathscr{B}$ is inverse-closed in $\mathscr{A}$.
c) The closure of an inverse-closed subalgebra of a Banach algebra is inverse-closed again.

Exercise 1.2.24. Let $\mathscr{A}$ be a unital involutive algebra. Show that an element a is invertible if and only if both a^*a and aa^* are invertible. Use this fact together with Proposition 1.2.36 and Exercise 1.2.12 to prove Theorem 1.2.38.

Exercise 1.2.25. Prove Theorem 1.2.39. (Hint: prove that $\sigma(r) = \{0\}$ and that $r \pm r^*$ have the property mentioned in the theorem whenever r has this property.)

1.3 Maximal ideals and representations

1.3.1 Maximal ideals and the radical

The ideals of an algebra are ordered with respect to the inclusion relation. A proper left (right, resp. two-sided) ideal is called a *maximal* left (right, resp. two-sided) ideal if it is not properly contained in any other proper left (right, resp. two-sided) ideal.

Proposition 1.3.1 (Krull's lemma)**.** *Every proper left (right, two-sided) ideal of a unital algebra $\mathscr{A}$ is contained in a maximal left (right, two-sided) ideal of $\mathscr{A}$.*

Proof. Let $\mathscr{J}$ be a left ideal of $\mathscr{A}$, and let Λ stand for the set of all proper left ideals of $\mathscr{A}$ which contain $\mathscr{J}$, ordered with respect to inclusion. If Λ' is a totally ordered subset of Λ, then $\mathscr{J}' := \cup_{\mathscr{J} \in \Lambda'} \mathscr{J}$ is a left ideal of $\mathscr{A}$, which is proper (it does not contain the identity). Thus, every linearly ordered subset of Λ is bounded above. By Zorn's lemma, Λ has a maximal element. Evidently, every maximal element of Λ is a maximal left ideal of $\mathscr{A}$ which contains $\mathscr{J}$. ∎

Let $\mathscr{A}$ be an algebra with identity e. The intersection of all maximal left ideals of $\mathscr{A}$ is called the *radical* of $\mathscr{A}$ and will be denoted by $\mathscr{R}_{\mathscr{A}}$. The radical of an algebra is a left ideal of this algebra. An algebra is called *semi-simple* if its radical is $\{0\}$.

Maximal ideals and the radical of an algebra are closely related with the invertible elements of that algebra.

Proposition 1.3.2. *Let $\mathscr{A}$ be an algebra with identity $e \neq 0$. An element of $\mathscr{A}$ is left invertible if and only if it is not contained in some maximal left ideal of $\mathscr{A}$.*

Proof. Let $a \in \mathscr{A}$ be left invertible, and let $\mathscr{J}$ be a maximal left ideal of $\mathscr{A}$ which contains a. Then $ba = e$ for some $b \in \mathscr{A}$ and, thus, $e \in \mathscr{J}$. The latter is impossible since $\mathscr{J}$ is proper. Conversely, if a is not left invertible, then $\mathscr{A}a$ is a left ideal of $\mathscr{A}$ which contains a and which is proper (because $e \notin \mathscr{A}a$). By Krull's lemma, $\mathscr{A}a$ is contained in some maximal left ideal of $\mathscr{A}$.   ∎

Proposition 1.3.3. *The following conditions are equivalent for an element r of an algebra $\mathscr{A}$ with identity e:*

  (i) *r is in the radical of $\mathscr{A}$;*
  (ii) *$e - ar$ is left invertible for every $a \in \mathscr{A}$;*
 (iii) *$e - arb$ is invertible for every $a, b \in \mathscr{A}$;*
 (iv) *$e - ra$ is right invertible for every $a \in \mathscr{A}$;*
  (v) *r is in the intersection of all maximal right ideals of $\mathscr{A}$.*

Proof. (i) $\Rightarrow$ (ii): Let $r \in \mathscr{R}_{\mathscr{A}}$, but suppose that $e - ar$ is not left invertible for some $a \in \mathscr{A}$. Then $\mathscr{A}(e - ar)$ lies in a proper left ideal of $\mathscr{A}$ which is contained in some maximal left ideal $\mathscr{J}$ by Krull's lemma. Since $\mathscr{A}$ is unital, one has $e - ar \in \mathscr{J}$. Moreover, $ar \in \mathscr{J}$ since r is in the radical and, consequently, in $\mathscr{J}$. Thus, $e \in \mathscr{J}$, in contradiction of the properness of $\mathscr{J}$.

(ii) $\Rightarrow$ (i): Let $e - ar$ be left invertible for every $a \in \mathscr{A}$. If $r \notin \mathscr{R}_{\mathscr{A}}$, then there is a maximal left ideal $\mathscr{J}$ with $r \notin \mathscr{J}$. The set $\mathscr{L} := \{l - ar : l \in \mathscr{J}, a \in \mathscr{A}\}$ is a left ideal of $\mathscr{A}$ which contains $\mathscr{J}$ properly (because $r \in \mathscr{L} \setminus \mathscr{J}$). The maximality of $\mathscr{J}$ implies $\mathscr{L} = \mathscr{A}$, hence, there are elements $l \in \mathscr{L}$ and $a \in \mathscr{A}$ such that $l + ar = e$. By hypothesis, $l = e - ar$ is a left invertible element which lies in a maximal left ideal. This contradicts Proposition 1.3.2.

(i), (ii) $\Rightarrow$ (iii): Let $r \in \mathscr{R}_{\mathscr{A}}$ and $a \in \mathscr{A}$. Then $e - ar$ is left invertible. We claim that it is right invertible, too. Let $e + b$ be a left inverse of $e - ar$, i.e., $(e + b)(e - ar) = e$ or, equivalently, $b = (a + ba)r$. Since $r \in \mathscr{R}_{\mathscr{A}}$ and the radical is a left ideal, we conclude that b belongs to the radical, too. Consequently, $e + cb$ is left invertible for all $c \in \mathscr{A}$. In particular, $e + b$ is left invertible, and $e + b$ is also right invertible by its definition. Thus, $e + b$ is invertible, and so is $e - ar$.

Now let $a, b \in \mathscr{A}$ and $r \in \mathscr{R}_{\mathscr{A}}$. Then $e - bar$ is invertible as we have just seen, and Lemma 1.2.1 implies the invertibility of $e - arb$. This settles assertion (iii). The implication (iii) $\Rightarrow$ (ii) is obvious, and the remaining implications follow as before thanks to the left-right symmetry of assertion (iii).   ∎

Corollary 1.3.4. *Every nilpotent element of a commutative algebra belongs to the radical of this algebra.*

Proof. If $r^n = 0$ for some n, then one can see immediately that $e + ar + a^2 r^2 + \ldots + a^{n-1} r^{n-1}$ is the inverse of $e - ar$.   ∎

Now we turn to the context of Banach algebras.

Theorem 1.3.5. *Let $\mathscr{A}$ be a unital Banach algebra over $\mathbb{K}$. Then:*

(i) *the closure of a proper (left, right, two-sided) ideal of $\mathscr{A}$ is a proper (left, right, two-sided) ideal;*
(ii) *every (left, right, two-sided) maximal ideal is closed;*
(iii) *the radical is a closed ideal of $\mathscr{A}$.*

Proof. Let $\mathscr{J}$ be a proper (left, right, or two-sided) ideal of $\mathscr{A}$. The closure of $\mathscr{J}$ is an ideal of the same type due to the continuity of the algebraic operations. As $\mathscr{G}_{\mathscr{A}}$ is open and $\mathscr{G}_{\mathscr{A}} \cap \mathscr{J} = \emptyset$ by Proposition 1.3.2, the closure of $\mathscr{J}$ is properly contained in $\mathscr{A}$. This proves assertion (i). Assertions (ii) and (iii) are immediate consequences of (i). ∎

    The following proposition states that the elements of the radical are, in a certain sense, "inessential" to the invertibility of other elements of the algebra. Thus, as long as one is concerned with invertibility properties only, the radical can be factored out, and one can work in a semi-simple algebra.

    An element s of an algebra $\mathscr{A}$ with identity is said to be a *perturbation for invertibility* if $a - s$ is invertible for every invertible element $a \in \mathscr{A}$.

Proposition 1.3.6. *Let $\mathscr{A}$ be a Banach algebra over $\mathbb{K}$ with identity e. Then the set of the perturbations for invertibility in $\mathscr{A}$ coincides with the radical of $\mathscr{A}$.*

Proof. If s is in the radical of $\mathscr{A}$ and $a \in \mathscr{A}$ is invertible, then $e - a^{-1}s$ is invertible by Proposition 1.3.3. Thus, $a - s = a(e - a^{-1}s)$ is invertible, and s is a perturbation for invertibility.

    For the reverse inclusion, we first verify that the set $\mathscr{S}$ of all perturbations for invertibility is a left ideal in $\mathscr{A}$. It is immediate that $\mathscr{S}$ is a linear space. Further, if a is invertible, then as is in $\mathscr{S}$ again. Indeed, if b is invertible, then $b - as = a(a^{-1}b - s)$ is also invertible. For arbitrary a, choose a complex number λ such that both $a + \lambda e$ and $a - \lambda e$ are invertible (which is possible because the spectrum of a is bounded) and, hence, both $(a + \lambda e)s$ and $(a - \lambda e)s$ belong to $\mathscr{S}$. Then the linearity of $\mathscr{S}$ implies that $as = \frac{a + \lambda e}{2}s + \frac{a - \lambda e}{2}s$ belongs to $\mathscr{S}$, too, whence the left ideal property of $\mathscr{S}$ follows. Now it is evident that $e - as$ is left invertible for all $a \in \mathscr{A}$. By Proposition 1.3.3 again, $s \in \mathscr{R}_{\mathscr{A}}$. ∎

    Due to the maximality, quotients of algebras by maximal ideals cannot contain non-trivial proper ideals. The converse is also true. This observation is the basis of a remarkable correspondence between maximal ideals and multiplicative functionals which holds in the commutative setting and which will be the subject of Section 1.3.3.

Theorem 1.3.7. *A proper closed ideal $\mathscr{M}$ of a unital Banach algebra $\mathscr{A}$ over $\mathbb{K}$ is maximal if and only if the quotient algebra $\mathscr{A}/\mathscr{M}$ is simple.*

Proof. Let $\mathscr{M}$ be a proper closed ideal of the algebra $\mathscr{A}$ with identity e, and let Φ be the canonical homomorphism from $\mathscr{A}$ onto $\mathscr{A}/\mathscr{M}$. Suppose there exists a non-trivial proper ideal $\mathscr{J}$ of $\mathscr{A}/\mathscr{M}$. It is easy to check that $\Phi^{-1}(\mathscr{J})$ is an ideal of $\mathscr{A}$

which properly contains $\mathcal{M}$, but does not coincide with $\mathcal{A}$. Thus, if $\mathcal{M}$ is maximal, then $\mathcal{A}/\mathcal{M}$ is simple.

Conversely, if $\mathcal{M}$ is not maximal, then there exists a proper ideal $\mathcal{J}$ of $\mathcal{A}$ such that $\mathcal{M}$ is properly contained in $\mathcal{J}$. Then $\Phi(\mathcal{J})$ is an ideal of $\mathcal{A}/\mathcal{M}$ such that $\Phi(\mathcal{J}) \neq 0 + \mathcal{M}$ because $\mathcal{J} \neq \mathcal{M}$ and $\Phi(\mathcal{J}) \neq \mathcal{A}/\mathcal{M}$ because $e + \mathcal{M} \notin \Phi(\mathcal{J})$. Thus, $\Phi(\mathcal{J})$ is a non-trivial ideal of $\mathcal{A}/\mathcal{M}$, and $\mathcal{A}/\mathcal{M}$ is not simple.   ∎

The following is a consequence of various results about ideals and invertibility obtained so far.

Theorem 1.3.8. *Let $\mathcal{A}$ be a commutative unital Banach algebra over $\mathbb{K}$, and let $\mathcal{M}$ be a maximal ideal of $\mathcal{A}$. Then the quotient algebra $\mathcal{A}/\mathcal{M}$ is a field. In the case $\mathbb{K} = \mathbb{C}$, the quotient $\mathcal{A}/\mathcal{M}$ is isometrically isomorphic to $\mathbb{C}$.*

Proof. If $\mathcal{M}$ is a maximal ideal, the quotient algebra $\mathcal{A}/\mathcal{M}$ is simple by Theorem 1.3.7. Thus, every non-zero element of $\mathcal{A}/\mathcal{M}$ is invertible in $\mathcal{A}/\mathcal{M}$ by Proposition 1.3.2. This settles the first assertion, and the second one results from the Gelfand-Mazur theorem 1.2.11.   ∎

1.3.2 Representations

In this section, we summarize some basic notions and results from the representation theory of Banach and C^*-algebras. Albeit most notions apply to Banach algebras, the strongest results will follow only for C^*-algebras. More detailed expositions as well as proofs can be found in every textbook on C^*-algebras.

A *representation* of an algebra $\mathcal{A}$ over a field $\mathbb{F}$ is a pair (X, π) consisting of a linear space X over $\mathbb{F}$ and an algebra homomorphism π from $\mathcal{A}$ into the algebra $L(X)$ of all linear operators on X. If the space X is evident from π, sometimes it is omitted, and we talk simply a representation π. The representation (X, π) is *faithful* if the kernel of π consists of the zero element only. In this case, π is an algebra isomorphism from $\mathcal{A}$ onto a subalgebra of $L(X)$.

Example 1.3.9. Let $\mathcal{A}$ be a unital algebra. Every element $a \in \mathcal{A}$ determines a linear operator $L_a : \mathcal{A} \to \mathcal{A}$ by $L_a(x) := ax$. The mapping $L : a \mapsto L_a$ is a representation of $\mathcal{A}$, the so-called *left regular representation*. More generally, if $\mathcal{J}$ is a left ideal of $\mathcal{A}$ then $\mathcal{A}/\mathcal{J}$ becomes a linear space in a natural way. Let $\Phi : \mathcal{A} \to \mathcal{A}/\mathcal{J}$ denote the canonical linear mapping $a \mapsto a + \mathcal{J}$. To each element $a \in \mathcal{A}$, we associate an operator $L_a^{\mathcal{J}} : \mathcal{A}/\mathcal{J} \to \mathcal{A}/\mathcal{J}$ via $L_a^{\mathcal{J}}(\Phi(x)) := \Phi(ax)$. The homomorphism $L^{\mathcal{J}} : \mathcal{A} \to L(\mathcal{A}/\mathcal{J})$, $a \mapsto L_a^{\mathcal{J}}$ is called the *left regular representation of $\mathcal{A}$ induced by $\mathcal{J}$*. See Exercise 1.3.6 for some details of these constructions.   □

Let $\mathscr{A}$ be an algebra and (X, π) a representation of $\mathscr{A}$. A subspace K of X is *invariant* for π if

$$\pi(a)K \subseteq K \quad \text{for all} \ a \in \mathscr{A}.$$

The subspaces $\{0\}$ and X are invariant for every representation. A non-zero representation (X, π) of an algebra $\mathscr{A}$ is called *algebraically irreducible* if $\{0\}$ and X are the only invariant subspaces for π.

Lemma 1.3.10 (Schur). *Let (X, π) be an algebraically irreducible representation of the algebra $\mathscr{A}$ and $B \neq 0$ a linear operator on X. If $B\pi(a) = \pi(a)B$ for all $a \in \mathscr{A}$, then B is invertible.*

Proof. The condition $B\pi(a) = \pi(a)B$ implies that $\operatorname{Ker} B$ and $\operatorname{Im} B$ are invariant subspaces for π. Since π is irreducible, and since $\operatorname{Ker} B \neq X$ and $\operatorname{Im} B \neq \{0\}$ by hypothesis, we have $\operatorname{Ker} B = \{0\}$ and $\operatorname{Im} B = X$. Thus, B is injective and surjective, hence invertible.   ∎

In the category of Banach algebras, an appropriate notion of a representation is defined as follows. Let $\mathscr{A}$ be a complex Banach algebra. A *representation* of $\mathscr{A}$ is a pair (X, π) where X is a complex Banach space and π is an algebra homomorphism from $\mathscr{A}$ into the algebra $\mathscr{L}(X)$ of the bounded linear operators on X. Note that the continuity of π is not required. A non-zero representation (X, π) of a Banach algebra $\mathscr{A}$ is called *topologically irreducible* if $\{0\}$ and X are the only *closed* subspaces of X which are invariant for π.

Sometimes, the continuity of π follows automatically. One instance is provided by the following theorem, a proof of which can be found in [10, Chapter 3, Section 25].

Theorem 1.3.11. *If (X, π) is an algebraically irreducible representation of a Banach algebra $\mathscr{A}$, then π is continuous.*

Example 1.3.12. Let $\mathscr{J}$ be a closed left ideal of a Banach algebra $\mathscr{A}$. Then the left regular representation $L^{\mathscr{J}}$ is continuous and has norm 1. Moreover, every left regular representation of $\mathscr{A}$ induced by a *maximal* left ideal $\mathscr{J}$ is algebraically irreducible. The proof is left as an exercise (Exercise 1.3.7).   □

We will need the following corollary of Schur's lemma.

Corollary 1.3.13. *Let $\mathscr{A}$ be a unital Banach algebra and $\mathscr{J}$ a maximal left ideal of $\mathscr{A}$, and let $L^{\mathscr{J}} : \mathscr{A} \to \mathscr{L}(\mathscr{A}/\mathscr{J})$ be the induced left regular representation. Further let B be a (not necessarily bounded) linear operator on $\mathscr{A}/\mathscr{J}$. If $BL_a^{\mathscr{J}} = L_a^{\mathscr{J}} B$ for all $a \in \mathscr{A}$, then B is a scalar multiple of the identity operator.*

Proof. Let $x \in \mathscr{A}/\mathscr{J}$ and choose $a \in \mathscr{A}$ such that $\|a\| \leq 2\|\Phi(a)\|$ and $\Phi(a) = x$. Then

$$\|Bx\| = \left\|BL_a^{\mathscr{J}}\, \Phi(e)\right\| = \left\|L_a^{\mathscr{J}}\, B\Phi(e)\right\|$$
$$\leq \|a\|\|B\Phi(e)\| \leq 2\|\Phi(a)\|\|B\Phi(e)\| = 2\|B\Phi(e)\|\|x\|.$$

Thus B is bounded. Being a bounded operator on a Banach space, B has a non-empty spectrum by Theorem 1.2.10. Choose any point λ in the spectrum of B. By Exercise 1.3.7, the left regular representation $L^{\mathscr{J}}$ is algebraically irreducible. So we can apply Schur's lemma to $\lambda I - B$. Since $\lambda I - B$ is not invertible, it must be 0. Thus, $B = \lambda I$.   ∎

Definition 1.3.14. A Banach algebra is *primitive* if it contains a maximal left ideal for which the induced left regular representation is injective.

Lemma 1.3.15. *A Banach algebra $\mathscr{A}$ is primitive if and only if there exists a maximal left ideal in $\mathscr{A}$ that does not contain non-trivial maximal ideals of $\mathscr{A}$.*

Proof. First let $\mathscr{J}$ be a maximal left ideal of $\mathscr{A}$ which induces an injective left regular representation. Let $\Phi : \mathscr{A} \to \mathscr{A}/\mathscr{J}$ denote the canonical linear mapping $a \mapsto a + \mathscr{J}$. Further let $\mathscr{I} \subset \mathscr{J}$ be an ideal and $a \in \mathscr{I}$. Then $L_a^{\mathscr{J}}(\Phi(x)) = \Phi(ax) = 0$ for any $x \in \mathscr{A}$, i.e., $L_a^{\mathscr{J}}$ is the zero operator. Since $L^{\mathscr{J}}$ is an isomorphism, $a = (L^{\mathscr{J}})^{-1}(L_a^{\mathscr{J}}) = 0$. Thus, $\mathscr{I}$ is the zero ideal.

For the reverse direction assume that the left ideal $\mathscr{J}$ does not contain proper non-trivial ideals. If $L^{\mathscr{J}}(a) = 0$, then $ax \in \mathscr{J}$ for all elements $x \in \mathscr{A}$. Define $\mathscr{I} := \{b \in \mathscr{A} : b\mathscr{A} \subset \mathscr{J}\}$. The set $\mathscr{I}$ is clearly an ideal of $\mathscr{A}$ contained in $\mathscr{J}$ and $a \in \mathscr{J}$. This implies $a = 0$, whence the injectivity of $L^{\mathscr{J}}$.   ∎

In the context of C^*-algebras and their homomorphism, by a *representation* of a C^*-algebra $\mathscr{A}$ one means a pair (H, π) where now H is a Hilbert space and π is a *-homomorphism from $\mathscr{A}$ into the algebra $\mathscr{L}(H)$. By Theorem 1.1.5, π is bounded (and even a contraction). In this setting, if K is a closed subspace of H and P_K denotes the orthogonal projection from H onto K, invariance of K for the representation (H, π) just means that

$$P_K \pi(a) P_K = \pi(a) P_K \quad \text{for all } a \in \mathscr{A} \tag{1.12}$$

which turns out to be equivalent to

$$P_K \pi(a) = \pi(a) P_K \quad \text{for all } a \in \mathscr{A}. \tag{1.13}$$

Indeed, (1.12) follows from (1.13) by multiplication by P_K, and (1.12) implies (1.13):

$$P_K \pi(a) = (\pi(a^*)P_K)^* = (P_K \pi(a^*)P_K)^* = P_K \pi(a)P_K = \pi(a)P_K.$$

Theorem 1.3.16. *The following assertions are equivalent for a representation* (H, π) *of a C^*-algebra $\mathscr{A}$:*

(i) (H, π) *is algebraically irreducible;*
(ii) (H, π) *is topologically irreducible;*
(iii) *every vector $x \in H \setminus \{0\}$ is algebraically cyclic, i.e., $\pi(\mathscr{A})x$ coincides with H;*
(iv) *every vector $x \in H \setminus \{0\}$ is topologically cyclic, i.e., $\pi(\mathscr{A})x$ is dense in H.*

1.3.3 Multiplicative linear functionals

Let $\mathscr{A}$ be an algebra over a field $\mathbb{F}$. An algebra homomorphism $\phi : \mathscr{A} \to \mathbb{F}$ (where the field $\mathbb{F}$ is considered as an algebra over itself) is also called a *multiplicative linear functional* or a *character* of $\mathscr{A}$. The homomorphism $\mathscr{A} \to \mathbb{F} : a \mapsto 0$ is called the *trivial* or the *zero character*. Trivial characters are usually excluded in what follows.

Proposition 1.3.17. *Let $\mathscr{A}$ be an algebra with identity e and ϕ a non-zero multiplicative linear functional over $\mathscr{A}$. Then:*

(i) $\phi(e) = 1;$
(ii) $\phi(a) \in \sigma(a)$ *for each element $a \in \mathscr{A}$;*
(iii) *the kernel of ϕ is a maximal ideal of $\mathscr{A}$.*

Proof. For the first assertion, choose $a \in \mathscr{A}$ such that $\phi(a) \neq 0$. Then $\phi(a) = \phi(ea) = \phi(e)\phi(a)$, whence $\phi(e) = 1$. For (ii), let $a \in \mathscr{A}$. Then $\phi(a)e - a$ is in the kernel of ϕ, which is a proper ideal by (i). So $\phi(a)e - a$ cannot be invertible. Finally, the kernel of a non-zero multiplicative linear functional is a hyperplane of codimension 1 in the linear space $\mathscr{A}$. So, as an ideal, it must be maximal. ∎

Theorem 1.3.18. *Non-zero multiplicative linear functionals over a unital Banach algebra over $\mathbb{K}$ are bounded and have norm 1.*

Proof. Let $\mathscr{A}$ be a Banach algebra with identity e. Suppose that there is an element $a \in \mathscr{A}$ such that $\|a\| = 1$ and $|\phi(a)| > 1$. Then $\phi(a)e - a$ is invertible, and

$$1 = \phi(e) = \phi\left((\phi(a)e - a)(\phi(a)e - a)^{-1}\right) = \phi\left((\phi(a)e - a)\right)\phi\left((\phi(a)e - a)^{-1}\right) = 0.$$

This contradiction implies that

$$\|\phi\| = \sup_{\|a\|=1} |\phi(a)| \leq 1.$$

As $\phi(e) = 1$, one has $\|\phi\| = 1$. ∎

There is a close relationship between maximal ideals and characters of a commutative complex Banach algebra.

Theorem 1.3.19. *Let $\mathscr{A}$ be a commutative unital Banach algebra over $\mathbb{C}$. The kernel of a non-trivial character of $\mathscr{A}$ is a maximal ideal of $\mathscr{A}$ and, vice versa, any non-trivial maximal ideal in $\mathscr{A}$ is the kernel of one and only one non-trivial character of $\mathscr{A}$.*

Proof. The first part of the assertion is a particular case of Proposition 1.3.17 (iii). For the second part, let $\mathscr{M}$ be a maximal ideal in $\mathscr{A}$. Then the quotient algebra $\mathscr{A}/\mathscr{M}$ is isometrically isomorphic to the field $\mathbb{C}$ by Theorem 1.3.8. Identifying $\mathscr{A}/\mathscr{M}$ with $\mathbb{C}$, one can consider the canonical homomorphism $\Phi : \mathscr{A} \to \mathscr{A}/\mathscr{M}$ as a non-trivial character with the given maximal ideal as kernel. Thus, there exist characters with kernel $\mathscr{M}$. That there is only one such character will follow easily from the following observation which we formulate as a separate lemma.   ∎

We call two linear functionals ϕ_1 and ϕ_2 on an algebra over $\mathbb{F}$ *proportional* if there exists a non-zero constant $\alpha \in \mathbb{F}$ such that $\phi_1 = \alpha \phi_2$.

Lemma 1.3.20. *Two linear functionals defined on an algebra over $\mathbb{F}$ have the same kernel if and only if they are proportional.*

Proof. Let ϕ_1 and ϕ_2 be functionals from an $\mathbb{F}$-algebra $\mathscr{A}$ into $\mathbb{F}$ which have the same kernel $\mathscr{N}$. Choose any element $c \in \mathscr{A}$ which is not in the kernel. This element spans a one-dimensional subspace of $\mathscr{A}$, which together with $\mathscr{N}$ spans the whole algebra $\mathscr{A}$. Let $a \in \mathscr{A}$. Write a as $\alpha c + n$ with $\alpha \in \mathbb{F}$ and $n \in \mathscr{N}$. Then $\phi_1(a) = \alpha \phi_1(c)$ and $\phi_2(a) = \alpha \phi_2(c)$, whence

$$\phi_2(a) = \alpha \phi_2(c) = \alpha \frac{\phi_2(c)}{\phi_1(c)} \phi_1(c) = \frac{\phi_2(c)}{\phi_1(c)} \phi_1(a).$$

The proof in the reverse direction is immediate.   ∎

Theorem 1.3.21. *Let $\mathscr{A}$ be a unital commutative Banach algebra over $\mathbb{C}$. An element $a \in \mathscr{A}$ is invertible if and only if $\phi(a) \neq 0$ for every non-zero character ϕ of $\mathscr{A}$.*

Proof. If a is invertible, then $1 = \phi(e) = \phi(a)\phi(a^{-1})$ for every non-zero character ϕ of $\mathscr{A}$. Thus, $\phi(a) \neq 0$. If a is not invertible, then a belongs to a maximal ideal $\mathscr{M}$ in $\mathscr{A}$ by Proposition 1.3.2. Let ϕ be the multiplicative functional which has $\mathscr{M}$ as its kernel. Then $\phi(a) = 0$.   ∎

Corollary 1.3.22. *The set of the non-zero characters of a commutative unital complex Banach algebra is not empty.*

Proof. If the algebra is a field, one has the identity functional. If not, then the algebra possesses at least one non-invertible non-zero element due to the Gelfand-Mazur theorem. Hence, there is a non-zero maximal ideal by Proposition 1.3.2 which in turn implies the existence non-trivial character via Theorem 1.3.19. ∎

Another consequence of Theorem 1.3.21 concerns spectra in singly generated Banach algebras.

Theorem 1.3.23. *Let $\mathscr{A}$ be a unital Banach algebra over $\mathbb{C}$ and let $a \in \mathscr{A}$. Further let $\mathscr{B}$ refer to the algebra* alg$\{a\}$, *i.e., to the smallest closed subalgebra of $\mathscr{A}$ which contains a and the identity element. Then $\rho_{\mathscr{B}}(a)$ is connected.*

Proof. Contrary to what we want to show, suppose $\sigma_{\mathscr{A}}(a)$ has a hole $B \subset \rho_{\mathscr{A}}(a)$, and let $\mu_0 \in B$. Let p be a polynomial. By the maximum modulus principle and the spectral mapping theorem for polynomials (Proposition 1.2.5),

$$
\begin{aligned}
|p(\mu_0)| &\leq \max\{|p(\mu)| : \mu \in \partial B\} \\
&\leq \max\{|p(\mu)| : \mu \in \sigma_{\mathscr{A}}(a)\} \\
&= \max\{|\mu| : \mu \in \sigma_{\mathscr{A}}(p(a))\} = r(p(a)) \leq \|p(a)\|. \quad (1.14)
\end{aligned}
$$

This estimate shows that if p_1 and p_2 are polynomials with $p_1(a) = p_2(a)$, then $p_1(\mu_0) = p_2(\mu_0)$. Therefore, one can define a functional ϕ on the set of polynomials of a by

$$
\phi(p(a)) := p(\mu_0)
$$

which is evidently linear and multiplicative. By (1.14), this functional is continuous. So it can be continuously extended onto the closure of the set of polynomials, which coincides with $\mathscr{B}$. Since $a = p(a)$ for the polynomial $p(z) := z$, one has $\phi(a) = \mu_0$, whence $\mu_0 \in \sigma_{\mathscr{B}}(a)$. This contradiction proves the assertion. ∎

We end this section with the notion of the joint spectrum of a collection of elements belonging to a commutative Banach algebra $\mathscr{A}$. Let $M_{\mathscr{A}}$ represent the set of maximal ideals of $\mathscr{A}$ and ϕ_x denote the character associated with the maximal ideal $x \in M_{\mathscr{A}}$.

Given a set $A := \{a_1, \ldots, a_n\} \subset \mathscr{A}$, the *joint spectrum* of A is the set

$$
\sigma_{\mathscr{A}}(a_1, \ldots, a_n) := \{(\phi_x(a_1), \ldots, \phi_x(a_n)) : x \in M_{\mathscr{A}}\} \subset \mathbb{C}^n.
$$

This notion coincides with the usual notion of spectrum, in the case A contains a single element. The proof of the next proposition is trivial.

Proposition 1.3.24. *Let $\mathscr{A}$ be a unital commutative Banach algebra over $\mathbb{C}$, and $a_1, \ldots, a_n \in \mathscr{A}$. Then the joint spectrum $\sigma_{\mathscr{A}}(a_1, \ldots, a_n)$ coincides with the set of points $(\lambda_1, \ldots, \lambda_n) \in \mathbb{C}^n$, such that the smallest closed ideal of $\mathscr{A}$ which contains $\lambda_i e - a_i, i = 1, \ldots, n$, is proper.*

1.3.4 Exercises

Exercise 1.3.1. Consider the algebra $\mathbb{R}^{2\times 2}$. Find its left, right and two-sided ideals. What is its radical? Do the same for the subalgebra of $\mathbb{R}^{2\times 2}$ which consists of upper triangular matrices.

Exercise 1.3.2. For $n > 1$, determine the characters of the algebra $\mathbb{R}^{n\times n}$ and of the algebra of upper triangular matrices in $\mathbb{R}^{n\times n}$.

Exercise 1.3.3. Show that for each algebra $\mathscr{A}$, the quotient algebra $\mathscr{A}/\mathscr{R}_{\mathscr{A}}$ is semi-simple and that the radical of $\mathscr{R}_{\mathscr{A}}$ is $\mathscr{R}_{\mathscr{A}}$.

Exercise 1.3.4. Let $\mathscr{A}$ be a Banach algebra with identity. Show that $a \in \mathscr{A}$ is invertible if and only if $a + \mathscr{R}_{\mathscr{A}} \in \mathscr{A}/\mathscr{R}_{\mathscr{A}}$ is invertible.

Exercise 1.3.5. Let $\mathscr{A}$ be a commutative unital algebra, and let p be a non-constant polynomial, all zeros of which are simple. Show that if $p(a) = 0$ and $p(a+r) = 0$ for certain elements $a \in \mathscr{A}$ and $r \in \mathscr{R}_{\mathscr{A}}$, then $r = 0$.

Exercise 1.3.6. Let $\mathscr{A}$ be a unital Banach algebra. Consider the left regular representation $a \mapsto L_a$ of $\mathscr{A}$ in Example 1.3.9, and denote it by v.

a) Show that $L_a \in \mathscr{L}(\mathscr{A})$.
b) Show that $\mathscr{A}_1 := \{L_a : a \in \mathscr{A}\}$ is a subalgebra of $\mathscr{L}(\mathscr{A})$.
c) Prove that $\mathscr{A}_1$ is closed in $\mathscr{L}(\mathscr{A})$ with respect to the norm $\|L_a\| := \sup_{\|x\|\leq 1} \|ax\|$.
d) Conclude that the algebras $\mathscr{A}$ and $\mathscr{A}_1$ are isometrically isomorphic.
e) Prove that $v : \mathscr{A} \to \mathscr{L}(\mathscr{A})$ has norm 1.
f) Prove that if $F \subset E$ is a linear subspace of E such that $L_a(F) \subset F$ for all $a \in \mathscr{A}$ then $F = E$ or $F = \{0\}$. (Hint: check that the invariant subspaces for the representation defined in Example 1.3.9 are the left ideals of $\mathscr{A}$.) This result means that v is an algebraic irreducible representation of $\mathscr{A}$.

Exercise 1.3.7. Show that the left regular representation of a Banach algebra $\mathscr{A}$ induced by a *maximal* left ideal $\mathscr{J}$ is algebraically irreducible.

Exercise 1.3.8. Determine the maximal ideals of algebras which are generated by one idempotent (Exercise 1.2.17). Do the same for algebras which are generated by a flip (Exercise 1.2.18).

Exercise 1.3.9. Let l_∞ refer to the Banach space of all bounded sequences $u : \mathbb{N} \to \mathbb{C}$ with the supremum norm, and let l_∞^c and l_∞^0 denote the subspaces of l_∞ of all convergent sequences and of all convergent sequences with limit zero, respectively. Provide l_∞ with pointwise defined multiplication.

a) Show that l_∞ is a commutative Banach algebra and l_∞^c a closed subalgebra of l_∞.
b) Show that l_∞^0 is a proper closed ideal of l_∞.
c) Find a multiplicative linear functional in l_∞^c with kernel l_∞^0 and conclude that l_∞^0 is a maximal ideal in l_∞^c. Is l_∞^0 also a maximal ideal in l_∞? (Hint: use Theorem 1.3.8.)
d) Determine all maximal ideals of l_∞^c.

Exercise 1.3.10. Determine the maximal ideals of the Wiener algebra discussed in Example 1.1.12.

Exercise 1.3.11. Give an example of a real algebra $\mathscr{A}$ and of a maximal ideal $\mathscr{J}$ of that algebra for which $\mathscr{A}/\mathscr{J}$ is *not* isomorphic to $\mathbb{R}$.

1.4 Some examples of Banach algebras

Here we present some concrete examples of Banach algebras. The first two and the last example will provide us with a frame in which to study invertibility problems, whereas the other examples will serve to illustrate the general theory of what follows.

1.4.1 $\mathscr{L}(X)$, $\mathscr{K}(X)$, and $\mathscr{L}(X)/\mathscr{K}(X)$

Let X be a Banach space. We have already met the Banach algebra $\mathscr{L}(X)$ of all bounded linear operators on X. This algebra is unital, with the identity operator I as the unit element. The set $\mathscr{K}(X)^1$ of the compact operators is a closed (two-sided) ideal of X which is proper if X has infinite dimension, whereas the operators of finite rank form an ideal which is non-closed in general. In the case that $X = H$ is a Hilbert space, $\mathscr{L}(H)$ is a C^*-algebra, and $\mathscr{K}(H)$ is a symmetric closed ideal of that algebra. If, moreover, H is separable and has infinite dimension, then $\mathscr{K}(H)$ is the only non-trivial closed ideal of $\mathscr{L}(H)$.

For each operator $A \in \mathscr{L}(X)$, we let Im $A := AX$ denote its range and Ker $A := \{x \in X : Ax = 0\}$ its kernel. By definition, an operator $A \in \mathscr{L}(X)$ is invertible in $\mathscr{L}(X)$ if there exists an operator $B \in \mathscr{L}(X)$ such that $AB = I$ and $BA = I$. The first of these conditions implies that Im $A = X$, i.e., that A is onto, whereas the second condition implies that Ker $A = \{0\}$, i.e., A is one-to-one. Thus, A is invertible as a mapping. Conversely, if A is invertible as a mapping, then a theorem by Banach establishes that the inverse mapping B is bounded again; thus, A is invertible in the algebra $\mathscr{L}(X)$.

If X has infinite dimension, it makes sense to consider operators which are "almost invertible" in the sense that their kernel has *finite* dimension and that the addition of a *finite*-dimensional space to their range already yields all of X. Operators with these properties are called *Fredholm operators*. Equivalently, an operator $A \in \mathscr{L}(X)$ is a Fredholm operator if

- Im A is closed,
- dim Ker $A < \infty$, and
- dim Coker $A < \infty$

---

[1] When the space X is evident from the context, we will sometimes write $\mathscr{L}$ and $\mathscr{K}$ in place of $\mathscr{L}(X)$ and $\mathscr{K}(X)$, respectively.

where Coker $A = X/\overline{\text{Im}\,A}$. The *Fredholm index* (or *index* for short) ind A of a Fredholm operator A is the integer

$$\text{ind}\,A := \dim \text{Ker}\,A - \dim \text{Coker}\,A. \tag{1.15}$$

Below we collect several important properties of Fredholm operators (for details see [150]). Note the remarkable stability properties of the index under small and compact perturbations. Neither the kernel dimension of A nor the cokernel dimension behave particularly well under such perturbations, but their difference does.

Let $A, B \in \mathscr{L}(X)$ be Fredholm operators. Then:

- there exists an operator $R \in \mathscr{L}(X)$ and finite rank operators K_1, K_2 such that $AR = I + K_1$ and $RA = I + K_2$; conversely, if there exist operators $R_1, R_2 \in \mathscr{L}(X)$ and compact operators K_1, K_2 on X such that $R_1 A = I + K_1$ and $AR_2 = I + K_2$, then A is a Fredholm operator;
- if $\|K\|$ is small enough then $A + K$ is a Fredholm operator and ind $A = $ ind $(A + K)$; thus, the set of all Fredholm operators is open in $\mathscr{L}(X)$, and the index is a continuous function on this set;
- if K is compact then $A + K$ is a Fredholm operator and ind $A = $ ind $(A + K)$;
- AB is a Fredholm operator and ind $(AB) = $ ind $A + $ ind B;
- the Banach adjoint A^* of A is a Fredholm operator and ind $A^* = -$ind A. The same holds for the Hilbert space adjoint if X is a Hilbert space.

The quotient $\mathscr{L}(X)/\mathscr{K}(X)$ is called the *Calkin algebra* of X. The first assertion in the above list states that the coset $A + \mathscr{K}(X)$ is invertible in the Calkin algebra if and only if A is a Fredholm operator. This fact is also known as Calkin's theorem. Thus, the Fredholm property is equivalent to invertibility in a suitable Banach algebra.

1.4.2 Sequences of operators

For later reference, we collect some facts on the convergence of sequences of linear bounded operators on a Banach space. In some places in what follows we will have to deal with generalized sequences, which are defined on an unbounded subset of the real line as the interval $[1, \infty[$. Hence, we will state the facts below for such generalized sequences. In general, this will not cause any changes when compared with common sequences, with one exception: It is no longer true that strongly convergent generalized sequences are bounded.

In the remainder of this section, let $\mathbb{I}$ be an unbounded subset of the positive semi-axis $\mathbb{R}^+$ and let X be a Banach space. By a *generalized sequence* in X we mean a function $n \mapsto x_n$ from $\mathbb{I}$ to X, which we will often write as $(x_n)_{n \in \mathbb{I}}$. As usual, the sequence $(x_n)_{n \in \mathbb{I}}$ is called *bounded* if $\sup_{n \in \mathbb{I}} \|x_n\| < \infty$, and the sequence $(x_n)_{n \in \mathbb{I}}$ is said to converge to $x \in X$ if, for every $\varepsilon > 0$, there is an n_0 such that $\|x_n - x\| < \varepsilon$ for all $n \in \mathbb{I}$ with $n \geq n_0$. A convergent generalized sequence is not necessarily bounded, but it is eventually bounded in the following sense: The generalized sequence $(x_n)_{n \in \mathbb{I}}$ in X is *eventually bounded* if there is an $n_0 \in \mathbb{I}$ such that the sequence $(\|x_n\|)_{n \geq n_0}$ is bounded. Clearly, if $\mathbb{I} = \mathbb{Z}^+$, then every eventually bounded sequence is bounded. If the specification of $\mathbb{I}$ is evident from the context, we will often omit the word *generalized* and speak simply about sequences.

Now we turn to (generalized) sequences of operators, that is, to functions $n \mapsto A_n$ from $\mathbb{I}$ to $\mathscr{L}(X)$, which we write as $(A_n)_{n \in \mathbb{I}}$. In accordance with the above notation, the sequence $(A_n)_{n \in \mathbb{I}}$ is said to be *bounded* if the sequence $(\|A_n\|)_{n \in \mathbb{I}}$ is bounded, and it is called *eventually bounded* if the sequence $(\|A_n\|)_{n \in \mathbb{I}}$ has this property. Further, we say that a sequence $(A_n)_{n \in \mathbb{I}}$ in $\mathscr{L}(X)$ converges to $A \in \mathscr{L}(X)$ as $n \to \infty$:

- *in the norm* if $\|A_n - A\|_{\mathscr{L}(X)} \to 0$;
- *strongly* if $\|(A_n - A)u\|_X \to 0$ for every $u \in X$;
- *weakly* if $\langle v, (A_n - A)u \rangle \to 0$ for every pair $u \in X$ and $v \in X^*$.

Clearly, norm convergence implies strong convergence, and strong convergence implies weak convergence. We will use the symbols "$\rightrightarrows$", "$\to$" and "$\rightharpoonup$" to indicate convergence in the norm, strong convergence and weak convergence, respectively. It is in general not true that the adjoint sequence of a strongly convergent sequence is strongly convergent again. Thus we call a sequence (A_n) *-strongly convergent* if the sequence (A_n) converges strongly on X and the sequence (A_n^*) converges strongly on X^*.

Lemma 1.4.1. *Let an operator sequence (A_n) be eventually bounded.*

(i) *If $\langle v, A_n u \rangle \to 0$ for all u and v belonging to a dense subset of X and X^*, respectively, then the sequence (A_n) converges weakly to zero.*
(ii) *If $\|A_n u\| \to 0$ for each u belonging to a dense subset of X, then the sequence (A_n) converges strongly to zero.*

Proof. We will prove assertion (i) only. The proof of (ii) is similar. Choose $n_0 \in \mathbb{I}$ such that the sequence $(A_n)_{n \geq n_0}$ is bounded, and set $M := \sup_{n \geq n_0} \|A_n\|$. Let $\varepsilon > 0$. For $u \in X$ and $v \in X^*$, choose u_ε in the dense subset of X and v_ε in the dense subset of X^* such that $\|u - u_\varepsilon\| < \varepsilon$ and $\|v - v_\varepsilon\| < \varepsilon$. Then, for $n \geq n_0$,

$$
\begin{aligned}
|\langle v, A_n u \rangle| &\leq |\langle v - v_\varepsilon, A_n u \rangle| + |\langle v_\varepsilon, A_n u \rangle| \\
&\leq |\langle v - v_\varepsilon, A_n u \rangle| + |\langle v_\varepsilon, A_n(u - u_\varepsilon) \rangle| + |\langle v_\varepsilon, A_n u_\varepsilon \rangle| \\
&\leq \|v - v_\varepsilon\| \|A_n u\| + \|v_\varepsilon\| \|A_n\| \|u - u_\varepsilon\| + |\langle v_\varepsilon, A_n u_\varepsilon \rangle| \\
&\leq \varepsilon M \|u\| + \varepsilon M(\|v\| + \varepsilon) + |\langle v_\varepsilon, A_n u_\varepsilon \rangle| \\
&\leq \varepsilon M(\|u\| + \|v\| + \varepsilon) + |\langle v_\varepsilon, A_n u_\varepsilon \rangle|.
\end{aligned}
$$

Now let $n_1 \geq n_0$ be such that $|\langle v_\varepsilon, A_n u_\varepsilon \rangle| < \varepsilon$ for $n \geq n_1$. Then

$$
|\langle v, A_n u \rangle| \leq \varepsilon M(\|u\| + \|v\| + \varepsilon) + \varepsilon \quad \text{for } n \geq n_1,
$$

whence the weak convergence of (A_n) to zero. ■

We next discuss some consequences of the uniform boundedness principle, which is one of the cornerstones of operator theory. For a detailed account, one can consult [26, Chapter III, Section 14], for instance. The second assertion of the following theorem is also known as the Banach-Steinhaus theorem.

Theorem 1.4.2 (Banach-Steinhaus).

(i) *Every weakly convergent generalized sequence of operators is eventually bounded.*
(ii) *Every strongly convergent generalized sequence $(A_n) \subseteq \mathscr{L}(X)$ is eventually bounded, the strong limit A of this sequence belongs to $\mathscr{L}(X)$, and*

$$\|A\| \leq \liminf_{n \to +\infty} \|A_n\|.$$

Proof. Again we prove assertion (i) only. Let $(A_n)_{n \in \mathbb{I}} \subseteq \mathscr{L}(X)$ be a weakly convergent generalized sequence. Assume this sequence is not eventually bounded. Then there exists, for every positive integer n, a number $t_n \in \mathbb{I}$ with $t_n \geq n$ and such that $\|A_{t_n}\| \geq n$. For every pair $u \in X$ and $v \in X^*$, the (common) sequence $(\langle v, A_{t_n} u \rangle)_{n \in \mathbb{N}} = (\langle A_{t_n}^* v, u \rangle)_{n \in \mathbb{N}}$ is convergent. Then this sequence is bounded, and the uniform boundedness principle implies the boundedness of the sequence $(A_{t_n}^* v)_{n \in \mathbb{N}}$ of linear functionals for every $v \in X^*$. Applying the uniform boundedness principle again, we get the boundedness of the sequence $(A_{t_n}^*)_{n \in \mathbb{N}}$, whence the boundedness of the sequence $(A_{t_n})_{n \in \mathbb{N}}$, a contradiction with the choice of t_n. ∎

Next we recall some results on products of convergent sequences of operators. Because we are working with algebraic structures, questions like "Is the limit of the product of sequences equal to the product of the limits?" are of great importance. The answer, as one can see in the following results, depends on the type of convergence. Let A, B, A_n and B_n be operators in $\mathscr{L}(X)$.

Lemma 1.4.3. *If $A_n \rightrightarrows A$ then $A_n^* \rightrightarrows A^*$, and the sequence (A_n) is eventually bounded. If also $B_n \rightrightarrows B$ then $A_n B_n \rightrightarrows AB$.*

Proof. The first assertion comes from the fact that $\|T\| = \|T^*\|$ for every operator $T \in \mathscr{L}(X)$. The eventual boundedness of the sequence (A_n) can be seen by writing $\|A_n\| \leq \|A_n - A\| + \|A\|$. Finally one has

$$\|AB - A_n B_n\| \leq \|A - A_n\| \|B\| + \|A_n\| \|B - B_n\|$$

which, together with the eventual boundedness, implies the last assertion. ∎

Lemma 1.4.4. *If the sequence (A_n) is eventually bounded and $B_n \to 0$, then $A_n B_n \to 0$. If $A_n \to A$ and $B_n \to B$, then $A_n B_n \to AB$.*

Proof. The first assertion follows from $\|A_n B_n u\| \leq \|A_n\| \|B_n u\|$ which holds for every $u \in X$. For the second one, note that (A_n) is eventually bounded by the Banach-Steinhaus theorem. Thus, the estimate

$$\|ABu - A_n B_n u\| \leq \|(A - A_n)Bu\| + \|A_n\| \|(B - B_n)u\|$$

implies the assertion. ∎

In contrast to the results above, the limit of the product of weakly convergent sequences is not, in general, equal to the product of their limits. However, the following still holds.

Lemma 1.4.5. *If $A_n \rightharpoonup A$ and $B \in \mathscr{L}(X)$, then $A_n B \rightharpoonup AB$ and $BA_n \rightharpoonup BA$. More generally, if $A_n \rightharpoonup A$ and $B_n^* \to B^*$, then $B_n A_n \rightharpoonup BA$.*

Proof. The first result is an immediate consequence of the definition of weak convergence. For the second assertion, note that for every pair $u \in X$ and $v \in X^*$, we have

$$\langle v, B_n A_n u \rangle = \langle B_n^* v, A_n u \rangle = \langle B^* v, A_n u \rangle - \langle (B^* - B_n^*)v, A_n u \rangle.$$

By hypothesis, $\langle B^* v, A_n u \rangle \to \langle B^* v, Au \rangle = \langle v, BAu \rangle$, whereas

$$|\langle (B^* - B_n^*)v, A_n u \rangle| \leq \|(B^* - B_n^*)v\| \|A_n\| \|u\| \to 0$$

due to the eventual boundedness of the A_n by Theorem 1.4.2. ∎

Lemma 1.4.6. *If K is compact on X and $A_n \rightharpoonup A$ weakly on X, then $KA_n \to KA$ strongly.*

Proof. Let $x \in X$ and put $x_n := (A - A_n)x$. Then the sequence $(x_n) \subset X$ converges weakly to zero, and we have to prove that the sequence (Kx_n) converges to zero in the norm of X. By the uniform boundedness principle, the sequence (x_n) is eventually bounded. Without loss of generality, we assume that $\|x_n\| \leq 1$ for all sufficiently large $n \in \mathbb{I}$. Since K is compact, there is a (common) subsequence (x_{n_k}) of (x_n) and a $y \in X$ such that $\|Kx_{n_k} - y\| \to 0$ as $k \to \infty$. Since $x_{n_k} \rightharpoonup 0$ weakly and K is continuous, one has $Kx_{n_k} \rightharpoonup K0 = 0$ weakly, whence $y = 0$. Since 0 is the only cluster point of the sequence (Kx_n) and this sequence is contained in a compact set, $\|Kx_n\| \to 0$. ∎

Lemma 1.4.7. *If K is compact on X and if $A_n \to A$ and $B_n^* \to B^*$ strongly, then $A_n K B_n \rightrightarrows AKB$.*

Proof. Consider first the case when $B_n = I$ for all $n \in \mathbb{I}$. Let V be the image of the closed unit ball $X_1 := \{u \in X : \|u\| \leq 1\}$ under the mapping K. Then V is relatively compact, and for each $\varepsilon > 0$ there exists a finite ε-net in V, i.e., a finite set of elements $v_1, \ldots, v_m \in V$ such that, given $v \in V$, one finds a v_i with $\|v - v_i\| < \varepsilon$. Choose $u_1, \ldots, u_m \in X_1$ such that $Kx_i = v_i$ for $i = 1, \ldots, m$. Then, for every $u \in X_1$,

$$\|(A_nK - AK)u\| \leq \min_{1 \leq i \leq m} \left(\|(A_n - A)(Ku - Ku_i)\| + \|(A_n - A)Ku_i\| \right)$$
$$\leq \varepsilon(\|A_n\| + \|A\|) + \min_{1 \leq i \leq m} \|(A_n - A)Ku_i\|. \tag{1.16}$$

Since the sequence (A_n) is eventually bounded by the Banach-Steinhaus theorem, the estimate (1.16) implies that $\|A_nK - AK\| \to 0$. Passing to adjoint operators, one concludes that the assertion also holds in the case $B_n^* \to B^*$ strongly and $A_n = I$ for all $n \in \mathbb{I}$. Now the general assertion follows from

$$\|A_nKB_n - AKB\| \leq \|A_nKB_n - A_nKB\| + \|A_nKB - AKB\|$$
$$\leq \|A_n\| \|KB_n - KB\| + \|A_nK - AK\| \|B\|$$

and the eventual boundedness of the sequence (A_n) again.   ∎

1.4.3 Algebras of continuous functions

Let X be a topological space. The set $C(X)$ of all complex-valued continuous functions on X is a $*$-algebra with respect to pointwise operations and pointwise involution, and the function $x \mapsto 1$ serves as the identity element of $C(X)$. If X is a compact Hausdorff space, then the functions in $C(X)$ are bounded, and $C(X)$ becomes a C^*-algebra on defining a norm by

$$\|f\| := \sup_{x \in X} |f(x)|.$$

If K is a closed subset of X, then $\{f \in C(X) : f|_K = 0\}$ is a closed ideal of $C(X)$. The following theorem states that every closed ideal of $C(X)$ can be obtained in this way. Given a closed ideal $\mathscr{I}$ of $C(X)$, set $X(\mathscr{I}) := \cap_{f \in \mathscr{I}} f^{(-1)}(0)$. Being the intersection of closed sets, $X(\mathscr{I})$ is a closed subset of X.

Theorem 1.4.8. *Let X be a compact Hausdorff space and $\mathscr{I}$ a closed ideal of $C(X)$. Then*

$$\mathscr{I} = \{f \in C(X) : f|_{X(\mathscr{I})} = 0\}.$$

Thus, there is a one-to-one correspondence between the closed subsets of X and the closed ideals of $C(X)$.

Proof. The inclusion $\mathscr{I} \subseteq \{f \in C(X) : f|_{X(\mathscr{I})} = 0\}$ is evident. For the reverse inclusion, let $f \in C(X)$ be a function which vanishes on $X(\mathscr{I})$. Given $\varepsilon > 0$, there is an open neighborhood $U(\mathscr{I})$ of $X(\mathscr{I})$ such that $|f(x)| < \varepsilon$ for all $x \in U(\mathscr{I})$. Further, for every $x \in X \setminus X(\mathscr{I})$, one can choose a function g_x in $\mathscr{I}$ with $g_x(x) = 1$. The functions $h_x := g_x \overline{g_x}$ belong to the ideal $\mathscr{I}$, too, they are real-valued, and $h_x(x) = 1$. Let $U_x := \{y \in X : h_x(y) > 1/2\}$. The open sets $U(\mathscr{I})$ and U_x with $x \in X \setminus X(\mathscr{I})$ cover the compact set X; hence, there is a finite subcovering, say

$$X = U(\mathscr{I}) \cup U_{x_1} \cup \ldots \cup U_{x_n}.$$

Write U_i in place of U_{x_i} for brevity, and let $1 = f_I + f_1 + \ldots + f_n$ be a partition of the identity with respect to this covering, i.e., f_I and the f_i are non-negative continuous functions with supp $f_{\mathscr{I}} \subseteq U(\mathscr{I})$ and supp $f_i \subseteq U_i$. The function h_i is greater than or equal to $1/2$ on the closure $\overline{U}_i$ of U_i, and thus, the restriction of h_i onto $\overline{U}_i$ can be inverted, and the inverse is continuous on $\overline{U}_i$ again. The Tietze-Uryson extension theorem ([160, Theorem IV.11]) implies the existence of a continuous function r_i on X which coincides with $(h_i|_{\overline{U}_i})^{-1}$ on $\overline{U}_i$. Hence,

$$f_i = f_i h_i (h_i|_{\overline{U}_i})^{-1} = f_i h_i r_i$$

which shows that f_i belongs to $\mathscr{I}$ for all i. Because $f = f \cdot 1 = f f_I + f f_1 + \ldots + f f_n$, we have

$$\|f - f f_1 - \ldots - f f_n\| = \|f f_I\| \le \|f|_{U(\mathscr{I})}\| < \varepsilon.$$

Thus, f can be approximated as closely as desired by functions in $\mathscr{I}$, which implies $f \in \mathscr{I}$ due to the closedness of $\mathscr{I}$. ∎

The correspondence between the closed subsets of X and the closed ideals of $C(X)$ is monotone: The greater the ideal $\mathscr{I}$, the smaller the set $X(\mathscr{I})$. An immediate consequence is the following characterization of the maximal ideals of $C(X)$.

Corollary 1.4.9. *Let X be a compact Hausdorff space and $x \in X$. Then $\{f \in C(X) : f(x) = 0\}$ is a maximal ideal of $C(X)$, and every maximal ideal of $C(X)$ is of this form.*

Thus, the non-trivial characters of $C(X)$ are exactly the mappings $f \mapsto f(x)$ with fixed $x \in X$. The following proposition describes the topological divisors of zero of the algebra $C(X)$.

Proposition 1.4.10. *Let X be a compact Hausdorff space. A function $f \in C(X)$ is a topological divisor of zero of the algebra $C(X)$ if and only if it has a zero on X.*

Proof. If f has no zero on X, then f is invertible. Thus, f cannot be a topological divisor of zero due to Proposition 1.2.26.

Now suppose that $f(x) = 0$ for some $x \in X$. Given $n \in \mathbb{N}$, choose an open neighborhood U_n of x with $f(y) < 1/n$ for $y \in U_n$, and choose a continuous function $f_n : X \to [0, 1]$ which takes the value 1 at x and vanishes outside U_n. (The existence of a function with these properties is guaranteed by the Tietze-Uryson extension theorem again.) Then $\|f_n\| = 1$ and $\|f f_n\| \le 1/n$. Thus, f is a topological divisor of zero. ∎

For a compact Hausdorff space X and $1 \le p < \infty$, consider the Lebesgue space $L^p(X)$ of all measurable functions on X such that

$$\|f\|_p := \left(\int_X |f(x)|^p \, dx \right)^{1/p}$$

is finite. Consider also the space $L^\infty(X)$ of all essentially bounded measurable functions on X. The set $L^p(X)$ equipped with the above defined norm is a Banach space. Each function $f \in L^\infty(X)$ gives rise to an operator on $L^p(X)$, which is the operator $u \mapsto fu$ *of multiplication by* f. We abbreviate this operator by fI. The following result is valid for operators of multiplication by arbitrary essentially bounded functions on Lebesgue spaces $L^p(X)$ with $p \geq 1$. For simplicity, we prove it for continuous functions only.

Proposition 1.4.11. *Let X be a compact Hausdorff space. Then the algebras*

$$\mathscr{A}_1 := C(X),$$
$$\mathscr{A}_2 := \{fI : f \in C(X)\} \subset \mathscr{L}(L^p(X))$$

and

$$\mathscr{A}_3 := (\mathscr{A}_2 + \mathscr{K}(L^p(X)))/\mathscr{K}(L^p(X))$$
$$= \{fI + \mathscr{K}(L^p(X)) : f \in C(X)\} \subset \mathscr{L}(L^p(X))/\mathscr{K}(L^p(X))$$

are isometrically isomorphic. Moreover,

$$\|f\|_\infty = \|fI\|_{\mathscr{L}(L^p(X))} = \|fI + \mathscr{K}(L^p(X))\|_{\mathscr{L}(L^p(X))/\mathscr{K}(L^p(X))}$$

for each function $f \in C(X)$.

Proof. Let $f \in C(X)$, and abbreviate the coset $fI + \mathscr{K}(L^p(X))$ to $\hat{f}$. Then

$$\|\hat{f}\|_{\mathscr{L}(L^p(X))/\mathscr{K}(L^p(X))} \leq \|fI\|_{\mathscr{L}(L^p(X))} \leq \|f\|_\infty,$$

with the first inequality coming from the definition of the quotient norm and the second one following by a straightforward estimate. It remains to verify that $\|f\|_\infty \leq \|\hat{f}\|$. Assume that $\|\hat{f}\| < \|f\|_\infty$. Since X is compact, there exists an $x_0 \in X$ with $|f(x_0)| = \|f\|_\infty$. Consider the function $f(x_0) - f$. This function is not invertible as a continuous function; but the corresponding coset in the Calkin algebra is invertible, as one easily gets by writing

$$f(x_0) - \hat{f} = \|f\|_\infty \left(\frac{f(x_0)}{\|f\|_\infty} I - \frac{\hat{f}}{\|f\|_\infty} \right)$$

and employing a Neumann series argument (note that the norm of $\hat{f}/\|f\|_\infty$ is less than 1). Thus, $(f(x_0) - f)I$ is a Fredholm operator. The set of all Fredholm operators is open in $\mathscr{L}(L^p(X))$ as mentioned in Section 1.4.1. Thus, if $f_0 \in C(X)$ is a function which is zero in a neighborhood of x_0 and for which $\|f(x_0) - f - f_0\|$ is sufficiently small, then $f_0 I$ is a Fredholm operator, too. This is impossible because, being zero in a neighborhood of x_0, the multiplication operator $f_0 I$ has an infinite-dimensional kernel. This contradiction proves the assertion. ∎

1.4.4 Singular integral operators with continuous coefficients

Let $\mathbb{T} := \{z \in \mathbb{C} : |z| = 1\}$ be the complex unit circle, which we provide with counterclockwise orientation and normalized Lebesgue measure, and let $1 < p < \infty$. We are now going to examine a closed subalgebra of $\mathscr{L}(L^p(\mathbb{T}))$ which is generated by the operators of multiplication by continuous functions and by the Cauchy singular integral operator S on $\mathbb{T}$. In the following two chapters, this algebra will serve as an example to illustrate various concepts. Algebras of this type can be introduced in much more general contexts: weighted Lebesgue spaces, curves with complicated topology, coefficients with various kinds of discontinuities, etc. We will study these algebras in detail in a quite general (but not too involved) setting in Chapter 4.

The *singular integral operator* S on $\mathbb{T}$ is formally defined by

$$(Su)(t) := \frac{1}{\pi \mathbf{i}} \int_{\mathbb{T}} \frac{u(s)}{s-t} \, ds, \quad t \in \mathbb{T}.$$

In general, this integral does not exist in the common sense of an improper integral. Rather one has to interpret this integral as a *Cauchy principal value integral*, i.e., one defines

$$(Su)(t) := \lim_{\varepsilon \to 0} \frac{1}{\pi \mathbf{i}} \int_{\mathbb{T} \setminus \mathbb{T}_{t,\varepsilon}} \frac{u(s)}{s-t} \, ds, \quad t \in \mathbb{T},$$

where $\mathbb{T}_{t,\varepsilon}$ is the part of $\mathbb{T}$ within the ε-disk centered at the point t. For $p \neq 2$, it is by no means a triviality to prove that $(Su)(t)$ exists for $u \in L^p(\mathbb{T})$ almost everywhere on $\mathbb{T}$, that Su belongs to $L^p(\mathbb{T})$ again, and that S is a bounded operator on that space. The interested reader is referred to [120, Section II.2] where Riesz' elegant proof of the boundedness of S on $L^p(\mathbb{T})$ for $p \neq 2$ is presented. Let us also mention that S is not well defined on the spaces $L^1(\mathbb{T})$, $L^\infty(\mathbb{T})$ and $C(\mathbb{T})$ (but spaces of Hölder continuous functions will work).

In the case $p = 2$, there is a simple way to check all these facts. For $n \in \mathbb{Z}$, let p_n refer to the function $z \mapsto z^n$ on $\mathbb{T}$. It is an easy exercise in complex function theory to show that

$$(Sp_n)(t) = \begin{cases} p_n & \text{if } n \geq 0, \\ -p_n & \text{if } n < 0. \end{cases} \tag{1.17}$$

From (1.17), and since $\{p_n\}_{n \in \mathbb{Z}}$ forms an orthogonal basis of $L^2(\mathbb{T})$, it becomes evident that S acts as an isometry on the dense subspace of $L^2(\mathbb{T})$ which is spanned by the trigonometric polynomials. Thus, S can be continuously extended to an isometry on all of $L^2(\mathbb{T})$. We denote this extension by S again. It is also evident from (1.17) that $S^2 = I$. Thus, S is a self-adjoint and unitary operator.

The property $S^2 = I$ (which holds for general $1 < p < \infty$) implies that the operators $P := (I+S)/2$ and $Q := (I-S)/2$ satisfy $P^2 = P$ and $Q^2 = Q$, i.e., they are projections, and they are even orthogonal projections if $p = 2$. Moreover,

$$P + Q = I, \quad P - Q = S \quad \text{and} \quad PQ = QP = 0.$$

By a *singular integral operator (SIO) with continuous coefficients* on $L^p(\mathbb{T})$, we mean an operator of the form

$$A = cI + dS + T = fP + gQ + T$$

where $c, d \in C(\mathbb{T})$, T is a compact operator, and $f := c + d$ and $g := c - d$. It is not hard to examine the smallest closed subalgebra $\mathcal{A}_p$ of $\mathcal{L}(L^p(\mathbb{T}))$ which contains all singular integral operators of that form. We left this as an exercise (Exercise 1.4.6). Our primary goal is to study the Fredholm properties of the operator A. Since the Fredholmness of an operator is not influenced by a compact perturbation, we can assume that $T = 0$.

First we show that the commutators $fS - SfI$, $fP - PfI$ and $fQ - QfI$ are compact for each continuous function f. It is clearly sufficient to prove the compactness of the first of these commutators. From

$$fS - SfI = (P+Q)(f(P-Q) - (P-Q)f)(P+Q) = 2(QfP - PfQ)$$

we further conclude that it is sufficient to check the compactness of QfP and PfQ. In case f is a trigonometric polynomial, the compactness of these operators is evident (both operators have finite rank). The case of a general continuous function follows from the fact that every continuous function can be uniformly approximated by trigonometric polynomials and

$$\max\{\|QfP\|, \|PfQ\|\} \leq \|P\| \|Q\| \|f\|_\infty$$

by Proposition 1.4.11. We summarize these facts in the following proposition. Recall that $1 < p < \infty$ throughout this section.

Proposition 1.4.12.

(i) *The singular integral operator S is well defined and bounded on $L^p(\mathbb{T})$, and $S^2 = I$. In case $p = 2$, the operator S is moreover self-adjoint and unitary.*

(ii) *For every continuous function f on $\mathbb{T}$, the operators*

$$fS - SfI, \quad fP - PfI, \quad fQ - QfI, \quad PfQ \quad \text{and} \quad QfP$$

are compact on $L^p(\mathbb{T})$.

The following lemma will be needed in the proof of the Fredholm criterion.

Lemma 1.4.13.

(i) *Let $f_1P + g_1Q + T_1$ and $f_2P + g_2Q + T_2$ be singular integral operators on $L^p(\mathbb{T})$ with continuous coefficients f_1, f_2, g_1, g_2 and with compact operators T_1, T_2. Then their product is of the form $f_1f_2P + g_1g_2Q + T$ with a compact operator T.*

(ii) *Let $f, g \in C(\mathbb{T})$ and $T \in \mathcal{K}(L^p(\mathbb{T}))$. Then*

$$\max\{\|f\|_\infty, \|g\|_\infty\} \leq \|fP + gQ + T\|. \tag{1.18}$$

Proof. The first assertion follows easily from Proposition 1.4.12. For the second assertion, let $u(s) := s$ for $s \in \mathbb{T}$ and consider the isometries U and U^{-1} on $L^p(\mathbb{T})$ acting by $v \mapsto uv$ and $v \mapsto u^{-1}v$, respectively. One has $U^n f U^{-n} = fI$ for every continuous function f. Further, $U^n P U^{-n} \to 0$ strongly as $n \to +\infty$ and $U^n P U^{-n} \to I$ strongly as $n \to -\infty$, as one can easily check. Finally, the weak convergence of the sequences (U^n) and (U^{-n}) to zero and Lemma 1.4.6 imply that $U^n T U^{-n} \to 0$ strongly as $n \to \pm\infty$. Thus,

$$U^n(fP + gQ + T)U^{-n} \to gI \quad \text{strongly as } n \to +\infty,$$

and, analogously,

$$U^{-n}(fP + gQ + T)U^n \to fI \quad \text{strongly as } n \to +\infty.$$

Since,

$$\|U^n(fP + gQ + T)U^{-n}\| = \|U^{-n}(fP + gQ + T)U^n\| = \|fP + gQ + T\|,$$

assertion (1.18) follows from the Banach-Steinhaus theorem. ∎

Here is the aforementioned Fredholm criterion for singular integral operators.

Theorem 1.4.14. *The singular integral operator $A := fP + gQ + T$ with $f, g \in C(\mathbb{T})$ and T compact is Fredholm on $L^p(\mathbb{T})$ if and only if*

$$(fg)(s) \neq 0 \quad \text{for all} \quad s \in \mathbb{T}. \tag{1.19}$$

Proof. Let (1.19) be satisfied. Consider the operator $B := f^{-1}P + g^{-1}Q$. Using Lemma 1.4.13 (i) one easily checks that $AB - I$ and $BA - I$ are compact operators. Thus, A is a Fredholm operator.

Conversely, let condition (1.19) be violated. Then at least one of the functions f and g has a zero on $\mathbb{T}$. Let $f(s_0) = 0$, for instance. Then the function f is a topological divisor of zero of the algebra $C(\mathbb{T})$ due to Proposition 1.4.10. By definition, there is a sequence (f_n) in $C(\mathbb{T})$ with $\|f_n\|_\infty = 1$ such that $\|f_n f\|_\infty \to 0$. Consider the singular integral operators $A_n := f_n P$. For the cosets of A and A_n modulo compact operators one finds

$$\|(A_n + \mathscr{K}(L^p(\mathbb{T})))(A + \mathscr{K}(L^p(\mathbb{T})))\| = \|f_n fP + \mathscr{K}(L^p(\mathbb{T}))\| \leq \|f_n f\|_\infty \|P\| \to 0$$

as well as $\|A_n + \mathscr{K}(L^p(\mathbb{T}))\| \geq \|f_n\| = 1$ due to (1.18). Thus, $A + \mathscr{K}(L^p(\mathbb{T}))$ is a topological divisor of zero in the Calkin algebra, which cannot be invertible in that algebra. Hence, A fails to be Fredholm. ∎

The statement of this theorem can also be formulated as follows: the singular integral operator $A := fP + gQ + T$ with continuous coefficients is Fredholm on $L^p(\mathbb{T})$ if and only if each of the operators $A_s := f(s)P + g(s)Q$ with $s \in \mathbb{T}$ is invertible on that space. The equivalence between the invertibility of the operators A_s (which are singular integral operators with constant coefficients) and the condition $f(s)g(s) \neq 0$ in (1.19) was the subject of Exercise 1.2.17.

An obvious interpretation of the operators A_s is that they are obtained by *freezing* the coefficients at the point $s \in \mathbb{T}$. This idea is widely used in the theory of partial differential equations. So, the operator A_s can also be viewed as a *local representative* of the operator A at the point s. Note that the local representative is independent of the compact operator T.

It seems to be promising to try the same idea for more involved objects, say for singular integral operators with coefficients which are allowed to have jump discontinuities. There are several questions which come immediately to attention:

- Which operators should be considered as local representatives?
- Does the Fredholmness of A imply the invertibility of its local representatives?

and, the basic question:

- Does the invertibility of all local representatives imply the Fredholmness of A?

Indeed, we will see in the forthcoming chapter that there are several general methods – so-called local principles – which allow one to answer these questions in the affirmative. These principles are formulated in the context of Banach algebras. They can be considered as non-commutative generalizations of the (commutative) Gelfand theory. Therefore we open Chapter 2 with a section on classical Gelfand theory. We will also see that the use of local principles is by no means limited to Fredholm problems. In fact they are (in principle) applicable to all invertibility problems in Banach algebras. Some applications to numerical analysis will be given later on.

1.4.5 Algebras of matrix sequences

For each sequence δ of positive integers, let $\mathscr{F}^\delta$ stand for the set of all bounded sequences (A_n) of matrices $A_n \in \mathbb{C}^{\delta(n) \times \delta(n)}$. We think of $\mathbb{C}^{k \times k}$ as provided with the spectral norm, i.e., with the operator norm induced by the Euclidean norm on the complex Hilbert space $\mathbb{C}^k$. With respect to the operations

$$(A_n) + (B_n) := (A_n + B_n), \quad \alpha(A_n) := (\alpha A_n), \quad (A_n)(B_n) := (A_n B_n),$$

the involution $(A_n)^* := (A_n^*)$ and the norm $\|(A_n)\|_{\mathscr{F}} := \sup_n \|A_n\|$, the set $\mathscr{F}^\delta$ becomes a C^*-algebra with identity (I_n) where I_n stands for the $\delta(n) \times \delta(n)$ identity matrix. We refer to $\mathscr{F}^\delta$ as the *algebra of matrix sequences* and to δ as its *dimension function*. Thus, the algebra of matrix sequences with constant dimension function $\delta = 1$ is $l^\infty(\mathbb{N})$.

Let $\mathscr{F}^\delta$ be the algebra of matrix sequences with dimension function δ. The set $\mathscr{G}^\delta$ of all sequences in $\mathscr{F}^\delta$ which tend to zero in the norm forms a closed ideal of $\mathscr{F}^\delta$. We call $\mathscr{G}^\delta$ the *ideal of the zero sequences associated with* $\mathscr{F}^\delta$. In particular, the ideal $\mathscr{G}^\delta$ with the constant dimension function $\delta = 1$ is $c_0(\mathbb{N})$.

The special choice of the dimension function is often not of importance; so as a convention we will usually simply write $\mathscr{F}$ and $\mathscr{G}$ in place of $\mathscr{F}^\delta$ and $\mathscr{G}^\delta$.

Our main interest lies in the quotient algebra $\mathscr{F}/\mathscr{G}$. Here is what can be said about the norm and the invertibility of a coset $(A_n) + \mathscr{G}$ in $\mathscr{F}/\mathscr{G}$.

Proposition 1.4.15. *For* $(A_n) \in \mathscr{F}$,

$$\|(A_n) + \mathscr{G}\|_{\mathscr{F}/\mathscr{G}} = \limsup_{n \to \infty} \|A_n\|. \tag{1.20}$$

Proof. Let $(A_n) \in \mathscr{F}$. Then, for every sequence $(G_n) \in \mathscr{G}$,

$$\limsup \|A_n\| = \limsup \|A_n + G_n\| \le \sup \|A_n + G_n\| = \|(A_n) + (G_n)\|_{\mathscr{F}}$$

whence the estimate $\limsup \|A_n\| \le \|(A_n) + \mathscr{G}\|$. For the reverse inequality, let $\varepsilon > 0$, and choose $n_0 \in \mathbb{N}$ such that $\|A_n\| \le \limsup_{n \to \infty} \|A_n\| + \varepsilon$ for all $n \ge n_0$. Set

$$G_n := \begin{cases} -A_n & \text{if } n < n_0, \\ 0 & \text{if } n \ge n_0. \end{cases}$$

The sequence (G_n) belongs to the ideal $\mathscr{G}$, and

$$\|(A_n) + \mathscr{G}\| \le \|(A_n) + (G_n)\| = \|(0, \dots, 0, A_{n_0}, A_{n_0+1}, \dots)\|$$
$$= \sup_{n \ge n_0} \|A_n\| \le \limsup \|A_n\| + \varepsilon.$$

Letting ε go to zero yields the desired result. $\blacksquare$

The importance of the quotient algebra $\mathscr{F}/\mathscr{G}$ in numerical analysis rests on the following observation, also known as Kozak's theorem . It allows one to consider the problem of stability of a sequence as an invertibility problem in a suitably chosen Banach algebra. Here, as usual in the context of numerical analysis, a sequence (A_n) is called *stable* if the entries A_n are invertible for all sufficiently large n and if the norms of their inverses are uniformly bounded. There is a close relation between the notions of stability and applicability of an approximation method; we will pick up this topic at the beginning of Chapter 6.

Theorem 1.4.16 (Kozak). *A sequence* $(A_n) \in \mathscr{F}$ *is stable if and only if its coset* $(A_n) + \mathscr{G}$ *is invertible in* $\mathscr{F}/\mathscr{G}$.

It is easily seen that the results of this section remain valid for algebras formed by bounded sequences (A_n) of bounded operators A_n which act on Hilbert spaces H_n (with some evident modifications).

1.4.6 Exercises

Exercise 1.4.1. Let H be a separable infinite-dimensional Hilbert space. Prove that $\mathscr{K}(H)$ is the only non-trivial closed ideal of $\mathscr{L}(H)$. (Hint: prove that every non-zero ideal of $\mathscr{L}(H)$ contains operators of rank one.)

Exercise 1.4.2. Let $V : l^2(\mathbb{Z}^+) \to l^2(\mathbb{Z}^+)$ denote the shift operator considered in Exercise 1.2.7. Set $V_n := V^n$ and $V_{-n} := (V^*)^n$ for each positive integer n and define $V_0 := I$. Show that each operator V_n (with $n \in \mathbb{Z}$) is Fredholm and determine its kernel, cokernel and index.

Exercise 1.4.3. Verify (1.17).

Exercise 1.4.4. Let X denote a compact Hausdorff space, X_0 a closed subset of X and $\mathscr{J}(X_0) := \{f \in C(X) : f|_{X_0} = 0\}$. Show that $C(X)/\mathscr{J}(X_0) \cong C(X_0)$.

Exercise 1.4.5. Show that Theorem 1.4.8 does not hold for commutative Banach algebras in general. That is, there is no one-to-one correspondence between the closed ideals of an algebra $\mathscr{B}$ and the closed subsets of $M_{\mathscr{B}}$. Hint: consider the algebra $C^1[0, 1]$ of all continuously differentiable functions as a Banach algebra under the norm $\|f\|_{C^1[0, 1]} := \|f\|_\infty + \|f'\|_\infty$. Let $x \in [0, 1]$. Show that both

$$\mathscr{J}_0 := \{f \in C^1[0, 1] : f(x) = 0\} \quad \text{and} \quad \mathscr{J}_1 := \{f \in C^1[0, 1] : f(x) = f'(x) = 0\}$$

are closed ideals of $C^1[0, 1]$.

Exercise 1.4.6. Let $\mathscr{A}_p$ denote the smallest closed subalgebra of $\mathscr{L}(L^p(\mathbb{T}))$ which contains all singular integral operators $aP + bQ + T$ with a, b continuous and T compact.

a) Show that each operator $A \in \mathscr{A}_p$ can be written as $A = aP + bQ + T$ with a, b continuous and T compact. Moreover, the functions a, b and the compact operator T are uniquely determined by A.

b) Prove that the mapping

$$\mathscr{A}_p \to C(\mathbb{T}) \times C(\mathbb{T}), \quad aP + bQ + T \mapsto (a, b)$$

is a continuous algebra homomorphism. What is its kernel?

c) Conclude that the mapping

$$\mathscr{A}_p/\mathscr{K}(L^p(\mathbb{T})) \to C(\mathbb{T}) \times C(\mathbb{T}), \quad aP + bQ + \mathscr{K}(L^p(\mathbb{T})) \mapsto (a, b)$$

is well defined, and that it is an injective continuous algebra homomorphism.

d) Conclude that the algebras $\mathscr{A}_p/\mathscr{K}(L^p(\mathbb{T}))$ and $C(\mathbb{T}) \times C(\mathbb{T})$ are topologically isomorphic.

e) Consider the smallest closed subalgebra $\widetilde{\mathscr{A}}_p$ of $\mathscr{L}(L^p(\mathbb{T}))$ which contains all operators $aP + bQ$ with a and b continuous. Prove that $\mathscr{K}(L^p(\mathbb{T})) \subset \widetilde{\mathscr{A}}_p$ and, thus, $\widetilde{\mathscr{A}}_p = \mathscr{A}_p$. (A proof will be given later on in Theorem 4.1.5.)

Exercise 1.4.7. Let X be a Banach space. Show that $A \in \mathcal{L}(X)$ has a finite-dimensional kernel and a closed image if and only if there is a number $c > 0$ and a compact operator $K \in \mathcal{K}(X)$ such that

$$\|Ax\| + \|Kx\| \geq c\|x\|$$

for all $x \in X$.

Exercise 1.4.8. Let the notation be as in Exercise 1.4.6, but let $p = 2$.

a) Let $a \in C(\mathbb{T})$. Show that the matrix representation of the operator $PaP : \operatorname{Im} P \to \operatorname{Im} P$ is given by the matrix

$$\begin{bmatrix} a_0 & a_{-1} & a_{-2} & a_{-3} & \cdots \\ a_1 & a_0 & a_{-1} & a_{-2} & \ddots \\ a_2 & a_1 & a_0 & a_{-1} & \ddots \\ a_3 & a_2 & a_1 & a_0 & \ddots \\ \vdots & \ddots & \ddots & \ddots & \ddots \end{bmatrix} \tag{1.21}$$

where

$$a_k := \frac{1}{2\pi} \int_0^{2\pi} a(e^{i\theta}) e^{-ik\theta}\, d\theta, \quad k \in \mathbb{Z}.$$

is the kth Fourier coefficient of a. The matrix (1.21) induces a bounded linear operator on $l^2(\mathbb{Z}^+)$ which is called the *Toeplitz operator* with generating function a. We denote it by $T(a)$.

b) Conclude that the smallest closed subalgebra of $\mathcal{L}(\operatorname{Im} P)$ which contains all operators PaP with $a \in C(\mathbb{T})$ is isometrically isomorphic to the smallest closed subalgebra of $\mathcal{L}(l^2(\mathbb{Z}^+))$ which contains all Toeplitz operators $T(a)$ with $a \in C(\mathbb{T})$. The latter algebra is called the *Toeplitz algebra*. We denote it by $\mathcal{T}(C)$.

c) Show that $\mathcal{T}(C)$ coincides with the smallest closed subalgebra of $\mathcal{L}(l^2(\mathbb{Z}^+))$ which contains the shift operators V and V_{-1} from Exercise 1.4.2.

d) Show that $\mathcal{K}(l^2(\mathbb{Z}^+)) \subset \mathcal{T}(C)$ and that the quotient algebra $\mathcal{T}(C)/\mathcal{K}(l^2(\mathbb{Z}^+))$ is isometrically isomorphic to the algebra $C(\mathbb{T})$. Conclude that the Toeplitz operator $T(a)$ is Fredholm if and only if the function a has no zero on $\mathbb{T}$.

e) Prove that the index of the Fredholm Toeplitz operator $T(a)$ is equal to the negative winding number of the function a around the origin. (Hints: Start with a rational function a without zeros and poles on $\mathbb{T}$. Let κ be the winding number of a. Write a as $a(t) = a_-(t)t^\kappa a_+(t)$ where the rational functions $a_+^{\pm 1}$ allow analytic continuation into the interior of the unit disk whereas the functions $a_-^{\pm 1}$ can be continued analytically into the exterior of the unit disk. Verify that $T(a) = T(a_-)V_\kappa T(a_+)$, with the outer factors being invertible. Use Exercise 1.4.2.)

f) Using this result, propose a way to determine the index of the Fredholm singular integral operator $aP + bQ$ with continuous coefficients.

Exercise 1.4.9. This exercise is aimed at providing a counter-example to the converse of Lemma 1.2.33. Write $\mathscr{K}$ for $\mathscr{K}(l^2(\mathbb{Z}^+))$ and consider the smallest closed subalgebra $\mathscr{B}$ of $\mathscr{T}(C)$ which contains the operators I and V and the ideal $\mathscr{K}$.

a) Prove that $\mathscr{B}$ is inverse-closed in $\mathscr{T}(C)$. (Hints: Any element of $\mathscr{T}(C)$ can be written as $T(a) + K$ with $a \in C(\mathbb{T})$ and K compact. By Coburn's theorem [21, Theorem 2.38], a Fredholm Toeplitz operator with index zero is invertible.)

b) Check that $\mathscr{B}/\mathscr{K}$ is not inverse-closed in $\mathscr{T}(C)/\mathscr{K}$.

Exercise 1.4.10. Let $\mathscr{B}$ be a closed and inverse-closed subalgebra of the sequence algebra $\mathscr{F}$ which contains the ideal $\mathscr{G}$ of the zero sequences. Prove that $\mathscr{B}/\mathscr{G}$ is inverse closed in $\mathscr{F}/\mathscr{G}$. (Hint: if (A_τ) is a sequence in $\mathscr{B}$ for which $(A_\tau) + \mathscr{G}$ is invertible in $\mathscr{F}/\mathscr{G}$, then there are sequences (B_τ) and (C_τ) in $\mathscr{F}$ and (G_τ) and (H_τ) in $\mathscr{G}$ such that $B_\tau A_\tau = I - G_\tau$ and $A_\tau C_\tau = I - H_\tau$ for all $\tau > 0$; build a Neumann series argument, by modifying the sequences above.)

Exercise 1.4.11. Prove that the set $\mathscr{F}$ of all bounded matrix sequences (introduced in Section 1.4.5) is a unital C^*-algebra and that $\mathscr{G}$ is a closed ideal of that algebra.

Exercise 1.4.12. Prove Kozak's theorem 1.4.16.

1.5 Notes and comments

The concept of operator algebra was introduced for the first time in 1913 by Frigyes Riesz [163]. From 1929, von Neumann developed, in a series of papers [198, 199, 200], also with Murray [122, 123, 124], the theory of the algebras that were given his name. The other pioneering contribution came from Israel Gelfand who, beginning with his Phd thesis in 1938, developed the theory of what he called *normed rings*, that included the von Neumann algebras as a particular case [59, 60, 61, 62]. Gelfand showed that these algebras could be characterized in an abstract way, as Banach algebras with involution with some additional axioms. Together with Naimark he proved that any commutative C^*-algebra with identity is isomorphic to the algebra of continuous complex-valued functions defined on the compact space of the maximal ideals of the C^*-algebra, and that any C^*-algebra could be represented as an algebra of operators acting on a Hilbert space [63]. There were other important early contributors like Köthe [101] and Jacobson [90] who developed concepts behind the radical of an algebra. We should also mention Shilov and his work on commutative Banach algebras [180] (the notion of a Shilov boundary and the Shilov idempotent theorem are due to him) and Segal [178] who also contributed to representation theory (the "S" in the "GNS-construction") and to decomposition theory of C^*-algebras. Finally, a most influential paper for the development of the topic of this book is Gohberg's small note [65] from 1952 where he first applied the then new theory of Banach algebras (normed rings) to algebras of singular integral operators.

From the mid-nineteen-forties research on algebras expanded enormously and proceeded along four main streams, each fertilizing the others. By ascending order of structure there was the abstract algebra theory stream that studied rings, finite-dimensional algebras and PI-algebras, whose contribution to this book will be seen in the next chapter; there was the complete normed algebra stream, where the elements of the algebra can be seen as operators acting on a Banach space, which is the main subject of this book; there was the C^*-algebra stream, where the operators act on a Hilbert space; and finally there was the von Neumann algebra stream, where the C^*-algebra must be closed in the strong-operator topology. There were many other streams, particularizations and generalizations crossing the main streams like module theory, commutative algebras, non-unital algebras, etc. Depending on the structure of elements of the algebra and the underlying structure of the space where the operators act, the problems to solve are different, and different concepts need to be introduced.

In the eighties, the field of operator algebras experienced a new dramatic evolution, associated to application in other areas like non-commutative geometry, K-theory, quantum field theory. Another new area of application developed at that time was numerical analysis, which is also covered in this book, in Chapter 6.

At the current time, any (even basic) work on operator algebras that aspires to be encyclopedic would need dozens of volumes. The most any work on this area can aspire to, is to conduct the reader through a path in the immense field of operator algebras. Each book has thus a particular tone, and it is difficult to choose one over another. The book the reader has in his hands now follows and explores the path leading to non-commutative local principles. For more comprehensive and detailed introductions to the topic of Banach algebras, we refer to [4, 10, 32, 41, 67, 84, 134, 135, 162, 202]. For C^*-algebras there are good books like [3, 8, 24, 34, 39, 91, 92, 121, 137, 192]. The books from Kadison and Ringrose have a very interesting collection of exercises, a couple of which we adapted for Chapters 1 and 2. From the above books we can recommend in particular (authors' personal taste) [3, 4, 41, 121, 202], whereas [32] and [8] provide nice overviews on the present theory of Banach and C^*-algebras, respectively.

Chapter 2
Local principles

Before we start our walk through the world of local principles, it is useful to give a general idea of what a local principle should be. A local principle will allow us to study invertibility properties of an element of an algebra by studying the invertibility properties of a (possibly large) family of (hopefully) simpler objects. These simpler objects will usually occur as homomorphic images of the given element. To make this more precise, consider a unital algebra $\mathscr{A}$ and a family $\mathscr{W} = (W_t)_{t \in T}$ of unital homomorphisms $W_t : \mathscr{A} \to \mathscr{B}_t$ from $\mathscr{A}$ into certain unital algebras $\mathscr{B}_t$. We say that $\mathscr{W}$ forms a *sufficient family of homomorphisms for* $\mathscr{A}$ if the following implication holds for every element $a \in \mathscr{A}$:

$$W_t(a) \text{ is invertible in } \mathscr{B}_t \text{ for every } t \in T \implies a \text{ is invertible in } \mathscr{A}$$

(the reverse implication is satisfied trivially). Equivalently, the family $\mathscr{W}$ is sufficient if and only if

$$\sigma_{\mathscr{A}}(a) \subseteq \cup_{t \in T} \sigma_{\mathscr{B}_t} \qquad \text{for all } a \in \mathscr{A}$$

(again with the reverse inclusion holding trivially). In case the family $\mathscr{W}$ is a singleton, $\{W\}$ say, then $\mathscr{W}$ is sufficient if and only if W is a symbol mapping in the sense of Section 1.2.1.

Every sufficient family $(W_t)_{t \in T}$ of homomorphisms for $\mathscr{A}$ gives rise to a symbol mapping W from $\mathscr{A}$ into the direct product $\Pi_{t \in T} \mathscr{B}_t$ of the algebras $\mathscr{B}_t$ via

$$W : a \mapsto (t \mapsto W_t(a)).$$

Since symbol mappings preserve spectra, they preserve spectral radii. In the C^*-case (i.e., if all occurring algebras are C^* and the homomorphisms are symmetric), this implies that they also preserve norms of self-adjoint elements (Proposition 1.2.36 (i)), and hence, by the C^*-axiom, the norms of arbitrary elements. Thus, a symmetric symbol mapping $W : \mathscr{A} \to \mathscr{B}$ between C^*-algebras is nothing but a *-isomorphism between $\mathscr{A}$ and a C^*-subalgebra of $\mathscr{B}$. For general (in particular, Banach) algebras there is a clear distinction between symbol mappings and isomorphisms. Note that $(W_t)_{t \in T}$ being sufficient only implies that $\cap_{t \in T} \text{Ker } W_t \subseteq \mathscr{R}_{\mathscr{A}}$.

S. Roch et al., *Non-commutative Gelfand Theories*, Universitext,
DOI 10.1007/978-0-85729-183-7_2, © Springer-Verlag London Limited 2011

So, in case of an algebraic environment, "local principle" will mean a process to construct sufficient families of homomorphisms. All local principles encountered in this chapter will use commutativity properties for this construction. It is either the commutativity of the algebra itself (classical Gelfand theory), the presence of a (sufficiently large) center (Allan's local principle and its relatives due to Douglas, Gohberg-Krupnik, and Simonenko), or the appearance of a completely new identity which replaces the common $ab = ba$ (Krupnik's principle for *PI*-algebras) which will be employed.

Unless stated otherwise, all algebras and homomorphisms occurring in this chapter are complex.

2.1 Gelfand theory

The Gelfand transform for commutative Banach algebras which we are going to discuss in this section, will not only provide us with the first and simplest local principle; it will also serve as a model for other local principles. Gelfand's theory associates with every commutative Banach algebra $\mathscr{B}$ with identity a subalgebra of the algebra $C(X)$ of all continuous functions on an appropriately chosen compact Hausdorff space X depending on the internal structure of the algebra. The Gelfand transform is then a homomorphism from $\mathscr{B}$ into $C(X)$, which is a symbol mapping for $\mathscr{B}$.

2.1.1 The maximal ideal space

Let $\mathscr{B}$ be a commutative unital Banach algebra, and let $M_{\mathscr{B}}$ denote the set of all maximal ideals of $\mathscr{B}$. Let x be a maximal ideal[1] of $\mathscr{B}$. Then the quotient algebra $\mathscr{B}/x$ is isomorphic to $\mathbb{C}$ by Theorem 1.3.8. Thus, to each $x \in M_{\mathscr{B}}$ and each $b \in \mathscr{B}$, there is associated a complex number, $\widehat{b}(x)$, which is the image of the coset $b + x$ under the above mentioned isomorphism.

The mapping $b \mapsto \widehat{b}(x)$ is a multiplicative linear functional on $\mathscr{B}$ the kernel of which is x. By Theorem 1.3.19, there are no other multiplicative linear functionals with the same kernel on $\mathscr{B}$, and there is a one-to-one correspondence between the non-trivial multiplicative linear functionals on $\mathscr{B}$ and the maximal ideals of $\mathscr{B}$: the kernel of every non-trivial multiplicative linear functional is a maximal ideal, and every maximal ideal is the kernel of a uniquely determined non-trivial multiplicative linear functional.

---

[1] The notation "x" for a maximal ideal may seem strange at first. But if one remembers Section 1.4.3 and the relation between the maximal ideals and the multiplicative linear functionals of the algebra $C(X)$, this choice of notation becomes clear.

Definition 2.1.1. Given an element $b \in \mathscr{B}$, the complex-valued function

$$\widehat{b} : M_{\mathscr{B}} \to \mathbb{C}, \quad x \mapsto \widehat{b}(x)$$

is called the *Gelfand transform of the element* $b \in \mathscr{B}$, and the resulting mapping

$$\widehat{} : \mathscr{B} \to C(M_{\mathscr{B}}), \quad b \mapsto \widehat{b}$$

is called the *Gelfand transform on* $\mathscr{B}$.

The set $M_{\mathscr{B}}$ can be made into a topological space by the requirement that all Gelfand transforms of elements of $\mathscr{B}$ become continuous functions on it:

Definition 2.1.2. The *Gelfand topology* on $M_{\mathscr{B}}$ is the coarsest topology on $M_{\mathscr{B}}$ that makes all Gelfand transforms $\widehat{b}$ with $b \in \mathscr{B}$ continuous. The set $M_{\mathscr{B}}$ provided with the Gelfand topology is referred to as the *maximal ideal space* of the Banach algebra $\mathscr{B}$.

Equivalently, the sets $\widehat{b}^{-1}(U)$ with b and U running through $\mathscr{B}$ and the open subsets of $\mathbb{C}$, respectively, form a sub-basis of the Gelfand topology.

2.1.2 Classical Gelfand theory

The following two theorems, also known as *Gelfand's representation theorem*, are the central results of classical Gelfand theory. They claim that every complex commutative unital semi-simple Banach algebra is isomorphic to an algebra of complex functions, defined in a certain compact Hausdorff space.

Theorem 2.1.3. *Let $\mathscr{B}$ be a commutative unital Banach algebra. Then:*

(i) *the Gelfand transform is a continuous homomorphism of norm 1;*
(ii) *the set $\widehat{\mathscr{B}}$ is a subalgebra of $C(M_{\mathscr{B}})$ which separates the points of $M_{\mathscr{B}}$ and contains the identity of $C(M_{\mathscr{B}})$;*
(iii) *the element $b \in \mathscr{B}$ is invertible if and only if $\widehat{b}(x) \neq 0$ for all $x \in M_{\mathscr{B}}$;*
(iv) *the kernel of the Gelfand transform is the radical of $\mathscr{B}$. Thus, the Gelfand transform is an isomorphism between $\mathscr{B}$ and $\widehat{\mathscr{B}}$ if and only if $\mathscr{B}$ is semi-simple;*
(v) *the spectrum of $b \in \mathscr{B}$ coincides with the image of $\widehat{b}$, and $r(b) = \|\widehat{b}\|_{\infty}$.*

Proof. The definition of the Gelfand topology in $M_{\mathscr{B}}$ guarantees the continuity of each function $\widehat{b}$. As $\widehat{b_1 + b_2} = \widehat{b_1} + \widehat{b_2}$, $\widehat{\lambda b_1} = \lambda \widehat{b_1}$ and $\widehat{b_1 b_2} = \widehat{b_1} \widehat{b_2}$, for $b_1, b_2 \in \mathscr{B}$ and $\lambda \in \mathbb{C}$, the Gelfand transform is a homomorphism. By the definition of the quotient norm, one also has

$$\|\widehat{b}\|_{\infty} = \sup_{x \in M_{\mathscr{B}}} |\widehat{b}(x)| \leq \|b\|, \tag{2.1}$$

which implies the continuity of the Gelfand transform and that its norm is less than or equal to 1. That $\widehat{\mathscr{B}}$ is a subalgebra of $C(M_{\mathscr{B}})$ is obvious, because the Gelfand transform is a homomorphism. To prove the rest of point (ii) let us suppose that we have two maximal ideals $x_1 \neq x_2$. Choosing an element $b_1 \in x_1$ such that $b_1 \notin x_2$ we obtain $\widehat{b}_1(x_1) = 0$ but $\widehat{b}_1(x_2) \neq 0$. We have also $\widehat{e}(x) = 1$ for all $x \in M_{\mathscr{B}}$, and using point (i) we can now deduce that the norm of $\widehat{}$ is 1. The third point is a direct consequence of Theorem 1.3.21. The proof of point (iv) is easy: just remember that $\widehat{b}(x) = 0$ for all $x \in M_{\mathscr{B}}$, if and only if $b \in x$ for all $x \in M_{\mathscr{B}}$.

Finally, regarding (v), we have that $\lambda \in \sigma(b) \Leftrightarrow \lambda e - b$ is not invertible, which is equivalent to $\widehat{\lambda e - b}(x) = 0$ for some $x \in M_{\mathscr{B}}$ by point (iii) above. This is equivalent to $\lambda - \widehat{b}(x) = 0$ for some $x \in M_{\mathscr{B}}$, that is $\lambda \in \widehat{b}(M_{\mathscr{B}})$. By the definition of spectral radius we have $r(b) = \sup_{x \in M_{\mathscr{B}}} |\widehat{b}(x)| = \|\widehat{b}\|_{\infty}$. ∎

In general, neither equality holds in the estimate (2.1), nor is the Gelfand transform $\widehat{} : \mathscr{B} \to C(M_{\mathscr{B}})$ injective or surjective. Examples are provided in Exercises 2.1.1 and 2.1.2.

Theorem 2.1.4. *The maximal ideal space $M_{\mathscr{B}}$ of a commutative Banach algebra with identity is a compact Hausdorff space; thus normal.*

Proof. We prepare the proof by recalling some facts from functional analysis. Let X be a Banach space with Banach dual X^*. For each $b \in X$, define a function

$$f_b : X^* \to \mathbb{C}, \quad \phi \mapsto \phi(b).$$

The w^*-topology on X^* is, by definition, the weakest topology on X^* for which all functions f_b with $b \in X$ are continuous. The restriction of the w^*-topology to the closed unit ball

$$E^* := \{\phi : \phi \in X^* \text{ and } \|\phi\| \leq 1\}$$

of X^* makes E^* a compact subset of a Hausdorff space in the w^*-topology (see, for instance, [184, Theorem 49-A]).

Now let $\mathscr{B}$ be a commutative Banach algebra with identity e, $\mathscr{B}^*$ its Banach dual space, and E^* the closed unit ball of $\mathscr{B}^*$, provided with its w^*-topology. Every non-zero multiplicative functional on $\mathscr{B}$ has norm 1 and can thus be considered as an element of E^*, which implies an embedding of the maximal ideal space $M_{\mathscr{B}}$ into E^*. It turns out that the restriction of the w^*-topology on E^* to $M_{\mathscr{B}}$ coincides with the Gelfand topology. Indeed, the restriction of f_b to $M_{\mathscr{B}}$ coincides with the Gelfand transform $\widehat{b}$, because of $f_b(x) = f_b(\phi_x) = \phi_x(b) = \widehat{b}(x)$.

Since E^* is a compact subset of a Hausdorff space with respect to the w^*-topology, it remains only to prove that $M_{\mathscr{B}}$ is a closed subset of E^*. Let $\mathbb{I}$ be a directed set and $(\phi_\alpha)_{\alpha \in \mathbb{I}}$ a net in $M_{\mathscr{B}}$ which converges to $\phi \in E^*$ in the w^*-topology. Then

$$\phi(e) = \lim_\alpha \phi_\alpha(e) = \lim_\alpha 1 = 1$$

and

$$\phi(b_1 b_2) = \lim_\alpha \phi_\alpha(b_1 b_2)$$
$$= \lim_\alpha \phi_\alpha(b_1)\phi_\alpha(b_2) = \lim_\alpha \phi_\alpha(b_1)\lim_\alpha \phi_\alpha(b_2) = \phi(b_1)\phi(b_2)$$

which shows that $\phi \in M_\mathscr{B}$ again. ∎

Remark 2.1.5. Using the terminology given at the beginning of the chapter, assertions (i) and (iii) of Theorem 2.1.3 can be rephrased as follows: The Gelfand transform $\widehat{} : \mathscr{B} \to C(M_\mathscr{B})$ is a symbol mapping for $\mathscr{B}$, and the family of all homomorphisms $W_x : b \mapsto \widehat{b}(x)$ of $\mathscr{B}$ with $x \in M_\mathscr{B}$ is a sufficient family. □

The following result provides a criterion for the injectivity of the Gelfand transform.

Proposition 2.1.6. *Let $\mathscr{B}$ be a commutative unital Banach algebra. The following conditions are equivalent, for any $b \in \mathscr{B}$:*

(i) $\|b^2\| = \|b\|^2$;
(ii) $r(b) = \|b\|$;
(iii) $\|\widehat{b}\| = \|b\|$.

Proof. Condition (i) implies $\|b^{2^k}\| = \|b\|^{2^k}$ for any natural k. Then the formula for the spectral radius (Theorem 1.2.12) gives

$$r(b) = \lim_{n\to\infty} \|b^n\|^{\frac{1}{n}} = \lim_{k\to\infty} \|b^{2^k}\|^{\frac{1}{2^k}} = \lim_{k\to\infty} \|b\| = \|b\|,$$

which is (ii). Conversely, if $\lambda \in \sigma(b)$ then $\lambda^2 \in \sigma(b^2)$ by Exercise 1.2.5. Hence, $\|b^2\| = r(b^2) = r(b)^2 = \|b\|^2$, showing that (ii) implies (i). The equivalence between (ii) and (iii) comes from (v) in Theorem 2.1.3. ∎

For normal elements in a C^*-algebra, condition (i) in the above proposition is always satisfied and we have $\|a^2\| = \|a\|^2$ and $r(a) = \|a\|$, as was seen in Proposition 1.2.36 (i). In a commutative C^*-algebra, all elements are normal. Thus, the Gelfand transform acts as an isometry on commutative C^*-algebras. One can say even more in this case.

Theorem 2.1.7 (Gelfand-Naimark). *Let $\mathscr{B}$ be a commutative unital C^*-algebra. Then the Gelfand transform is an (isometric) *-isomorphism from $\mathscr{B}$ onto the algebra $C(M_\mathscr{B})$.*

Proof. Let us first verify that the Gelfand transform is a *-homomorphism. If $h \in \mathscr{B}$ is self-adjoint, then $\sigma(h) \subset \mathbb{R}$ by Proposition 1.2.36, implying that $\mathrm{Im}\,\widehat{h} \subset \mathbb{R}$ by Theorem 2.1.3 (iii). Consequently, $\overline{\widehat{h}} = \widehat{h} = \widehat{h^*}$. Now let $b \in \mathscr{B}$ be arbitrary. Write b as $b = h + ik$ with h, k self-adjoint. Then

$$\widehat{b^*} = \widehat{(h - ik)} = \widehat{h} - \widehat{ik} = \overline{\widehat{h}} - i\overline{\widehat{k}} = \overline{\widehat{h + ik}} = \overline{\widehat{b}}.$$

Thus, the Gelfand transform is symmetric, and by Propositions 1.2.36 and 2.1.6, it is also an isometry. We are left with proving the surjectivity. The image $\widehat{\mathscr{B}}$ is a closed self-adjoint subalgebra of $C(M_{\mathscr{B}})$ by Theorem 1.1.6, and it separates the points of $M_{\mathscr{B}}$ and contains the constant functions. By the Stone-Weierstrass theorem (see for example [171, Section 5.7]), $\widehat{\mathscr{B}}$ coincides with $C(M_{\mathscr{B}})$.   ∎

The theorem above justifies thinking of elements of commutative unital C^*-algebras as continuous functions on a compact Hausdorff space, and we shall use this henceforth.

2.1.3 The Shilov boundary

Let X be a compact Hausdorff space. For each function $a \in C(X)$ and each closed non-empty subset E of X we set

$$\|a_{|E}\|_\infty := \max_{x \in E} |a(x)|.$$

Let $\mathscr{C}$ be a subset of $C(X)$. A closed subset F of X is called a *maximizing set for $\mathscr{C}$* if

$$\|a_{|X}\|_\infty = \|a_{|F}\|_\infty \quad \text{for all } a \in \mathscr{C}.$$

Lemma 2.1.8. *Let $\mathscr{C}$ be a subalgebra of $C(X)$ which contains the constant functions and separates the points of X. Then the intersection of all maximizing sets for $\mathscr{C}$ is a maximizing set for $\mathscr{C}$.*

Proof. Let S denote the intersection of all maximizing sets for $\mathscr{C}$. We claim that every point $x_0 \in X \setminus S$ has an open neighborhood U such that $F \setminus U$ is a maximizing set for $\mathscr{C}$ whenever F is a maximizing set for $\mathscr{C}$.

Indeed, since $x_0 \notin S$, there is a maximizing set F_0 for $\mathscr{C}$ which does not contain x_0. Since $\mathscr{C}$ contains the constant functions and separates the points of X, for each $y \in F_0$, there is a function $a_y \in \mathscr{C}$ with $a_y(x_0) = 0$ and $a_y(y) = 2$. Each set $U_y := \{x \in X : |a_y(x)| > 1\}$ is an open neighborhood of y. Thus, the compact set F_0 can be covered by a finite number of sets of the form U_y. We denote the corresponding functions in $\mathscr{C}$ by $a_1, \ldots, a_r$. Thus, $a_k(x_0) = 0$ for $k = 1, \ldots, r$, and for each $y \in F_0$ there is a $k \in \{1, \ldots, r\}$ such that $a_k(y) > 1$. Let

$$U := \{x : |a_k(x)| < 1 \text{ for every } k = 1, \ldots, r\}.$$

Then U is an open neighborhood of x_0 and $U \cap F_0 = \emptyset$.

Now let F be a maximizing set for $\mathscr{C}$, and suppose that the set $F \setminus U$ is not maximizing for $\mathscr{C}$. Then there is a function $a \in \mathscr{C}$ with

$$\|a_{|X}\|_\infty = 1 > \|a_{|F \setminus U}\|_\infty.$$

Let $M := \max\{\|a_{k|X}\|_\infty : k = 1, \ldots, r\}$ and choose n such that $\|a_{|F \setminus U}\|_\infty^n < 1/M$. Then $\|a^n a_{k|F \setminus U}\|_\infty < 1$ for every $k = 1, \ldots, r$, and one also has $|a^n(x)a_k(x)| < 1$ for all $x \in U$ and $k = 1, \ldots, r$. Since F is maximizing, this implies

$$\|a^n a_{k|X}\|_\infty = \|a^n a_{k|F}\|_\infty < 1 \quad \text{for every } k = 1, \ldots, r.$$

Since F_0 is a maximizing set, there is a point $y \in F_0$ with $a(y) = 1$. For this point, one gets

$$|a_k(y)| = |a^n(y)a_k(y)| \le \|a^n a_{k|X}\|_\infty < 1,$$

which implies that $y \in F_0 \cap U$. Hence, $F_0 \cap U$ is not empty; a contradiction. This contradiction proves our claim.

Having the claim at our disposal, the proof of the lemma can be completed as follows. Let $a \in \mathscr{C}$, and let $K := \{x \in X : |a(x)| = \|a_{|X}\|_\infty\}$. We have to show that $S \cap K$ is not empty. Contrary to what we want to show, assume that $S \cap K = \emptyset$. Then each point $x_0 \in K$ has an open neighborhood U given by the claim. Since K is compact, it is covered by a finite number of these neighborhoods, say $U_1, \ldots, U_n$. The set X is maximizing, and so are the sets

$$X \setminus U_1, \ X \setminus (U_1 \cup U_2), \ \ldots, \ X \setminus (U_1 \cup \ldots \cup U_n) =: E,$$

say. Since $E \cap K = \emptyset$ one obtains $\|a_{|E}\|_\infty < \|a_{|X}\|_\infty$ which contradicts the maximality of E. ∎

Now let $\mathscr{A}$ be a commutative Banach algebra with identity e. Then the algebra $\mathscr{C}$ of all Gelfand transforms of elements of $\mathscr{A}$ contains the constant functions and separates the points of the maximal ideal space $M_\mathscr{A}$. By the above lemma, the intersection of all maximizing sets for $\mathscr{C}$ is a maximizing set for $\mathscr{C}$. This intersection is called the *Shilov boundary of* $M_\mathscr{A}$. We denote it by $\partial_S M_\mathscr{A}$.

Equivalently, a point $x_0 \in M_\mathscr{A}$ belongs to $\partial_S M_\mathscr{A}$ if and only if, for each open neighborhood $U \subset M_\mathscr{A}$ of x_0, there exists an $a \in \mathscr{A}$ such that

$$\|\widehat{a}_{|M_\mathscr{A} \setminus U}\|_\infty < \|\widehat{a}_{|U}\|_\infty.$$

Theorem 2.1.9. *Let $\mathscr{A}$ and $\mathscr{B}$ be commutative unital Banach algebras.*

(i) *If* $\mathsf{W} : \mathscr{A} \to \mathscr{B}$ *is a unital homomorphism which preserves spectral radii, i.e., if*

$$r_\mathscr{B}(\mathsf{W}(a)) = r_\mathscr{A}(a) \quad \text{for all } a \in \mathscr{A},$$

then $\partial_S M_\mathscr{A} \subseteq \mathsf{W}^*(\partial_S M_\mathscr{B})$ *(with* W^* *referring to the dual mapping of* W*).*

(ii) *Now let $\mathscr{A}$ be a unital closed subalgebra of $\mathscr{B}$. Then each maximal ideal in the Shilov boundary of $M_\mathscr{A}$ is contained in some maximal ideal of $\mathscr{B}$.*

Proof. (i) We think of the elements of the maximal ideal space $M_\mathscr{B}$ as non-trivial multiplicative functionals, and thus as elements of the Banach dual $\mathscr{B}^*$. Since W is a homomorphism, its dual W^* sends multiplicative functionals on $\mathscr{B}$ to multiplicative functionals on $\mathscr{A}$. Let $e_{\mathscr{A}}$ and $e_{\mathscr{B}}$ denote the identity elements of $\mathscr{A}$ and $\mathscr{B}$, respectively. From

$$W^*(\varphi)(e_{\mathscr{A}}) = \varphi(W(e_{\mathscr{A}})) = \varphi(e_{\mathscr{B}}) = 1$$

one deduces that the image of a non-trivial multiplicative functional φ on $\mathscr{B}$ is non-trivial again. Thus, W^* maps $M_{\mathscr{B}}$ into $M_{\mathscr{A}}$.

The Shilov boundary $\partial_S M_{\mathscr{B}}$ is a compact subset of $M_{\mathscr{B}}$, and $W^* : M_{\mathscr{B}} \to M_{\mathscr{A}}$ is continuous by the definition of the Gelfand topology. Hence, $W^*(\partial_S M_{\mathscr{B}})$ is a compact subset of $M_{\mathscr{A}}$. Since $M_{\mathscr{A}}$ is a compact Hausdorff space, the set $W^*(\partial_S M_{\mathscr{B}})$ is closed in $M_{\mathscr{A}}$. The assertion will follow once we have shown that this set is maximizing for the algebra of all Gelfand transforms of elements of $\mathscr{A}$. Let $a \in \mathscr{A}$. Then

$$\|\widehat{a}_{|W^*(\partial_S M_{\mathscr{B}})}\|_\infty = \|\widehat{W(a)}_{|\partial_S M_{\mathscr{B}}}\|_\infty = r_{\mathscr{B}}(W(a)) = r_{\mathscr{A}}(a)$$

with r referring to the spectral radius. Thus, $W^*(\partial_S M_{\mathscr{B}})$ is maximizing and includes the Shilov boundary of the maximal ideal space of $\mathscr{A}$.

(ii) Consider the inclusion map $W : \mathscr{A} \to \mathscr{B}$. Then W is a homomorphism which preserves spectral radii (use the formula for the spectral radius to check this). Thus, by part (i) of this theorem, every maximal ideal in the Shilov boundary of $\mathscr{A}$ arises as the restriction of some maximal ideal (in the Shilov boundary) of $\mathscr{B}$.   ∎

2.1.4 Example: SIOs with continuous coefficients

We return now to the SIOs introduced in Section 1.4.4, but will apply Gelfand's representation theorem. Let $\mathscr{B}$ be the smallest closed subalgebra of $\mathscr{L}(L^p(\mathbb{T}))$ which contains all singular integral operators of the form

$$A = cI + dS + K = fP + gQ + K$$

where $c, d \in C(\mathbb{T})$, K is compact, and $f := c + d$ and $g := c - d$. From Proposition 1.4.12 we infer that the Calkin image $\mathscr{B}^{\mathscr{K}} := \mathscr{B}/\mathscr{K}$ of $\mathscr{B}$ is a commutative and unital Banach algebra which is, consequently, subject to Gelfand's representation theorem. We start with identifying the maximal ideal space of $\mathscr{B}^{\mathscr{K}}$.

Proposition 2.1.10. *All proper ideals of $\mathscr{B}^{\mathscr{K}}$ are contained in ideals of the form*

$$\mathscr{I}_{P,X_0} := \{fP + gQ + \mathscr{K} : f(X_0) = 0\} \quad or \quad \mathscr{I}_{Q,X_0} := \{fP + gQ + \mathscr{K} : g(X_0) = 0\}$$

with a certain subset X_0 of $\mathbb{T}$.

Proof. Suppose there is an ideal $\mathscr{I}$ of $\mathscr{B}^{\mathscr{K}}$, which does not have the claimed property. Then, for every $x \in \mathbb{T}$, there are functions f_x and $g_x \in C(\mathbb{T})$ such that

$f_x(x) \neq 0$, $g_x(x) \neq 0$, and $f_x P + g_x Q + \mathcal{K} \in \mathcal{I}$. Consequently,

$$A_x := (f_x P + g_x Q + \mathcal{K})(\overline{f}_x P + \overline{g}_x Q + \mathcal{K}) = |f_x|^2 P + |g_x|^2 Q + \mathcal{K} \in \mathcal{I}.$$

The functions $|f_x|^2$ and $|g_x|^2$ are positive in a certain open neighborhood U_x of x, and the collection of all of these neighborhoods covers $\mathbb{T}$. By compactness, one can extract a finite subcovering $\mathbb{T} = U_{x_1} \cup \ldots \cup U_{x_n}$, say. It is easy to see that then the operator $A = \sum_{k=1}^n A_{x_k} \in \mathcal{I}$ is invertible in $\mathscr{B}^{\mathcal{K}}$, which is a contradiction. ∎

Proposition 2.1.10 implies that the maximal ideals of $\mathscr{B}^{\mathcal{K}}$ are necessarily of the form

$$\mathscr{I}_{P,x_0} := \{fP + gQ + \mathcal{K} : f(x_0) = 0\} \text{ and } \mathscr{I}_{Q,x_0} := \{fP + gQ + \mathcal{K} : g(x_0) = 0\}$$

with $x_0 \in \mathbb{T}$. These ideals are closed by Theorem 1.3.5. Thus, there is a bijection between the maximal ideal space of $\mathscr{B}^{\mathcal{K}}$ and two copies of the unit circle $\mathbb{T}$ or, more precisely, between the maximal ideal space of $\mathscr{B}^{\mathcal{K}}$ and the product $\mathbb{T} \times \{0, 1\}$. A closer look shows that this bijection is even a homeomorphism, which allows one to identify the maximal ideal space of $\mathscr{B}^{\mathcal{K}}$ with $\mathbb{T} \times \{0, 1\}$, provided with the usual product topology. Under this identification, the Gelfand transform of an element $A + \mathcal{K} = fP + gQ + \mathcal{K}$ is given by

$$\widehat{A + \mathcal{K}}(x, n) = \begin{cases} f(x) & \text{if } n = 0, \\ g(x) & \text{if } n = 1. \end{cases}$$

Thus, the coset $A + \mathcal{K}$ is invertible (equivalently, the operator A is Fredholm) if and only if $f(x) \neq 0$ and $g(x) \neq 0$ for all $x \in \mathbb{T}$. It is also easy to see that the radical of $\mathscr{B}^{\mathcal{K}}$ is $\{0\}$.

These results remain valid for curves Γ other than the unit circle, provided that the singular integral operator S_Γ on that curve satisfies $S_\Gamma^2 = I$ and that the commutator of S_Γ with every operator of multiplication by a continuous function on Γ is compact. Examples of such curves will be discussed in Chapter 4.

2.1.5 Exercises

Exercise 2.1.1. Let $\mathscr{B}$ be the algebra of all matrices $\begin{bmatrix} a & b \\ 0 & a \end{bmatrix}$ with complex entries a, b. Determine the maximal ideal space of $\mathscr{B}$ and the Gelfand transform on $\mathscr{B}$. Conclude that the Gelfand transform is not injective. (Evidently, the reason for being not injective is that $\mathscr{B}$ has a non-trivial radical.)

Exercise 2.1.2. Let $\mathbb{A}$ refer to the disk algebra introduced in Example 1.2.23. Characterize $\mathbb{A}$ as the set of all functions $f \in C(\mathbb{T})$ which possess an analytic continuation into the open unit disk $\mathbb{D} := \{z \in \mathbb{C} : |z| < 1\}$. Show that $\mathbb{A}$ is a commutative Banach algebra with identity and that every character of $\mathbb{A}$ is of the form $f \mapsto f(z)$ with some fixed point $z \in \mathbb{D} \cup \mathbb{T}$. Conclude that the maximal ideal space of $\mathbb{A}$ is homeomorphic to the closed unit disk $\mathbb{D} \cup \mathbb{T}$ with its standard (Euclidean) topology. Show further that the Gelfand transform of $f \in \mathbb{A}$ coincides with the analytic continuation of f into the interior of the unit disk. Thus, the image of the Gelfand transform on $\mathbb{A}$ is a proper subset of $C(M_{\mathbb{A}}) = C(\mathbb{D} \cup \mathbb{T})$.

Exercise 2.1.3. A Banach algebra $\mathcal{B}$ with identity e is *singly generated* if there is an element $b \in \mathcal{B}$ such that the smallest closed subalgebra of $\mathcal{B}$ which contains e and b coincides with $\mathcal{B}$. In this case, b is called a *generator* of $\mathcal{B}$. Prove that the maximal ideal space of a singly generated (by b, say) Banach algebra is homeomorphic to the spectrum $\sigma_{\mathcal{B}}(b)$. Suggestion: under this homeomorphism, the point $\lambda \in \sigma_{\mathcal{B}}(b)$ corresponds to the smallest closed ideal of $\mathcal{B}$ which contains $b - \lambda e$ (see [37, 15.3.6]).

Exercise 2.1.4. We know from Exercise 1.4.8 that the Toeplitz algebra $\mathcal{T}(C)$ contains the ideal $\mathcal{K}(l^2(\mathbb{Z}^+))$ of the compact operators and that the quotient algebra $\mathcal{T}(C)/\mathcal{K}(l^2(\mathbb{Z}^+))$ is commutative. Identify the maximal ideal space of this quotient algebra. Show that the Gelfand transform of $T(a) + \mathcal{K}(l^2(\mathbb{Z}^+))$ can be identified with $a \in C(\mathbb{T})$.

Exercise 2.1.5. Let $\mathcal{B}$ be a commutative Banach algebra with identity, generated by the elements $\{b_1, \ldots, b_n\}$. Show that $M_{\mathcal{B}}$ is homeomorphic to the joint spectrum $\sigma_{\mathcal{B}}(b_1, \ldots, b_n)$ in $\mathcal{B}$.

Exercise 2.1.6. Review Exercise 1.3.9. Let M_{l_∞} be the space of multiplicative linear functionals of l_∞, with the w^* topology. Given $u \in l_\infty$, define $\widehat{u}(\phi)$ as $\phi(u)$ for $\phi \in M_{l_\infty}$. Note that M_{l_∞} is a compact Hausdorff space.

a) Show that M_{l_∞} is extremely disconnected (i.e., the closure of every open set is open).
b) Let $\phi_n(u) := u_n$ for $n \in \mathbb{N}$ and $u \in l_\infty$. Show that the subset $\{\phi_n : n \in \mathbb{N}\}$ of M_{l_∞} is homeomorphic to $\mathbb{N}$ and that, consequently, $\mathbb{N}$ can be identified with a subset of M_{l_∞}. Show that $\mathbb{N}$ is dense in M_{l_∞}.
c) Show that the one point subset $\{\phi_n\}$ of M_{l_∞} is open in M_{l_∞}.
d) Show that $\phi(u) = 0$ when $\phi \in M_{l_\infty} \setminus \mathbb{N}$ and $u \in l_\infty^0$.

Exercise 2.1.7. Determine the Shilov boundaries of the maximal ideal spaces of the disk algebra $\mathbb{A}$ and of the C^*-algebra $C(X)$ where X is a compact Hausdorff space. (Hint: use the Tietze-Uryson extension theorem.)

2.2 Allan's local principle

As we have seen, Gelfand theory associates with each unital commutative Banach algebra $\mathcal{B}$ a compact Hausdorff space $M_{\mathcal{B}}$, called the *maximal ideal space* of $\mathcal{B}$, and with every element $b \in \mathcal{B}$ a continuous function $\widehat{b} : M_{\mathcal{B}} \to \mathbb{C}$, called the *Gelfand transform* of b, such that the mapping

$$\frown : \mathscr{B} \to C(M_{\mathscr{B}}), \quad b \mapsto \widehat{b}$$

becomes a contractive algebra homomorphism which preserves spectra. Allan's local principle is a generalization of classical Gelfand theory to unital Banach algebras which are close to commutative algebras in the sense that their centers are non-trivial.

2.2.1 Central subalgebras

Let $\mathscr{A}$ be a unital Banach algebra. Recall that the center of $\mathscr{A}$ is the set $\text{Cen}\,\mathscr{A}$ of all elements $a \in \mathscr{A}$ such that $ab = ba$ for all $b \in \mathscr{A}$. Evidently, $\text{Cen}\,\mathscr{A}$ is a closed commutative subalgebra of $\mathscr{A}$ which contains the identity element. A *central* subalgebra of $\mathscr{A}$ is a closed subalgebra $\mathscr{B}$ of the center of $\mathscr{A}$ which contains the identity element. Thus, $\mathscr{B}$ is a commutative Banach algebra with compact maximal ideal space $M_{\mathscr{B}}$. For each maximal ideal x of $\mathscr{B}$, consider the smallest closed two-sided ideal $\mathscr{I}_x$ of $\mathscr{A}$ which contains x, and let Φ_x refer to the canonical homomorphism from $\mathscr{A}$ onto the quotient algebra $\mathscr{A}/\mathscr{I}_x$.

In contrast to the commutative setting where $\mathscr{B}/x \cong \mathbb{C}$ for all $x \in M_{\mathscr{B}}$, the quotient algebras $\mathscr{A}/\mathscr{I}_x$ will depend on $x \in M_{\mathscr{B}}$ in general. Moreover, it can happen that $\mathscr{I}_x = \mathscr{A}$ for certain maximal ideals x. In this case we *define* that $\Phi_x(a)$ is invertible in $\mathscr{A}/\mathscr{I}_x$ and that $\|\Phi_x(a)\| = 0$ for each $a \in \mathscr{A}$.

2.2.2 Allan's local principle

The proof of Allan's local principle is based on the following observation.

Proposition 2.2.1 (Allan). *Let $\mathscr{B}$ be a central subalgebra of the unital Banach algebra $\mathscr{A}$. If $\mathscr{M}$ is a maximal left, right, or two-sided ideal of $\mathscr{A}$, then $\mathscr{M} \cap \mathscr{B}$ is a (two-sided) maximal ideal of $\mathscr{B}$.*

Proof. For definiteness, let $\mathscr{M}$ be a maximal left ideal of $\mathscr{A}$. Then $\mathscr{M} \cap \mathscr{B}$ is a proper (since $e \in \mathscr{B} \setminus \mathscr{M}$) closed two-sided ideal of $\mathscr{B}$. The maximality of $\mathscr{M} \cap \mathscr{B}$ will follow once we have shown that

$$\text{for all } z \in \mathscr{B} \setminus \mathscr{M}, \text{ there is a } \lambda \in \mathbb{C} \setminus \{0\} \text{ with } z - \lambda e \in \mathscr{M}. \tag{2.2}$$

Indeed, let $\mathscr{I}$ be a two-sided ideal of $\mathscr{B}$ with $\mathscr{M} \cap \mathscr{B} \subset \mathscr{I}$ and $\mathscr{M} \cap \mathscr{B} \neq \mathscr{I}$. Choose $z \in \mathscr{I} \setminus (\mathscr{M} \cap \mathscr{B}) \subset \mathscr{B} \setminus \mathscr{M}$. According to (2.2), there is a $\lambda \in \mathbb{C}$ and an $l \in \mathscr{M} \cap \mathscr{B}$ with $e = \lambda^{-1} z + l$. Hence, $e \in \mathscr{I}$, whence $\mathscr{I} = \mathscr{B}$ and the maximality of $\mathscr{M} \cap \mathscr{B}$.

We are left with verifying (2.2). In a first step we show that every element $z \in \mathscr{B} \setminus \mathscr{M}$ has a unique inverse modulo $\mathscr{M}$. The set $\mathscr{I}_z := \{l + az : l \in \mathscr{M}, a \in \mathscr{A}\}$ is a left ideal of $\mathscr{A}$ which contains $\mathscr{M}$ properly (since $z \notin \mathscr{M}$). Since $\mathscr{M}$ is maximal, we must have $\mathscr{I}_z = \mathscr{A}$. Hence, $e \in \mathscr{I}_z$, and there is an $a \in \mathscr{A}$ with $az - e, za - e \in \mathscr{M}$ (note that $z \in \operatorname{Cen} A$). Thus, a is an inverse of z modulo $\mathscr{M}$.

Further, $\mathscr{K}_z := \{a \in \mathscr{A} : az \in \mathscr{M}\}$ is a proper (since $e \notin \mathscr{K}_z$) left ideal of $\mathscr{A}$ which contains $\mathscr{M}$. Since $\mathscr{M}$ is maximal, we have $\mathscr{K}_z = \mathscr{M}$. In particular, if a_1 and a_2 are inverses modulo $\mathscr{M}$ of z, then $a_1 - a_2 \in \mathscr{M}$. Thus, the inverses modulo $\mathscr{M}$ of z determine a unique element of the quotient space $\mathscr{A}/\mathscr{M}$.

Contrary to (2.2), suppose that $z - \lambda e \notin \mathscr{M}$ for all $\lambda \in \mathbb{C}$. Let $y^\pi(\lambda)$ denote the (uniquely determined) coset of $\mathscr{A}/\mathscr{M}$ containing the inverses modulo $\mathscr{M}$ of $z - \lambda e$. Then $y^\pi : \mathbb{C} \to \mathscr{A}/\mathscr{M}$ is an analytic function. Indeed, let $\lambda_0 \in \mathbb{C}$, and let $y_0 \in y^\pi(\lambda_0)$ be an inverse modulo $\mathscr{M}$ of $z - \lambda_0 e$. Then, for $|\lambda - \lambda_0| < 1/\|y_0\|$, the element $e - (\lambda - \lambda_0)y_0$ is invertible in $\mathscr{A}$, and it is readily verified that $y_0[e - (\lambda - \lambda_0)y_0]^{-1}$ is an inverse modulo $\mathscr{M}$ of $z - \lambda e$. Thus, for $|\lambda - \lambda_0| < 1/\|y_0\|$,

$$y^\pi(\lambda) = y_0[e - (\lambda - \lambda_0)y_0]^{-1} + \mathscr{M},$$

which implies the asserted analyticity. If $|\lambda| > 2\|z\|$, then $z - \lambda e$ is actually invertible in $\mathscr{A}$ and

$$\|y^\pi(\lambda)\| \leq \|(z - \lambda e)^{-1}\| = \frac{1}{|\lambda|}\|(e - z/\lambda)^{-1}\| \leq \frac{1}{|\lambda|}\frac{1}{1 - \|z\|/|\lambda|} < \frac{1}{\|z\|}.$$

Therefore, y^π is bounded, whence $y^\pi(\lambda) = 0$ for all $\lambda \in \mathbb{C}$ by Liouville's theorem. In particular, $y^\pi(0) = 0$. Thus, there is a $y_0 \in \mathscr{M}$ with $y_0 z - e \in \mathscr{M}$, whence $e \in \mathscr{M}$. This is impossible since $\mathscr{M}$ is a proper ideal of $\mathscr{A}$. This contradiction implies that there is a $\lambda \in \mathbb{C}$ with $z - \lambda e \in \mathscr{M}$. Since $z \notin \mathscr{M}$, one also has $\lambda \neq 0$. ∎

Before tackling Allan's principle, let us recall that a function $f : M_{\mathscr{B}} \to \mathbb{R}$ is said to be *upper semi-continuous* at $x_0 \in M_{\mathscr{B}}$ if, for each $\varepsilon > 0$, there exists a neighborhood $U_\varepsilon \subset M_{\mathscr{B}}$ of x_0 such that $f(x) < f(x_0) + \varepsilon$ for any $x \in U_\varepsilon$. The function f is said to be upper semi-continuous on $M_{\mathscr{B}}$ if it is upper semi-continuous at each point of $M_{\mathscr{B}}$. Every upper semi-continuous function defined on a compact set attains its supremum.

Theorem 2.2.2 (Allan's local principle). *Let $\mathscr{B}$ be a central subalgebra of the unital Banach algebra $\mathscr{A}$. Then:*

(i)   *an element $a \in \mathscr{A}$ is invertible if and only if the cosets $\Phi_x(a)$ are invertible in $\mathscr{A}/\mathscr{I}_x$ for each $x \in M_{\mathscr{B}}$;*
(ii) *the mapping $M_{\mathscr{B}} \to \mathbb{R}^+$, $x \mapsto \|\Phi_x(a)\|$ is upper semi-continuous for every $a \in \mathscr{A}$;*
(iii) *$\|a\| \geq \max_{x \in M_{\mathscr{B}}} \|\Phi_x(a)\|$;*
(iv) *$\cap_{x \in M_{\mathscr{B}}} \mathscr{I}_x$ lies in the radical of $\mathscr{A}$.*

Proof. To prove (i) we show that $a \in \mathscr{A}$ is left invertible if and only if $\Phi_x(a)$ is left invertible for all $x \in M_{\mathscr{B}}$. The proof for the right invertibility is analogous.

Clearly, $\Phi_x(a)$ is left invertible if a is so. To verify the reverse implication assume the contrary, i.e., suppose $\Phi_x(a)$ to be left invertible in $\mathscr{A}/\mathscr{I}_x$ for all $x \in M_{\mathscr{B}}$

but let a have no left inverse in $\mathscr{A}$. Denote by $\mathscr{M}$ a maximal left ideal of $\mathscr{A}$ which contains the set $\mathscr{I} := \{ba : b \in \mathscr{A}\}$ (note that $e \notin \mathscr{I}$). Put $x = \mathscr{M} \cap \mathscr{B}$. By Proposition 2.2.1, x is a maximal ideal of $\mathscr{B}$. We claim that $\mathscr{I}_x \subseteq \mathscr{M}$. Indeed, if $l = \sum_{k=1}^{n} a_k x_k b_k$ where $x_k \in x$ and $a_k, b_k \in \mathscr{A}$, then $l = \sum_{k=1}^{n} a_k b_k x_k$ (because $\mathscr{B}$ is central), and hence $l \in \mathscr{M}$ (because $\mathscr{M}$ is a left ideal). Thus, $\mathscr{I}_x \subseteq \mathscr{M}$. By our assumption, $\Phi_x(a)$ is left invertible in $\mathscr{A}/\mathscr{I}_x$, that is, there exists a $b \in \mathscr{A}$ with $ba - e \in \mathscr{I}_x$, and since $\mathscr{I}_x \subseteq \mathscr{M}$ we have $ba - e \in \mathscr{M}$. On the other hand, $ba \in \mathscr{I} \subseteq \mathscr{M}$. This implies that $e \in \mathscr{M}$ which contradicts the maximality of $\mathscr{M}$.

(ii) Let $x \in M_{\mathscr{B}}$ and $\varepsilon > 0$. We have to show that there is a neighborhood U of x such that
$$\|\Phi_y(a)\| < \|\Phi_x(a)\| + \varepsilon \quad \text{for all } y \in U.$$

Choose elements $a_1, \dots, a_n \in \mathscr{A}$ and $x_1, \dots, x_n \in x$ with
$$\left\| a + \sum_{j=1}^{n} a_j x_j \right\| < \|\Phi_x(a)\| + \varepsilon/2, \tag{2.3}$$

set $\delta := \sum_{i=1}^{n} \|a_i\| + 1$, and define an open neighborhood $U \subset M_{\mathscr{B}}$ of x by
$$U := \{y \in M_{\mathscr{B}} : |\widehat{x_j}(y)| < \varepsilon/(2\delta) \text{ for } j = 1, \dots, n\}.$$

Let $y \in U$ and set $y_j := x_j - \widehat{x_j}(y)\,e$. Then $\widehat{y_j}(y) = \widehat{x_j}(y) - \widehat{x_j}(y)\widehat{e}(y) = 0$, whence $y_j \in y$ and
$$\|\Phi_y(a)\| \leq \left\| a + \sum_{j=1}^{n} a_j y_j \right\|. \tag{2.4}$$

The estimates (2.3) and (2.4) give
$$\begin{aligned}
\|\Phi_y(a)\| - \|\Phi_x(a)\| &\leq \left\| a + \sum a_j y_j \right\| - \left\| a + \sum a_j x_j \right\| + \varepsilon/2 \\
&\leq \left\| \sum a_j(y_j - x_j) \right\| + \varepsilon/2 \\
&= \left\| \sum \widehat{x_j}(y) a_j \right\| + \varepsilon/2 < \varepsilon,
\end{aligned}$$

whence the upper semi-continuity of $y \mapsto \|\Phi_y(a)\|$ at x.

(iii) By definition, $\|a\| \geq \|\Phi_x(a)\|$ for any $x \in M_{\mathscr{B}}$, which implies that $\|a\| \geq \sup_{x \in M_{\mathscr{B}}} \|\Phi_x(a)\|$. The supremum in this estimate is actually a maximum due to the compactness of $M_{\mathscr{B}}$.

(iv) Let $k \in \cap_{x \in M_{\mathscr{B}}} \mathscr{I}_x$. Then, for any $a \in \mathscr{A}$, $\Phi_x(e - ak) = \Phi_x(e)$ and by (i) above $e - ak$ is invertible. Thus, by Proposition 1.3.3, k belongs to the radical of $\mathscr{A}$. ∎

2.2.3 Local invertibility and local spectra

As consequences of the upper semi-continuity, we mention the following properties of local invertibility and local spectra.

Let X be a locally compact Hausdorff space with one-point compactification $X \cup \{x_\infty\}$, and let M be a mapping from X into the set of all compact subsets of the complex plane. For each net $y := (y_t)_{t \in T}$ in X with limit x_∞, consider the set $L(y)$ of all limits of convergent nets $(\lambda_t)_{t \in T}$ with $\lambda_t \in M(y_t)$. The *limes superior* (also called the *partial limiting set*) $\limsup_{x \to x_\infty} M(x)$ is defined as the union of all sets $L(y)$, where the union is taken over all nets y in X tending to x_∞. For $X = \mathbb{Z}^+$, this definition coincides with that one given before Proposition 1.2.16. Below we apply this definition when Y is a compact Hausdorff space, $x_\infty \in Y$ and $X := Y \setminus \{x_\infty\}$.

Proposition 2.2.3. *Let the hypothesis be as in Theorem 2.2.2.*

(i) *If $\Phi_x(a)$ is invertible in $\mathscr{A}/\mathscr{I}_x$, then there is a neighborhood U of $x \in M_{\mathscr{B}}$ as well as a neighborhood V of $a \in \mathscr{A}$ such that $\Phi_y(c)$ is invertible in $\mathscr{A}/\mathscr{I}_y$ and*

$$\|\Phi_y(c)^{-1}\| \leq 4\|\Phi_x(a)^{-1}\| \quad \text{for all } y \in U \text{ and } c \in V.$$

(ii) *For all $a \in \mathscr{A}$ and $x \in M_{\mathscr{B}}$,*

$$\limsup_{y \to x} \sigma(\Phi_y(a)) \subseteq \sigma(\Phi_x(a)).$$

The number 4 in the estimate in assertion (i) can be replaced by any constant greater than 1.

Proof. (i) Let $\Phi_x(a)$ be invertible and choose $b \in \mathscr{A}$ such that

$$\Phi_x(ab - e) = \Phi_x(ba - e) = 0.$$

By Theorem 2.2.2 (ii), the mappings

$$y \mapsto \|\Phi_y(ab - e)\| \quad \text{and} \quad y \mapsto \|\Phi_y(ba - e)\|$$

are upper semi-continuous on the maximal ideal space of $\mathscr{B}$. Hence,

$$\|\Phi_y(ab - e)\| < 1/4 \quad \text{and} \quad \|\Phi_y(ba - e)\| < 1/4$$

for all maximal ideals y in a certain neighborhood U' of x. Further, let V stand for the set of all elements $c \in \mathscr{A}$ with $\|c - a\| < (4\|b\|)^{-1}$. Then

$$\Phi_y(c)\Phi_y(b) = \Phi_y(e) + \Phi_y(cb - e) \quad \text{and} \quad \Phi_y(b)\Phi_y(c) = \Phi_y(e) + \Phi_y(bc - e)$$

with

$$\|\Phi_y(cb - e)\| \leq \|\Phi_y(ab - e)\| + \|\Phi_y((c - a)b)\| \leq 1/4 + \|c - a\|\,\|b\| < 1/2$$

and, analogously, $\|\Phi_y(bc-e)\| \leq 1/2$. Since $\Phi(e)$ is the identity element in $\mathscr{A}/\mathscr{I}_y$, a Neumann series argument implies that $\Phi_y(c)$ is invertible in $\mathscr{A}/\mathscr{I}_y$ and that

$$\|\Phi_y(c)^{-1}\| \leq 2\|\Phi_y(b)\| \quad \text{for all } y \in U' \text{ and } c \in V.$$

Employing the upper semi-continuity once more, one finally gets

$$\|\Phi_y(b)\| \leq 2\|\Phi_x(b)\| = 2\|\Phi_x(a)^{-1}\|$$

for all y in a neighborhood $U \subseteq U'$ of x.

(ii) Let $\lambda \in \limsup_{y \to x} \sigma(\Phi_y(a))$. By definition of the limes superior, there is a net $(y_t)_{t \in T} \in M_{\mathscr{B}}$ with $y_t \to x$ and numbers $\lambda_t \in \sigma(\Phi_{y_t}(a))$ with $\lambda_t \to \lambda$. Consider the elements $a - \lambda_t e$, which converge to $a - \lambda e$ in the norm of $\mathscr{A}$. If the coset $\Phi_x(a - \lambda e)$ was invertible, then the local cosets $\Phi_{y_t}(a - \lambda_t e)$ would be invertible for all sufficiently large t by part (i) of this proposition. This is impossible due to the choice of the points λ_t. Consequently, $\lambda \in \sigma(\Phi_x(a))$. ∎

2.2.4 Localization over central C^*-algebras

We will now have a closer look at Allan's local principle in the case that the central subalgebra $\mathscr{B}$ of $\mathscr{A}$ is a C^*-algebra. Thus, we let $\mathscr{A}, \mathscr{B}, M_{\mathscr{B}}, \mathscr{I}_x$ and Φ_x be as in the previous subsection, but in addition we assume that there is an involution on the central subalgebra $\mathscr{B}$ of $\mathscr{A}$ which makes $\mathscr{B}$ to a C^*-algebra with respect to the norm inherited from $\mathscr{A}$. In this context, we will obtain a nice expression for the local norm and a canonical representation for the elements in the local ideals. Further we will briefly discuss some issues related to inverse-closedness.

Proposition 2.2.4. *Let $\mathscr{A}$ be a unital Banach algebra and let $\mathscr{B}$ be a central C^*-subalgebra of $\mathscr{A}$ which contains the identity. Then, for each $a \in \mathscr{A}$ and $x \in M_{\mathscr{B}}$,*

$$\|\Phi_x(a)\| = \inf_b \|ab\|$$

where the infimum is taken over all $b \in \mathscr{B}$ with $0 \leq b \leq 1$ which are identically 1 in a certain neighborhood of x.

Proof. Denote the infimum on the right-hand side by q. If $b \in \mathscr{B}$ is identically 1 in some neighborhood of x, then $a(b-1)$ belongs to the local ideal $\mathscr{I}_x$, whence

$$\|\Phi_x(a)\| \leq \|a + a(b-1)\| = \|ab\|.$$

Taking the infimum over all b with the properties mentioned above gives the estimate $\|\Phi_x(a)\| \leq q$.

For the reverse estimate, let $\varepsilon > 0$. Choose functions $b_1, \ldots, b_n \in \mathscr{B}$ which vanish at x and non-zero elements $a_1, \ldots, a_n \in \mathscr{A}$ such that

$$\|a + a_1 b_1 + \ldots + a_n b_n\| \leq \|\Phi_x(a)\| + \varepsilon.$$

If b is any function in $\mathscr{B}$ with $0 \leq b \leq 1$ which is identically 1 in a certain neighborhood of x, then

$$\begin{aligned}
\|ab\| &\leq \|(a + \sum a_i b_i)b\| + \|\sum a_i b_i b\| \\
&\leq \|a + \sum a_i b_i\| + \sum \|a_i\| \|b_i b\| \\
&\leq \|\Phi_x(a)\| + \varepsilon + \sum \|a_i\| \|b_i b\|.
\end{aligned}$$

The quantity on the right-hand side becomes smaller than $\|\Phi_x(a)\| + 2\varepsilon$ if b is chosen such that $\|b_i b\| < \varepsilon/(n\|a_i\|)$ for all i. Since $\varepsilon > 0$ is arbitrary, $\|ab\| \leq \|\Phi_x(a)\|$, whence the estimate $q \leq \|\Phi_x(a)\|$.   ∎

Here is the aforementioned result on the structure of elements in the local ideals.

Proposition 2.2.5. *Let $\mathscr{A}$ be a Banach algebra with identity e, $\mathscr{B}$ be a central C^*-subalgebra of $\mathscr{A}$ which contains e, and $x \in M_{\mathscr{B}}$. Then*

$$\mathscr{I}_x = \{ca : a \in \mathscr{A} \text{ and } c \in x\}. \tag{2.5}$$

Proof. By definition, the ideal $\mathscr{I}_x$ is the closure in $\mathscr{A}$ of the set of all finite sums

$$\sum_{j=1}^{n} c_j a_j \quad \text{with } c_j \in x \text{ and } a_j \in \mathscr{A}.$$

We claim that each sum of this form can be written as a single product ca with $c_j \in x$ and $a_j \in \mathscr{A}$. Clearly, it is sufficient to prove this fact for $n = 2$. Let $c_1, c_2 \in x$ and $a_1, a_2 \in \mathscr{A}$. We identify the elements of $\mathscr{B}$ with the corresponding Gelfand transforms. Define $c := \sqrt{|c_1| + |c_2|}$. Then $c \in x$, and the point x belongs to the set $N_c := \{y \in M_{\mathscr{B}} : c(y) = 0\}$. For $j = 1, 2$, put

$$g_j(y) := \begin{cases} c_j(y)/c(y) & \text{if } y \notin N_c, \\ 0 & \text{if } y \in N_c. \end{cases}$$

For $y \notin N_c$, one has

$$|g_j(y)| = \frac{|c_j(y)|}{|c(y)|} = \frac{|c_j(y)|}{|c_1(y)| + |c_2(y)|} |c(y)| \leq |c(y)|.$$

Since the estimate $|g_j(y)| \leq |c(y)|$ holds for $y \in N_c$ as well, the functions g_j are continuous. Thus, they can be identified with elements of $\mathscr{B}$. Since $c_j = c g_j$, it follows that $c_1 a_1 + c_2 a_2 = c(g_1 a_1 + g_2 a_2)$ as desired. Hence, the set on the right-hand side of (2.5) forms an ideal of $\mathscr{A}$. We abbreviate this ideal by $\mathscr{I}'_x$ for a moment.

Next we are going to prove that $\mathscr{I}'_x$ is a closed ideal. Let d be in the closure of $\mathscr{I}'_x$. Given any convergent series $\sum_{k=1}^{\infty} \varepsilon_k$ of positive numbers, there are elements

$a_k \in \mathscr{A}$ and $c_k \in x$ such that $\|d - c_k a_k\| < \varepsilon_k/2$ for every k. For each k, there is an open neighborhood $U_k \subset M_{\mathscr{B}}$ of x such that $\|a_k\|\|c_k(y)\| < \varepsilon_k/2$ for all $y \in U_k$. Without loss of generality, one can assume that $\overline{U_{k+1}} \subset U_k$ for every k. Further, let f_k be elements of $\mathscr{B}$ with $0 \le f_k \le 1$ and such that $f_k|_{\overline{U_{k+1}}} = 1$ and $f_k|_{M_{\mathscr{B}} \setminus U_k} = 0$ for all k. Then

$$\|c_k f_k\| = \|c_k f_k\|_\infty \le \sup_{y \in U_k} |c_k(y)|,$$

whence

$$\|f_k d\| \le \|d - c_k a_k\|\|f_k\| + \|a_k\|\|c_k f_k\| < \varepsilon_k.$$

Consequently, the series $d + \sum_{k=1}^\infty f_k d$ is absolutely convergent. Let $d_\infty \in \mathscr{A}$ denote the limit of that series. The estimate

$$0 \le \left(1 + \sum_{k=1}^n f_k(y)\right)^{-1} - \left(1 + \sum_{k=1}^{n+m} f_k(y)\right)^{-1}$$

$$\le \begin{cases} (1 + \sum_{k=1}^n f_k(y))^{-1} & \text{for } y \in U_{n+1} \\ 0 & \text{for } y \in M_{\mathscr{B}} \setminus U_{n+1} \end{cases} \le \frac{1}{n+1}$$

shows that $(1 + \sum_{k=1}^n f_k)^{-1}$ converges in $\mathscr{B}$ as $n \to \infty$ to some element c. Clearly, $c(x) = 0$, whence $c \in x$. Since $d = c d_\infty$ by construction, $\mathscr{I}_x'$ is closed. Thus, $\mathscr{I}_x'$ is a closed ideal of $\mathscr{A}$ which contains the ideal x of $\mathscr{B}$. Since $\mathscr{I}_x$ is the smallest ideal of $\mathscr{A}$ with these properties, the assertion follows. ∎

Finally we will show that every central C^*-subalgebra of a Banach algebra is inverse-closed. Recall that the algebra $\mathscr{B}$ is inverse-closed in $\mathscr{A}$ if every element $b \in \mathscr{B}$ which is invertible in $\mathscr{A}$ possesses an inverse in $\mathscr{B}$. The following definitions make sense in the context of the general local principle (i.e., without assuming the C^*-property of $\mathscr{B}$). Also Lemma 2.2.6 holds in the general context.

Write the maximal ideal space $M_{\mathscr{B}}$ as $M_{\mathscr{B}}^0 \cup M_{\mathscr{B}}^+$ where $M_{\mathscr{B}}^0$ collects those maximal ideals x of $\mathscr{B}$ for which $\mathscr{I}_x \equiv \mathscr{A}$ and where $M_{\mathscr{B}}^+$ is the complement of $M_{\mathscr{B}}^0$ in $M_{\mathscr{B}}$. The set $M_{\mathscr{B}}^0$ is open in $M_{\mathscr{B}}$. Indeed, if $x \in M_{\mathscr{B}}^0$, then $\Phi_x(0)$ is invertible by definition. By Proposition 2.2.3, $\Phi_y(0)$ is invertible for all y in a certain open neighborhood U of x. This is only possible if $y \in M_{\mathscr{B}}^0$.

Lemma 2.2.6. *If $M_{\mathscr{B}}^0 = \emptyset$, then $\mathscr{B}$ is inverse-closed in $\mathscr{A}$.*

Proof. Assume $\mathscr{B}$ is not inverse-closed. Then there is an element $b \in \mathscr{B}$ which is invertible in $\mathscr{A}$ but not in $\mathscr{B}$. Hence, b is contained in some maximal ideal x of $\mathscr{B}$. Then $b \in \mathscr{I}_x$, and $b + \mathscr{I}_x$ is 0 in $\mathscr{A}/\mathscr{I}_x$. But on the other hand, $b + \mathscr{I}_x$ is invertible in $\mathscr{A}/\mathscr{I}_x$. This is possible only if $x \in M_{\mathscr{B}}^0$. Hence, the component $M_{\mathscr{B}}^0$ of $M_{\mathscr{B}}$ is not empty. ∎

Corollary 2.2.7. *Let $\mathscr{A}$ be a Banach algebra with identity e and let $\mathscr{B}$ be a central C^*-subalgebra of $\mathscr{A}$ which contains e. Then:*

(i) *$M_{\mathscr{B}}^0$ is empty;*
(ii) *$\mathscr{B}$ is inverse-closed in $\mathscr{A}$.*

Proof. Let $x \in M_{\mathscr{B}}^0$. Then $\Phi_x(e) = 0$. Proposition 2.2.4 implies that $\inf \|eb\| = \inf \|b\| = 0$, with the infimum taken over all $b \in \mathscr{B}$ with $0 \le b \le 1$ which are identically 1 in a certain neighborhood of x. This is impossible since the Gelfand transform acts as an isometry on $\mathscr{B}$ by the Gelfand-Naimark theorem, whence $\|b\| \ge |\widehat{b}(x)| = 1$. The second assertion follows via Lemma 2.2.6. ∎

The most general result regarding inverse-closedness of C^*-algebras in Banach algebras was obtained by Goldstein [78]. By a C^*-subalgebra $\mathscr{B}$ of a Banach algebra $\mathscr{A}$ we mean a (not necessarily closed) subalgebra of $\mathscr{A}$ which carries the structure of a C^*-algebra, i.e., there is an involution and a norm on $\mathscr{B}$ which make $\mathscr{B}$ into a C^*-algebra with respect to the operations inherited from $\mathscr{A}$.

Theorem 2.2.8 (Goldstein). *Let $\mathscr{A}$ be a unital Banach algebra, and let $\mathscr{B}$ be a (not necessarily closed) C^*-subalgebra of $\mathscr{A}$ which contains the identity. Then $\mathscr{B}$ is inverse-closed in $\mathscr{A}$.*

We wish to add a related result. It is well known that the maximal ideal space of a singly generated unital Banach algebra is homeomorphic to the spectrum of its generating element (see Exercise 2.1.3).

Proposition 2.2.9. *Let $\mathscr{A}$ be a unital Banach algebra and let $\mathscr{B}$ be a central closed subalgebra of $\mathscr{A}$ which contains the identity and which is singly generated by an element b and the identity. Identify the maximal ideal space of $\mathscr{B}$ with $\sigma_{\mathscr{B}}(b)$. Then $M_{\mathscr{B}}^+ = \sigma_{\mathscr{A}}(b)$.*

Proof. If $x \in M_{\mathscr{B}}^+$, then

$$\Phi_x(b - xe) = \widehat{b}(x)\Phi_x(e) - x\Phi_x(e) = 0\Phi_x(e) = \Phi_x(0)$$

is not invertible in $\mathscr{A}_x$. (Note that if $M_{\mathscr{B}}$ is identified with $\sigma_{\mathscr{B}}(b)$ then the Gelfand transform of b is the identity mapping on $\sigma_{\mathscr{B}}(b)$.) By Allan's local principle, $b - xe$ is not invertible in $\mathscr{A}$, whence $x \in \sigma_{\mathscr{A}}(b)$.

Conversely, let $x \in \sigma_{\mathscr{A}}(b)$. Then $b - xe$ is not invertible in $\mathscr{A}$, and Allan's local principle implies the existence of a point $y \in \sigma_{\mathscr{B}}(b) = M_{\mathscr{B}}$ such that $\Phi_y(b - xe)$ is not invertible. On the other hand, the element

$$\Phi_z(b - xe) = \widehat{b}(z)\Phi_z(e) - x\Phi_z(e) = (z - x)\Phi_z(e)$$

is invertible for every $z \ne x$. Thus, $y = x$, i.e., $\Phi_x(b - xe)$ is not invertible. This implies that $x \in M_{\mathscr{B}}^+$. ∎

2.2.5 Douglas' local principle and sufficient families

We are now going to specialize Allan's local principle to the context of C^*-algebras. Allan's local principle provides us with a sufficient family of homomorphisms. We will first discuss sufficient families in the context of C^*-algebras.

Let $\mathscr{A}$ be a unital C^*-algebra, $(\mathscr{B}_t)_{t \in T}$ a family of unital C^*-algebras, and $\mathscr{W} := (W_t)_{t \in T}$ a family of unital *-homomorphisms $W_t : \mathscr{A} \to \mathscr{B}_t$. Further, let $\mathscr{F}$ stand for the direct product of the C^*-algebras $\mathscr{B}_t$ with $t \in T$, and let W denote the *-homomorphism

$$W : \mathscr{A} \to \mathscr{F}, \quad a \mapsto (t \mapsto W_t(a)). \tag{2.6}$$

Besides sufficient families of homomorphisms it will be convenient to introduce families of homomorphisms which are sufficient in a weaker sense. The family $\mathscr{W}$ is called *weakly sufficient* if the implication

$$W_t(a) \text{ is invertible in } \mathscr{B}_t \text{ for every } t \in T \text{ and } \sup_{t \in T} \|W_t(a)^{-1}\| < \infty$$
$$\Rightarrow \quad a \text{ is invertible in } \mathscr{A}$$

holds for every $a \in \mathscr{A}$.

Theorem 2.2.10. *Let $\mathscr{A}$ be a unital C^*-algebra. The following conditions are equivalent for a family $\mathscr{W} = (W_t)_{t \in T}$ of unital *-homomorphisms $W_t : \mathscr{A} \to \mathscr{B}_t$:*

(i) *the family $\mathscr{W}$ is weakly sufficient;*
(ii) *if $W_t(a) = 0$ for every $t \in T$, then $a = 0$;*
(iii) *for every $a \in \mathscr{A}$, $\|a\| = \sup_{t \in T} \|W_t(a)\|$;*
(iv) *the homomorphism (2.6) is a symbol mapping for $\mathscr{A}$.*

Proof. For the implication (i) $\Rightarrow$ (ii), let $a \in \mathscr{A}$ be an element such that $W_t(a) = 0$ for all $t \in T$, and let b be an arbitrary invertible element of $\mathscr{A}$. Then $W_t(b)$ is invertible for all $t \in T$, and the norms $\|W_t(b)^{-1}\|$ are uniformly bounded by $\|b^{-1}\|$. Consequently, the elements $W_t(a+b)$ are invertible for all $t \in T$, and the norms of their inverses are uniformly bounded, too. By (i), the element $a + b$ is invertible. Hence, a belongs to the radical of $\mathscr{A}$ which consists of the zero element only.

If hypothesis (ii) is satisfied, then the gluing mapping (2.6) is a *-homomorphism with kernel $\{0\}$. Hence, W is an isometry, which is equivalent to (iii). The implication (iii) $\Rightarrow$ (iv) follows since every isometry is a symbol mapping.

Finally, for the implication (iv) $\Rightarrow$ (i), let $a \in \mathscr{A}$ be an element for which the operators $W_t(a)$ are invertible for all $t \in T$ and the norms $\|W_t(a)^{-1}\|$ are uniformly bounded. Then $W(a)$ is invertible in the direct product $\mathscr{F}$. Due to the inverse-closedness of C^*-algebras, $W(a)$ is also invertible in $W(\mathscr{A})$. Since W is a symbol mapping, a is invertible in $\mathscr{A}$. ∎

In the proof of the next result, the concept of the square root of a positive element will be used. The square root of a positive element in a C^*-algebra is defined via the continuous functional calculus, which is an immediate corollary of the Gelfand-Naimark theorem (Theorem 2.1.7).

Theorem 2.2.11. *Let $\mathscr{A}$ be a unital C^*-algebra. The following conditions are equivalent for a family $\mathscr{W} = (W_t)_{t \in T}$ of unital $*$-homomorphisms $W_t : \mathscr{A} \to \mathscr{B}_t$:*

   (i) *the family $\mathscr{W}$ is sufficient;*
  (ii) *for every $a \in \mathscr{A}$, there is a $t \in T$ such that $\|W_t(a)\| = \|a\|$.*

Thus, a weakly sufficient family $\mathscr{W}$ is sufficient if and only if the supremum in Theorem 2.2.10 (iii) is attained.

Proof. (i) $\Rightarrow$ (ii): Suppose there is an $a \in \mathscr{A}$ such that

$$\|W_t(a)\| < \sup_{s \in T} \|W_s(a)\| \quad \text{for all } t \in T. \tag{2.7}$$

Since

$$\begin{aligned}
\|W_t(a)\|^2 &= \|W_t(a)^* W_t(a)\| \\
&= \|(W_t(a)^* W_t(a))^{1/2} (W_t(a)^* W_t(a))^{1/2}\| \\
&= \|(W_t(a)^* W_t(a))^{1/2}\|^2 = \|W_t((a^* a)^{1/2})\|^2,
\end{aligned}$$

one can assume without loss of generality that the element a in (2.7) is self-adjoint and positive. Since the norm of a self-adjoint element b coincides with its spectral radius $r(b)$, (2.7) can be rewritten as

$$r(W_t(a)) < \sup_{s \in T} r(W_s(a)) \quad \text{for all } t \in T. \tag{2.8}$$

Denote the supremum on the right-hand side of (2.8) by M and set $c := a - Me$. The elements $W_t(c) = W_t(a) - Me_t$ are invertible for all $t \in T$ since $r(W_t(a)) < M$. Thus, $c = a - Me$ is invertible by hypothesis (i). Since the set of the invertible elements is open, we get the invertibility of $a - me$ for all $m \in \mathbb{R}$ belonging to some neighborhood U of M. On the other hand, by the definition of the supremum, there is an $s_U \in T$ such that $m_U := r(W_{s_U}(a))$ belongs to U. The element $W_{s_U}(a) - m_U e_{s_U}$ is not invertible, because the spectral radius of a positive element belongs to the spectrum of that element. Hence, $a - m_U e$ is not invertible. This contradiction proves the assertion.

(ii) $\Rightarrow$ (i): Let $a \in \mathscr{A}$ be not invertible. We claim that there is a $t \in T$ such that $W_t(a)$ is not invertible.

If a is not invertible, then at least one of the elements aa^* or a^*a is not invertible, say a^*a for definiteness. Since a^*a is non-negative, a clear application of the Gelfand-Naimark theorem yields

$$\left\| \|a^*a\|e - a^*a \right\| = \|a^*a\|. \tag{2.9}$$

Set $b := \|a^*a\|e - a^*a$, and choose $t \in T$ such that $\|W_t(b)\| = \|b\|$. Then (2.9) implies

$$\left\| \|a^*a\|e_t - W_t(a^*a) \right\| = \|W_t(b)\| = \|b\| = \|a^*a\|.$$

Since $\|W_t(a^*a)\| \leq \|a^*a\|$, one can apply the Gelfand-Naimark theorem once more to get the non-invertibility of $W_t(a^*a)$ and, thus, of $W_t(a)$. ∎

Theorem 2.2.12 (Douglas' local principle). *Let $\mathscr{A}$ be a unital C^*-algebra and $\mathscr{B}$ a symmetric closed subalgebra of the center of $\mathscr{A}$ which contains the identity element. Then the assertions of Allan's local principle can be completed as follows:*

(i) $M^0_{\mathscr{B}} = \emptyset$;
(ii) $\|a\| = \max_{x \in M_{\mathscr{B}}} \|\Phi_x(a)\|$ *for each* $a \in \mathscr{A}$;
(iii) $\cap_{x \in M_{\mathscr{B}}} \mathscr{I}_x = \{0\}$.

Proof. Assertion (i) is Corollary 2.2.7 (i). Assertion (ii) follows from Theorem 2.2.11, and (iii) is a consequence of the semi-simplicity of C^*-algebras. ∎

2.2.6 Example: SIOs with piecewise continuous coefficients

Consider the algebra $PC(\mathbb{T})$ of all *piecewise continuous functions* on $\mathbb{T}$, that is, the algebra of all functions $a : \mathbb{T} \to \mathbb{C}$ which possess finite one-sided limits $a(x^{\pm})$ at each point of $\mathbb{T}$. For definiteness, let $a(x^+)$ refer to the limit of a at x, taken in the clockwise direction. The algebra $PC(\mathbb{T})$ is naturally embedded in $L^{\infty}(\mathbb{T})$ and contains $C(\mathbb{T})$. Further, let $\mathscr{A} := \mathrm{alg}(S, PC(\mathbb{T}))$ stand for the smallest closed subalgebra of $\mathscr{L}(L^p(\mathbb{T}))$ which contains the singular integral operator S, all operators of multiplication by a piecewise continuous function, and the ideal $\mathscr{K}$ of the compact operators.

We are interested in the subalgebra $\mathscr{A}^{\mathscr{K}} := \mathscr{A}/\mathscr{K}$ of the Calkin algebra. Proposition 1.4.12 implies that the set $C(\mathbb{T}) + \mathscr{K} = \{fI + \mathscr{K} : f \in C(\mathbb{T})\}$ is a central subalgebra of $\mathscr{A}^{\mathscr{K}}$, and this algebra is isometrically isomorphic to the algebra $C(\mathbb{T})$ by Proposition 1.4.11. The maximal ideal space of $C(\mathbb{T}) + \mathscr{K}$ is homeomorphic to $\mathbb{T}$, and the maximal ideal corresponding to $x \in \mathbb{T}$ is $\{(fI) + \mathscr{J} : f \in C(\mathbb{T}),\ f(x) = 0\}$, as was seen in Section 1.4.3.

Let $\mathscr{I}_x$ denote the smallest closed two-sided ideal in $\mathscr{A}^{\mathscr{K}}$ which contains the maximal ideal x of $C(\mathbb{T}) + \mathscr{K}$. Allan's local principle transfers the invertibility problem in $\mathscr{A}^{\mathscr{K}}$ to a family of invertibility problems, one in each local algebra $\mathscr{A}^{\mathscr{K}}_x := \mathscr{A}^{\mathscr{K}}/\mathscr{I}_x$. Let $\Phi^{\mathscr{K}}_x$ stand for the canonical homomorphism from $\mathscr{A}$ to $\mathscr{A}^{\mathscr{K}}_x$. The next results will give some clues about the nature of the local algebras $\mathscr{A}^{\mathscr{K}}_x$.

Lemma 2.2.13. *If $c \in PC(\mathbb{T})$ is continuous at x and $c(x) = 0$, then $\Phi^{\mathscr{K}}_x(cI) = 0$.*

Proof. Given $\varepsilon > 0$, choose $f_\varepsilon \in C(\mathbb{T})$ such that $0 \leq f_\varepsilon < 1$ except at x, where $f_\varepsilon(x) = 1$, and that the support of f_ε is contained in the interval $[xe^{-i\varepsilon}, xe^{+i\varepsilon}]$ of $\mathbb{T}$. It is easy to see that $\Phi^{\mathscr{K}}_x(f_\varepsilon I)$ is the identity in the local algebra. Consequently,

$$\|\Phi^{\mathscr{K}}_x(cI)\| = \|\Phi^{\mathscr{K}}_x(cI)\Phi^{\mathscr{K}}_x(f_\varepsilon I)\| = \|\Phi^{\mathscr{K}}_x(cf_\varepsilon I)\| \leq \|cf_\varepsilon\|_{L^{\infty}},$$

and the last norm can be as small as desired by choosing ε small enough.   ∎

For $x \in \mathbb{T}$, define

$$\chi_x(t) := \begin{cases} 0 & \text{if } t \in]xe^{-i\pi}, x[, \\ 1 & \text{if } t \in [x, xe^{-i\pi}]. \end{cases}$$

Proposition 2.2.14. *Let $x \in \mathbb{T}$. Every local algebra $\mathscr{A}_x^{\mathscr{K}}$ is unital, and it is generated by the identity element and by two idempotents, namely $\Phi_x^{\mathscr{K}}(\chi_x I)$ and $\Phi_x^{\mathscr{K}}(P)$.*

Proof. It is evident that $\Phi_x^{\mathscr{K}}(I)$ is the identity element of $\mathscr{A}_x^{\mathscr{K}}$ and that $\Phi_x^{\mathscr{K}}(S) = \Phi_x^{\mathscr{K}}(2P - I)$ is a linear combination of the identity element and the idempotent $\Phi_x^{\mathscr{K}}(P)$. Now let $a \in PC(\mathbb{T})$. Then, by Lemma 2.2.13,

$$\Phi_x^{\mathscr{K}}(aI) = \Phi_x^{\mathscr{K}}\left(a(x^-)(1 - \chi_x)I + a(x^+)\chi_x I\right)$$
$$= a(x^-)\Phi_x^{\mathscr{K}}(I) + \left(a(x^+) - a(x^-)\right)\Phi_x^{\mathscr{K}}(\chi_x I),$$

representing $\Phi_x^{\mathscr{K}}(aI)$ as a linear combination of the identity element and the idempotent $\Phi_x^{\mathscr{K}}(\chi_x I)$. Since $\Phi_x^{\mathscr{K}}(S)$ together with all cosets $\Phi_x^{\mathscr{K}}(aI)$ generate the algebra $\mathscr{A}_x^{\mathscr{K}}$, the assertion follows.   ∎

We would like to emphasize once more that the local algebras $\mathscr{A}_x^{\mathscr{K}}$ are generated by two idempotents (and the identity). Algebras generated by (two or more) idempotents appear frequently as local algebras, and Chapter 3 will be devoted to a detailed study of them.

2.2.7 Exercises

Exercise 2.2.1. Show that the family $\{\delta_t\}_{t \in [0,1]}$ of homomorphisms $\delta_t : f \mapsto f(t)$ is sufficient for $C[0, 1]$, whereas $\{\delta_t\}_{t \in [0,1[}$ is weakly sufficient but not sufficient.

Exercise 2.2.2. Describe the center of the algebra $\mathbb{C}^{2 \times 2}$. More generally, describe the center of the algebra $\mathscr{L}(E)$ when E is a Banach space.

Exercise 2.2.3. Describe the center of the algebra $M_2([0, 1], \mathbb{C})$ of the functions $f : [0, 1] \rightarrow \mathbb{C}^{2 \times 2}$. What is the result of localization of $M_2([0, 1], \mathbb{C})$ over its center via Allan's local principle?

Exercise 2.2.4. Localize $C(\mathbb{T})$ with respect to the disk algebra $\mathbb{A}$.

Exercise 2.2.5. Let $\mathscr{T}(PC)$ stand for the smallest closed subalgebra of $\mathscr{L}(l^2(\mathbb{Z}^+))$ which contains all Toeplitz operators with piecewise continuous functions.

(i) Show that $\mathscr{T}(C)/\mathscr{K}(l^2(\mathbb{Z}^+))$ is a central subalgebra of $\mathscr{T}(PC)/\mathscr{K}(l^2(\mathbb{Z}^+))$.
(ii) Using Allan's local principle, localize the algebra $\mathscr{T}(PC)/\mathscr{K}(l^2(\mathbb{Z}^+))$ over the maximal ideal space $\mathbb{T}$ of $\mathscr{T}(C)/\mathscr{K}(l^2(\mathbb{Z}^+))$ (recall Exercise 2.1.4 in

this connection). For $x \in \mathbb{T}$, write Φ_x for the canonical homomorphism from $\mathscr{T}(PC)$ onto the associated local algebra at x. Show that

$$\Phi_x(T(a)) = \Phi_x(a(x^-)T(1 - \chi_x) + a(x^+)T(\chi_x))$$

with the notation as in Section 2.2.6. Thus, the local algebra at x is singly generated by $\Phi_x(T(\chi_x))$ (and the identity element).

(iii) Show that the spectrum of $\Phi_x(T(\chi_x))$ is the interval $[0, 1]$.

(iv) Conclude that the Toeplitz operator $T(a)$ with $a \in PC(\mathbb{T})$ is Fredholm on $l^2(\mathbb{Z}^+)$ if and only if the function

$$\hat{a} : \mathbb{T} \times [0, 1] \to \mathbb{C}, \quad (x, t) \mapsto a(x^-)(1 - t) + a(x^+)t \qquad (2.10)$$

has no zeros.

(v) Conclude from Douglas' local principle that the algebra $\mathscr{T}(PC)/\mathscr{K}(l^2(\mathbb{Z}^+))$ is commutative.

(vi) Show that there is a bijection between the maximal ideal space of the quotient algebra $\mathscr{T}(PC)/\mathscr{K}(l^2(\mathbb{Z}^+))$ and $\mathbb{T} \times [0, 1]$ and that, under the identification of these two sets, the Gelfand transform of the coset $T(a) + \mathscr{K}(l^2(\mathbb{Z}^+))$ is given by (2.10).

Warning: the maximal ideal space of $\mathscr{T}(PC)/\mathscr{K}(l^2(\mathbb{Z}^+))$ and the product $\mathbb{T} \times [0, 1]$ coincide as sets, but the Gelfand topology on $\mathbb{T} \times [0, 1]$ is quite different from the common (Euclidean) product topology. For details see [21, Section 4.88].

Exercise 2.2.6. Let $\mathscr{A}$ be a unital Banach algebra and $p \in \mathscr{A}$ a non-trivial idempotent in the center of $\mathscr{A}$. Then $\text{alg}\{p\}$ and $\text{alg}\{e - p\}$ are maximal ideals of $\text{alg}\{e, p\}$ which we denote by 0 and 1. Show that the local algebras $\mathscr{A}_0$ and $\mathscr{A}_1$ can be identified with $(e - p)\mathscr{A}(e - p)$ and $p\mathscr{A}p$, respectively. (Note that $e - p$ and p are considered as the identity elements of these algebras.)

2.3 Norm-preserving localization

2.3.1 Faithful localizing pairs

Allan's local principle replaces the question of whether an element a in a unital Banach algebra $\mathscr{A}$ is invertible by a whole variety of "simpler" invertibility problems in local algebras $\mathscr{A}/\mathscr{I}_x$. The transition from $\mathscr{A}$ to its local algebras perfectly respects spectral properties: An element in $\mathscr{A}$ is invertible *if and only if* all local representatives of that element are invertible. But if one is interested in the structure of the algebra rather than in the spectra of its elements then this localization can fail. The point is that the intersection of the local ideals $\mathscr{I}_x$ can contain non-zero elements, in which case some structural information gets lost in the process of localization. This cannot happen in the case that $\mathscr{A}$ is a C^*-algebra which is localized over one of its central C^*-subalgebras (compare Douglas' local principle). In this section we are going to establish an effective mixture between Allan and Douglas (likewise, between Banach and C^*-algebras) which combines the advantages of Allan's principle (broad applicability) with those of Douglas' principle (no loss of structural information).

Let $\widehat{b}$ again refer to the Gelfand transforms of the element b of a commutative C^*-algebra.

Definition 2.3.1. Let $\mathscr{A}$ be a unital Banach algebra and $\mathscr{B}$ a subalgebra of $\mathscr{A}$. We say that $(\mathscr{A}, \mathscr{B})$ is a *faithful localizing pair*[2] if:

a) $\mathscr{B}$ is contained in the center of $\mathscr{A}$ and includes the identity element of $\mathscr{A}$;
b) there is an involution $b \mapsto b^*$ in $\mathscr{B}$ that turns $\mathscr{B}$ into a C^*-algebra;
c) $\|a(b_1 + b_2)\| \leq \max\{\|ab_1\|, \|ab_2\|\}$ for all elements $a \in \mathscr{A}$ and $b_1, b_2 \in \mathscr{B}$ with $\operatorname{supp}\widehat{b}_1 \cap \operatorname{supp}\widehat{b}_2 = \emptyset$.

Of course, c) is the striking condition in Definition 2.3.1. If, also, the "outer" algebra $\mathscr{A}$ is C^*, then this condition is satisfied automatically.

Proposition 2.3.2. *Let $\mathscr{A}$ be a unital C^*-algebra and $\mathscr{B}$ be a central and unital C^*-subalgebra of $\mathscr{A}$. Then $(\mathscr{A}, \mathscr{B})$ is a faithful localizing pair.*

Proof. Let $a \in \mathscr{A}$, and let b_1, b_2 be elements of $\mathscr{B}$ such that $\operatorname{supp}\widehat{b}_1 \cap \operatorname{supp}\widehat{b}_2 = \emptyset$. Further, let $r : \mathscr{A} \to \mathbb{R}^+$ denote the spectral radius function. Taking into account that $b_1 b_2 = 0$ one gets

$$
\begin{aligned}
\|a(b_1 + b_2)\|^2 &= r\left((b_1 + b_2)(b_1^* + b_2^*)a^* a\right) \\
&= \lim_{n \to \infty} \|(b_1 + b_2)^n (b_1^* + b_2^*)^n (a^* a)^n\|^{1/n} \\
&= \lim_{n \to \infty} \|(b_1^*)^n b_1^n (a^* a)^n + (b_2^*)^n b_2^n (a^* a)^n\|^{1/n} \\
&\leq \lim_{n \to \infty} \left(\|b_1^* a^* a b_1\|^n + \|b_2^* a^* a b_2\|^n\right)^{1/n} \\
&= \max\{\|ab_1\|^2, \|ab_2\|^2\},
\end{aligned}
$$

which is the assertion. ∎

---

[2] Formerly we used the notation "$\mathscr{A}$ is KMS with respect to $\mathscr{B}$" instead of "$(\mathscr{A}, \mathscr{B})$ is a faithful localizing pair", and we called c) the "KMS-property" of $\mathscr{A}$. We changed this notation to avoid confusion with a standard abbreviation in C^*-theory where KMS stands for "Kubo/Martin/Schwinger" (and not for political reasons as one might guess: our "KMS" was named after the town "Karl-Marx-Stadt" where the material presented in this section was developed in the eighties; since 1990 this town has again borne its historic name "Chemnitz").

2.3.2 Local norm estimates

Let $(\mathscr{A}, \mathscr{B})$ be a faithful localizing pair. In accordance with Allan's local principle, we localize $\mathscr{A}$ over $\mathscr{B}$ and get local ideals $\mathscr{I}_x$ and local homomorphisms $\Phi_x(a)$ for every x in the maximal ideal space $M_{\mathscr{B}}$ of $\mathscr{B}$. In the present setting it seems to be more convenient to write $\widehat{a}(x)$ instead of $\Phi_x(a)$.

Here is the main result on faithful (norm-preserving) localization.

Theorem 2.3.3. *Let $\mathscr{A}$ be a unital Banach algebra and $\mathscr{B}$ be a central and unital C^*-subalgebra of $\mathscr{A}$. Then $(\mathscr{A}, \mathscr{B})$ is a faithful localizing pair if and only if*

$$\|a\| = \max_{x \in M_{\mathscr{B}}} \|\widehat{a}(x)\|. \tag{2.11}$$

Proof. Let $(\mathscr{A}, \mathscr{B})$ be a faithful localizing pair. By assertion (iii) of Theorem 2.2.2, $\max_{x \in M_{\mathscr{B}}} \|\widehat{a}(x)\| \leq \|a\|$. It remains to verify the reverse inequality. Let a be in $\mathscr{A}$. As a consequence of Proposition 2.2.4, given $x \in M_{\mathscr{B}}$ and $\varepsilon > 0$, there is a b in $C(M_{\mathscr{B}})$ such that $0 \leq \widehat{b} \leq 1$, the support of b is contained in some open neighborhood $U(x)$ of x, and $\|ba\| < \|\widehat{a}(x)\| + \varepsilon$. Consequently, each x in $M_{\mathscr{B}}$ possesses an open neighborhood $U(x)$ such that $\|ba\| < \max_{x \in M_{\mathscr{B}}} \|\widehat{a}(x)\| + \varepsilon$ whenever $b \in C(M_{\mathscr{B}})$, $0 \leq \widehat{b} \leq 1$, and $\operatorname{supp} \widehat{b} \subseteq U(x)$. Choose a finite number $U_1, \ldots, U_n$ of these neighborhoods which cover $M_{\mathscr{B}}$, fix any (large) positive integer m, and let $k \in \{1, \ldots, m\}$. Further, let $e = f_1 + \cdots + f_n$ be a partition of unity subordinate to the covering $M_{\mathscr{B}} = \cup_{i=1}^n U_i$ (i.e., every $\widehat{f}_i$ is a continuous function with values in $[0, 1]$ and support in U_i), and put

$$V_{ki}^m := \left\{ x \in M_{\mathscr{B}} : \widehat{f}_i(x) \geq \frac{k+1}{n(m+1)}, \, \widehat{f}_{i+1}(x) \leq \frac{k}{n(m+1)}, \ldots, \widehat{f}_n(x) \leq \frac{k}{n(m+1)} \right\}$$

for $i = 1, \ldots, n-1$, and

$$V_{kn}^m := \left\{ x \in M_{\mathscr{B}} : \widehat{f}_n(x) \geq \frac{k+1}{n(m+1)} \right\}.$$

A straightforward check shows that the sets $V_{k1}^m, \ldots, V_{kn}^m$ are closed and pairwise disjoint, that $V_{ki}^m \subset U_i$ for $i = 1, \ldots, n$, and that each x in $M_{\mathscr{B}}$ belongs to at most n of the sets $G_k^m := M_{\mathscr{B}} \setminus \cup_{i=1}^n V_{ki}^m$. Now let $\widehat{g}_{k1}^m, \ldots, \widehat{g}_{kn}^m$ be any functions in $C(M_{\mathscr{B}})$ such that $\widehat{g}_{ki}^m |_{V_{ki}^m} = 1$, $\operatorname{supp} \widehat{g}_{ki}^m \cap \operatorname{supp} \widehat{g}_{kj}^m = \emptyset$ whenever $i \neq j$, $\operatorname{supp} \widehat{g}_{ki}^m \subset U_i$, and $0 \leq \widehat{g}_{ki}^m \leq 1$. Finally, put $\widehat{g}_k^m = \widehat{g}_{k1}^m + \cdots + \widehat{g}_{kn}^m$. Since $(\mathscr{A}, \mathscr{B})$ is a faithful localizing pair, we have

$$\|g_k^m a\| = \|(g_{k1}^m + \cdots + g_{kn}^m)a\| \leq \max_i \|g_{ki}^m a\| < \max_{x \in M_{\mathscr{B}}} \|\widehat{a}(x)\| + \varepsilon$$

(for the last '<' recall that $\operatorname{supp} \widehat{g}_{ki}^m \subset U_i$). Hence,

$$\|(g_1^m + \cdots + g_m^m)a\| \leq m \left(\max_{x \in M_{\mathscr{B}}} \|\widehat{a}(x)\| + \varepsilon \right). \tag{2.12}$$

Put $\widehat{h}_k^m := 1 - \widehat{g}_k^m$. Then $0 \le \widehat{h}_k^m \le 1$ and $\operatorname{supp} \widehat{h}_k^m \subseteq G_k^m$, and one has

$$\|(g_1^m + \cdots + g_m^m)a\| = \|ma - (h_1^m + \cdots + h_m^m)a\| \ge m\|a\| - \|h_1^m + \cdots + h_m^m\|\,\|a\|.$$

Because $\operatorname{supp}(\widehat{h}_1^m + \cdots + \widehat{h}_m^m) \subset \cup_{k=1}^m G_k^m$, and since each $x \in M_{\mathscr{B}}$ belongs to at most n of the sets G_k^m, it follows that $\widehat{h}_1^m(x) + \cdots + \widehat{h}_m^m(x) \le n$ for all x in $M_{\mathscr{B}}$. This implies

$$\|(g_1^m + \cdots + g_m^m)a\| \ge (m-n)\|a\|. \tag{2.13}$$

Combining (2.12) and (2.13) we arrive at the inequality

$$\|a\| \le \frac{m}{m-n}\left(\max_{x \in M_{\mathscr{B}}} \|\widehat{a}(x)\| + \varepsilon\right).$$

Letting m go to infinity and ε go to zero we obtain the desired inequality.

To prove the reverse implication, let $a \in \mathscr{A}$ and let $b_1, b_2 \in \mathscr{B}$ such that $\operatorname{supp} \widehat{b}_1 \cap \operatorname{supp} \widehat{b}_2 = \emptyset$. Then

$$\|a(b_1 + b_2)\|^2 = \max_{x \in M_{\mathscr{B}}} \left\|\widehat{a}(x)\left(\widehat{b}_1(x) + \widehat{b}_2(x)\right)\right\|.$$

For each $x \in M_{\mathscr{B}}$, only one of the values $\widehat{b}_1(x)$ and $\widehat{b}_2(x)$ can be different from 0. Consequently,

$$\left\|\widehat{a}(x)\left(\widehat{b}_1(x) + \widehat{b}_2(x)\right)\right\| = \max\left\{\left\|\widehat{a}(x)\widehat{b}_1(x)\right\|, \left\|\widehat{a}(x)\widehat{b}_2(x)\right\|\right\},$$

and the result follows.   ∎

Besides the norm computation aspect, faithful localizing pairs are advantageous to investigate local enclosement properties.

Theorem 2.3.4. *Let $\mathscr{A}$ be a unital Banach algebra and $\mathscr{B}$ be a subalgebra of $\mathscr{A}$ such that $(\mathscr{A}, \mathscr{B})$ forms a faithful localizing pair. Further, let $\mathscr{C}$ be a closed linear subset of $\mathscr{A}$ such that $bc \in \mathscr{C}$ whenever $b \in \mathscr{B}$ and $c \in \mathscr{C}$ (i.e., $\mathscr{C}$ is a $\mathscr{B}$-module). Let $a \in \mathscr{A}$ and assume that, for each $x \in M_{\mathscr{B}}$, there is an $a_x \in \mathscr{C}$ such that $\widehat{(a - a_x)}(x) = (\widehat{a} - \widehat{a}_x)(x) = 0$. Then $a \in \mathscr{C}$.*

Proof. Let $x \in M_{\mathscr{B}}$ and $\varepsilon > 0$. By Theorem 2.2.2 (b), there is a neighborhood $U(x)$ of x such that

$$\left\|\widehat{(a - a_x)}(y)\right\| < \varepsilon$$

for all $y \in U(x)$. Choose a finite number $U(x_1), \ldots, U(x_n)$ of these neighborhoods which cover $M_{\mathscr{B}}$, and let $e = f_1 + \cdots + f_n$ with $f_k \in \mathscr{B}$ be a partition of unity subordinate to this covering. Put $b_\varepsilon := \sum_{k=1}^n f_k a_{x_k}$. Then

$$\|a - b_\varepsilon\| = \left\| \sum_{k=1}^{n} f_k(a - a_{x_k}) \right\|$$

$$= \sup_{y \in M_{\mathscr{B}}} \left\| \sum_{k=1}^{n} \widehat{f_k}(y) \, (\widehat{a} - \widehat{a_{x_k}})(y) \right\| \qquad \text{(Theorem 2.3.3)}$$

$$= \sup_{y \in M_{\mathscr{B}}} \left\| \sum_{k:y \in U(x_k)} \widehat{f_k}(y) \, (\widehat{a - a_{x_k}})(y) \right\|$$

$$\leq \sup_{y \in M_{\mathscr{B}}} \sum_{k:y \in U(x_k)} \widehat{f_k}(y)\varepsilon = \varepsilon.$$

The algebraic properties of $\mathscr{C}$ ensure that $b_\varepsilon \in \mathscr{C}$ for every ε. Since $\mathscr{C}$ is closed and $\|a - b_\varepsilon\| < \varepsilon$ for every $\varepsilon > 0$, we have $a \in \mathscr{C}$. ∎

As an application of the above result we will get a complete description of the image $\widetilde{\mathscr{A}}$ of $\mathscr{A}$ in the direct product $\mathscr{F}$ of the local algebras $\mathscr{A}/\mathscr{I}_x$ under the mapping

$$\mathscr{A} \to \mathscr{F}, \quad a \mapsto (x \mapsto \Phi_x(a)), \tag{2.14}$$

provided $(\mathscr{A}, \mathscr{B})$ is a faithful localizing pair. To that end, we will have to introduce the concept of *semi-continuity with respect to a given set*. Let X be a compact Hausdorff space. To each point $x \in X$, we associate a Banach algebra $\mathscr{A}_x$ with unit element e_x. Denote the direct product of the algebras $\mathscr{A}_x$ by $\mathscr{F}$. Evidently, this product can be identified with the set of all bounded functions f on X which take at $x \in X$ a value $f(x) \in \mathscr{A}_x$. In particular, if g is a continuous complex-valued function on X, then the function

$$x \mapsto g(x)e_x \tag{2.15}$$

belongs to $\mathscr{F}$. The set $\mathscr{D}$ of all functions of the form (2.15) is a closed subalgebra of the center of $\mathscr{F}$, and it is easy to check that $(\mathscr{F}, \mathscr{D})$ forms a faithful localizing pair. Let $\mathscr{E}$ be a subset of $\mathscr{F}$ which is subject to the following conditions:

(i) $\mathscr{D} \subseteq \mathscr{E}$;
(ii) given $x \in X$ and $a \in \mathscr{A}_x$, there is an $f \in \mathscr{E}$ such that $f(x) = a$;
(iii) the function $x \mapsto \|f(x)\|_{\mathscr{A}_x}$ is upper semi-continuous on X for each $f \in \mathscr{E}$;
(iv) $\mathscr{E}$ is a (not necessarily closed) algebra.

We call a function $g \in \mathscr{F}$ *semi-continuous with respect to* $\mathscr{E}$ if, for each $x_0 \in X$, each $f \in \mathscr{E}$ with $f(x_0) = g(x_0)$, and each $\varepsilon > 0$, there is a neighborhood $U = U(x_0, f, g, \varepsilon)$ of x_0 such that $\|g(x) - f(x)\|_{\mathscr{A}_x} < \varepsilon$ for all $x \in U$. Thus, a function g is semi-continuous with respect to $\mathscr{E}$ if it behaves locally as a function in $\mathscr{E}$.

It is easy to show that the set of all functions which are semi-continuous with respect to $\mathscr{E}$ forms a closed subalgebra of $\mathscr{F}$ which we will denote by $\mathscr{F}(\mathscr{E})$. The following result can be considered as a generalization of the classical Weierstrass theorem (see Exercise 2.3.2).

Theorem 2.3.5. *The smallest closed subalgebra of $\mathscr{F}$ which contains $\mathscr{E}$ coincides with $\mathscr{F}(\mathscr{E})$.*

Proof. The inclusion $\mathrm{clos}\,\mathscr{E} \subseteq \mathscr{F}(\mathscr{E})$ is an immediate consequence of the fact that, by property (iii) of $\mathscr{E}$, each function in $\mathscr{E}$ is semi-continuous with respect to $\mathscr{E}$. For the reverse inclusion, let $b \in \mathscr{F}(\mathscr{E})$. By property (ii), there is an $f \in \mathscr{E}$ such that $f(x) = b(x)$ or, equivalently, $\widehat{(f - b)}(x) = 0$. Thus, the algebras $\mathscr{F}(\mathscr{E})$ and $\mathscr{D}$ and the $\mathscr{D}$-module $\mathrm{clos}\,\mathscr{E}$ satisfy the assumptions imposed in Theorem 2.3.4 on the algebras $\mathscr{A}$ and $\mathscr{B}$ and the $\mathscr{B}$-module $\mathscr{C}$. Thus, the conclusion follows immediately from that theorem.   ∎

Let $(\mathscr{A}, \mathscr{B})$ be a faithful localizing pair. As a consequence of the preceding theorem, we get a description of the image $\widetilde{\mathscr{A}}$ of $\mathscr{A}$ in the direct product $\mathscr{F}$ of the local algebras $\mathscr{A}/\mathscr{I}_x$ under the mapping (2.14).

Corollary 2.3.6. *$\widetilde{\mathscr{A}}$ coincides with the closed subalgebra of $\mathscr{F}$ which consists of all functions $f \in \mathscr{F}$ which are semi-continuous with respect to $\widetilde{\mathscr{A}}$.*

Indeed, by Theorem 2.2.2 (b), $\widetilde{\mathscr{A}}$ satisfies the assumptions made for the set $\mathscr{E}$ in Theorem 2.3.5 (with X being the maximal ideal space of $\mathscr{B}$).

Remark 2.3.7. There are more general concepts of faithful localization: in [81], the C^*-compatible norm in Definition 2.3.1 (a) is allowed to be different from the original norm and also (b) is substituted by a more general condition. The price one has to pay is that (2.11) is no longer an equality. Rather, one has

$$C_1 \|a\| \leq \max_{x \in M_{\mathscr{B}}} \|\widehat{a}(x)\| \leq C_2 \|a\| \quad \text{for all } a \in \mathscr{A}$$

with certain constants C_1, C_2 independent of a. Thus, the exact norm computation aspect of Theorem 2.3.3 gets lost, but the local enclosement Theorem 2.3.4 remains valid without changes.   □

2.3.3 Exercises

Exercise 2.3.1. Show that (with the notation of the previous subsection) the set of all functions which are semi-continuous with respect to $\mathscr{E}$ forms a closed subalgebra of $\mathscr{F}$.

Exercise 2.3.2. Let $\mathscr{F} := C[0, 1]$ and $\mathscr{E} := \mathbb{C}$ (considered as constant functions). Describe the functions in $\mathscr{F}$ which are semi-continuous with respect to $\mathscr{E}$. Use Theorem 2.3.5 to derive the classical Weierstrass theorem.

2.4 Gohberg-Krupnik's local principle

The local principle by Gohberg and Krupnik has several advantages: its formulation as well as the proofs of its main results are quite elementary, and it works equally well in the case of complex and real algebras.

2.4.1 Localizing classes

Let $\mathscr{A}$ be a (real or complex) Banach algebra with identity e. A subset $M \subset \mathscr{A}$ is called a *localizing class* if it does not contain the element 0 and if for arbitrary elements $f_1, f_2 \in M$, there exists a third element $f \in M$ such that $f_j f = f f_j = f$ for $j = 1, 2$.

Let M be a localizing class. Two elements $a, b \in \mathscr{A}$ are said to be *M-equivalent from the left* (resp. from the right) if

$$\inf_{f \in M} \|(a - b)f\| = 0 \qquad (\text{resp.} \inf_{f \in M} \|f(a - b)\| = 0).$$

Finally, an element $a \in \mathscr{A}$ is called *M-invertible from the left* (resp. from the right) if there exist elements $b \in \mathscr{A}$ and $f \in M$ such that $baf = f$ (resp. $fab = f$).

Proposition 2.4.1. *Let M be a localizing class, and let a_1 and a_2 be elements of $\mathscr{A}$ which are M-equivalent from the left (resp. from the right). Then a_1 is M-invertible from the left (resp. from the right) if and only if a_2 is so.*

Proof. Let a_1 be M-invertible from the left. Choose elements $b_1 \in \mathscr{A}$ and $f \in M$ such that $b_1 a_1 f = f$. Since a_1 and a_2 are M-equivalent from the left, there is a $g \in M$ such that $\|(a_1 - a_2)g\| < \|b_1\|^{-1}$. Let $h \in M$ be such that $fh = gh = h$. Then

$$\begin{aligned}
b_1 a_2 h &= b_1 a_1 h - b_1(a_1 - a_2)h \\
&= b_1 a_1 fh - b_1(a_1 - a_2)gh \\
&= fh - b_1(a_1 - a_2)gh \\
&= h - b_1(a_1 - a_2)gh.
\end{aligned}$$

Set $u := b_1(a_1 - a_2)g$. Then the above identity can be rewritten as

$$b_1 a_2 h = (e - u)h.$$

Since $\|u\| < 1$, the element $e - u$ is invertible in $\mathscr{A}$. Setting $b_2 := (e - u)^{-1}b_1$, one obtains $b_2 a_2 h = h$. Thus, a_2 is M-invertible from the left. The proof for right M-invertibility is similar. ∎

Definition 2.4.2. Let T be a topological space. A system $\{M_\tau\}_{\tau \in T}$ of localizing classes is said to be:

- *covering* if from each choice $\{f_\tau\}_{\tau \in T}$ of elements $f_\tau \in M_\tau$ one can select a finite number of elements $\{f_{\tau_1}, \ldots, f_{\tau_m}\}$ the sum of which is invertible in $\mathscr{A}$;
- *overlapping* if each M_τ is a bounded set in $\mathscr{A}$, if $f \in M_{\tau_0}$ for some $\tau_0 \in T$ implies $f \in M_\tau$ for all τ in some open neighborhood of τ_0, and if the elements of $F :=$ $\cup_{\tau \in T} M_\tau$ commute pairwise.

Let $\{M_\tau\}_{\tau \in T}$ be an overlapping system of localizing classes. The *commutant* of F is the set $\mathrm{Com} F := \{a \in \mathscr{A} : af = fa \, \forall f \in F\}$. It is easy to verify that $\mathrm{Com} F$ is a closed subalgebra of $\mathscr{A}$. For $\tau \in T$, let Z^τ denote the set of all elements in $\mathrm{Com} F$ which are M_τ-equivalent to zero both from the left and from the right.

Lemma 2.4.3. *The set Z^τ is a proper closed two-sided ideal of* $\mathrm{Com} F$.

Proof. The closedness of Z^τ follows easily from the boundedness of M_τ, and the ideal property of Z^τ can be also straightforwardly checked. The properness of Z^τ can be seen as follows. Suppose the identity element e of $\mathscr{A}$ belongs to Z^τ. Then there exists a sequence of $f_n \in M_\tau$ such that $\|f_n\| \to 0$ as $n \to \infty$. Since there exist non-zero elements $g_n \in M_\tau$ such that $f_n g_n = g_n$, it follows that $\|f_n\| \geq 1$, and we obtained a contradiction. ∎

For $a \in \mathrm{Com} F$, let a^τ denote the coset $a + Z^\tau$ of a in the quotient algebra $\mathrm{Com} F / Z^\tau$.

Proposition 2.4.4. *Let $\{M_\tau\}_{\tau \in T}$ be a system of localizing classes, with each M_τ a bounded set in $\mathscr{A}$. Let $\tau \in T$ and $a \in \mathrm{Com} F$. Then a is M_τ-invertible in $\mathrm{Com} F$ from the left (resp. from the right) if and only if a^τ is left (resp. right) invertible in $\mathrm{Com} F / Z^\tau$.*

Proof. Let a^τ be left invertible in $\mathrm{Com} F / Z^\tau$. Then there is a $b \in \mathrm{Com} F$ such that $ba - e \in Z^\tau$. This implies that ba is M_τ-equivalent from the left to e. Proposition 2.4.1 yields the M_τ-invertibility of ba, and thus of a, from the left. Conversely, if there is $b \in \mathrm{Com} F$ and $f \in M_\tau$ such that $baf = f$, then $(ba - e)f = 0$. Hence $ba - e \in Z^\tau$, and thus $b^\tau a^\tau = e$. The proof for right invertibility is similar. ∎

2.4.2 Gohberg-Krupnik's local principle

The following theorem is a very similar result to Theorem 2.2.2, with the ideals $\mathscr{I}_x$ and the maximal ideal space of the central subalgebra substituted respectively by the ideals Z^τ and the index set T of the system of localizing classes $\{M_\tau\}_{\tau \in T}$. But contrary to Allan's local principle (where complex function theoretic arguments are used in the proof of Proposition 2.2.1), the local principle by Gohberg-Krupnik is valid for real Banach algebras, too.

Theorem 2.4.5 (Gohberg-Krupnik). *Let $\mathscr{A}$ be a Banach algebra with identity and $\{M_\tau\}_{\tau \in T}$ a covering system of localizing classes, the elements of which belong to the center of $\mathscr{A}$. Further, let $a \in \mathscr{A}$ and, for every $\tau \in T$, let a_τ be an element of $\mathscr{A}$ which is M_τ-equivalent from the left to a.*

   (i) *The element a is left invertible in $\mathscr{A}$ if and only if a_τ is M_τ-invertible in $\mathscr{A}$ from the left for every $\tau \in T$.*
  (ii) *Suppose that each M_τ is a bounded set in $\mathscr{A}$. Then a is left invertible in $\mathscr{A}$ if and only if a^τ is left invertible in $\mathscr{A}/Z^\tau$ for all $\tau \in T$.*
 (iii) *If the system $\{M_\tau\}_{\tau \in T}$ is overlapping, then the function $T \to \mathbb{R}^+$, $\tau \mapsto \|a^\tau\|$ is upper semi-continuous.*
 (iv) *If $\mathscr{A}$ is a C^*-algebra, then the system $\{M_\tau\}_{\tau \in T}$ is overlapping. If moreover $M_\tau^* = M_\tau$ for all $\tau \in T$, then*

$$\|a\| = \sup_{\tau \in T} \|a^\tau\|.$$

The result remains valid if left *is replaced by* right *everywhere.*

Proof. We will give the proof in case of left equivalence and left invertibility. The proof for right equivalence and right invertibility is, of course, similar.

(i) If a is left invertible, then a is M_τ-invertible in $\mathscr{A}$ $(= \mathrm{Com}F)$ from the left for all $\tau \in T$. By Proposition 2.4.1, a_τ is M_τ-invertible from the left for all $\tau \in T$. Conversely, suppose a_τ is M_τ-invertible from the left for all $\tau \in T$. Again by Proposition 2.4.1, it follows that a is M_τ-invertible from the left for all $\tau \in T$. Thus there are $b_\tau \in \mathscr{A}$ and $f_\tau \in M_\tau$ such that $b_\tau a f_\tau = f_\tau$. Since $\{M_\tau\}_{\tau \in T}$ is covering, one can choose a finite number of elements $f_{\tau_1}, \ldots, f_{\tau_m}$ so that $\sum_{j=1}^{m} f_{\tau_j}$ is invertible. Put

$$s := \sum_{j=1}^{m} b_{\tau_j} f_{\tau_j}$$

to obtain

$$sa = \sum_{j=1}^{m} b_{\tau_j} f_{\tau_j} a = \sum_{j=1}^{m} b_{\tau_j} a f_{\tau_j} = \sum_{j=1}^{m} f_{\tau_j}.$$

Thus, $(\sum_{j=1}^{m} f_{\tau_j})^{-1} s$ is a left inverse of a.

(ii) If a^τ is left invertible in $\mathscr{A}/Z^\tau$ for all $\tau \in T$, then a is left invertible in $\mathscr{A}$ by Proposition 2.4.4. and part (i) above. The converse is trivial.

(iii) Let $\tau_0 \in T$ and $\varepsilon > 0$. Choose $z \in Z^\tau$ so that $\|a + z\| < \|a^{\tau_0}\| + \varepsilon/2$. Since z is M_{τ_0}-equivalent to zero from the left, there is an $f \in M_{\tau_0}$ such that $\|zf\| < \varepsilon/2$. Because $f \in M_{\tau_0}$ implies that $f \in M_\tau$ for all τ in some open neighborhood of τ_0 due to the overlapping property, we deduce that $f \in M_\tau$ for all τ in some open neighborhood U_{τ_0} of τ_0. Put $y := z - zf$. If $\tau \in U_{\tau_0}$, then there exists a $g \in M_\tau$ such that $fg = g$. Consequently, we have that $yg = zg - zfg = zg - zg = 0$. Since $y \in \mathrm{Com}F$ (by the definition of Z^τ and due to the overlapping property), it follows that $y \in Z^\tau$ for all $\tau \in U_{\tau_0}$. Hence, $\|a^\tau\| \le \|a + y\|$ for $\tau \in U_{\tau_0}$. Thus, if $\tau \in U_{\tau_0}$, then

$$\|a^\tau\| - \|a^{\tau_0}\| < \|a+y\| - \|a+z\| + \frac{\varepsilon}{2} \le \|y-z\| + \frac{\varepsilon}{2} = \|zf\| + \frac{\varepsilon}{2} < \varepsilon,$$

which proves the upper semi-continuity of the mapping $\tau \mapsto \|a^\tau\|$ at τ_0.

(iv) If $\mathrm{Com}F$ and $\mathrm{Com}F/Z^\tau$ are C^*-algebras then, indeed,

$$\begin{aligned}
\|a\|^2 = r(aa^*) &= \sup_{\tau \in T} r((aa^*)^\tau) \text{ by (ii)}\\
&= \sup_{\tau \in T} r(a^\tau (a^\tau)^*)) = \sup_{\tau \in T} \|a^\tau\|^2.
\end{aligned}$$

■

2.4.3 Exercises

Exercise 2.4.1. Study the algebra $\mathscr{A}^{\mathscr{K}}$ considered in the example of Section 2.2.6, now using Gohberg-Krupnik's local principle, instead of Allan's.

Exercise 2.4.2. Prove Allan's local principle in the special case of localization with respect to a unital central C^*-subalgebra (provided it exists) by means of Gohberg-Krupnik's local principle. Make sure that your proof does not employ properties of *complex* analytic functions. Derive a version of Allan's local principle for *real* Banach algebras. (Hint: see the proof of Theorem 1.21 in [151].)

2.5 Simonenko's local principle

Simonenko's local principle can be viewed as a particular realization of the two previously considered local principles by Allan-Douglas and Gohberg-Krupnik. It is well adapted to the study of Fredholm properties, and its formulation is quite intuitive. In particular, no Banach algebra "infrastructure" is involved.

2.5.1 Spaces and operators of local type

Let X be a locally compact Hausdorff topological space, and let μ be a non-negative (possibly infinite) measure which is defined on a σ-algebra over X which contains all Borel subsets of X. The characteristic function of a measurable subset U of X will be denoted by χ_U in what follows.

Definition 2.5.1. A Banach space E, the elements of which are (equivalence classes of) measurable functions $f : X \to \mathbb{C}$, is called an *ideal Banach space over* X if the characteristic function of every compact subset of X belongs to E and if, for every measurable function $f : X \to \mathbb{C}$ and every function $g \in E$, the inequality

$$|f(x)| \leq |g(x)| \quad \text{a.e. on } X \tag{2.16}$$

implies that $f \in E$ and $\|f\|_E \leq \|g\|_E$.

The archetypal examples of ideal Banach spaces are the Lebesgue spaces $L^p(X)$ with $1 \leq p \leq \infty$. Note also that an ideal Banach space over X contains every bounded measurable function on X with compact support, which follows immediately from the definition.

Lemma 2.5.2. *Let E be an ideal Banach space over X, and let a be a bounded measurable function on X. Then the product af belongs to E for every function $f \in E$. In particular, the operator $aI : E \to E$, $f \mapsto af$ of multiplication by a is well defined. This operator is bounded, and $\|aI\| = \|a\|_\infty$.*

Proof. For every $f \in E$, one has $|a(x)f(x)| \leq |\,\|a\|_\infty f(x)|$ almost everywhere on X, whence

$$\|af\|_E \leq \|\,\|a\|_\infty f\|_E = \|a\|_\infty \|f\|_E. \tag{2.17}$$

The left inequality in (2.17) shows that $af \in E$ (since $\|a\|_\infty f \in E$), and the inequality $\|af\|_E \leq \|a\|_\infty \|f\|_E$ implies the boundedness of aI and the estimate $\|aI\| \leq \|a\|_\infty$. For the reverse estimate, let $\varepsilon > 0$ and choose a compact subset U of X with $\mu(U) > 0$ such that

$$(\|a\|_\infty - \varepsilon)\chi_U(x) \leq |a(x)\chi_U(x)| \quad \text{a.e. on } X.$$

Then the second condition in Definition 2.5.1 implies that

$$(\|a\|_\infty - \varepsilon)\|\chi_U\|_E \leq \|a\chi_U\|_E \leq \|aI\|\,\|\chi_U\|_E$$

which gives the estimate $\|a\|_\infty - \varepsilon \leq \|aI\|$. Letting ε go to zero we obtain the assertion. ∎

Definition 2.5.3. Let E be an ideal Banach space over X. An operator $A \in \mathscr{L}(E)$ is said to be of *local type* if the operator $\chi_{F_1} A \chi_{F_2} I$ is compact for every choice of disjoint closed subsets F_1, F_2 of X. We denote the set of all operators of local type by $\Lambda(E)$.

Let $A \in \mathscr{L}(E)$. The norm of the coset $A + \mathscr{K}(E)$ considered as an element of the Calkin algebra $\mathscr{L}(E)/\mathscr{K}(E)$ is called the *essential norm* of A. We denote it by $|A|$. Thus,

$$|A| := \inf_{K \in \mathscr{K}(E)} \|A + K\|.$$

Two operators $A, B \in \mathscr{L}(E)$ are said to be *essentially equivalent* if $|A - B| = 0$, in which case we write $A \sim B$.

Definition 2.5.4. An ideal Banach space E over X is called a *Banach space of local type over X* if, for each pair $A, B \in \mathscr{L}(E)$ of operators of local type and for each pair F_1, F_2 of disjoint closed subsets of X,

$$|\chi_{F_1} A \chi_{F_1} I + \chi_{F_2} B \chi_{F_2} I| \leq \max\{|A|, |B|\}. \tag{2.18}$$

Example 2.5.5. The Lebesgue spaces $L^p(X) =: E$ are Banach spaces of local type for every $1 \leq p \leq \infty$. To see this, notice that

$$\|\chi_{F_1} A \chi_{F_1} I + \chi_{F_2} B \chi_{F_2} I\| \leq \max\{\|A\|, \|B\|\} \tag{2.19}$$

for each pair of operators $A, B \in \mathcal{L}(E)$ (not necessarily of local type) and for each pair F_1, F_2 of disjoint closed subsets of X. Thus,

$$
\begin{aligned}
|\chi_{F_1} A \chi_{F_1} I + \chi_{F_2} B \chi_{F_2} I| &= \inf_{M \in \mathcal{K}(E)} \|\chi_{F_1} A \chi_{F_1} I + \chi_{F_2} B \chi_{F_2} I + M\| \\
&\leq \inf_{K, L \in \mathcal{K}(E)} \|\chi_{F_1} A \chi_{F_1} I + \chi_{F_2} B \chi_{F_2} I + \chi_{F_1} K \chi_{F_1} I + \chi_{F_2} L \chi_{F_2} I\| \\
&= \inf_{K, L \in \mathcal{K}(E)} \|\chi_{F_1} (A+K) \chi_{F_1} I + \chi_{F_2} (B+L) \chi_{F_2} I\| \\
&\leq \inf_{K, L \in \mathcal{K}(E)} \max\{\|A+K\|, \|B+L\|\} \qquad \text{by (2.19)} \\
&= \max\{|A|, |B|\}.
\end{aligned}
$$

$\square$

We proceed with equivalent characterizations of operators of local type which will prove useful in what follows.

Theorem 2.5.6. *Let E be a Banach space of local type over X. The following conditions are equivalent for an operator $A \in \mathcal{L}(E)$:*

(i) *A is of local type;*
(ii) *for each function $f \in C(X)$, the commutator $AfI - fA$ is compact;*
(iii) *for all measurable sets F_1, F_2 of X with $\operatorname{clos} F_1 \subset \operatorname{int} F_2$, one has $\chi_{F_1} A \chi_{F_2} I \sim \chi_{F_1} A$ and $\chi_{F_2} A \chi_{F_1} I \sim A \chi_{F_1} I$.*

Proof. (i) $\Rightarrow$ (ii): Let $f \in C(X)$ be a real-valued function with $0 \leq f \leq 1$. For $j, n \in \mathbb{N}$ with $1 \leq j \leq 4n$, define the sets

$$
F_j^n := \begin{cases} \{x \in X : 0 \leq f(x) \leq \frac{1}{4n}\} & \text{if } j = 1, \\ \{x \in X : \frac{j-1}{4n} < f(x) \leq \frac{j}{4n}\} & \text{if } j \geq 2 \end{cases}
$$

and

$$
G_j^n := \{x \in X : \frac{j-2}{4n} \leq f(x) \leq \frac{j+1}{4n}\},
$$

and set $\chi_j := \chi_{F_j^n}$ as well as $\hat{\chi}_j := \chi_{G_j^n}$. Finally, let $\chi_0 = \chi_{4n+1} := 0$. Being preimages of closed intervals under a continuous function, the sets G_j^n are closed. For each operator $A \in \Lambda(E)$, one has

$$
|AfI - fA| = \left| \sum_{j=1}^{4n} \sum_{k=1}^{4n} (\chi_j A \chi_k fI - f \chi_j A \chi_k I) \right| = \left| \sum_{j=1}^{4n} \sum_{k=j-1}^{j+1} (\chi_j A \chi_k fI - f \chi_j A \chi_k I) \right|
$$

since $\operatorname{clos} F_j^n \cap \operatorname{clos} F_k^n = \emptyset$ for $|j-k| > 1$ and since A is of local type. A shift of the summation index yields

$$
\begin{aligned}
|AfI - fA| &= \left| \sum_{j=1}^{4n} \sum_{k=-1}^{1} (\chi_j A \chi_{j+k} fI - f\chi_j A \chi_{j+k} I) \right| \\
&= \left| \sum_{j=1}^{4n} \sum_{k=-1}^{1} \left(\chi_j A \chi_{j+k} \left(f - \tfrac{j-1}{4n} \right) I - \left(f - \tfrac{j-1}{4n} \right) \chi_j A \chi_{j+k} I \right) \right| \\
&\leq \sum_{k=-1}^{1} \left(\left| \sum_{j=1}^{4n} \chi_j A \chi_{j+k} \left(f - \tfrac{j-1}{4n} \right) I \right| + \left| \sum_{j=1}^{4n} \left(f - \tfrac{j-1}{4n} \right) \chi_j A \chi_{j+k} I \right| \right).
\end{aligned}
$$
$$(2.20)$$

Consider the first of the two inner sums on the right-hand side of (2.20). Taking into account that $G_j^n \cap G_k^n = \emptyset$ for $|j-k| \geq 4$ and $\chi_{j+k}\hat{\chi}_j = \chi_{j+k}$ for $k \in \{-1, 0, 1\}$ and employing the local property of E, we get for $k \in \{-1, 0, 1\}$,

$$
\begin{aligned}
\left| \sum_{j=1}^{4n} \chi_j A \chi_{j+k} \left(f - \tfrac{j-1}{4n} \right) I \right| &= \left| \sum_{j=1}^{4n} \hat{\chi}_j \chi_j A \chi_{j+k} \left(f - \tfrac{j-1}{4n} \right) \hat{\chi}_j I \right| \\
&\leq \sum_{r=1}^{4} \left| \sum_{j=0}^{n-1} \hat{\chi}_{4j+r} \chi_{4j+r} A \chi_{4j+r+k} \left(f - \tfrac{4j+r-1}{4n} \right) \hat{\chi}_{4j+r} I \right| \\
&\leq \sum_{r=1}^{4} \max_{0 \leq j \leq n-1} \left| \chi_{4j+r} A \chi_{4j+r+k} \left(f - \tfrac{4j+r-1}{4n} \right) I \right|.
\end{aligned}
$$

For $x \in F_{4j+r+k}^n$ one has

$$
\frac{4j+r+k-1}{4n} < f(x) \leq \frac{4j+r+k}{4n}
$$

which implies

$$
\frac{k}{4n} < f(x) - \frac{4j+r-1}{4n} \leq \frac{k+1}{4n}
$$

and, consequently,

$$
\left| \chi_{4j+r+k} \left(f - \frac{4j+r-1}{4n} \right) \right| \leq \frac{1}{2n}.
$$

Thus, the first of the inner sums in (2.20) can be estimated by

$$
\frac{1}{2n} \sum_{r=1}^{4} \max_{0 \leq j \leq n-1} |\chi_{4j+r} A| \leq \frac{2}{n} |A|,
$$

and a similar estimate for the second sum finally yields

$$
|AfI - fA| \leq 3(2/n + 2/n) = 12/n.
$$

Letting n go to infinity, we conclude that the operator $AfI - fA$ is compact for every continuous function $f : X \to [0, 1]$. The generalization to arbitrary $f \in C(X)$ is made by scaling and by writing f as the sum of its real, imaginary, positive and negative parts.

(ii) $\Rightarrow$ (iii): Let F_1, F_2 be measurable subsets of X with $\operatorname{clos} F_1 \subset \operatorname{int} F_2$. Then $\operatorname{clos} F_1$ and $\operatorname{clos}(X \setminus F_2)$ are disjoint. Since X is a locally compact Hausdorff space, there is a continuous function f on X such that

$$f(x) = \begin{cases} 0 & \text{if } x \in \operatorname{clos} F_1, \\ 1 & \text{if } x \in \operatorname{clos}(X \setminus F_2). \end{cases}$$

Then

$$\chi_{F_1} A - \chi_{F_1} A \chi_{F_2} I = \chi_{F_1} A \chi_{X \setminus F_2} I = \chi_{F_1} A f \chi_{X \setminus F_2} I \sim \chi_{F_1} f A \chi_{X \setminus F_2} I = 0.$$

The second relation follows in the same way.

(iii) $\Rightarrow$ (i): Let F_1, F_2 be disjoint closed subsets of X. Then $F_1 \subset X \setminus F_2$ which is open. Thus, by (iii),

$$\chi_{F_1} A \chi_{F_2} I = \chi_{F_1} A - \chi_{F_1} A \chi_{X \setminus F_2} I \sim 0$$

which finishes the proof of the theorem.   ∎

As the first consequence of the above result, we mention the following.

Theorem 2.5.7. *Let E be a Banach space of local type over X. Then $\Lambda(E)$ is a closed and inverse-closed subalgebra of $\mathscr{L}(E)$. The ideal $\mathscr{K}(E)$ of the compact operators is contained in $\Lambda(E)$, and the quotient algebra $\Lambda(E)/\mathscr{K}(E)$ is inverse-closed in the Calkin algebra $\mathscr{L}(E)/\mathscr{K}(E)$. In particular, if A is a Fredholm operator of local type, then each of its regularizers is of local type again.*

Proof. The proof of this theorem becomes straightforward if the equivalence between conditions (i) and (ii) in Theorem 2.5.6 is employed. For example, the identities

$$f(AB) - (AB)fI = (fA - Af)B + A(fB - BfI)$$

and

$$fA^{-1} - A^{-1}fI = A^{-1}(Af - fA)A^{-1}$$

show that $AB \in \Lambda(E)$ whenever A and B are in $\Lambda(E)$ and that $A^{-1} \in \Lambda(E)$ whenever $A \in \Lambda(E)$ is invertible in $\mathscr{L}(E)$. If (A_n) is a sequence of operators of local type which converges to $A \in \mathscr{L}(E)$ in the norm of $\mathscr{L}(E)$, then $fA - AfI$ is the norm limit of the sequence of compact operators $(fA_n - A_n fI)$, and hence compact. It is also evident that $\mathscr{K}(E) \subset \Lambda(E)$. Finally, let $A \in \Lambda(E)$ be a Fredholm operator. Then, there are compact operators K, L, as well as an operator $R \in \mathscr{L}(E)$ such that $AR = I + K$ and $RA = I + L$. For each $f \in C(X)$ one obtains

$$fR - RfI = (RA - L)fR - Rf(AR - K)$$
$$= R(Af - fA)R - LfR + RfK \in \mathcal{K}(E)$$

which shows that $R \in \Lambda(E)$, too.   ∎

2.5.2 Local equivalence and local norms

The following definitions of local equivalence and of local norms go back to Simonenko.

Definition 2.5.8. Let E be an ideal Banach space over X and let $x \in X$. Then the *local essential norm* of an operator $A \in \Lambda(E)$ at the point x is the quantity

$$|A|_x := \inf_U |\chi_U A|$$

where the infimum is taken over all open neighborhoods U of x. The operators A and B are called *locally equivalent* at x if $|A - B|_x = 0$. Local equivalence at the point x will be denoted by $A \overset{x}{\sim} B$.

Lemma 2.5.9. *Let X, E, A and x be as in the previous definition. Then the local essential norm $|A|_x$ coincides with each of the following quantities:*

(i) $\inf |fA|$, *where the infimum is taken over all continuous functions $f : X \to [0,1]$ which are identically 1 in a neighborhood of x;*
(ii) $\inf |fA|$, *where the infimum is taken over all continuous functions f on X with $f(x) = 1$.*

Proof. Let m_1 and m_2 denote the quantities defined in (i) and (ii), respectively. Evidently, $m_2 \leq m_1$. Further, given a neighborhood U of x, choose a neighborhood W of x with $\overline{W} \subset U$. Then there is a continuous function $f : X \to [0,1]$ which is identically 1 on W and vanishes outside U. Thus,

$$|fA| = |f\chi_U A| \leq |\chi_U A|,$$

whence the estimate $m_1 \leq |A|_x$. For the estimate $|A|_x \leq m_2$, let $\varepsilon > 0$ and choose a continuous function f with $f(x) = 1$ and $|fA| < m_2 + \varepsilon$. Since f is continuous, there is a neighborhood U of x such that $|f(y) - 1| < \varepsilon$ for all $y \in U$. With this neighborhood, one gets

$$|\chi_U A| \leq |\chi_U fA| + |\chi_U (1 - f)A| \leq |fA| + |\chi_U (1 - f)||A| \leq m_2 + \varepsilon + \varepsilon |A|.$$

Consequently, $|A|_x \leq m_2 + \varepsilon(1 + |A|)$. Letting ε go to zero yields the assertion. ∎

Remark 2.5.10. In combination with Theorem 2.5.6, the preceding lemma shows that

$$|A|_x = \inf_U |A\chi_U I|$$

with the infimum taken over all open neighborhoods of x. $\qquad\qquad\qquad\square$

Note that Theorem 2.5.6 offers another way to deal with local properties of operators of local type. The second characterization of $\Lambda(E)$ given in that theorem implies that $\mathscr{B} := C(X) + \mathscr{K}(E)$ is a central subalgebra of $\Lambda(E)/\mathscr{K}(E)$ which contains the identity element. The central subalgebra $\mathscr{B}$ is isometrically isomorphic to the algebra $C(X)$. Indeed, as in Proposition 1.4.11 (where the case $E = L^p(X)$ is considered) one gets that $\|fI + \mathscr{K}(E)\| = \|fI\|$ for each function $f \in C(X)$, and the equality $\|fI\| = \|f\|_\infty$ has been established in Lemma 2.5.2. Thus, the maximal ideal space of the commutative C^*-algebra $\mathscr{B}$ is homeomorphic to X, and one can apply Allan's local principle to localize the algebra $\mathscr{A} := \Lambda(E)/\mathscr{K}(E)$ over X. We let $\mathscr{J}_x$ denote the local ideal of $\mathscr{A}$ which is induced by $x \in X$ and write Φ_x for the canonical homomorphism $\mathscr{A} \to \mathscr{A}/\mathscr{J}_x$. Further we let Ψ_x stand for the homomorphism $\Phi_x \circ \pi$, where π refers to the canonical homomorphism $\Lambda(E) \to \Lambda(E)/\mathscr{K}(E)$. Thus,

$$\Psi_x : \Lambda(E) \to \mathscr{A}/\mathscr{J}_x, \; A \mapsto \Phi_x(A + \mathscr{K}(E)).$$

Proposition 2.5.11. *Let E be a Banach space of local type over X. Then, for each $A \in \Lambda(E)$ and each $x \in X$,*

$$\|\Psi_x(A)\| = |A|_x.$$

Proof. Let $A \in \Lambda(E)$ and $x \in X$. By Lemma 2.5.9, we have to show that

$$\|\Psi_x(A)\| = \inf |fA|, \qquad\qquad\qquad (2.21)$$

where the infimum is taken over all continuous functions $f : X \to [0,1]$ which are identically 1 in a neighborhood of x. We denote this infimum by q. If $f \in C(X)$ is identically 1 in some neighborhood of x, then the coset $(f-1)A + \mathscr{K}(E)$ belongs to $\mathscr{J}_x$. Thus,

$$\|\Psi_x(A)\| \le |A + (f-1)A| = |fA|.$$

Taking the infimum over all f with these properties we get $\|\Psi_x(A)\| \le q$.

For the reverse inequality, given $\varepsilon > 0$, choose functions $f_1, \dots, f_n \in C(X)$ which vanish at x and operators $B_1, \dots, B_n \in \Lambda(E)$ such that

$$|A + f_1 B_1 + \dots + f_n B_n| < \|\Psi_x(A)\| + \varepsilon.$$

If f is any function in $C(X)$ with $0 \le f \le 1$ and $f \equiv 1$ in some neighborhood of x, then

$$|fA| \leq \left| f\left(A + \sum f_i B_i\right) \right| + \left| \sum f f_i B_i \right|$$
$$\leq \left| A + \sum f_i B_i \right| + \sum \|f f_i\|_\infty |B_i|$$
$$\leq \|\Psi_x(A)\| + \varepsilon + \sum \|f f_i\|_\infty |B_i|.$$

The right-hand side of this estimate becomes smaller than $\|\Psi_x(A)\| + 2\varepsilon$ if f is chosen such that $\|f f_i\|_\infty < \varepsilon/(n|B_i|)$ for every i. Hence, $q \leq \|\Psi_x(A)\|$. ∎

Theorem 2.5.12. *Let E be a Banach space of local type over X and let $A \in \mathscr{L}(E)$ be an operator of local type. Then the essential norm of A can be expressed in terms of local norms by*

$$|A| = \max_{x \in X} |A|_x.$$

This theorem will be an immediate consequence of Theorem 2.3.3 and Proposition 2.5.11 once we have shown the following.

Theorem 2.5.13. $(\Lambda(E)/\mathscr{K}(E), C(X) + \mathscr{K}(E))$ *is a faithful localizing pair.*

Proof. We have to show that

$$|(f+g)A| \leq \max\{|fA|, |gA|\}$$

whenever $A \in \Lambda(E)$ and f and g are functions in $C(X)$ with disjoint supports $M, N \subseteq X$, respectively. Since A is of local type, we conclude from (2.18) that

$$\max\{|fA|, |gA|\} \geq |\chi_M f A \chi_M I + \chi_N g A \chi_N I|$$
$$= |fA\chi_M I + gA\chi_N I|$$
$$= |fA + gA - fA\chi_{X\setminus M} I - gA\chi_{X\setminus N} I|$$
$$= |fA + gA|.$$

Here we used that

$$fA\chi_{X\setminus M} I = (fA - Af)\chi_{X\setminus M} I + Af\chi_{X\setminus M} I = (fA - Af)\chi_{X\setminus M} I$$

and $gA\chi_{X\setminus N} I$ are compact operators. ∎

2.5.3 Local Fredholmness and Fredholmness

Definition 2.5.14. Let E be an ideal Banach space over X and $A \in \mathscr{L}(E)$. An operator $R_l \in \mathscr{L}(E)$ (resp. R_r) is called a *local left (resp. right) regularizer* of the operator A at the point $x \in X$ if there is a neighborhood U of the point x such that $R_l A \chi_U I \sim \chi_U I$ (resp. $\chi_U A R_r \sim \chi_U I$). An operator A is said to be *locally Fredholm at* $x \in X$ if it possesses both a local left and a local right regularizer at that point.

Proposition 2.5.15. *Let E be a Banach space of local type over X and let $A \in \mathcal{L}(E)$ be an operator of local type which is locally Fredholm at $x \in X$. Then A possesses local left and right regularizers at x which are* of local type.

Proof. Let U be an open neighborhood of x and let R_l, $R_r \in \mathcal{L}(E)$ be operators such that

$$R_l A \chi_U I \sim \chi_U I \quad \text{and} \quad \chi_U A R_r \sim \chi_U I.$$

Let g be a continuous function on X which is identically 1 in a neighborhood V of x and which has its support inside U. Since

$$g R_l g A \chi_V I \sim g R_l A g \chi_V I = g R_l A \chi_U g \chi_V I \sim g \chi_U g \chi_V I = \chi_V I,$$

the operator $g R_l g I$ is a local left regularizer of A at x, too. We claim that $g R_l g I$ is of local type. Let $f \in C(X)$. Then

$$
\begin{aligned}
f g R_l g I - g R_l g f I &= (f g R_l - g R_l f) g \chi_U I \\
&\sim (f g R_l - g R_l f) g \chi_U A R_r \\
&\sim (f g R_l - g R_l f) A g R_r \\
&\sim (f g R_l A - g R_l A f) g R_r \\
&= (f g R_l A \chi_U - g R_l A \chi_U f) g R_r \\
&\sim (f g - g f) \chi_U g R_r = 0,
\end{aligned}
$$

which proves the claim. Similarly one gets that $g R_r g I$ is a local right regularizer of A at x which is of local type. ∎

The main result on local Fredholmness reads as follows.

Theorem 2.5.16. *Let E be a Banach space of local type over X. An operator $A \in \mathcal{L}(E)$ of local type is Fredholm if and only if it is locally Fredholm at every point $x \in X$.*

The proof will follow immediately from Allan's local principle (Theorem 2.2.2) and from Theorem 2.5.7 once we have checked the following assertion. The notation is as in Section 2.5.2.

Proposition 2.5.17. *Let E be a Banach space of local type over X. The operator $A \in \Lambda(E)$ is locally Fredholm at $x \in X$ if and only if its local coset $\Psi_x(A)$ is invertible in $\mathcal{A} / \mathcal{J}_x$.*

Proof. If A is locally Fredholm at x, then there are operators R_l and R_r of local type as well as an open neighborhood U of x such that

$$R_l A \chi_U I \sim \chi_U I \quad \text{and} \quad \chi_U A R_r \sim \chi_U I.$$

Let f be a continuous function on X with $f(x) = 1$ and supp $f \subset U$. Then

$$R_l A f I \sim f I \quad \text{and} \quad f A R_r \sim f I.$$

Applying the local homomorphism Ψ_x to these identities and taking into account that $\Psi_x(f)$ is the identity element of $\mathscr{A}/\mathscr{J}_x$, one gets the invertibility of $\Psi_x(A)$. Conversely, let $\Psi_x(A)$ be invertible in $\mathscr{A}/\mathscr{J}_x$. Let $R \in \Lambda(E)$ be such that $\Psi_x(R)$ is the inverse of $\Psi_x(A)$. Then $\Psi_x(RA - I) = \Psi_x(AR - I) = 0$, whence $|RA - I|_x = 0$ by Proposition 2.5.11. By the definition of the local norm, there is an open neighborhood U of x such that $|R A \chi_U I - \chi_U I| < 1/2$. Equivalently, there is an operator $C \in \mathscr{L}(E)$ with $\|C\| < 1/2$ and a compact operator K such that $R A \chi_U I - \chi_U I = C + K$. Multiplying this equality from both sides by $\chi_U I$ we find

$$\chi_U R A \chi_U I = \chi_U I + \chi_U C \chi_U I + \chi_U K \chi_U I. \tag{2.22}$$

Since $\|\chi_U C \chi_U I\| < 1/2$, the operator $\chi_U I + \chi_U C \chi_U I$ (considered as an operator on the range of the projection $\chi_U I$) is invertible by Neumann series. Let D denote its inverse. Then (2.22) implies

$$D \chi_U R A \chi_U I = \chi_U I + K'$$

with a certain compact operator K'. Consequently, $D \chi_U R A \chi_U I \sim \chi_U I$, i.e., $D \chi_U R$ is a local left regularizer of A at x. The existence of a local right regularizer follows analogously. ∎

2.5.4 The envelope of an operator function

Let E be an ideal Banach space over X.

Definition 2.5.18. An operator function $X \to \Lambda(E)$, $x \mapsto A_x$ is said to possess an *envelope* if there is an operator $A \in \Lambda(E)$ such that $A \overset{x}{\sim} A_x$ for all $x \in X$. Each operator A with this property is called an envelope of the function $(A_x)_{x \in X}$.

Definition 2.5.19. The operator function $X \to \Lambda(E)$, $x \mapsto A_x$ is said to be locally semi-continuous if for every point $x_0 \in X$ and every $\varepsilon > 0$ there exists an open neighborhood U of the point x_0 such that every point $x \in U$ has an open neighborhood $V \subset U$ with

$$|(A_{x_0} - A_x)\chi_V I| < \varepsilon \tag{2.23}$$

Theorem 2.5.20. *Let E be a Banach space of local type over X. Then a bounded operator function $X \to \Lambda(E)$, $x \mapsto A_x$ possesses an envelope if and only if it is locally semi-continuous. The envelope is uniquely determined up to a compact operator, and*

$$|A| \le \sup_{x \in X} |A_x|$$

for each envelope A of the given operator function.

Proof. It easy to check that the existence of an envelope implies the local semi-continuity of the operator function. The reverse implication will be shown by having recourse to Theorem 2.3.5 and its application to faithful localizing pairs. Thus, we let $\mathscr{F}$ refer to the set of all bounded functions on X which take, at $x \in X$, a value in $\mathscr{A} / \mathscr{J}_x$ (the notation is as in Subsection 2.5.2). The set $\mathscr{F}$ becomes a Banach algebra by defining elementwise operations and the supremum norm. The Banach algebra $\mathscr{F}$ together with its subalgebra $\mathscr{D}$, consisting of all functions $x \mapsto f(x)\Psi_x(I)$ where $f \in C(X)$, forms a faithful localizing pair. Let $\mathscr{E}$ stand for the set of all functions in $\mathscr{F}$ of the form

$$x \mapsto \Psi_x(A) \quad \text{with} \quad A \in \Lambda(E).$$

Then $\mathscr{E}$ is a subalgebra of $\mathscr{F}$ which satisfies conditions (i) - (iv) on page 89 in Subsection 2.3.2. Moreover, $\mathscr{E}$ is closed. Indeed, let (f_n) with $f_n : x \mapsto \Psi_x(A_n)$ be a Cauchy sequence in $\mathscr{F}$. From

$$\|f_n - f_m\|_{\mathscr{F}} = \sup_{x \in X} \|\Psi_x(A_n - A_m)\| = |A_n - A_m|$$

(by Theorem 2.3.3 and Proposition 2.5.11) we conclude that $(A_n + \mathscr{K}(E))$ is a Cauchy sequence in $\mathscr{A}$. Let $A \in \Lambda(E)$ be such that $A + \mathscr{K}(E)$ is the limit of that sequence. Then $f : x \mapsto \Psi_x(A)$ is a function in $\mathscr{F}$, and $\|f_n - f\|_{\mathscr{F}} \to 0$.

Now let $X \to \Lambda(E)$, $x \mapsto A_x$ be a bounded and locally semi-continuous function on X, and let $x_0 \in X$. Let $A \in \Lambda(E)$ be an operator with $\Psi_{x_0}(A) = \Psi_{x_0}(A_{x_0})$. Due to Allan's local principle (Theorem 2.2.2), the function

$$X \to \mathbb{R}, \quad x \mapsto \|\Psi_x(A - A_{x_0})\|$$

is upper semi-continuous at x_0, i.e., given $\varepsilon > 0$ there is a neighborhood U_1 of x_0 such that $\|\Psi_x(A - A_{x_0})\| < \varepsilon$ for all $x \in U_1$. Further we conclude from (2.23) that there is a neighborhood U_2 of x_0 such that, for $x \in U_2$,

$$\|\Psi_x(A_x) - \Psi_x(A_{x_0})\| = |A_x - A_{x_0}|_x \leq |(A_x - A_{x_0})\chi_V I| < \varepsilon.$$

Hence, for all $x \in U := U_1 \cap U_2$,

$$\|\Psi_x(A_x) - \Psi_x(A)\| \leq \|\Psi_x(A_x) - \Psi_x(A_{x_0})\| + \|\Psi_x(A_{x_0}) - \Psi_x(A)\| \leq 2\varepsilon.$$

Hence, the function $x \mapsto \Psi_x(A_x)$ is semi-continuous with respect to $\mathscr{E}$ in the sense of Section 2.3.2. By Theorem 2.3.5, this function belongs to the closure of $\mathscr{E}$ in $\mathscr{F}$, which actually coincides with $\mathscr{E}$ as we have already seen. Thus, this function is of the form $x \mapsto \Psi_x(B)$ with a certain operator B of local type; in other words, B is an envelope for $(A_x)_{x \in X}$.

For the proof of the norm estimate, we take into account that $|A|_x = |A_x|_x$ by the definition of the local norm $|\cdot|_x$. Then

$$|A| = \max_{x \in X} |A|_x = \max_{x \in X} |A_x|_x \leq \sup_{x \in X} |A_x|$$

where the first equality comes from Theorem 2.5.12.   ∎

2.5.5 Exercises

Exercise 2.5.1. Prove that local equivalence is a reflexive, symmetric and transitive relation. Moreover,

(i) if A, B, A_x and B_x are local type operators with $A \overset{x}{\sim} A_x$ and $B \overset{x}{\sim} B_x$, then $AB \overset{x}{\sim} A_x B_x$;

(ii) if A and B are local type operators and $x \in X$, then $A \overset{x}{\sim} B$ if and only if for every $\varepsilon > 0$ there exists a neighborhood U of x such that

$$|(A-B)\chi_U I| < \varepsilon \quad \text{and} \quad |\chi_U (A-B)| < \varepsilon;$$

(iii) if A and B are local type operators and U_i, $i = 1, \dots, 8$ are neighborhoods of $x \in X$ such that $\chi_{U_1} A \chi_{U_2} I \overset{x}{\sim} \chi_{U_3} B \chi_{U_4} I$, then $\chi_{U_5} A \chi_{U_6} I \overset{x}{\sim} \chi_{U_7} B \chi_{U_8} I$;

(iv) if (A_i) and (B_i) are sequences such that $|A_i - A| \to 0$, $|B_i - B| \to 0$, and $A_i \overset{x}{\sim} B_i$ for every i, then $A \overset{x}{\sim} B$.

Exercise 2.5.2. Let E be a Banach space of local type over X and $A \in \mathcal{L}(E)$ a local type operator. Prove that

$$|A| = \sup_{x \in X} |A|_x.$$

Exercise 2.5.3. Show that the equality $|A| = \sup_{x \in X} |A_x|$ does not hold in general.

Exercise 2.5.4. Let A be an operator of local type, $x \in X$, W a neighborhood of x, and R_l (resp. R_r) a local left (resp. right) regularizer of A at x. Prove also that $R_l \chi_W I$ and $\chi_W R_l$ (resp. $R_r \chi_W I$ and $\chi_W R_r$) are local left (resp. right) regularizers of A at x.

Exercise 2.5.5. Let A be an operator of local type, $x \in X$, W a neighborhood of x, and R_l (resp. R_r) a local left (resp. right) regularizer of A at x, and let $f \in C(X)$ be a function which is identically 1 on W. Prove then that $R_l f I$ and $f R_l$ (resp. $R_r f I$ and $f R_r$) are local left (resp. right) regularizers of A at x.

Exercise 2.5.6. Show that if A is a local type operator which is locally Fredholm at a point $x \in X$, then A possesses a local type regularizer at that point.

Exercise 2.5.7. Let operators A and B be locally equivalent at $x \in X$ and assume that A possesses a left (resp. right) local regularizer at x. Show that then B also possesses a left (resp. right) local regularizer at x. Prove that if A, B and the left (right) local regularizer of A are local type operators, then the left (right) local regularizer of B has the same property.

Exercise 2.5.8. Let operators A and B be locally equivalent at $x \in X$, and let A be locally Fredholm at x. Show that B is also locally Fredholm at x.

Exercise 2.5.9. Let A be a local type operator which possesses a local left (right) local type regularizer for any $x \in X$. Prove that A possesses a global left (right) local type regularizer.

2.6 PI-algebras and QI-algebras

In this section we are going to present another generalization of Gelfand's transform, applicable to special classes of Banach algebras, the so called *PI-* and *QI-algebras*, where *PI* stands for *polynomial identity* and *QI* for *quasi identity*. These algebras are close to commutative algebras in the sense that the defining condition $ab = ba$ of a commutative algebra is replaced by another polynomial identity. For PI- and QI-algebras we will obtain sufficient families of *finite-dimensional* homomorphisms, that is, the algebras $\mathscr{B}_t$ defined in the introduction of the chapter will prove to be matrix algebras, $\mathscr{B}_t = \mathbb{C}^{l(t) \times l(t)}$, with $\sup l(t) < \infty$.

2.6.1 Standard polynomial identities

In this section, $\mathscr{A}$ denotes a unital algebra over an arbitrary field $\mathbb{F}$, and $\mathcal{P}$ is a polynomial of positive degree in n non-commuting variables with coefficients in $\mathbb{F}$. Each time the variables in $\mathcal{P}$ are replaced by elements $a_1, \ldots, a_n$ of the algebra, the result is an element of the algebra which we denote by $\mathcal{P}(a_1, \ldots, a_n)$.

Definition 2.6.1. Let $\mathcal{P}$ be a polynomial of positive degree in n non-commuting variables with coefficients in $\mathbb{F}$. An algebra $\mathscr{A}$ is said to satisfy the *polynomial identity* $\mathcal{P}$ if $\mathcal{P}(a_1, \ldots, a_n) = 0$ for every choice of elements $a_1, \ldots, a_n \in \mathscr{A}$. We then call $\mathscr{A}$ a $\mathcal{P}$-*algebra*. If $\mathscr{A}$ satisfies at least one non-trivial polynomial identity, it is called a *PI-algebra*.

Let Σ_n refer to the permutation group of the set $\{1, \ldots, n\}$. Polynomials of the form

$$\mathcal{P}(a_1, \ldots, a_n) = \sum_{\sigma \in \Sigma_n} \lambda_\sigma \, a_{\sigma(1)} \ldots a_{\sigma(n)} \qquad (2.24)$$

with coefficients $\lambda_\sigma \in \mathbb{F}$ are called *multilinear*. Considered as a mapping from $\mathscr{A}^n$ to $\mathscr{A}$, a multilinear polynomial is indeed linear in each component.

A polynomial $\mathcal{P}$ of positive degree in n non-commuting variables is called *alternating* if any repetition in the choice of elements $a_1, \ldots, a_n$ yields 0, that is, $\mathcal{P}(\ldots, a_j, \ldots, a_j, \ldots) = 0$. Finally, for $1 \le i \ne j \le n$, we let $\mathcal{P}_{ij}$ stand for the polynomial $\mathcal{P}$ with the variables at places i and j interchanged.

Lemma 2.6.2. *Let $\mathcal{P}$ be a multilinear polynomial. Then $\mathcal{P}$ is alternating if and only if $\mathcal{P}_{ij} = -\mathcal{P}$ for each choice of indices $1 \le i \ne j \le n$.*

Proof. Let $\mathcal{P}$ be alternating and $1 \le i \ne j \le n$. Then

$$\begin{aligned}
0 &= \mathcal{P}(\ldots, a_i + a_j, \ldots, a_i + a_j, \ldots) \\
&= \mathcal{P}(\ldots, a_i, \ldots, a_i, \ldots) + \mathcal{P}(\ldots, a_i, \ldots, a_j, \ldots) \\
&\quad + \mathcal{P}(\ldots, a_j, \ldots, a_i, \ldots) + \mathcal{P}(\ldots, a_j, \ldots, a_j, \ldots) \\
&= \mathcal{P}(\ldots, a_i, \ldots, a_j, \ldots) + \mathcal{P}(\ldots, a_j, \ldots, a_i, \ldots).
\end{aligned}$$

Conversely, by interchanging the positions where the repeated element a_j is present, we obtain

$$\mathcal{P}(\ldots, a_j, \ldots, a_j, \ldots) = -\mathcal{P}(\ldots, a_j, \ldots, a_j, \ldots),$$

which implies $\mathcal{P}(\ldots, a_j, \ldots, a_j, \ldots) = 0$. ∎

The following definition introduces the process of *multilinearization*. Applied to a (non-multilinear) polynomial, this process results in another polynomial, with one "new" variable and a lesser degree in one of the "old" variables. This process, taken repeatedly, finally allows a multilinear polynomial to be obtained from any polynomial.

Definition 2.6.3. Let $\mathcal{P}: \mathscr{A}^n \to \mathscr{A}$ be a function in n variables. For $1 \leq i \leq n$, we define the function $\Delta_i \mathcal{P}: \mathscr{A}^{n+1} \to \mathscr{A}$ by

$$\begin{aligned}
\Delta_i \mathcal{P}(a_1, \ldots, a_{n+1}) := {} & \mathcal{P}(a_1, \ldots, a_{i-1}, a_i + a_{n+1}, a_{i+1}, \ldots, a_n) \\
& - \mathcal{P}(a_1, \ldots, a_{i-1}, a_i, a_{i+1}, \ldots, a_n) \\
& - \mathcal{P}(a_1, \ldots, a_{i-1}, a_{n+1}, a_{i+1}, \ldots, a_n).
\end{aligned} \tag{2.25}$$

Lemma 2.6.4. *If an algebra $\mathscr{A}$ satisfies a polynomial identity of degree k, then it also satisfies a multilinear identity of degree $\leq k$.*

Proof. Let $\mathscr{A}$ satisfy a polynomial identity $\mathcal{P}$ of degree k in n variables. If $\mathcal{P}$ is not linear in the first variable (this happens if the degree of the first variable is greater than 1), then consider

$$\begin{aligned}
& \Delta_1 \mathcal{P}(a_1, \ldots, a_n, a_{n+1}) \\
& = \mathcal{P}_n(a_1 + a_{n+1}, a_2, \ldots, a_n) - \mathcal{P}(a_1, a_2, \ldots, a_n) - \mathcal{P}(a_{n+1}, a_2, \ldots, a_n).
\end{aligned}$$

Clearly, $\mathscr{A}$ also satisfies the polynomial identity $\Delta_1 \mathcal{P}$, and the degree of $\Delta_1 \mathcal{P}$ is not greater than that of $\mathcal{P}$. But the degree of the first variable in $\Delta_1 \mathcal{P}$ is strictly lower than the degree of the first variable in $\mathcal{P}$. Repeated application of this procedure to every nonlinear variable yields, after a finite number of steps, a multilinear identity of degree not greater than k which is also satisfied by $\mathscr{A}$. ∎

Lemma 2.6.5. *The matrix algebra $\mathbb{F}^{n \times n}$ over the field $\mathbb{F}$ does not satisfy a polynomial identity of degree less than $2n$.*

Proof. By the above lemma, we just have to check that $\mathbb{F}^{n \times n}$ does not satisfy any multilinear identity of degree less than $2n$. Suppose that $\mathbb{F}^{n \times n}$ satisfies a multilinear identity $\mathcal{P}_m$ of degree $m < 2n$. Let $E_{pq} \in \mathbb{F}^{n \times n}$ be the matrix with zeros at every entry except at the entry pq, which is 1. Inserting the matrices

$$a_i := \begin{cases} E_{\frac{i+1}{2}\,\frac{i+1}{2}} & \text{if } i \text{ is odd}, \\ E_{\frac{i}{2}\,\frac{i+2}{2}} & \text{if } i \text{ is even} \end{cases}$$

into (2.24), we get immediately that the coefficient associated with the identity permutation is zero. Rearranging the matrices above in $\mathcal{P}_m$, we then conclude that all coefficients must be zero. This contradiction proves the assertion.   ∎

We are now going to introduce a class of multilinear polynomials which will play a dominant role in what follows.

Definition 2.6.6. Let $\mathscr{A}$ be an algebra and $a_1, \ldots, a_n \in \mathscr{A}$. The *standard polynomial* of degree n is defined by

$$S_n(a_1, \ldots, a_n) := \sum_{\sigma \in \Sigma_n} \mathrm{sgn}\,\sigma \, a_{\sigma(1)} \cdots a_{\sigma(n)},$$

where $\mathrm{sgn}\,\sigma$ takes the value $+1$ if the permutation $\sigma \in \Sigma_n$ is even and -1 if it is odd.

The standard polynomials can also be defined recursively by $S_1(a_1) := a_1$ and

$$S_n(a_1, \ldots, a_n) = \sum_{i=1}^{n} (-1)^{i-1} a_i S_{n-1}(a_1, \ldots, \tilde{a}_i, \ldots, a_n)$$

or, equivalently,

$$S_n(a_1, \ldots, a_n) = \sum_{i=1}^{n} (-1)^{n-i} S_{n-1}(a_1, \ldots, \tilde{a}_i, \ldots, a_n) a_i$$

if $n > 1$ where the tilde indicates that the corresponding element is omitted.

It easy to see that standard polynomials and their scalar multiples are alternating. The next result shows the converse is also true.

Proposition 2.6.7. *Any multilinear alternating polynomial of degree n is a multiple of the standard polynomial S_n.*

Proof. Let $\mathcal{P}$ be a multilinear alternating polynomial of the form (2.24). Since every permutation is a composition of interchanges of variables, one has by Lemma 2.6.2, $\mathcal{P}(a_1, \ldots, a_n) = \mathrm{sgn}\,\sigma\, \mathcal{P}(a_{\sigma(1)}, \ldots, a_{\sigma(n)})$ for each permutation $\sigma \in \Sigma_n$. The coefficient of the monomial $a_{\sigma(1)} \cdots a_{\sigma(n)}$ in the polynomial on the left-hand side of this equality is λ_σ; its counterpart on the right-hand side is $\mathrm{sgn}\,\sigma \lambda_{id}$, with the identity permutation id. Hence, $\lambda_\sigma = \mathrm{sgn}\,\sigma \lambda_{id}$, which implies

$$\mathcal{P}(a_1, \ldots, a_n) = \lambda_{id} \sum_{\sigma \in \Sigma_n} \mathrm{sgn}\,\sigma a_{\sigma(1)} \cdots a_{\sigma(n)}.$$

Thus, $\mathcal{P} = \lambda_{id} S_n$.   ∎

The following fact will be used in the proof of the Amitsur-Levitzki theorem.

Corollary 2.6.8. *Let $\mathcal{P}(a_1, \ldots, a_{2n}) = \sum_{\sigma \in \Sigma_{2n}} \mathrm{sgn}\,\sigma [a_{\sigma_1}, a_{\sigma_2}] \ldots [a_{\sigma_{2n-1}}, a_{\sigma_{2n}}]$, where $[a,b]$ represents the commutator $ab - ba$. Then $\mathcal{P} = 2^n S_{2n}$.*

Proof. It is not difficult to see that $\mathcal{P}$ is a multilinear and alternating polynomial of degree $2n$. By the above proposition, $\mathcal{P}$ is a multiple of the standard polynomial S_{2n}. The polynomial $\mathcal{P}$ is the sum of $2^n(2n)!$ multilinear monomials, none of which cancel. Thus, the constant is 2^n. ∎

Let $\mathrm{tr}\, a$ denote the trace of the matrix a.

Lemma 2.6.9. *Let* $a_1, \ldots, a_{2k} \in \mathbb{F}^{n \times n}$. *Then* $\mathrm{tr}\left[S_{2k}(a_1, \ldots, a_{2k})\right] = 0$.

Proof. For $i = 1, \ldots, 2k$, let $\mathbf{a}_{2k}^i$ denote the $(2k-1)$-tuple $(a_1, \ldots, \tilde{a}_i, \ldots, a_{2k})$ where the tilde indicates that the corresponding element is omitted. Then

$$2\mathrm{tr}\left[S_{2k}(a_1, \ldots, a_{2k})\right]$$

$$= \mathrm{tr}\left[\sum_{i=1}^{2k}(-1)^{i-1} a_i S_{2k-1}(\mathbf{a}_{2k}^i)\right] + \mathrm{tr}\left[\sum_{i=1}^{2k}(-1)^{2k-i} S_{2k-1}(\mathbf{a}_{2k}^i) a_i\right]$$

$$= \sum_{i=1}^{2k}(-1)^{i-1}\mathrm{tr}\left[a_i S_{2k-1}(\mathbf{a}_{2k}^i) - S_{2k-1}(\mathbf{a}_{2k}^i)a_i\right] = 0,$$

since the trace of the commutator of two matrices is zero. ∎

Definition 2.6.10. The algebra $\mathscr{A}$ is said to *satisfy the standard identity of order* n if $S_n(a_1, \ldots, a_n) = 0$ for any $a_1, \ldots, a_n \in \mathscr{A}$. The family of all algebras with that property will be denoted by SI_n. Further, let SI_{2n}^m stand for the family of all algebras $\mathscr{A}$ with the property that $(S_{2n}(a_1, \ldots, a_{2n}))^m = 0$ for all choices of elements $a_1, \ldots, a_{2n} \in \mathscr{A}$. Finally, if $\mathscr{A}$ is a (real or complex) Banach algebra, then we call $\mathscr{A}$ a *QI-algebra* and write $\mathscr{A} \in SI_{2n}^\infty$ if there is a number n such that, for any choice of $a_1, \ldots, a_{2n} \in \mathscr{A}$,

$$\lim_{m \to \infty} \|(S_{2n}(a_1, \ldots, a_{2n}))^m\|^{1/m} = 0.$$

The standard polynomial of order 2 is $S_2(a_1, a_2) = a_1 a_2 - a_2 a_1$. Thus an algebra satisfies the standard identity of order 2 if and only if it is commutative.

In Lemma 2.6.5 we have seen that the algebra $\mathbb{F}^{n \times n}$ does not satisfy any polynomial identity of order less than $2n$. The next theorem states that it satisfies the standard identity of degree $2n$.

Theorem 2.6.11 (Amitsur-Levitzki). *The algebra* $\mathbb{F}^{n \times n}$ *satisfies the standard identity of degree* $2n$.

Proof. First let $\mathbb{F} = \mathbb{Q}$ be the field of the rational numbers and consider a matrix $a \in \mathbb{Q}^{n \times n}$. Newton's formula for the coefficients of the characteristic polynomial $\mathcal{P}_a$ of a,

$$\mathcal{P}_a(\lambda) := \det(\lambda I - a) = \lambda^n + \sum_{k=1}^{n} \alpha_k \lambda^{n-k} \qquad (2.26)$$

implies that the coefficients α_k are given as follows. Let Ω_k denote the set of all j-tuples $m = (m_1, \ldots, m_j)$ of integers with $1 \leq m_1 \leq m_2 \leq \cdots \leq m_j$ and $m_1 + m_2 +$

$\cdots + m_j = k$. Notice that both the k-tuple $(1, \ldots, 1)$ as well as the 1-tuple (k) belong to Ω_k. Then, for $k = 1, \ldots, n$,

$$\alpha_k = \sum_{m \in \Omega_k} q_m \operatorname{tr}(a^{m_1}) \ldots \operatorname{tr}(a^{m_j})$$

with certain rational numbers q_m (see also [170, Theorem 1.3.19]).

By the Cayley-Hamilton theorem, $\mathcal{P}_a(a) = 0$. We apply the multilinearization process to the mapping $\mathcal{P}: \mathbb{F}^{n \times n} \to \mathbb{F}$, $a \mapsto \mathcal{P}_a(a)$ to get the multivariable function

$$(\Delta \mathcal{P})(a_1, \ldots, a_n) := (\Delta_1)^{n-1} \mathcal{P}(a_1).$$

Clearly, $(\Delta \mathcal{P})(a_1, \ldots, a_n) = 0$. Since the trace is additive, we thereby arrive at the identity

$$0 = (\Delta \mathcal{P})(a_1, \ldots, a_n) = \sum_{\sigma \in \Sigma_n} a_{\sigma_1} \ldots a_{\sigma_n} +$$

$$+ \sum_{k=1}^{n} \sum_{m \in \Omega_k} \sum_{\sigma \in \Sigma_n} q_m \operatorname{tr}(a_{\sigma_1} \ldots a_{\sigma_{m_1}}) \ldots \operatorname{tr}(a_{\sigma_{m_1 + \ldots + m_{j-1}+1}} \cdots a_{\sigma_k}) a_{\sigma_{k+1}} \ldots a_{\sigma_n}.$$

For instance, for $n = 2$ one has

$$0 = \sum_{\sigma \in \Sigma_2} a_{\sigma_1} a_{\sigma_2} - \sum_{\sigma \in \Sigma_2} (\operatorname{tr} a_{\sigma_1}) a_{\sigma_2} + \frac{1}{2} \sum_{\sigma \in \Sigma_2} (\operatorname{tr} a_{\sigma_1})(\operatorname{tr} a_{\sigma_2}) - \frac{1}{2} \sum_{\sigma \in \Sigma_2} \operatorname{tr}(a_{\sigma_1} a_{\sigma_2})$$

due to

$$\mathcal{P}_a(a) = a^2 - (\operatorname{tr} a) a + \frac{1}{2} (\operatorname{tr} a)^2 - \frac{1}{2} \operatorname{tr}(a^2).$$

Back to general n. Given $2n$ matrices $a_1, \ldots, a_{2n} \in \mathbb{F}^{n \times n}$ and a permutation $\sigma' \in \Sigma_{2n}$, we replace each variable a_i in the above identity by $[a_{\sigma'_{2i-1}}, a_{\sigma'_{2i}}]$ and form the sum

$$0 = \sum_{\sigma' \in \Sigma_{2n}} \operatorname{sgn} \sigma' (\Delta \mathcal{P}) \left([a_{\sigma'_1}, a_{\sigma'_2}], \ldots, [a_{\sigma'_{2n-1}}, a_{\sigma'_{2n}}] \right).$$

Using Corollary 2.6.8, we write this identity as

$$0 = 2^n \mathcal{S}_{2n}(a_1, \ldots, a_{2n}) + \mathcal{P}(a_1, \ldots, a_{2n})$$

where $\mathcal{P}(a_1, \ldots, a_{2n})$ is a sum of terms of the form

$$q_m \operatorname{tr} \mathcal{S}_{2m_1}(a_{2\sigma_1 - 1}, \ldots, a_{2\sigma_{m_1}}) \times \cdots \times$$
$$\times \operatorname{tr} \mathcal{S}_{2m_j}(a_{2\sigma_{(m_1 + \cdots + m_{j-1}+1)}-1}, \ldots, a_{2\sigma_k}) \mathcal{S}_{2(n-k)}(a_{2\sigma_{k+1}-1}, \ldots, a_{2\sigma_n})$$

for some $\sigma \in \Sigma_n$. (Again, it is useful to consider the particular case $n = 2$ to understand this construction.) By Lemma 2.6.9, each of these terms is 0. Thus, $\mathcal{P}(a_1, \ldots a_{2n}) = 0$ and, consequently, $\mathcal{S}_{2n}(a_1, \ldots a_{2n}) = 0$.

Now let $\mathbb{F}$ be an arbitrary field, and $a_i \in \mathbb{F}^{n \times n}$. Each matrix a_i can be written as a linear combination $a_i = \sum_{j,k} a_{jk}^{(i)} E_{jk}$ where the E_{jk} are the unit matrices (i.e., the jk^{th} entry of E_{jk} is equal to 1, and the others are zero). Then, due to the multi-linearity of S_{2n}, the matrix $S_{2n}(a_1, \ldots, a_{2n})$ is a linear combination of the matrices $S_{2n}(E_{j_1 k_1}, \ldots, E_{j_{2n} k_{2n}})$ which are all zero by the first part of the proof. ∎

It is useful to observe that the argument used in the last part of the above proof also applies to an arbitrary commutative unital algebra $\mathscr{C}$ over $\mathbb{F}$. Thus the Amitsur-Levitzki theorem remains valid if the field $\mathbb{F}$ is replaced by a commutative unital algebra $\mathscr{C}$.

Having this result at our disposal, we can extend the results on matrix algebras from Section 1.1.4 as follows.

Theorem 2.6.12. *Let $\mathscr{A}$ be a unital algebra over a field $\mathbb{F}$, and $\mathscr{C}$ its center. Then $\mathscr{A}$ is isomorphic to $\mathscr{C}^{n \times n}$ if and only if*

(i) *$\mathscr{A} \in SI_{2n}$, and*
(ii) *$\mathscr{A}$ contains a unital subalgebra $\mathscr{A}_0$ which is isomorphic to $\mathbb{F}^{n \times n}$.*

Proof. If $\mathscr{A}$ is isomorphic to $\mathscr{C}^{n \times n}$, then $\mathscr{A} \in SI_{2n}$ by the Amitsur-Levitzki theorem. Hence, (i) is satisfied, and (ii) is obvious. Conversely, let $\mathscr{A}$ satisfy (i) and (ii). Then Theorem 1.1.17 establishes an isomorphism $w = [w_{jk}]_{j,k=1}^{n}$ between the algebras $\mathscr{A}$ and $\mathscr{D}^{n \times n}$, where $\mathscr{D} = w_{jk}(\mathscr{A})$ is a subalgebra of $\mathscr{A}$. Moreover, the centers of $\mathscr{A}$ and $\mathscr{D}$ coincide. It remains to prove the inclusion $w_{jk}(a) \in \mathscr{C}$ for all $1 \leq j, k \leq n$, and $a \in \mathscr{A}$. Since $b = \sum_i \sum_m e_{ii} b e_{mm}$ for all $b \in \mathscr{A}$ and $w_{jk}(a) = \sum_{i=1}^{n} e_{ij} a e_{ki}$ by definition, the commutator $w_{jk}(a) b - b w_{jk}(a)$ is zero only if

$$e_{ij} a e_{ki} b e_{mm} - e_{ii} b e_{mj} a e_{km} = 0$$

for all $a, b \in \mathscr{A}$ and $1 \leq i, j, k, m \leq n$. But the latter identity is a simple consequence of

$$\begin{aligned} & e_{ij} a e_{ki} b e_{mm} - e_{ii} b e_{mj} a e_{km} \\ &= e_{i1} S_{2n+1}(e_{1j} a e_{k1}, e_{1i} b e_{m1}, e_{12}, \ldots, e_{n-1,n}, e_{nn}, e_{n,n-1}, \ldots, e_{21}) e_{1m} \end{aligned}$$

and of the fact that each SI_{2n}-algebra also satisfies the standard identity of degree $2n + 1$. ∎

Corollary 2.6.13. *Let $\mathscr{A} \in SI_{2n}$ be a complex Banach algebra with center $\mathscr{C}$, which contains a unital subalgebra $\mathscr{A}_0$ isomorphic to $\mathscr{C}^{n \times n}$. Then the maximal ideal spaces of $\mathscr{A}$ and $\mathscr{C}$ are homeomorphic.*

2.6.2 Matrix symbols

Let $\mathscr{A}$ be a Banach algebra with identity e over the field $\mathbb{C}$ and $\mathscr{B} \subset \mathscr{A}$ be a subalgebra of $\mathscr{A}$. Let X be an arbitrary set and l a bounded function from X into the set of the positive integers. Set $n := \sup_{x \in X} l(x)$. Assume we are given a family $\{\mu_x\}_{x \in X}$ of representations of $\mathscr{A}$ having the property that $\mu_x(a) \in \mathbb{C}^{l(x) \times l(x)}$ for each $a \in \mathscr{A}$. If it is true for every element $b \in \mathscr{B}$ that b is invertible in $\mathscr{A}$ if and only if $\mu_x(b)$ is invertible for all $x \in X$, then we say that $\{\mu_x\}_{x \in X}$ generates a *matrix symbol of order n for $\mathscr{B}$ in $\mathscr{A}$*. The collection of all subalgebras $\mathscr{B}$ of $\mathscr{A}$ which possess a matrix symbol of order n for $\mathscr{B}$ in $\mathscr{A}$ is denoted by $\mathrm{IS}(n, \mathscr{A})$. In case $\mathscr{A}$ belongs to $\mathrm{IS}(n, \mathscr{A})$ we just say that $\mathscr{A}$ has a matrix symbol of order n.

Let $\mathscr{J}$ be a maximal left ideal of a Banach algebra $\mathscr{A}$, and write E for the linear space $\mathscr{A} / \mathscr{J}$ and $\Phi : \mathscr{A} \to E$ for the canonical linear mapping. Further, let $L^{\mathscr{J}}$ denote the left regular representation of $\mathscr{A}$ induced by $\mathscr{J}$, which was introduced in Example 1.3.12.

Lemma 2.6.14. *Let E_0 be a finite-dimensional linear manifold in E and let $x \in E \setminus E_0$. Then there is an $a \in \mathscr{A}$ such that $L_a^{\mathscr{J}}(E_0) = \{0\}$ and $L_a^{\mathscr{J}}(x) \neq 0$.*

Proof. The proof proceeds by induction with respect to the dimension of E_0. The assertion of the lemma is evidently true if $\dim E_0 = 0$. Suppose that it is true for $\dim E_0 = k$. Choose $y \notin E_0$ and set $E_1 := E_0 + \mathbb{C}y$. Further let $\mathscr{L} := \{L_a^{\mathscr{J}} : L_a^{\mathscr{J}}(E_0) = \{0\}\}$. Consider the set $\{L_a^{\mathscr{J}}(y) : L_a^{\mathscr{J}} \in \mathscr{L}\}$. This set is a nontrivial linear subspace of E which is invariant under all operators $L_b^{\mathscr{J}}$ with $b \in \mathscr{A}$. Since $L^{\mathscr{J}}$ is an algebraically irreducible representation (by Exercise 1.3.7), one has $\{L_a^{\mathscr{J}}(y) : L_a^{\mathscr{J}} \in \mathscr{L}\} = E$.

Now suppose the assertion of the lemma is not valid for E_1. Then there is a $z \in E \setminus E_1$ such that $L_a^{\mathscr{J}}(z) = 0$ for every a satisfying $L_a^{\mathscr{J}}(E_1) = \{0\}$. Consider the operator B that acts on E according to $Bx := L_a^{\mathscr{J}}(z)$, with a chosen such that $L_a^{\mathscr{J}}(y) = x$. It can easily be checked that B is correctly defined and satisfies the conditions of Corollary 1.3.13. Consequently, B is a scalar operator, i.e., $B = \lambda I$ with a complex λ, and for every $L_a^{\mathscr{J}} \in \mathscr{L}$ we have

$$L_a^{\mathscr{J}}(z) = B L_a^{\mathscr{J}}(y) = \lambda L_a^{\mathscr{J}}(y) \Leftrightarrow L_a^{\mathscr{J}}(z - \lambda y) = 0.$$

By hypothesis, if $L_a^{\mathscr{J}}(\xi) = 0$ for all $L_a^{\mathscr{J}} \in \mathscr{L}$, then $\xi \in E_0$. Thus $z - \lambda y \in E_0$, which contradicts the choice of $z \in E \setminus E_1$. ∎

Lemma 2.6.15. *Let $v_1, \ldots, v_n$ and $e_1, \ldots, e_n$ be elements of E, and suppose that the elements e_k are linearly independent. Then there is an element $a \in \mathscr{A}$ with $L_a^{\mathscr{J}} e_k = v_k$ for all $k = 1, \ldots, n$.*

Proof. It follows from Lemma 2.6.14 that, for each $k = 1, \ldots, n$, there is an $a_k \in \mathscr{A}$ such that $L_{a_k}^{\mathscr{J}}(e_k) \neq 0$ and $L_{a_k}^{\mathscr{J}}(e_m) = 0$ for $m \neq k$. Consider the linear manifold $E_k :=$

$\{L_x^{\mathcal{J}} L_{a_k}^{\mathcal{J}}(e_k) : x \in \mathscr{A}\}$. Since $L_a^{\mathcal{J}}(E_k) \subset E_k$ and $L_e^{\mathcal{J}} L_{a_k}^{\mathcal{J}}(e_k) \neq 0$, we have $E_k = E$ for every k. Hence, there are elements $x_k \in \mathscr{A}$ such that $L_{x_k}^{\mathcal{J}} L_{a_k}^{\mathcal{J}}(e_k) = L_{x_k a_k}^{\mathcal{J}}(e_k) = v_k$ and $L_{x_k a_k}^{\mathcal{J}}(e_m) = 0$ for $m \neq k$. The element $a := \sum_{k=1}^{n} x_k a_k$ has the desired property. ∎

Theorem 2.6.16 (Kaplansky). *Every primitive Banach algebra $\mathscr{A} \in SI_{2n}^\infty$ is isomorphic to $\mathbb{C}^{l \times l}$ for some $l \leq n$.*

Proof. If the algebra $\mathscr{A}$ is primitive, then it contains a maximal left ideal $\mathscr{J}$ for which the corresponding left regular representation $L^{\mathscr{J}} : \mathscr{A} \to \{L_a^{\mathscr{J}} : a \in \mathscr{A}\}$ is an isomorphism. We show that if $\mathscr{A} \in SI_{2n}^\infty$ then dim $\mathscr{A}/\mathscr{J} \leq n$. Set $E := \mathscr{A}/\mathscr{J}$ and suppose dim $E > n$. Let $e_1, \ldots, e_{n+1}$ be linearly independent elements in E. For $i, j, k = 1, \ldots, n+1$, set

$$v_k^{(ij)} := \delta_{jk} e_i$$

where δ_{jk} is the Kronecker symbol. By Lemma 2.6.15, there are elements $a_{i,j} \in \mathscr{A}$ such that $L_{a_{i,j}}^{\mathscr{J}} e_k = v_k^{(ij)}$. Then a simple computation yields

$$S_{2n}\left(L_{a_{n+1,n}}^{\mathscr{J}}, L_{a_{n,n-1}}^{\mathscr{J}}, \ldots, L_{a_{2,1}}^{\mathscr{J}}, L_{a_{1,2}}^{\mathscr{J}}, \ldots, L_{a_{n,n+1}}^{\mathscr{J}}\right) e_{n+1} = e_{n+1},$$

whence

$$\left\| S_{2n}^m\left(L_{a_{n+1,n}}^{\mathscr{J}}, L_{a_{n,n-1}}^{\mathscr{J}}, \ldots, L_{a_{2,1}}^{\mathscr{J}}, L_{a_{1,2}}^{\mathscr{J}}, \ldots, L_{a_{n,n+1}}^{\mathscr{J}}\right) \right\| \geq 1$$

for all m. Since $L^{\mathscr{J}}$ is continuous, this contradicts our assumption that $\mathscr{A} \in SI_{2n}^\infty$. Hence, dim $E \leq n$. But then, clearly, $\mathscr{L}(E) \equiv \mathbb{C}^{l \times l}$ with some $l \leq n$. ∎

The following theorem can be considered as an analog of the Gelfand theory for algebras satisfying a standard polynomial identity.

Theorem 2.6.17. *Let $\mathscr{A} \in SI_{2n}^\infty$ be a unital Banach algebra. Then:*

(i) *for each maximal ideal x of $\mathscr{A}$, the quotient algebra $\mathscr{A}_x := \mathscr{A}/x$ is isomorphic to $\mathbb{C}^{l \times l}$ with a certain $l = l(x)$ less than or equal to n;*

(ii) *an element $a \in \mathscr{A}$ is invertible if and only if the matrices $\Phi_x(a) \in \mathbb{C}^{l(x) \times l(x)}$ are invertible for all maximal ideals x of $\mathscr{A}$ (here we set $\Phi_x := \varphi_x \pi_x$ where π_x denotes the canonical homomorphism from $\mathscr{A}$ onto $\mathscr{A}_x$ and φ_x is an arbitrarily chosen isomorphism from $\mathscr{A}_x$ onto $\mathbb{C}^{l(x) \times l(x)}$, which exists by (i));*

(iii) *the radical of $\mathscr{A}$ coincides with the intersection of all maximal ideals of $\mathscr{A}$.*

Proof. (i) If the algebra itself is primitive then $\mathscr{A}$ is isomorphic to $\mathbb{C}^{l \times l}$ with some $l \leq n$ by Theorem 2.6.16. We will exclude this trivial case. For all x in $M_{\mathscr{A}}$, the quotient algebra $\mathscr{A}_x$ is primitive and belongs to SI_{2n}^∞. Then (i) follows from Theorem 2.6.16.

(ii) If $a \in \mathscr{A}$ is invertible, then $\Phi_x(a)$ is invertible for every $x \in M_{\mathscr{A}}$. To prove the reverse implication, suppose that the $\Phi_x(a)$ are invertible for every $x \in M_{\mathscr{A}}$. Then

$a + x$ is invertible in $\mathscr{A}_x$. Suppose that a is not left invertible. Then a belongs to a left maximal ideal $\mathscr{J}$ by Proposition 1.3.2. Let $L^{\mathscr{J}}$ be the left regular representation induced by $\mathscr{J}$, and set $\mathscr{I} := \mathrm{Ker}\, L^{\mathscr{J}}$. It is evident that $\mathscr{I}$ is an ideal which is contained in $\mathscr{J}$, that the quotient algebra $\mathscr{A}^{\mathscr{J}}$ is primitive, and $\mathscr{A}^{\mathscr{J}} \in SI_{2n}^{\infty}$. Theorem 2.6.16 implies that $\mathscr{A}^{\mathscr{J}}$ is isomorphic to $\mathbb{C}^{l \times l}$, with $l \leq n$, whence the maximality of $\mathscr{I}$. Since x is a subset of $\mathscr{J}$, the image $\mathscr{J}_x := \pi_x(\mathscr{J})$ is a left ideal again, and $\pi_x(a) \in \mathscr{J}_x$. So $\pi_x(a)$ cannot be invertible in $\mathscr{A}_x$. This contradicts the assumption. Hence, a is left invertible.

Let us prove now that a is also right invertible. Since a is left invertible, there is a $b \in \mathscr{A}$ such that $ba = e$. Thus, $\Phi_x(b)\Phi_x(a) = \Phi_x(e)$. Since $\Phi_x(a)$ is invertible in $\mathscr{A}_x$, it follows that $\Phi_x(a)\Phi_x(b) = \Phi_x(e)$ or, equivalently, $ab - e \in x$ for all $x \in M_{\mathscr{A}}$. Since each maximal left ideal contains a maximal ideal by Lemma 1.3.15, the element $r = ab - e$ belongs to the radical of $\mathscr{A}$. By Proposition 1.3.3, the element $ab = e + r$ is invertible, which implies the right invertibility of a.

(iii) The intersection of all maximal ideals belongs to the radical $\mathscr{R}_{\mathscr{A}}$ of $\mathscr{A}$. Conversely let $r \in \mathscr{R}_{\mathscr{A}}$. Then $r_x := r + x$ belongs to $\mathscr{R}_{\mathscr{A}_x}$. Since $\mathscr{A}_x$ is semi-simple, this implies that $r \in x$ for all maximal ideals x. ∎

Remark 2.6.18. The proof of Theorem 2.6.16 made use only of the multilinear property of S_{2n}. Thus the proof holds if instead of S_{2n} one has any multilinear polynomial. This in particular implies that a version of Theorem 2.6.17 is true if the algebra $\mathscr{A}$ is a PI-algebra, because of Lemma 2.6.4. □

Theorem 2.6.19. *Let $\mathscr{A}$ be a unital Banach algebra. The following assertions are equivalent:*

 (i) $\mathscr{A}/\mathscr{R}_{\mathscr{A}}$ *is a PI-algebra;*
 (ii) $\mathscr{A}$ *has a matrix symbol of order n;*
 (iii) $\mathscr{A}/\mathscr{R}_{\mathscr{A}} \in SI_{2n}$;
 (iv) $\mathscr{A} \in SI_{2n}^{\infty}$.

Proof. Suppose that $\mathscr{A}/\mathscr{R}_{\mathscr{A}}$ is a PI-algebra. Then, $\mathscr{A}/\mathscr{R}_{\mathscr{A}}$ satisfies a multilinear polynomial, and thus has a matrix symbol of some order n (see Remark 2.6.18). Assertion (ii) follows due to the equivalence of the invertibility of an element $a \in c\mathscr{A}$, and the invertibility of the coset $a + \mathscr{R}_{\mathscr{A}}$ in $\mathscr{A}/\mathscr{R}_{\mathscr{A}}$. We continue with the implication (ii) ⇒ (iii): Assume that there is a family of matrix-valued homomorphisms h_x on $\mathscr{A}$, labeled by the elements of some set X, such that an element $a \in \mathscr{A}$ is invertible in $\mathscr{A}$ if and only if the matrices $h_x(a)$ are invertible for all $x \in X$.

We claim that if $a \in \mathscr{A}$ is invertible, then $a + S_{2n}(a_1, \ldots, a_{2n})$ is invertible for each choice of elements $a_1, \ldots, a_{2n}$ of $\mathscr{A}$. Since

$$h_x(S_{2n}(a_1, \ldots, a_{2n})) = S_{2n}(h_x(a_1), \ldots, h_x(a_{2n}))$$

and all entries $h_x(a_k)$ are $l \times l$ matrices, Theorem 2.6.11 implies that

$$h_x\left(\mathcal{S}_{2n}(a_1, \ldots, a_{2n})\right) = 0$$

for all $x \in X$. Hence,

$$h_x(a) = h_x(a + \mathcal{S}_{2n}(a_1, \ldots, a_{2n}))$$

for every $x \in X$. Since the h_x constitute a matrix symbol, the claim follows. Then, by Proposition 1.3.3, $\mathcal{S}_{2n}(a_1, \ldots, a_{2n})$ is in the radical of $\mathcal{A}$, whence the assertion.

The implication (iii) $\Rightarrow$ (iv) is a consequence of the fact that elements of the radical have spectral radius 0 and of Theorem 1.2.12. (iii) also trivially implies (i). The final implication (iv) $\Rightarrow$ (ii) comes directly from Theorem 2.6.17.   ∎

Example 2.6.20. Let $\mathcal{A}_0$ denote the set of all bounded linear operators A on l^2, such that the coefficients of the matrix representation $(a_{ij})_{i,j=1}^{\infty}$ of A with respect to the standard basis satisfy the following conditions:

- $a_{ij} = 0$ whenever $i > j$;
- $a_{ij} = 0$ whenever $i < j$ with a finite number of exceptions;
- the limit $\lim_{i \to \infty} a_{ii}$ exists and is finite.

The set $\mathcal{A}_0$, provided with the operations inherited from $\mathcal{L}(l^2)$, forms an algebra. Let $\mathcal{A}$ be the closure of $\mathcal{A}_0$ in $\mathcal{L}(l^2)$. Then $\mathcal{A}$ is a QI-algebra in SI_2^{∞}, the mappings

$$\phi_n : \mathcal{A} \to \mathbb{C}, \quad \phi_n(A) := \begin{cases} a_{nn} & \text{if } n \in \mathbb{N}, \\ \lim_{i \to \infty} a_{ii} & \text{if } n = \infty. \end{cases} \tag{2.27}$$

are continuous homomorphisms, and an element $A \in \mathcal{A}$ is invertible in $\mathcal{A}$ if and only if $\phi_n(A) \neq 0$ for all $n \in \mathbb{N} \cup \{\infty\}$. The proof of these facts is left as an exercise.   □

We conclude this section with two results on the existence of matrix symbols for algebras with a non-trivial center.

Proposition 2.6.21. *Let $\mathcal{A}$ be a unital Banach algebra, $\mathcal{B}$ be a closed subalgebra and $\mathcal{C}$ a subalgebra of the center of $\mathcal{B}$. If*

$$\sup_{x \in M_\mathcal{C}} \dim \mathcal{B}_x =: m < \infty \tag{2.28}$$

then $\mathcal{B} \in \mathrm{IS}(n, \mathcal{A})$ for some $n \leq \sqrt{m}$.

Proof. By hypothesis, $\dim \mathcal{B}_x \leq m$ for any $x \in M_\mathcal{C}$. For $b_1, \ldots, b_{m+1} \in \mathcal{B}$, consider $\mathcal{S}_{m+1}(b_1 + x, \ldots, b_{m+1} + x)$. Let $e_1, \ldots, e_m$ be a basis of $\mathcal{B}_x$. Since every coset $b_i + x$ can be written as a linear combination of the basis elements $e_1, \ldots, e_m$, the multilinearity of the standard polynomial implies that $\mathcal{S}_{m+1}(b_1 + x, \ldots, b_{m+1} + x) = 0$. By Theorem 2.6.17, there exists a positive integer $k \leq \lceil \frac{m+1}{2} \rceil$ such that $\mathcal{B}_x$ has a matrix symbol of order k for all $x \in M_\mathcal{C}$. Consider the commutator

$$\mathscr{D} := \{a \in \mathscr{A} : ac = ca \text{ for every } c \in \mathscr{C}\}$$

of the algebra $\mathscr{C}$. It is easy to see that $\mathscr{D}$ is an inverse-closed Banach subalgebra of $\mathscr{A}$ which contains $\mathscr{B}$ and which has $\mathscr{C}$ in its center. Thus, the local algebras $\mathscr{D}_x$ are well defined for $x \in M_{\mathscr{C}}$. Since dim $\mathscr{B}_x < \infty$, the algebra $\mathscr{B}_x$ is inverse-closed in $\mathscr{D}_x$ (see Exercise 1.2.13). Thus, $\mathscr{B}_x \in \mathrm{IS}(k, \mathscr{D}_x)$ for all $x \in M_{\mathscr{C}}$.

We claim that $\mathscr{B} \in \mathrm{IS}(k, \mathscr{A})$. Indeed, let $\{\Phi_\tau^x\}_{\tau \in \mathscr{T}(x)}$ generate a matrix symbol for $\mathscr{B}_x$ in $\mathscr{D}_x$. For $b \in \mathscr{B}$, set $\Phi_{x,\tau}(b) := \Phi_\tau^x(b_x)$, which defines a homomorphism $\Phi_{x,\tau}$ on $\mathscr{B}$. Since $\mathscr{D}$ is inverse-closed in $\mathscr{A}$, the element $b \in \mathscr{B}$ is invertible in $\mathscr{A}$ if and only if it is invertible in $\mathscr{D}$. By Allan's local principle, b is invertible in $\mathscr{D}$ if and only if b_x is invertible in $\mathscr{D}_x$ for all $x \in M_{\mathscr{C}}$. Thus, b is invertible in $\mathscr{A}$ if and only if $\Phi_{x,\tau}(b)$ is invertible for all $x \in M_{\mathscr{C}}$ and $\tau \in \mathscr{T}(x)$, that is, the family $\{\Phi_{x,\tau}\}$ generates a matrix symbol for $\mathscr{B}$ in $\mathscr{A}$.

The homomorphisms $\Phi_{x,\tau} : \mathscr{A} \mapsto \mathbb{C}^{l(x) \times l(x)}$ obviously satisfy dim Im $\Phi_{x,\tau} \leq$ dim $\mathscr{B}_x$. Consequently, $l^2(x) \leq m$, and the result follows.   ∎

Corollary 2.6.22. *Let $\mathscr{A}$ be a unital Banach algebra, $\mathscr{B}$ a closed subalgebra of $\mathscr{A}$, and $\mathscr{B}^0$ be a dense subalgebra of $\mathscr{B}$. If $\mathscr{B}^0$ is an m-dimensional module over its center, then $\mathscr{B} \in \mathrm{IS}(n, \mathscr{A})$ with a certain $n \leq \sqrt{m}$.*

2.6.3 Exercises

Exercise 2.6.1. Prove that every finite-dimensional algebra $\mathscr{A}$ with dim $\mathscr{A} < n$ satisfies the standard identity of order n.

Exercise 2.6.2. Prove that the mappings ϕ_i in Example 2.6.20 are continuous homomorphisms and that an element $A \in \mathscr{A}$ is invertible in $\mathscr{A}$ if and only if $\phi_i(A) \neq 0$ for all $i \in \mathbb{N} \cup \{\infty\}$.

2.7 Notes and comments

The simplest local principle is the classical Gelfand theory for commutative Banach algebras. It was created by Gelfand in 1941. The original paper is [62]. The modern reader can choose between a few textbooks on Banach algebras which present nice introductions to Gelfand theory; see, for example, [162, 171, 202].

Classical Gelfand theory has found a variety of remarkable applications. For instance, it provides an elegant and relatively simple proof of the famous Wiener theorem which states that if f is a non-vanishing function with an absolutely convergent Fourier series expansion, then its inverse f^{-1} has the same property. It is thus nothing but natural that several attempts were made to establish non-commutative versions of Gelfand's theory.

Representation theory for C^*-algebras can be thought of as a non-commutative generalization of the Gelfand theory which carries much of its spirit: the maximal ideal space is replaced by the spectrum of the algebra (which coincides with the space of the primitive ideals in many cases), and the multiplicative functionals correspond to the irreducible representations.

The needs of operator theory, asymptotic spectral theory and numerical analysis show that one has to go beyond C^*-algebras. Already in 1942, Bochner and Phillips [9] generalized Wiener's theorem to functions with values in a Banach algebra. In their paper, they introduced a new tool which can be considered as the first appearance of a (non-commutative) local principle. When speaking about local principles in this text, we mean a generalization of Gelfand's theory to non-commutative Banach algebras which are not too far away from commutative algebras in the sense that they possess a large center, or that a "higher" commutator property is satisfied.

The idea to consider the cosets of an element a modulo the ideals $\mathscr{I}_x$ as a localization of a, as was done in Section 2.2, is quite old. The first reference we know is Glimm's paper [64] from 1960 where he introduced these ideals in the case that $\mathscr{A}$ is a C^*-algebra and $\mathscr{B}$ is the full center of that algebra, and where he already proved the upper semi-continuity of the mapping $x \mapsto \|a + \mathscr{I}_x\|$ (Theorem 2.2.2 (ii) above). In this setting, the ideals $\mathscr{I}_x$ are also known as *Glimm's ideals* in the literature.

The general version of this kind of localization (where $\mathscr{A}$ is a unital Banach algebra and $\mathscr{B}$ a closed central subalgebra) is known as Allan's local principle (Theorem 2.2.2). It appeared in its original form in [2]. Subsections 2.2.3 and 2.2.4 present some additional features related with local invertibility and inverse-closedness. Note, in connection with Subsection 2.2.3, that Allan's local principle can be used to study the continuity of the spectrum of elements in Banach algebras, as proposed in [50]. The combination of Allan's local principle with ideas presented in [9] leads to a piece of non-commutative Gelfand theory which is crucial in the study of one-sided invertibility of Banach algebra-valued holomorphic functions. The interested reader is directed to the monograph [128].

The material presented in Subsection 2.2.4 can be completed by a result which is of interest in connection with Theorem 2.2.8. The result belongs to Hulanicki [89] (see also [52] for a corrected proof) and reads as follows: Assume that $\mathscr{A}$ is an involutive Banach algebra with identity I_H and contained in $\mathscr{L}(H)$ for some Hilbert space H. If

$$r_{\mathscr{A}}(A) = r(A) = \|A\|_{op},$$

for all self-adjoint elements $A \in \mathscr{A}$, then $\mathrm{sp}_{\mathscr{A}} A = \mathrm{sp} A$ for all $A \in \mathscr{A}$. Moreover, the algebra $\mathscr{A}$ is symmetric, that is, $\mathrm{sp}_{\mathscr{A}} A^* A \subset [0, \infty[$ for all $A \in \mathscr{A}$.

Douglas' local principle (Theorem 2.2.12) is the specification of Allan's local principle to C^*-algebras. It appeared independently in [41]. Originally, Allan's local principle was aimed at the study of invertibility properties of holomorphic Banach algebra-valued functions, whereas Douglas' local principle was used in the study of invertibility properties of Toeplitz operators.

Proposition 2.2.5 goes back to Semenyuta and Khevelev ([179]). The proof presented here is an adaptation of the one presented in [14, Proposition 8.6].

The subject of Section 2.3 is perhaps the most important supplement to Allan's local principle: the concept of norm-preserving localization. It appeared for the first time in a paper by Krupnik [107] in the special context of Simonenko's local principle. This local principle is rather specific and works in the algebra of all bounded linear operators of local type acting on spaces $L^p(X, \mu)$, where X is a Hausdorff space of finite dimension. In [107], Krupnik proposed a sharper version which is true without any restriction to the dimension of X and which already offers norm-preserving localization in the Calkin image of the algebra of all operators of local type. What is called a faithful localizing pair in Subsection 2.3.1 first appeared in a joint paper of Böttcher, Krupnik and one of the authors in [18] under the name KMS-algebra. Theorem 2.3.3 is taken from that paper, whereas Theorems 2.3.4 and 2.3.5 are from [169].

Let us mention an independent circle of papers [33, 85, 194, 195] which were published at about the same time as Simonenko's and Allan's local principles appeared. The aim of these papers is to describe C^*-algebras as continuous fields C^*-algebras.

Gohberg-Krupnik's local principle was published in 1973 [70]. It is distinguished by its simplicity and broad applicability. Assertions (i) and (ii) in Theorem 2.4.5 are due to Gohberg and Krupnik, whereas (iii) appeared in [21]. Assertion (iv) is added by the authors.

As already mentioned, Simonenko's local principle, presented in Section 2.5, was originally formulated for operators of local type acting on Lebesgue spaces $L^p(X, \mu)$ with X a Hausdorff space of finite dimension. In the original paper [185, 186], Simonenko had already introduced the notion of an envelope and presented Theorem 2.5.20 under the assumption that the operator function $x \mapsto A_x$ is continuous. The equivalence of (i) and (ii) in Theorem 2.5.6 had already been mentioned by Seeley in his review of Simonenko's paper [185] (see Mathematical Reviews MR0179630). A partial case of Theorem 2.5.20 is contained in [189]. The proofs presented are in the spirit of [18] and [169].

Allan's local principle, completed by the concept of norm-preserving localization, can be regarded as a far-reaching generalization of Simonenko's local principle. For another generalization of Simonenko's principle we refer to Kozak [102].

Operators of local type on spaces $C^n(X)$ and Hölder spaces $H_\alpha(X)$ with $0 < \alpha \leq 1$ were studied in the Ph.D. thesis of Pöltz [146] and in the papers [116, 142, 143, 144, 145]. He constructed an analog of Simonenko's theory for the above mentioned spaces, none of which is of "local type" in the sense of Subsection 2.5.1. Unfortunately, the results of Pöltz are almost unknown even among the experts.

In Section 2.6 we mainly follow Krupnik's book [108] together with the paper [54] by Finck and two of the authors. There are several proofs of the Amitsur-Levitzki theorem, but unfortunately they are all rather technical. The proof presented here belongs to Razmyslov [159] (see also [42, 170]). In [108], a different proof can be found. The equivalence between (ii) and (iii) in Theorem 2.6.19 is a natural conclusion of both works [54, 108]. The last two results appeared in [75].

One basic question is left open in our exposition, namely whether it is possible to provide the set of the maximal ideals of a Banach PI-algebra with a topology which has (compactness) properties similar to the Gelfand topology in the commutative case. This questions seems to be delicate. Some partial results can be found in [108]. Regarding functional calculus for Banach PI-algebras, see [115].

Chapter 3
Banach algebras generated by idempotents

The goal of this chapter is to provide a possible tool to study the invertibility of the local cosets which arise after localization of convolution type operators. The basic observation is that the local algebras in some cases are generated by a finite number of elements p which are idempotent in the sense that $p^2 = p$. Under some additional conditions, it turns out that such algebras possess matrix-valued symbols. Thus, one can associate with every element of the algebra a matrix-valued function such that the element is invertible if and only if the associated function is invertible at every point. In this way, one gets an effective criterion for the invertibility of elements in local algebras.

A few words to the notions "idempotent" and "projection". In most cases, we use *idempotent* for elements p in an algebra which satisfy $p^2 = p$, whereas *projection* is reserved for a self-adjoint idempotent in an involutive algebra. This choice is in accordance, for example, with the use of these notions in "Shilov's idempotent theorem" and "Halmos' two projections theorem". On the other hand, it is also usual to call a bounded linear operator P on a Banach space a projection if $P = P^2$. We will, therefore, also not be too nitpicking and use the words "idempotent" and "projection" sometimes as synonyms.

3.1 Algebras generated by two idempotents

3.1.1 Motivation: Local algebras generated by two projections

To start with, we consider a simple and transparent situation. Let $C^{2\times2}[0,1]$ denote the C^*-algebra of all 2×2 matrices, the entries of which are continuous functions on the interval $[0,1]$. The functions

$$x \mapsto \begin{bmatrix} 1 & 0 \\ 0 & 0 \end{bmatrix}, \quad x \mapsto \begin{bmatrix} x & \sqrt{x(1-x)} \\ \sqrt{x(1-x)} & 1-x \end{bmatrix}, \quad x \in [0,1],$$

S. Roch et al., *Non-commutative Gelfand Theories*, Universitext,
DOI 10.1007/978-0-85729-183-7_3, © Springer-Verlag London Limited 2011

belong to $C^{2\times 2}[0,1]$, and they are projections, i.e., self-adjoint idempotents, in that algebra. Employing the Stone-Weierstrass theorem, it is not hard to see that the smallest C^*-subalgebra of $C^{2\times 2}[0,1]$ which contains these two projections coincides with the C^*-subalgebra of $C^{2\times 2}[0,1]$, the elements of which are diagonal at 0 and 1. The following remarkable theorem states that this structure is archetypal for all unital C^*-algebras which are generated by two projections p and r with $\sigma(prp) = [0,1]$.

Theorem 3.1.1 (Halmos). *Let $\mathscr{B}$ be a C^*-algebra with identity e, and let p and r be projections in $\mathscr{B}$ such that the smallest closed subalgebra of $\mathscr{B}$ which contains p, r and e coincides with $\mathscr{B}$. Further suppose that the spectrum of prp is $[0,1]$. Then $\mathscr{B}$ is *-isomorphic to the algebra of all continuous matrix-valued functions $[0,1] \to \mathbb{C}^{2\times 2}$ which are diagonal at 0 and 1. The isomorphism can be chosen in such a way that it maps e, p and r into*

$$x \mapsto \begin{bmatrix} 1 & 0 \\ 0 & 1 \end{bmatrix}, \quad x \mapsto \begin{bmatrix} 1 & 0 \\ 0 & 0 \end{bmatrix}, \quad and \quad x \mapsto \begin{bmatrix} x & \sqrt{x(1-x)} \\ \sqrt{x(1-x)} & 1-x \end{bmatrix},$$

respectively.

Before proceeding with generalizations of Halmos' two projections theorem, we would like to illustrate how this theorem can be employed in the context of Section 2.2.6. In this section, we analyzed the algebra $\mathscr{A} = \text{alg}(S, PC(\mathbb{T}))$, which we here consider as a subalgebra of $\mathscr{L}(L^2(\mathbb{T}))$. Localization via Allan's local principle yields, for every $t \in \mathbb{T}$, a local algebra $\mathscr{A}_t^{\mathscr{K}}$ which is generated by the local cosets $\Phi_t^{\mathscr{K}}(\chi_t I)$ and $\Phi_t^{\mathscr{K}}(P)$ and by the identity element $\Phi_t^{\mathscr{K}}(I)$. Since these cosets are projections, every local algebra $\mathscr{A}_t^{\mathscr{K}}$ is a particular example of a unital C^*-algebra generated by two projections. In order to apply Halmos' theorem to these local algebras one has to know if

$$\sigma_{\mathscr{A}_t^{\mathscr{K}}}(\chi_t P \chi_t I) = [0,1]. \tag{3.1}$$

Fortunately, this fact is well known. A proof using properties of Toeplitz operators can be found in [82, Propositions 4.18 and 4.41]. It is also not hard to give direct proof of (3.1) using a homogenization argument. Indeed, in Section 4.2.5 we will compute this local spectrum, but for the singular integral projection acting on the real line. The translation of this result to the singular integral projection on the unit circle and, thus, the derivation of (3.1), can be done via the mapping

$$(Bf)(z) := \frac{1}{z-1} f\left(\mathbf{i}\frac{z+1}{z-1}\right)$$

which is a linear homeomorphism from $L^2(\mathbb{R})$ onto $L^2(\mathbb{T})$ and has the property that $B^{-1}SB = -S_{\mathbb{R}}$. The details are left to the reader.

Thus, Halmos' two projections theorem implies that the singular integral operator $aP + bQ$ with $Q = I - P$ is locally invertible at $t \in \mathbb{T}$ if and only if the associated matrix function

$$x \mapsto \begin{bmatrix} a_+ & 0 \\ 0 & a_- \end{bmatrix} \begin{bmatrix} x & \sqrt{x(1-x)} \\ \sqrt{x(1-x)} & 1-x \end{bmatrix}$$

$$+ \begin{bmatrix} b_+ & 0 \\ 0 & b_- \end{bmatrix} \begin{bmatrix} 1-x & -\sqrt{x(1-x)} \\ -\sqrt{x(1-x)} & x \end{bmatrix}$$

$$= \begin{bmatrix} a_+x + b_+(1-x) & (a_+ - b_+)\sqrt{x(1-x)} \\ (a_- - b_-)\sqrt{x(1-x)} & a_-(1-x) + b_-x \end{bmatrix},$$

with $a_\pm$ referring to the one-sided limits $a(t^\pm)$, is invertible on $[0, 1]$. In combination with the local principle, this gives the desired criterion for the Fredholmness of the singular integral operator $aP + bQ$ with piecewise continuous coefficients and, more generally, for the Fredholmness of any operator in $\mathscr{A}(S, PC(\mathbb{T}))$.

There is another convenient approach to studying local algebras via homogenization techniques. We will consider this alternate approach in detail in the next chapter. Let us mention already here that it is a striking advantage of the approach via the two projections theorem that it also applies to the local invertibility of singular integral operators on more general spaces (actually on arbitrary L^p-spaces with Muckenhoupt weights over Carleson curves) and also to a wealth of further local invertibility problems (arising in numerical analysis, for example). One disadvantage of this approach (again in comparison to homogenization) is that it only yields criteria for (local) invertibility. Hence, using this approach, one only gets information on the structure of the local algebras up to elements in the radical.

3.1.2 A general two idempotents theorem

It is evidently desirable to have at one's disposal an analog of Halmos' theorem which holds for unital Banach algebras generated by two idempotents. One potential application of such a theorem could be the study of the Fredholm property of singular integral operators on L^p-spaces for $p \neq 2$. We shall see in this section that there is indeed a general *two idempotents theorem* which associates with every element of a Banach algebra (generated by the identity element and by two idempotents p and r)[1] a 2×2 matrix-valued function such that the element is invertible if and only if the associated function is invertible.

There are at least two approaches to attack the general two idempotents theorem, each of which is based on one of the following observations:

- Every algebra $\mathscr{B}$ generated by the identity e and by two idempotents p and r has the element $c := prp + (e-p)(e-r)(e-p)$ in its center. This observation offers a way to apply Allan's local principle to analyse the algebra $\mathscr{B}$ by localization over the spectrum of c.

---

[1] The projection represented by p has obviously nothing to do with the index of the L^p-space. Both are represented by the same letter, but is clear what is meant when the symbol appears.

- The algebra $\mathscr{B}$ satisfies the standard identity S_4. This fact renders the algebra $\mathscr{B}$ accessible to Krupnik's generalization of the Gelfand theory to PI-algebras.

For an application of Allan's principle we refer to Section 3.2. In this section, we will follow the second way. The basis for this approach is provided by the following proposition and its corollary.

Proposition 3.1.2. *Let $\mathscr{A}$ be an algebra which is generated by the identity element e and by the idempotents p and r. Then $\mathscr{A}$ satisfies the standard polynomial identity S_4.*

Proof. One easily checks that the element

$$c := prp + (e-p)(e-r)(e-p) = e - p - r + pr + rp$$

belongs to the center of the algebra $\mathscr{A}$:

$$pc = prp = cp \quad \text{and} \quad rc = rpr = cr.$$

Further, each element of $\mathscr{A}$ can be written as

$$h_1(c)e + h_2(c)p + h_3(c)r + h_4(c)pr \tag{3.2}$$

with polynomials $h_1, \ldots, h_4$. Thus, the algebra $\mathscr{A}$ is a module over its center of dimension not greater than four. To see this, note that the generating elements e, p, r of $\mathscr{A}$ are of the form (3.2) and that the set of all elements of $\mathscr{A}$ which are of the form (3.2) is a subalgebra of $\mathscr{A}$. The latter follows from the multiplication table

	p	r	pr
p	p	pr	pr
r	$c - e + p + r - pr$	r	cr
pr	cp	pr	cpr

Since S_4 is multilinear, and since all polynomials $h_i(c)$ belong to the center of $\mathscr{A}$, it remains to show that

$$S_4(b_1, b_2, b_3, b_4) = 0 \quad \text{whenever} \quad \{b_1, b_2, b_3, b_4\} \subseteq \{e, p, r, pr\}.$$

This can be easily done: If two of the b_i coincide, then clearly, $S_4(b_1, b_2, b_3, b_4) = 0$. Otherwise, one of the b_i must be the identity element, say $b_1 = e$. In this case one has $S_4(e, a_2, a_3, a_4) = 0$ for arbitrary elements a_2, a_3, a_4 of $\mathscr{A}$. ∎

Corollary 3.1.3. *Let $\mathscr{B}$ be a Banach algebra which is generated (as a Banach algebra) by the identity element e and by the idempotents p and r. Then $\mathscr{B}$ satisfies the standard polynomial identity S_4.*

This follows immediately from the preceding proposition and the continuity of the standard polynomial S_4.

Now let $\mathscr{B}$ be a Banach algebra which is generated by the identity element e and by the idempotents p and r. From Corollary 3.1.3 and Theorem 2.6.17 we infer that, for each maximal ideal x of $\mathscr{B}$,

$$\text{either} \quad \mathscr{B}/x \cong \mathbb{C}^{1\times 1} \quad \text{or} \quad \mathscr{B}/x \cong \mathbb{C}^{2\times 2}.$$

For $i = 1, 2$, let M_i denote the set of all maximal ideals x of $\mathscr{B}$ with $\mathscr{B}/x \cong \mathbb{C}^{i\times i}$. As in Theorem 2.6.17, for each $x \in M_i$, we choose an isomorphism ξ_x from $\mathscr{B}/x$ onto $\mathbb{C}^{i\times i}$, and we define

$$\Phi_x : \mathscr{B} \to \mathbb{C}^{i\times i}, \quad a \mapsto \xi_x(a+x).$$

By Theorem 2.6.17 again, the family $\{\Phi_x\}_{x\in M_i, i=1,2}$ forms a sufficient family of homomorphisms of the algebra $\mathscr{B}$.

We are going to describe the elements of this family up to similarity, that is, the possible form of the representations. Since homomorphisms map idempotents to idempotents, the restriction of each homomorphism Φ_x with $x \in M_1$ onto the set $\{e, p, r\}$ coincides with one of the following mappings $G_0, \ldots, G_3 : \{e, p, r\} \to \mathbb{C}$ given by

$$
\begin{aligned}
G_0(e) &= 1, \; G_0(p) = 0, \; G_0(r) = 0, \\
G_1(e) &= 1, \; G_1(p) = 1, \; G_1(r) = 0, \\
G_2(e) &= 1, \; G_2(p) = 0, \; G_2(r) = 1, \\
G_3(e) &= 1, \; G_3(p) = 1, \; G_3(r) = 1.
\end{aligned}
$$

Conversely, every continuous homomorphism $\Phi : \mathscr{B} \to \mathbb{C}$ is uniquely determined by its action on the set $\{e, p, r\}$ of the generators of $\mathscr{B}$. Thus, there are at most four elements in M_1.

Now let $x \in M_2$. The idempotent $\Phi_x(p)$ must have 0 *and* 1 as its eigenvalues (otherwise $\Phi_x(p)$ is the zero or the identity matrix, and the range of Φ_x is a commutative algebra which cannot coincide with $\mathbb{C}^{2\times 2}$). Thus, there is an invertible matrix $B_x \in \mathbb{C}^{2\times 2}$ such that

$$\Phi_x(p) = B_x^{-1} \begin{bmatrix} 1 & 0 \\ 0 & 0 \end{bmatrix} B_x.$$

We replace Φ_x by the mapping

$$\tilde{\Phi}_x : \mathscr{A} \to \mathbb{C}^{2\times 2}, \quad a \mapsto B_x \Phi_x(a) B_x^{-1}$$

which is also a homomorphism and which has the property that $\tilde{\Phi}_x(a)$ is invertible if and only if $\Phi_x(a)$ is invertible. Consider

$$\tilde{\Phi}_x(r) =: \begin{bmatrix} \alpha & \beta \\ \gamma & \delta \end{bmatrix}$$

with complex numbers α, β, γ and δ. We must have $\beta\gamma \neq 0$ since otherwise

$$\mathrm{alg}\{\tilde{\Phi}_x(e), \tilde{\Phi}_x(p), \tilde{\Phi}_x(r)\}$$

is an algebra of (upper or lower) triangular matrices which cannot coincide with $\mathbb{C}^{2\times 2}$. From the idempotent property of $\tilde{\Phi}_x(r)$ one then easily concludes that

$$\tilde{\Phi}_x(r) = \begin{bmatrix} \alpha & \varepsilon_x\sqrt{\alpha(1-\alpha)} \\ \varepsilon_x^{-1}\sqrt{\alpha(1-\alpha)} & 1-\alpha \end{bmatrix}$$

with a non-zero complex number ε_x. Here, $\sqrt{\alpha(1-\alpha)}$ denotes a complex number with $(\sqrt{\alpha(1-\alpha)})^2 = \alpha(1-\alpha)$. Define

$$\Psi_x : \mathscr{B} \to \mathbb{C}^{2\times 2}, \quad a \mapsto \begin{bmatrix} 1 & 0 \\ 0 & \varepsilon_x \end{bmatrix} \tilde{\Phi}_x(a) \begin{bmatrix} 1 & 0 \\ 0 & \varepsilon_x^{-1} \end{bmatrix}.$$

Then Ψ_x is an isomorphism, $\Psi_x(a)$ is invertible if and only if $\Phi_x(a)$ is invertible, and

$$\Psi_x(e) = \begin{bmatrix} 1 & 0 \\ 0 & 1 \end{bmatrix}, \ \Psi_x(p) = \begin{bmatrix} 1 & 0 \\ 0 & 0 \end{bmatrix}$$

and

$$\Psi_x(r) = \begin{bmatrix} \alpha & \sqrt{\alpha(1-\alpha)} \\ \sqrt{\alpha(1-\alpha)} & 1-\alpha \end{bmatrix}$$

with a certain $\alpha \in \mathbb{C}\setminus\{0,1\}$. Thus, if $x \in M_2$, then Φ_x is equivalent (up to similarity) to a homomorphism Ψ_x, the restriction of which onto the set $\{e, p, r\}$ coincides with one of the mappings $F_\alpha : \{e, p, r\} \to \mathbb{C}^{2\times 2}$ given by

$$F_\alpha(e) = \begin{bmatrix} 1 & 0 \\ 0 & 1 \end{bmatrix}, \qquad F_\alpha(p) = \begin{bmatrix} 1 & 0 \\ 0 & 0 \end{bmatrix}, \qquad F_\alpha(r) = \begin{bmatrix} \alpha & \sqrt{\alpha(1-\alpha)} \\ \sqrt{\alpha(1-\alpha)} & 1-\alpha \end{bmatrix},$$

with $\alpha \in \mathbb{C}\setminus\{0,1\}$.

Finally, one has to decide which of the mappings G_m with $m \in \{0,1,2,3\}$ and F_α with $\alpha \in \mathbb{C}\setminus\{0,1\}$ are indeed restrictions of homomorphisms of $\mathscr{B}$. This can be done by means of the *indicator elements*

$$b := p + 2r \quad \text{and} \quad c = e - p - r + pr + rp.$$

One easily checks that

$$G_m(b) := G_m(p) + 2G_m(r) = m \quad \text{for} \quad m = 0,1,2,3$$

and that none of the numbers $0,1,2,3$ is in the spectrum of $F_\alpha(b) := F_\alpha(p) + 2F_\alpha(q)$ if $\alpha \notin \{0,1\}$. Since every point in the spectrum of b must be obtained as a point in the spectrum of $\Phi_x(b)$ with $x \in M_1$ or $\Psi_x(b)$ with $x \in M_2$, we conclude that each point m in

$$\sigma(b) \cap \{0,1,2,3\}$$

must be obtained by a one-dimensional representation, and since only $G_m(b)$ yields m, we even have a one-to-one correspondence between M_1 and $\sigma(b) \cap \{0, 1, 2, 3\}$.

Similarly, for c one easily checks that

$$G_m(c) := G_m(e) - G_m(p) - G_m(r) + G_m(p)G_m(r) + G_m(r)G_m(p)$$

is in $\{0, 1\}$ for each choice of $m \in \{0, 1, 2, 3\}$. Thus the points in $\sigma(c) \setminus \{0, 1\}$ can only be obtained by two-dimensional representations $\Psi_x(c)$. From

$$F_\alpha(c) := F_\alpha(e) - F_\alpha(p) - F_\alpha(r) + F_\alpha(p)F_\alpha(r) + F_\alpha(r)F_\alpha(p) = \begin{bmatrix} \alpha & 0 \\ 0 & \alpha \end{bmatrix}$$

we conclude that each point $\alpha \in \sigma(c) \setminus \{0, 1\}$ induces one of the mappings F_α and that, conversely, each of the mappings F_α (consequently, each of the Ψ_x) can contribute only one point to $\sigma(c) \setminus \{0, 1\}$. Thus, there is a one-to-one correspondence between M_2 and $\sigma(c) \setminus \{0, 1\}$.

Finally we consider what happens if one of the points 0 or 1 belongs to $\sigma(c)$ but is not isolated in $\sigma(c)$. Suppose that 0 has this property. Then there is a sequence $(x_n) \subset \sigma(c) \setminus \{0\}$ such that $x_n \to 0$ as $n \to \infty$. We determine the spectrum of $F_{x_n}(b)$, i.e. the solutions of the equation

$$\det \begin{bmatrix} 2x_n + 1 - \lambda & 2\sqrt{x_n(1 - x_n)} \\ 2\sqrt{x_n(1 - x_n)} & 2(1 - x_n) - \lambda \end{bmatrix}$$
$$= (2x_n + 1 - \lambda)(2 - 2x_n - \lambda) - 4x_n(1 - x_n) = 0.$$

Since the roots of a polynomial depend continuously on the coefficients, the solutions λ_n and μ_n of the above equations tend to the solutions of the equation $(1 - \lambda)(2 - \lambda) = 0$ which comes from the above equation by letting x_n converge to 0. Thus, $\lambda_n \to 1$ and $\mu_n \to 2$, whence $1, 2 \in \sigma(b)$. In the same fashion one proves that $0, 3 \in \sigma(b)$ if 1 is in $\sigma(c)$ and not isolated in $\sigma(c)$.

We summarize these facts in the following theorem. In our terminology, this theorem describes algebras generated by two idempotents. But in what follows, we will refer to this theorem as the "two projections theorem", since it is a generalization of Halmos' theorem which is always referred to as "two projections theorem" in the literature.

Theorem 3.1.4 (two projections theorem). *Let $\mathscr{A}$ be a Banach algebra with identity e, and let p and r be idempotents in $\mathscr{A}$. The smallest closed subalgebra of $\mathscr{A}$ which contains p, r and e will be denoted by $\mathscr{B}$. Then:*

(i) *for each $x \in \sigma_{\mathscr{B}}(e - p - r + pr + rp) \setminus \{0, 1\}$, the mapping*

$$F_x : \{e, p, r\} \to \mathbb{C}^{2 \times 2},$$

given by

$$F_x(e) = \begin{bmatrix} 1 & 0 \\ 0 & 1 \end{bmatrix}, \quad F_x(p) = \begin{bmatrix} 1 & 0 \\ 0 & 0 \end{bmatrix}, \quad F_x(r) = \begin{bmatrix} x & \sqrt{x(1-x)} \\ \sqrt{x(1-x)} & 1-x \end{bmatrix},$$

where $\sqrt{x(1-x)}$ denotes any number with $(\sqrt{x(1-x)})^2 = x(1-x)$, extends to a continuous algebra homomorphism from $\mathscr{B}$ onto $\mathbb{C}^{2 \times 2}$ which we denote also by F_x;

(ii) for each $m \in \sigma_{\mathscr{B}}(p+2r) \cap \{0, 1, 2, 3\}$, the mapping

$$G_m : \{e, p, r\} \to \mathbb{C},$$

given by

$$G_0(e) = 1, \; G_0(p) = G_0(r) = 0, \quad G_1(e) = G_1(p) = 1, \; G_1(r) = 0,$$

$$G_2(e) = G_2(r) = 1, \; G_2(p) = 0, \quad G_3(e) = G_3(p) = G_3(r) = 1$$

extends to a continuous algebra homomorphism from $\mathscr{B}$ onto $\mathbb{C}$;

(iii) an element $a \in \mathscr{B}$ is invertible in $\mathscr{B}$ if and only if the matrices $F_x(a)$ are invertible for all $x \in \sigma_{\mathscr{B}}(e-p-r+pr+rp) \setminus \{0, 1\}$ and if the numbers $G_m(a)$ are non-zero for all $m \in \sigma_{\mathscr{B}}(p+2r) \cap \{0, 1, 2, 3\}$;

(iv) if 0 and 1 are not isolated points in the spectrum of c then each of the homomorphisms G_m, $m = 0, 1, 2, 3$, occurs.

Theorem 3.1.4 can be completed as follows.

Corollary 3.1.5. Let $\mathscr{B}_p$ and $\mathscr{B}_{e-p}$ be the Banach algebras $\{pbp : b \in \mathscr{B}\}$ and $\{(e-p)b(e-p) : b \in \mathscr{B}\}$, respectively.

(i) $\sigma_{\mathscr{B}}(c) \setminus \{0,1\} = \sigma_{\mathscr{B}_p}(prp) \setminus \{0,1\} = \sigma_{\mathscr{B}_{e-p}}((e-p)(e-r)(e-p)) \setminus \{0,1\}$.

(ii) If $\{0,1\} \subset \sigma_{\mathscr{B}_p}(prp)$ then $\sigma_{\mathscr{B}}(c) = \sigma_{\mathscr{B}_p}(prp)$. The same is true if one replaces prp by $(e-p)(e-r)(e-p)$.

(iii) If the closure of one of the sets $\sigma_{\mathscr{B}}(c) \setminus \{0,1\}$, $\sigma_{\mathscr{B}_p}(prp) \setminus \{0,1\}$, $\sigma_{\mathscr{B}_{e-p}}((e-p)(e-r)(e-p)) \setminus \{0,1\}$ contains the points 0 and 1, then these sets coincide.

Proof. Take $\alpha \in \sigma_{\mathscr{B}}(c) \setminus \{0,1\}$. A simple computation gives

$$F_\alpha(c) = \begin{bmatrix} \alpha & 0 \\ 0 & \alpha \end{bmatrix}, F_\alpha(prp) = \begin{bmatrix} \alpha & 0 \\ 0 & 0 \end{bmatrix}, \text{ and } F_\alpha((e-p)(r-p)(e-p)) = \begin{bmatrix} 0 & 0 \\ 0 & \alpha \end{bmatrix}.$$

Then assertion (i) of Theorem 3.1.4 immediately implies (i). For assertion (ii), note that

$$\sigma_{\mathscr{B}}(c) = \sigma_{\mathscr{B}_p}(prp) \cup \sigma_{\mathscr{B}_{e-p}}((e-p)(e-r)(e-p))$$

which, in combination with (i), gives (ii). Assertion (iii) is obvious. ∎

3.1.3 Exercises

Exercise 3.1.1. Show that an element p of a C^*-algebra is a projection if and only if $p = p^* p$.

Exercise 3.1.2. Consider a piecewise constant function a on $\mathbb{T}$ which takes only the values 0 and 1. Assume that the sets $\{t \in \mathbb{T} : a(t) = 0\}$ and $\{t \in \mathbb{T} : a(t) = 1\}$ have positive measure. Show that $\sigma(PaP) = [0, 1]$, and describe the smallest closed subalgebra of $\mathscr{L}(L^2(\mathbb{T}))$ which is generated by aI and by the singular integral operator S defined in Section 1.4.4.

Exercise 3.1.3. Describe the set of all idempotent 2×2 matrices.

Exercise 3.1.4. Determine the smallest number l such that $\mathbb{C}^{2 \times 2}$ is generated by l idempotents. Answer the same question for $\mathbb{C}^{n \times n}$ with $n > 2$.

Exercise 3.1.5. Let the notation be as in the two projections theorem. For each subset M of $\{0, 1, 2, 3\}$, provide an example where $\sigma_{\mathscr{B}}(p + 2r) = M$. Thus, all possible combinations of one-dimensional representations occur. Prove the corresponding assertion for the two-dimensional representations.

3.2 An N idempotents theorem

When trying to derive generalizations of the two projections theorem discussed in Section 3.1 to algebras generated by three and more idempotents, one soon realizes that, in general, such algebras do not possess a matrix-valued symbol at all. On the other hand, in specific applications one often has additional relations between the generating idempotents of the algebra which then guarantee the existence of a matrix symbol. Here we will consider one version of an N idempotents theorem which is motivated by properties of singular integral operators on composed curves.

3.2.1 Algebras generated by three idempotents

As we have seen, every algebra generated by two idempotents possesses a matrix symbol of order 2. This simple picture changes substantially when one more idempotent is added to the generators. There are several examples which illustrate that, in general, algebras generated by three idempotents cannot possess a matrix symbol of a fixed order. So it is an easy exercise to show that, for every $n \geq 3$, the algebra $\mathbb{C}^{n \times n}$ is generated by three idempotent matrices (see also [109]). Moreover, one has the following characterization of algebras generated by three idempotents. Recall that a topological space is *separable* if it possesses a countable dense subset.

Theorem 3.2.1. *Every separable Banach algebra is isomorphic to a subalgebra of a Banach algebra generated by three idempotents.*

The proof is in [12]. The converse of the theorem is evidently true: A Banach algebra which is generated by three idempotents is separable.

Theorem 3.2.1 indicates that the category of all Banach algebras generated by three idempotents is extremely large, and that these algebras will usually show a rather involved structure. There is therefore no hope to derive some substantial results for this case. But, as we will see in the following sections, there are many particular situations where some additional relations between the generators of the algebra exist, and these additional relations render the algebra accessible.

It turns out that these additional conditions have to be rather strong in order to guarantee the existence of matrix symbols. Indeed, even if the three idempotents are orthogonal projections on a Hilbert space, and even if two of them commute, a matrix symbol does not need to exist. For a particular situation where this happens see Example 5.2.3 in Chapter 5.

3.2.2 Choice of the additional conditions

Let $\mathscr{A}$ be a Banach algebra with identity element e, and let $p_1, \ldots, p_{2N}$ be a partition of the identity into non-zero idempotents, i.e., we suppose that $p_i \neq 0$ for all i,

$$p_i \cdot p_j = \delta_{ij} p_i \quad \text{for all } i, j \tag{3.3}$$

where δ_{ij} is the Kronecker delta, and

$$p_1 + p_2 + \ldots + p_{2N} = e. \tag{3.4}$$

Further, let P be an idempotent element of $\mathscr{A}$, set $Q := e - P$ and $p_{2N+1} := p_1$, and suppose that the conditions

$$P(p_{2i-1} + p_{2i})P = (p_{2i-1} + p_{2i})P \tag{3.5}$$

and

$$Q(p_{2i} + p_{2i+1})Q = (p_{2i} + p_{2i+1})Q \tag{3.6}$$

hold for all $i = 1, \ldots, N$. In what follows it will be convenient to define the idempotents p_k for every integer k by $p_k := p_l$ with $l \in \{1, \ldots, 2N\}$ whenever $k - l$ is divisible by $2N$. It is clear that then (3.5) and (3.6) hold for all integers i.

The algebra $\mathscr{B}$ we are interested in is the smallest closed subalgebra of $\mathscr{A}$ which contains the set $\{p_i\}_{i=1}^{2N}$ as well as the element P. Observe that $\mathscr{B}$ contains the identity element e due to (3.4) and that P and Q are complementary idempotents. Indeed, adding the identities (3.5) for $i = 1, \ldots, N$ and taking into account (3.4) yields $P^2 = P$. Thus, $\mathscr{B}$ is actually an algebra generated by the $2N + 1$ idempotents $p_1, \ldots, p_{2N}$ and P.

Example 3.2.2. If $N = 1$, then the partition $\{p_i\}$ consists of two elements p_1 and p_2 with $p_2 = e - p_1$. Moreover, the axioms (3.5) and (3.6) reduce to $P^2 = P$ and $Q^2 = Q$, respectively. Thus, Banach algebras generated by two idempotents are particular cases of the situation described above.   □

Example 3.2.3. Let $\mathscr{A}$ denote the algebra of all continuous functions on $[0,1]$ which take values in $\mathbb{C}^{4\times 4}$. Define idempotents $p_1, \ldots, p_4$ and P in $\mathscr{A}$ by

$$p_1(t) := \mathrm{diag}\{1,0,0,0\}, \quad p_2(t) := \mathrm{diag}\{0,1,0,0\},$$
$$p_3(t) := \mathrm{diag}\{0,0,1,0\}, \quad p_4(t) := \mathrm{diag}\{0,0,0,1\}$$

and

$$P(t) := \begin{bmatrix} t & t-1 & t-1 & t-1 \\ -t & 1-t & 1-t & 1-t \\ t & t & t & t-1 \\ -t & -t & -t & 1-t \end{bmatrix}$$

for $t \in [0,1]$. Then conditions (3.3) and (3.4) obviously hold with $N = 2$. A straightforward computation shows that also (3.5) and (3.6) are fulfilled.   □

The original motivation to consider (3.3)–(3.6) as the additional conditions needed for an N projections theorem came from singular integral operators with piecewise constant coefficients on composed curves. We will briefly discuss the related issues after the proof of Proposition 4.2.19 in Chapter 4.

3.2.3 The N projections theorem

If conditions (3.3)–(3.6) are satisfied then the algebra $\mathscr{B}$ possesses a matrix symbol of order N. For a precise statement of this result, set

$$X := \sum_{i=1}^{N} (p_{2i-1}Pp_{2i-1} + p_{2i}Qp_{2i}) \tag{3.7}$$

and

$$Y := \sum_{i=1}^{N} (p_{2i-1}P + p_{2i}Q) + \sum_{i=1}^{2N} (2i-1)p_i. \tag{3.8}$$

These elements will play the role of indicator elements, as the elements $e - p - r + pr + rp$ and $p + 2r$ did in the two projections theorem.

Theorem 3.2.4 (N projections theorem). *Let $\mathscr{A}$ be a Banach algebra with identity e, and let $p_1, \ldots, p_{2N}$ and P be non-zero elements of $\mathscr{A}$ satisfying (3.3)–(3.6). Further, let $\mathscr{B}$ stand for the smallest closed subalgebra of $\mathscr{A}$ containing the elements P and $p_1, \ldots, p_{2N}$. Then the following assertions hold.*

(i) *If $x \in \sigma_{\mathscr{B}}(X) \setminus \{0, 1\}$, then the mapping $F_x : \{P, p_1, \ldots, p_{2N}\} \to \mathbb{C}^{2N\times 2N}$ given by*

$$F_x(p_i) := \text{diag}\,(0, \ldots, 0, 1, 0, \ldots, 0),$$

with the 1 standing at the ith place, and

$$F_x(P) := \text{diag}\,(1, -1, 1, -1, \ldots, 1, -1) \times$$

$$\begin{bmatrix} x & x-1 & x-1 & x-1 & \cdots & x-1 & x-1 \\ x & x-1 & x-1 & x-1 & \cdots & x-1 & x-1 \\ x & x & x & x-1 & \cdots & x-1 & x-1 \\ x & x & x & x-1 & \cdots & x-1 & x-1 \\ \vdots & \vdots & \vdots & \vdots & \ddots & \vdots & \vdots \\ x & x & x & x & \cdots & x & x-1 \\ x & x & x & x & \cdots & x & x-1 \end{bmatrix}$$

extends to a continuous algebra homomorphism from $\mathscr{B}$ onto $\mathbb{C}^{2N \times 2N}$.

(ii) If $m \in \sigma_{\mathscr{B}}(Y) \cap \{1, \ldots, 4N\}$, then the mapping $G_m : \{P, p_1, \ldots, p_{2N}\} \to \mathbb{C}$ defined by

$$G_{4k}(p_i) := \begin{cases} 1 & \text{if } i = 2k, \\ 0 & \text{if } i \neq 2k, \end{cases} \qquad\qquad G_{4k}(P) := 0,$$

$$G_{4k-1}(p_i) := \begin{cases} 1 & \text{if } i = 2k, \\ 0 & \text{if } i \neq 2k, \end{cases} \qquad\qquad G_{4k-1}(P) := 1,$$

$$G_{4k-2}(p_i) := \begin{cases} 1 & \text{if } i = 2k-1, \\ 0 & \text{if } i \neq 2k-1, \end{cases} \qquad\qquad G_{4k-2}(P) := 1,$$

$$G_{4k-3}(p_i) := \begin{cases} 1 & \text{if } i = 2k-1, \\ 0 & \text{if } i \neq 2k-1, \end{cases} \qquad\qquad G_{4k-3}(P) := 0,$$

where $k = 1, \ldots, N$, extends to a continuous algebra homomorphism from $\mathscr{B}$ onto $\mathbb{C}$.

(iii) An element $b \in \mathscr{B}$ is invertible in $\mathscr{B}$ if and only if the matrices $F_x(b)$ are invertible for all $x \in \sigma_{\mathscr{B}}(X) \setminus \{0, 1\}$ and if the numbers $G_m(b)$ are non-zero for all $m \in \sigma_{\mathscr{B}}(Y) \cap \{1, \ldots, 4N\}$.

(iv) An element $b \in \mathscr{B}$ is invertible in $\mathscr{A}$ if and only if the matrices $F_x(b)$ are invertible for all $x \in \sigma_{\mathscr{A}}(X) \setminus \{0, 1\}$ and if the numbers $G_m(b)$ are non-zero for all $m \in \sigma_{\mathscr{A}}(Y) \cap \{1, \ldots, 4N\}$.

We split the proof of this theorem into several steps.

3.2.4 Algebraic structure of $\mathscr{B}$

We start with examining the smallest (not necessarily closed) subalgebra $\mathscr{B}^0$ which contains the idempotents p_i for $1 \leq i \leq 2N$ and the idempotent P.

Proposition 3.2.5. *The element X defined by* (3.7) *lies in the center of $\mathscr{B}^0$.*

Proof. Evidently, X commutes with each of the idempotents p_i. It remains to show that $PX = XP$. Let us first prove that

$$X = \sum_{i=1}^{N} \left((p_{2i} + p_{2i+1})Qp_{2i}Q + (p_{2i-1} + p_{2i})Pp_{2i-1}P \right). \tag{3.9}$$

Since the p_j form a partition of identity, it is sufficient to prove that

$$p_j X = p_j \sum_{i=1}^{N} \left((p_{2i} + p_{2i+1})Qp_{2i}Q + (p_{2i-1} + p_{2i})Pp_{2i-1}P \right)$$

for $j = 1, \ldots, 2N$ or, equivalently, that

$$p_{2i}Qp_{2i} = p_{2i}Qp_{2i}Q + p_{2i}Pp_{2i-1}P \tag{3.10}$$

and

$$p_{2i-1}Pp_{2i-1} = p_{2i-1}Qp_{2i-2}Q + p_{2i-1}Pp_{2i-1}P \tag{3.11}$$

for $i = 1, \ldots, N$. For (3.10) observe that

$$
\begin{aligned}
p_{2i}Qp_{2i}Q + p_{2i}Pp_{2i-1}P &= p_{2i}Q - p_{2i}Pp_{2i}Q + p_{2i}Pp_{2i-1}P \\
&= p_{2i}Q - p_{2i}Pp_{2i} + p_{2i}Pp_{2i}P + p_{2i}Pp_{2i-1}P \\
&= p_{2i}Q - p_{2i}Pp_{2i} + p_{2i}P(p_{2i-1} + p_{2i})P \\
&= p_{2i}Q - p_{2i}Pp_{2i} + p_{2i}P \\
&= p_{2i} - p_{2i}Pp_{2i} = p_{2i}(P + Q)p_{2i} - p_{2i}Pp_{2i} \\
&= p_{2i}Qp_{2i},
\end{aligned}
$$

and (3.11) follows analogously. Thus (3.9) holds. Further, axioms (3.5) and (3.6) imply

$$Q(p_{2i-1} + p_{2i})P = P(p_{2i} + p_{2i+1})Q = 0 \quad \text{for} \quad i = 1, \ldots, N, \tag{3.12}$$

and the axioms (3.5), (3.6) together with the identities (3.9), (3.12) yield

$$
\begin{aligned}
PX &= P \cdot \sum_{i=1}^{N} \left((p_{2i} + p_{2i+1})Qp_{2i}Q + (p_{2i-1} + p_{2i})Pp_{2i-1}P \right) \\
&= \sum_{i=1}^{N} (p_{2i-1} + p_{2i})Pp_{2i-1}P,
\end{aligned}
$$

and similarly $XP = \sum_{i=1}^{N} (p_{2i-1} + p_{2i})Pp_{2i-1}P$. Hence, $PX = XP$. ∎

Proposition 3.2.6. *Considered as a module over its center, the algebra $\mathscr{B}^0$ is generated by the $(2N)^2$ elements $(p_i)_{i=1}^{2N}$ and $(p_i P p_j)_{i,j=1}^{2N}$ with $i \neq j$. In other words, given $a \in \mathscr{B}^0$, there are polynomials R_{ij} in X such that*

$$a = \sum_{i=1}^{2N} R_{ii}(X) p_i + \sum_{\substack{i,j=1 \\ i \neq j}}^{2N} R_{ij}(X) p_i P p_j. \tag{3.13}$$

Proof. Let $\mathscr{B}^1$ denote the set of all elements of $\mathscr{B}^0$ which can be written in the form (3.13). First we show that the generating elements of $\mathscr{B}^0$ belong to $\mathscr{B}^1$. This is evident for the idempotents p_i. Since further

$$p_i P p_i = p_i P p_i \cdot p_i = \begin{cases} X \cdot p_i & \text{if } i \text{ is odd,} \\ (e - X) \cdot p_i & \text{if } i \text{ is even,} \end{cases} \tag{3.14}$$

the assertion for P can be obtained as follows:

$$P = \sum_{i,j=1}^{2N} p_i P p_j = \sum_{i=1}^{N} p_{2i} P p_{2i} + \sum_{i=1}^{N} p_{2i-1} P p_{2i-1} + \sum_{\substack{i,j=1 \\ i \neq j}}^{2N} p_i P p_j$$

$$= \sum_{i=1}^{N} (e - X) p_{2i} + \sum_{i=1}^{N} X p_{2i-1} + \sum_{\substack{i,j=1 \\ i \neq j}}^{2N} p_i P p_j.$$

In the second step we are going to show that $\mathscr{B}^1$ is actually an algebra. Since the generating elements of $\mathscr{B}^0$ belong to $\mathscr{B}^1$, this fact implies that $\mathscr{B}^0 = \mathscr{B}^1$.

Evidently, the set $\mathscr{B}^1$ is closed under addition. Its closedness under multiplication will follow once we have shown that the product of each pair of elements $(p_i)_{i=1}^{2N}$ and $(p_i P p_j)_{i,j=1}^{2N}$ with $i \neq j$ is in $\mathscr{B}^1$ again. This is obvious if one of these elements is p_i. Consider the product $p_i P p_j \cdot p_k P p_l$ with $i \neq j$ and $k \neq l$. This product is 0 (which is in $\mathscr{B}^1$) if $j \neq k$, and it is equal to $p_i P p_j P p_l$ in the case $j = k$. If j is even (say, $j = 2n$) then

$$p_i P p_{2n} P p_l = p_i P (p_{2n-1} + p_{2n}) P p_l - p_i P p_{2n-1} P p_l$$

$$= p_i (p_{2n-1} + p_{2n}) P p_l - p_i P p_{2n-1} P p_l \tag{3.15}$$

by axiom (3.5), whereas in the case that j is odd (say, $j = 2n - 1$),

$$p_i P p_{2n-1} P p_l = p_i P (p_{2n-2} + p_{2n-1}) P p_l - p_i P p_{2n-2} P p_l$$

$$= p_i P (p_{2n-2} + p_{2n-1}) p_l - p_i P p_{2n-2} P p_l \tag{3.16}$$

by (3.12). The first items in (3.15) and (3.16) belong to $\mathscr{B}^1$. Indeed, they are either 0 or equal to $p_i P p_l$ (depending on j). If $i \neq l$ then $p_i P p_l \in \mathscr{B}^1$ by definition, whereas the inclusion $p_i P p_i \in \mathscr{B}^1$ follows from (3.14). Thus, identities (3.15) and (3.16) reduce the question whether $p_i P p_j P p_l$ belongs to $\mathscr{B}^1$ to the question whether

$p_i P p_{j-1} P p_l$ belongs to $\mathscr{B}^1$. Repeating this argument, we finally arrive at an element of the form $p_i P p_i P p_l$. This element belongs to $\mathscr{B}^1$ since

$$p_i P p_i P p_l = p_i P p_i \cdot p_i P p_l = \begin{cases} X \cdot p_i P p_l & \text{if } i \text{ is odd,} \\ (e - X) \cdot p_i P p_l & \text{if } i \text{ is even} \end{cases}$$

and by (3.14).   ∎

Let us have a closer look at the products $p_i P p_j \cdot p_j P p_l$ in the case $i \neq j$ and $j \neq l$.

Proposition 3.2.7. *The following equalities hold:*

(i) *if $l > j > i$ or $j > i > l$ or $i > l > j$, then*

$$p_i P p_j P p_l = (-1)^{j-1}(X - e)\, p_i P p_l;$$

(ii) *if $l > i > j$ or $j > l > i$ or $i > j > l$, then*

$$p_i P p_j P p_l = (-1)^{j-1} X\, p_i P p_l;$$

(iii) *if $i = l$ and $i \neq j$, then*

$$p_i P p_j P p_i = (-1)^{j-i} X (X - e)\, p_i.$$

Proof. Let $j \notin \{i, l\}$. Then

$$p_i P p_j P p_l = p_i P(p_{j-i} + p_j)P p_l - p_i P p_{j-1} P p_l. \tag{3.17}$$

If, moreover, $j - 1 \notin \{i, l\}$, then we conclude from (3.5) and (3.12) that $p_i P(p_{j-i} + p_j)P p_l = 0$, whence

$$p_i P p_j P p_l = -p_i P p_{j-1} P p_l. \tag{3.18}$$

Suppose now that the conditions of assertion (i) are satisfied. Then there is a smallest positive integer k such that (all computations are done modulo $2N$) $j \notin \{i, l\}$, $j - 1 \notin \{i, l\}, \ldots, j - (k-1) \notin \{i, l\}$, but $j - k = i$. Consequently, a repeated application of (3.18) gives

$$p_i P p_j P p_l = (-1)^{k-1} p_i P p_{j-(k-1)} P p_l$$

whence, by virtue of (3.17),

$$\begin{aligned} p_i P p_j P p_l &= (-1)^{k-1}(p_i P(p_{j-k} + p_{j-(k-1)})P p_l - p_i P p_{j-k} P p_l) \\ &= (-1)^{k-1}(p_i P(p_i + p_{i+1})P p_l - p_i P p_i P p_l). \end{aligned}$$

Note that our assumptions imply that $l \neq i$ and $l \neq i + 1$ (otherwise $j - (k-1)$ would be equal to l). Thus,

$$p_i P p_j P p_l = \begin{cases} (-1)^{k-1}(p_i(p_i + p_{i+1})P p_l - p_i P p_i P p_l) & \text{if } i \text{ is odd} \\ (-1)^{k-1}(p_i P(p_i + p_{i+1})p_l - p_i P p_i P p_l) & \text{if } i \text{ is even} \end{cases}$$

$$= \begin{cases} (-1)^{k-1}(p_i P p_l - p_i P p_i P p_l) & \text{if } i \text{ is odd} \\ (-1)^{k-1}(-p_i P p_i P p_l) & \text{if } i \text{ is even} \end{cases}$$

$$= \begin{cases} (-1)^{k-1}(e - X)p_i P p_l & \text{if } i \text{ is odd} \\ (-1)^{k-1}(-1)(e - X)p_i P p_l & \text{if } i \text{ is even,} \end{cases}$$

where we have taken (3.14) into account. Replacing k by $j - i$ yields assertion (i). The proofs for (ii) and (iii) are analogous. ∎

3.2.5 Localization, and identification of the local algebras

The element X belongs to the center of the algebra $\mathscr{B}^0$ by Proposition 3.2.5 and thus to the center of $\mathscr{B}$. Hence, the smallest closed subalgebra $\mathscr{C}$ of $\mathscr{B}$ which contains both the identity element e and the element X is in the center of $\mathscr{B}$, and this fact offers the possibility of localizing $\mathscr{B}$ over $\mathscr{C}$ by the local principle of Allan and Douglas. The maximal ideal space of the singly generated Banach algebra $\mathscr{C}$ is homeomorphic to the spectrum $\sigma_{\mathscr{C}}(X)$ of its generator, where the homeomorphism identifies the point $x \in \sigma_{\mathscr{C}}(X)$ with the smallest closed ideal of $\mathscr{C}$ which contains $X - xe$ (see Exercise 2.1.3). For each $x \in \sigma_{\mathscr{C}}(X)$, we introduce the ideal $\mathscr{I}_x$ of $\mathscr{B}$ in accordance with the local principle. Further we let $\mathscr{B}_x := \mathscr{B}/\mathscr{I}_x$ denote the local algebra associated with x and write Φ_x for the canonical homomorphism from $\mathscr{B}$ onto $\mathscr{B}_x$.

From Proposition 2.2.9 we infer that the ideal $\mathscr{I}_x$ coincides with $\mathscr{B}$ if and only if x belongs to $\sigma_{\mathscr{C}}(X) \setminus \sigma_{\mathscr{B}}(X)$. Thus, by the local principle, an element $b \in \mathscr{B}$ is invertible if and only if the cosets $b + \mathscr{I}_x$ are invertible for all $x \in \sigma_{\mathscr{B}}(X)$. Our next goal is the explicit description of the local algebras $\mathscr{B}_x$.

Proposition 3.2.8. *If $x \in \sigma_{\mathscr{B}}(X) \setminus \{0, 1\}$, then $\mathscr{B}_x$ is isomorphic to $\mathbb{C}^{2N \times 2N}$.*

Proof. Consider the image $\Phi_x(\mathscr{B}^0)$ of the algebra $\mathscr{B}^0$ in $\mathscr{B}_x$. Since each element of $\mathscr{B}^0$ can be written in the form (3.13), and since $\Phi_x(X) = x\Phi_x(e)$ by definition, it follows that $\Phi_x(R(X)) = R(x)\Phi_x(e)$ for each polynomial R. Consequently, every element of $\Phi_x(\mathscr{B}^0)$ is a complex linear combination of the elements

$$\Phi_x(p_i) \quad (i = 1, \ldots, 2N) \quad \text{and} \quad \Phi_x(p_i P p_j) \quad (i, j = 1, \ldots, 2N, i \neq j). \quad (3.19)$$

Conversely, every linear combination of the elements (3.19) is in $\Phi_x(\mathscr{B}^0)$. Thus, $\Phi_x(\mathscr{B}^0)$ is a linear space of dimension less than or equal to $(2N)^2$. In particular, $\Phi_x(\mathscr{B}^0)$ is closed in $\mathscr{B}_x$. On the other hand, $\mathscr{B}^0$ is dense in $\mathscr{B}$. Hence, $\Phi_x(\mathscr{B}^0)$ is dense in $\Phi_x(\mathscr{B}) = \mathscr{B}_x$, whence $\mathscr{B}_x = \Phi_x(\mathscr{B}^0)$.

We claim that the dimension of $\mathscr{B}_x$ is exactly $(2N)^2$ and that the elements (3.19) form a basis of this space. Given $i, j = 1, \ldots, 2N$, define elements $a_{ij} \in \mathscr{B}_x$ by

$$a_{ij} = \begin{cases} (-1)^{i-1}(x-1)^{-1}\Phi_x(p_iPp_j) & \text{if } i < j, \\ (-1)^{i-1}x^{-1}\Phi_x(p_iPp_j) & \text{if } i > j, \\ \Phi_x(p_i) & \text{if } i = j. \end{cases}$$

This definition makes sense since $x \neq 0$ and $x \neq 1$. Proposition 3.2.7 implies that

$$a_{ij}a_{kl} = \delta_{jk} \cdot a_{il} \quad \text{for all} \quad 1 \leq i, j, k, l \leq 2N. \tag{3.20}$$

Indeed, in the case $j = k$ and $j > i > l$ we find

$$\begin{aligned} a_{ij}a_{jl} &= (-1)^{i-1}(x-1)^{-1}\Phi_x(p_iPp_j) \cdot (-1)^{j-1}x^{-1}\Phi_x(p_jPp_l) \\ &= (-1)^{i-1}(-1)^{j-1}x^{-1}(x-1)^{-1}\Phi_x(p_iPp_jPp_l) \\ &= (-1)^{i-1}(-1)^{j-1}x^{-1}(x-1)^{-1}\Phi_x((-1)^{j-1}(X-e)p_iPp_l) \\ &= (-1)^{i-1}x^{-1}\Phi_x(p_iPp_l) = a_{il}. \end{aligned}$$

The other cases can be treated analogously. Now suppose the elements a_{ij} are linearly dependent. Then there are complex numbers c_{ij} with

$$\sum_{i,j=1}^{2N} c_{ij}a_{ij} = 0 \tag{3.21}$$

and $c_{i_0 j_0} \neq 0$ for a certain pair $i_0 j_0$. Multiplying (3.21) by a_{ki_0} from the left and by $a_{j_0 k}$ from the right yields $c_{i_0 j_0}a_{ki_0}a_{i_0 j_0}a_{j_0 k} = c_{i_0 j_0}a_{kk} = 0$, whence $a_{kk} = 0$ for all $k = 1, \ldots, 2N$. Consequently,

$$\Phi_x(e) = \Phi_x\left(\sum_{k=1}^{2N} p_k\right) = \sum_{k=1}^{2N} a_{kk} = \Phi_x(0)$$

which is impossible since $\mathscr{I}_x \neq \mathscr{B}$ for $x \in \sigma_{\mathscr{B}}(X)$ by Proposition 2.2.9. Thus, the elements $(a_{ij})_{i,j=1}^{2N}$ are linearly independent. Therefore, also the elements (3.19) are linearly independent, and both sets of elements form a basis of $\mathscr{B}_x$. Finally, it is immediate from (3.20) that the mapping

$$\Psi_x : (a_{ij})_{i,j=1}^{2N} \to \mathbb{C}^{2N \times 2N}, \quad a_{ij} \mapsto E_{ij},$$

where E_{ij} refers to the $2N \times 2N$ matrix whose ij entry is 1 and all other entries of which are zero, extends to an algebra isomorphism from $\mathscr{B}_x$ onto $\mathbb{C}^{2N \times 2N}$. ∎

The following corollary identifies the images of the generating elements of the algebra $\mathscr{B}$ under the homomorphism $F_x := \Psi_x \circ \Phi_x : \mathscr{B} \to \mathbb{C}^{2N \times 2N}$.

Corollary 3.2.9. *Let $x \in \sigma_{\mathscr{B}}(X) \setminus \{0, 1\}$. Then*

$$F_x(p_i) = \text{diag}(0, \ldots, 0, 1, 0, \ldots, 0), \tag{3.22}$$

the 1 standing at the ith place, and

$$F_x(P) = \mathrm{diag}\,(1, -1, 1, -1, \ldots, 1, -1) \times$$

$$
\begin{bmatrix}
x & x-1 & x-1 & x-1 & \cdots & x-1 & x-1 \\
x & x-1 & x-1 & x-1 & \cdots & x-1 & x-1 \\
x & x & x & x-1 & \cdots & x-1 & x-1 \\
x & x & x & x-1 & \cdots & x-1 & x-1 \\
\vdots & \vdots & \vdots & \vdots & \ddots & \vdots & \vdots \\
x & x & x & x & \cdots & x & x-1 \\
x & x & x & x & \cdots & x & x-1
\end{bmatrix}.
\tag{3.23}
$$

Proof. For (3.22) recall that $\Phi_x(p_i) = a_{ii}$, and to get (3.23) observe that

$$
F_x(P) = F_x\left(\sum_{i,j=1}^{2N} p_i P p_j\right)
$$

$$
= (\Psi_x \circ \Phi_x)\left(\sum_{\substack{i,j=1 \\ i<j}}^{2N} p_i P p_j\right) + (\Psi_x \circ \Phi_x)\left(\sum_{\substack{i,j=1 \\ i>j}}^{2N} p_i P p_j\right) + F_x\left(\sum_{i=1}^{2N} p_i P p_i\right)
$$

$$
= \Psi_x\left(\sum_{\substack{i,j=1 \\ i<j}}^{2N} (-1)^{i-1}(x-1)a_{ij}\right) + \Psi_x\left(\sum_{\substack{i,j=1 \\ i>j}}^{2N} (-1)^{i-1}x a_{ij}\right)
$$

$$
+ F_x\left(\sum_{i=1}^{2N} p_i P p_i\right)
$$

and take into account (3.14). $\blacksquare$

Our next object is the local algebras $\mathscr{B}_x$ associated with the points in $\sigma_{\mathscr{B}}(X) \cap \{0, 1\}$. These algebras will not be identified completely; we will only show that all irreducible representations are one-dimensional and compute them.

Proposition 3.2.10. *If $x \in \sigma_{\mathscr{B}}(X) \cap \{0, 1\}$, then $\mathscr{B}_x$ is an SI_2^{N+1}-algebra.*

Proof. Instead of working with the polynomial

$$S_2^{N+1}(a, b) = (ab - ba)^{N+1}$$

in two variables, which is nonlinear, we consider the polynomial

$$\mathcal{P}(a_1, b_1, \ldots, a_{N+1}, b_{N+1}) := \prod_{k=1}^{N+1} (a_k b_k - b_k a_k)$$

in $2(N+1)$ variables, which is linear in each variable. Note that if $\mathscr{B}_x$ is a $\mathcal{P}$-algebra, then it is also an SI_2^{N+1}-algebra. Since $\mathscr{B}_x$ is a linear space and $\mathcal{P}$ is multilinear, it remains to prove that

$$\prod_{k=1}^{N+1}(a_kb_k - b_ka_k) = 0$$

for all choices of cosets a_k, b_k among the basis elements

$$\Phi_x(p_i) \quad (i = 1, \ldots, 2N) \quad \text{and} \quad \Phi_x(p_iPp_j) \quad (i, j = 1, \ldots, 2N, i \neq j)$$

of the algebra $\mathcal{B}_x$. Proposition 3.2.7 implies that each commutant $a_kb_k - b_ka_k$ can be written in the form $c_k\Phi_x(p_{i_k}Pp_{j_k})$, where $i_k, j_k \in \{1, \ldots, N+1\}$ and $c_k \in \mathbb{C}$ can be zero. Hence,

$$\prod_{k=1}^{N+1}(a_kb_k - b_ka_k) = c\,\Phi_x\left(\prod_{k=1}^{N+1} p_{i_k}Pp_{j_k}\right).$$

Since the partition $\{p_i\}$ of identity consists of $2N$ elements, there are two of the elements p_{i_k} and p_{j_k} with $k = 1, \ldots, N+1$ which coincide. Thus, $\prod_{k=1}^{N+1} p_{i_k}Pp_{j_k}$ contains at least one subproduct of the form $p_iPp_{l_1}Pp_{l_2}\ldots Pp_{l_r}Pp_i$ with $r \geq 1$. Invoking Proposition 3.2.7 once more, one easily gets $\Phi_x(p_iPp_{l_1}Pp_{l_2}\ldots Pp_{l_r}Pp_i) = 0$. Thus,

$$\Phi_x\left(\prod_{k=1}^{N+1} p_{i_k}Pp_{j_k}\right) = 0$$

for $x \in \sigma_{\mathcal{B}}(X) \cap \{0, 1\}$.   ∎

By the extended version of Theorem 2.6.17, the algebras $\mathcal{B}_x$ possess matrix symbols of order 1, i.e., scalar-valued symbols. Since each algebra homomorphism $\Psi : \mathcal{B}_x \to \mathbb{C}$ gives rise to an algebra homomorphism $\Psi \circ \Phi_x : \mathcal{B} \to \mathbb{C}$, we have to determine the one-dimensional representations of the algebra $\mathcal{B}$.

Clearly, each homomorphism $G : \mathcal{B} \to \mathbb{C}$ maps idempotents to idempotents. Thus, if $p \in \mathcal{B}$ is idempotent, then $G(p)$ is either 0 or 1. Moreover, since $G(e) = 1$ for each non-zero homomorphism G, we conclude that there is an i_0 such that $G(p_{i_0}) = 1$ and $G(p_i) = 0$ for all $i \neq i_0$. Hence, the restriction of a non-zero homomorphism $G : \mathcal{B} \to \mathbb{C}$ to the set $\{P, p_1, p_2, \ldots, p_{2N}\}$ coincides with one of the following mappings G_n with $n \in \{1, 2, \ldots, 4N\}$:

$$G_{4m}(p_i) = \begin{cases} 1 & \text{if} \quad i = 2m, \\ 0 & \text{if} \quad i \neq 2m, \end{cases} \qquad\qquad G_{4m}(P) = 0,$$

$$G_{4m-1}(p_i) = \begin{cases} 1 & \text{if} \quad i = 2m, \\ 0 & \text{if} \quad i \neq 2m, \end{cases} \qquad\qquad G_{4m-1}(P) = 1,$$

$$\tag{3.24}$$

$$G_{4m-2}(p_i) = \begin{cases} 1 & \text{if} \quad i = 2m-1, \\ 0 & \text{if} \quad i \neq 2m-1, \end{cases} \qquad\qquad G_{4m-2}(P) = 1,$$

$$G_{4m-3}(p_i) = \begin{cases} 1 & \text{if} \quad i = 2m-1, \\ 0 & \text{if} \quad i \neq 2m-1, \end{cases} \qquad\qquad G_{4m-3}(P) = 0,$$

where $m = 1, \ldots, N$. Set

$$Y := \sum_{i=1}^{N} (p_{2i-1}P + p_{2i}Q) + \sum_{i=1}^{2N} (2i-1)p_i.$$

Proposition 3.2.11. *Let* $m \in \sigma_{\mathscr{B}}(Y) \cap \{1, 2, \ldots, 4N\}$. *Then the mapping*

$$G_m : \{P, p_1, p_2, \ldots, p_{2N}\} \to \mathbb{C},$$

given by (3.24), *extends to an algebra homomorphism from* $\mathscr{B}$ *onto* $\mathbb{C}$.

Proof. First note that if G_m extends to an algebra homomorphism, then

$$G_m(Y) = m. \tag{3.25}$$

We claim that, for $m \in \sigma_{\mathscr{B}}(Y) \cap \{1, 2, \ldots, 4N\}$ and $x \in \sigma_{\mathscr{B}}(X) \setminus \{0, 1\}$,

$$m \notin \sigma_{\mathscr{B}_x}(\Phi_x(Y)). \tag{3.26}$$

By Corollary 3.2.9, what we have to prove to justfy the claim is that the $2N \times 2N$ matrices

$$(\Psi_x \circ \Phi_x)(Y) - \mathrm{diag}\,(m, m, \ldots, m)$$

$$= \begin{bmatrix}
x & x-1 & x-1 & x-1 & \cdots & x-1 & x-1 \\
x & x & x-1 & x-1 & \cdots & x-1 & x-1 \\
x & x & x & x-1 & \cdots & x-1 & x-1 \\
x & x & x & x & \cdots & x-1 & x-1 \\
\vdots & \vdots & \vdots & \vdots & \ddots & \vdots & \vdots \\
x & x & x & x & \cdots & x & x-1 \\
x & x & x & x & \cdots & x & x
\end{bmatrix}$$

$$+ \mathrm{diag}\,(1-m, 3-m, \ldots, 4N-1-m)$$

are invertible. For this goal we compute the determinant of a general $M \times M$ matrix of the form

$$\begin{bmatrix}
x+\lambda_1 & x-1 & x-1 & x-1 & \cdots & x-1 & x-1 \\
x & x+\lambda_2 & x-1 & x-1 & \cdots & x-1 & x-1 \\
x & x & x+\lambda_3 & x-1 & \cdots & x-1 & x-1 \\
x & x & x & x+\lambda_4 & \cdots & x-1 & x-1 \\
\vdots & \vdots & \vdots & \vdots & \ddots & \vdots & \vdots \\
x & x & x & x & \cdots & x+\lambda_{M-1} & x-1 \\
x & x & x & x & \cdots & x & x+\lambda_M
\end{bmatrix} \tag{3.27}$$

with complex entries. Consider x as being variable and denote the determinant of the matrix (3.27) by $D(x)$. Subtracting the first row in (3.27) from all other rows, and then the last column from all other columns, one gets a matrix the $1, N$ entry of which is $x - 1$ whereas all other entries are independent of x. Thus, $D(x)$ is a first order polynomial in x. Further, since $D(0) = \prod_{i=1}^{M} \lambda_i$ and $D(1) = \prod_{i=1}^{M}(1 + \lambda_i)$, one has

$$D(x) = x \prod_{i=1}^{M}(1 + \lambda_i) + (1 - x) \prod_{i=1}^{M} \lambda_i. \tag{3.28}$$

Now let $m \in \{1, \ldots 4N\}$, $M = 2N$, and $\lambda_i = 2i - 1 - m$ for $i = 1, \ldots, 2N$. If m is odd, then one of the numbers λ_i is equal to zero, but $\prod_{i=1}^{M}(1 + \lambda_i) \neq 0$. If m is even, then one of the numbers $1 + \lambda_i$ is zero, but $\prod_{i=1}^{M} \lambda_i \neq 0$. Hence, in any case,

$$x \prod_{i=1}^{M}(1 + \lambda_i) + (1 - x) \prod_{i=1}^{M} \lambda_i \neq 0$$

whenever $x \notin \{0, 1\}$. This proves our claim (3.26).

Now the assertion can be obtained as follows. Let $m \in \sigma_{\mathscr{B}}(Y) \cap \{1, 2, \ldots, 4N\}$. Then, by the local principle,

$$m \in \cup_{x \in \sigma_{\mathscr{B}}(X)} \sigma_{\mathscr{B}_x}(\Phi_x(Y))$$

whereas, by (3.26),

$$m \notin \cup_{x \in \sigma_{\mathscr{B}}(X) \setminus \{0,1\}} \sigma_{\mathscr{B}_x}(\Phi_x(Y)).$$

Hence,

$$m \in \cup_{x \in \sigma_{\mathscr{B}}(X) \cap \{0,1\}} \sigma_{\mathscr{B}_x}(\Phi_x(Y)).$$

The algebra $\mathscr{B}_x$ with $x \in \sigma_{\mathscr{B}}(X) \cap \{0, 1\}$ possesses a scalar-valued symbol by Proposition 3.2.10 and Theorem 2.6.17. Thus, if $m \in \sigma_{\mathscr{B}_{x_0}}(\Phi_{x_0}(Y))$ with a certain $x_0 \in \sigma_{\mathscr{B}}(X) \cap \{0, 1\}$ then there is an algebra homomorphism G' from $\mathscr{B}_{x_0}$ onto $\mathbb{C}$ with $G'(\Phi_{x_0}(Y)) = m$. Then $G := G' \circ \Phi_{x_0}$ is an algebra homomorphism from $\mathscr{B}$ onto $\mathbb{C}$ with $G(Y) = m$. The restriction of G to the set $\{P, p_1, \ldots, p_{2N}\}$ coincides with one of the mappings G_n introduced in (3.24) and, by (3.25), this restriction is just G_m. In other words, G_m extends to an (evidently continuous) algebra homomorphism from $\mathscr{B}$ onto $\mathbb{C}$. ∎

For $m \in \sigma_{\mathscr{B}}(Y) \cap \{1, 2, \ldots, 4N\}$, we denote the extension of G_m by G_m again. One easily checks that

$$G_m(X) = \begin{cases} 0 & \text{if } m \text{ is odd,} \\ 1 & \text{if } m \text{ is even.} \end{cases}$$

Thus, if $0 \in \sigma_{\mathscr{B}}(X)$ and m is odd, then the local ideal $\mathscr{I}_0$ lies in the kernel of G_m and consequently, for each $a \in \mathscr{B}$ the number $G_m(a)$ depends on the coset $\Phi_0(a)$ only. This shows that the quotient mapping

$$G'_m : \mathscr{B}_0 \to \mathbb{C}, \quad \Phi_0(a) \mapsto G_m(a)$$

is well defined, and this mapping is an algebra homomorphism from $\mathscr{B}_0$ onto $\mathbb{C}$. Analogously, if $0 \in \sigma_{\mathscr{B}}(X)$ and m is even, then

$$G'_m : \mathscr{B}_1 \to \mathbb{C}, \quad \Phi_1(a) \mapsto G_m(a)$$

is a correctly defined and non-trivial algebra homomorphism.

Proposition 3.2.12. *If $0 \in \sigma_{\mathscr{B}}(X)$, then the mappings G_m with m an odd number in $\sigma_{\mathscr{B}}(Y) \cap \{1, 2, \ldots, 4N\}$ constitute a scalar-valued symbol for $\mathscr{B}_0$. If $1 \in \sigma_{\mathscr{B}}(X)$, then the mappings G_m with even m in $\sigma_{\mathscr{B}}(Y) \cap \{1, 2, \ldots, 4N\}$ constitute a scalar-valued symbol for $\mathscr{B}_1$.*

Proof. The mappings G'_m with m odd (even) are the *only* non-trivial algebra homomorphisms from $\mathscr{B}_0$ (resp. $\mathscr{B}_1$) into $\mathbb{C}$. Since the algebras $\mathscr{B}_0$ (resp. $\mathscr{B}_1$) *possess* a scalar-valued symbol by Theorem 2.6.17 and Proposition 3.2.10, we conclude that for all $a \in \mathscr{B}$ the coset $\Phi_0(a)$ (resp. $\Phi_1(a)$) is invertible whenever all $G'_m(\Phi_0(a)) = G_m(a)$ with m odd (resp. even) are invertible. $\blacksquare$

After these preparations, we are now in a position to prove the N projections theorem:

Proof of Theorem 3.2.4. The proof of assertions (i), (ii) and (iii) is immediate from Allan's local principle in combination with the description of the local algebras given in the preceding results. The continuity of the mappings F_x and G_m is a consequence of a general result by Johnson (see, e.g., [84, Chapter 6, Theorem 2.65]) which states that an algebra homomorphism from a Banach algebra onto a semi-simple Banach algebra must be continuous.

For a proof of assertion (iv) recall that the algebra $\mathscr{B}^0$ is a $(2N)^2$-dimensional module over its center. Thus, Corollary 2.6.22 tells us that there is a set $\{v_t\}, t \in T$, of representations of $\mathscr{B}$ such that $\mathrm{Im}\, v_t = \mathbb{C}^{l \times l}$ with $l = l(t) \leq 2N$ and such that an element b of $\mathscr{B}$ is invertible in $\mathscr{A}$ if and only if $\det v_t(b) \neq 0$ for all $t \in T$. The very same arguments as in the proof of assertion (iii) imply that each of these representations is of the form F_x (with an $x \in \mathbb{C} \setminus \{0, 1\}$) as defined in Corollary 3.2.9, or G_m (with $m \in \{1, 2, \ldots, 2N\}$) as defined after Proposition 3.2.10. Hence, there exist two sets $\xi = \xi(\mathscr{A}, \mathscr{B}) \subset \mathbb{C} \setminus \{0, 1\}$ and $\mu = \mu(\mathscr{A}, \mathscr{B}) \subseteq \{1, 2, \ldots, 2N\}$ such that

$$\sigma_{\mathscr{A}}(b) = \cup_{x \in \xi} \sigma(F_x(b)) \cup \{G_m(b) : m \in \mu\} \tag{3.29}$$

for all $b \in \mathscr{B}$. We claim that $\xi = \sigma_{\mathscr{A}}(X) \setminus \{0, 1\}$ and $\mu = \sigma_{\mathscr{A}}(Y) \cap \{1, \ldots, 4N\}$.

Since $G_m(X) \in \{0, 1\}$ and $\xi \cap \{0, 1\} = \emptyset$, one has

$$\sigma_{\mathscr{A}}(X) \setminus \{0, 1\} = \cup_{x \in \xi} \sigma(F_x(X)) \cup \{G_m(X) : m \in \mu\} \setminus \{0, 1\} = \cup_{x \in \xi} \{x\} = \xi. \tag{3.30}$$

For the second claim note that, for any $\lambda \in \mathbb{C}$, the matrix $F_x(Y - \lambda e)$ coincides with the matrix (3.25), with the λ_i in (3.25) replaced by $2i - 1 - \lambda$. It follows from the explicit form (3.28) of the determinant of this matrix that every eigenvalue λ of $F_x(Y)$ solves the equation

$$x\prod_{i=1}^{2N}(2i-\lambda)+(1-x)\prod_{i=1}^{2N}(2i-1-\lambda)=0.$$

But, if $x\notin\{0,1\}$, then $\sigma(F_x(Y))\cap\{1,2,\ldots,2N\}=\emptyset$. Thus,

$$\sigma_{\mathscr{A}}(Y)\cap\{1,\ldots,4N\}=\{G_m(Y):m\in\mu\}\cap\{1,\ldots,4N\}=\{m\}_{m\in\mu}=\mu. \quad (3.31)$$

Now assertion (iv) follows immediately from (3.29), (3.30) and (3.31). ∎

Observe that assertion (iv) is evident in the case that the algebra $\mathscr{B}$ is inverse-closed in $\mathscr{A}$. However, this is not always satisfied as the following example indicates.

Example 3.2.13. Consider the algebra $\mathscr{A}$ of all continuous 2×2 matrix functions on the complex unit circle $\mathbb{T}$. Let t denote the identical mapping of $\mathbb{T}$. Then

$$P=\begin{bmatrix} t & 1-t \\ t & 1-t \end{bmatrix}, \quad p_1=\begin{bmatrix} 1 & 0 \\ 0 & 0 \end{bmatrix}, \quad \text{and} \quad p_2=\begin{bmatrix} 0 & 0 \\ 0 & 1 \end{bmatrix}$$

are elements of $\mathscr{A}$ which satisfy the assumptions of Theorem 3.2.4 with $N=1$. The function

$$X:=p_1Pp_1+p_2(e-P)p_2=\begin{bmatrix} t & 0 \\ 0 & t \end{bmatrix}$$

is invertible in $\mathscr{A}$ but not invertible in $\mathscr{B}$ since the latter algebra consists of matrix functions holomorphic in the unit disk only. □

In this connection, let us emphasize an evident consequence of assertions (iii) and (iv) of the previous theorem.

Corollary 3.2.14. *If $\sigma_{\mathscr{B}}(X)=\sigma_{\mathscr{A}}(X)$ and $\sigma_{\mathscr{B}}(Y)=\sigma_{\mathscr{A}}(Y)$, then the algebra $\mathscr{B}$ is inverse-closed in $\mathscr{A}$.*

The following additional information is often useful. The mappings F_0 and F_1 are formally defined as in Theorem 3.2.4 for $x=0,1$.

Proposition 3.2.15.

(i) *If $0\notin\sigma_{\mathscr{B}}(X)$ and $1\notin\sigma_{\mathscr{B}}(X)$, then*
$\sigma_{\mathscr{B}}(Y)\cap\{1,\ldots,4N\}=\emptyset.$

(ii) *If $0\in\sigma_{\mathscr{B}}(X)$ and $1\in\sigma_{\mathscr{B}}(X)$, and if none of these points is isolated in $\sigma_{\mathscr{B}}(X)$, then the family (F_x) with $x\in\sigma_{\mathscr{B}}(X)$ is a matrix symbol for $\mathscr{B}$.*

(iii) *If $0\notin\sigma_{\mathscr{A}}(X)$ and $1\notin\sigma_{\mathscr{A}}(X)$ then $\sigma_{\mathscr{A}}(Y)\cap\{1,\ldots,4N\}=\emptyset.$*

(iv) *If $0\in\sigma_{\mathscr{A}}(X)$ and $1\in\sigma_{\mathscr{A}}(X)$, and if both points are not isolated in $\sigma_{\mathscr{A}}(X)$, then the family (F_x) with $x\in\sigma_{\mathscr{A}}(X)$ is a matrix symbol for the invertibility of the elements of $\mathscr{B}$ in the algebra $\mathscr{A}$.*

Proof. (i) Note that $G_m(X)\in\{0,1\}$ in any case. Thus, if $\sigma_{\mathscr{B}}(X)\cap\{0,1\}=\emptyset$, then one-dimensional representations cannot exist.

(ii) The function $x \mapsto F_x(Y)$ is continuous on $\sigma_{\mathscr{B}}(X)$. Due to the continuous dependence of the eigenvalues of a matrix on the matrix itself (see [88, Appendix D]), one has

$$\sigma(F_0(Y)) = \limsup_{x \to 0} \sigma(F_x(Y))$$

and consequently,

$$\sigma(F_0(Y)) = \limsup_{x \to 0} \sigma(\Phi_x(Y)). \tag{3.32}$$

We further know from Proposition 2.2.3 (ii) that

$$\limsup_{x \to 0} \sigma(\Phi_x(Y)) \subseteq \sigma(\Phi_0(Y)). \tag{3.33}$$

From (3.32) and (3.33) we obtain $\sigma(F_0(Y)) \subseteq \sigma(\Phi_0(Y))$. Analogously, $\sigma(F_1(Y)) \subseteq \sigma(\Phi_1(Y))$. Hence,

$$\sigma(F_0(Y)) \cup \sigma(F_1(Y)) \subseteq \sigma(\Phi_0(Y)) \cup \sigma(\Phi_1(Y)) \subseteq \sigma_{\mathscr{B}}(Y).$$

A simple computation gives

$$\sigma(F_0(Y)) \cup \sigma(F_1(Y)) = \{1, 2, \dots, 4N\},$$

whence

$$\sigma_{\mathscr{B}}(Y) \cap \{1, 2, \dots, 4N\} = \{1, 2, \dots, 4N\}.$$

In other words, all possible one-dimensional representations occur.

It remains to observe that, for each $a \in \mathscr{B}$, the matrices $F_0(a)$ and $F_1(a)$ are triangular and that the diagonal of $F_0(a)$ equals $(G_1(a), G_3(a), \dots, G_{2N-1}(a))$, while the diagonal of $F_1(a)$ is $(G_2(a), G_4(a), \dots, G_{2N}(a))$.

The proof of assertions (iii) and (iv) can be given in a completely analogous manner. ∎

3.2.6 Other indicator elements

The elements X and Y indicate which matrix representations of the algebra $\mathscr{B}$ actually appear. While X is distinguished by the fact that it belongs to the center of $\mathscr{B}$, there is some latitude to choose Y. For example, one can show that the element

$$Z := P + \sum_{i=1}^{2N} 2i p_i$$

can substitute Y in the determination of all one-dimensional representations. Indeed, consider the mappings K_m given by $K_{2i}(P) = 0$, $K_{2i+1}(P) = 1$,

$$K_{2i}(p_i) = \begin{cases} 1 & \text{if} \quad i = j, \\ 0 & \text{if} \quad i \neq j, \end{cases} \qquad K_{2i+1}(p_i) = \begin{cases} 1 & \text{if} \quad i = j, \\ 0 & \text{if} \quad i \neq j, \end{cases}$$

where $i = 1, \dots, 2N$. The following is the analog of Proposition 3.2.11 and can be proved in the same way.

Proposition 3.2.16. *If $m \in \sigma_{\mathscr{B}}(Z) \cap \{2, 3, 4, \dots, 4N+1\}$ then the complex-valued mapping K_m defined on $\{P, p_1, \dots, p_{2N}\}$ extends to an algebra homomorphism from $\mathscr{B}$ onto $\mathbb{C}$.*

The following observation is often helpful in order to determine the spectrum of X. For $i = 1, 2, \dots, 2N$, let $\mathscr{B}_i$ denote the algebra $p_i \mathscr{B} p_i = \{p_i b p_i, b \in \mathscr{B}\}$.

Proposition 3.2.17. *If $\{0, 1\} \subseteq \sigma_{\mathscr{B}_i}(p_i X p_i)$ for some i then $\sigma_{\mathscr{B}}(X) = \sigma_{\mathscr{B}_i}(p_i X p_i)$.*

Proof. Since $\{p_i\}$ is a partition of identity and X is in the center of $\mathscr{B}$, we have

$$\sigma_{\mathscr{B}}(X) = \cup_{j=1}^{2N} \sigma_{\mathscr{B}_j}(p_j X p_j). \tag{3.34}$$

We claim that

$$\sigma_{\mathscr{B}_j}(p_j X p_j) \setminus \{0, 1\} = \sigma_{\mathscr{B}_k}(p_k X p_k) \setminus \{0, 1\} \tag{3.35}$$

for all $j, k = 1, \dots, 2N$. Indeed, let $\lambda \notin \sigma_{\mathscr{B}_j}(p_j X p_j)$. Then there is an a in $\mathscr{B}$ such that

$$p_j a p_j (p_j X p_j - \lambda p_j) = p_j.$$

Multiplying this identity from the left-hand side by $p_k P p_j$ and from the right-hand side by $p_j P p_k$ with some $k \neq j$ one gets

$$p_k P p_j a p_j (p_j X p_j - \lambda p_j) p_j P p_k = p_k P p_j P p_k$$

which can be written by means of Proposition 3.2.7 as

$$p_k P p_j a p_j P p_k (p_k X p_k - \lambda p_k) = (-1)^{j-k} X(X - I) p_k.$$

The element $p_k X p_k$ lies in the center of the algebra $\mathscr{B}_k$. Thus, localization of $\mathscr{B}_k$ over its smallest closed subalgebra which contains p_k and $p_k X p_k$ via Allan's local principle yields that, at the point $\mu \in \sigma_{\mathscr{B}_k}(p_k X p_k)$,

$$(\mu - \lambda)\Omega_\mu(p_k P p_j a p_j P p_k) = (-1)^{j-k} \mu(\mu - 1)\Omega_\mu(I),$$

where Ω_μ refers to the canonical homomorphism from $\mathscr{B}_k$ onto its local algebra at μ. Thus, if $\mu \notin \{0, 1\}$ then $\mu - \lambda \neq 0$ and, hence, $\lambda \notin \sigma_{\mathscr{B}_k}(p_k X p_k) \setminus \{0, 1\}$. This gives (3.35) which, in combination with (3.34), proves the assertion. ∎

3.2.7 The two projections theorem revisited

Let $N = 1$ in Theorem 3.2.4. Then, as we have already seen, the partition $\{p_i\}$ consists of two elements p_1 and p_2 with $p_2 = e - p_1$, and the axioms (3.5) and (3.6) reduce to $P^2 = P$ and $Q^2 = Q$, respectively. Thus, $\mathscr{B}$ is the (general) algebra generated by two idempotents P and p_1 and the identity, with no further relations between the generators.

Obviously, there are some differences between the specification of Theorem 3.2.4 to the case $N = 1$ and Theorem 3.1.4. In the case $N = 1$, set $p := p_1$ and $r := P$ in Theorem 3.1.4. The first difference concerns the indicator element for the one-dimensional representations. In Theorem 3.1.4, it is the element $p + 2r$, whereas this role is played by the element $Y = pr + (e - p)(e - r) + p + 3(e - p) = 2pr + 4e - 3p - r$ in Theorem 3.2.4, which seems to be much more complicated. But if Y is replaced by the element Z from the preceding subsection, then $Z = r + 2p + 4(e - p) = r - 2p + 4e$ which is as simple as $p + 2r$.

The second difference concerns the explicit form of the 2×2 matrices. In Theorems 3.1.4 and 3.2.4, the matrices associated with the idempotent r at the point $x \in \sigma_{\mathscr{B}}(X) \setminus \{0, 1\}$ are

$$\begin{bmatrix} x & \sqrt{x(1-x)} \\ \sqrt{x(1-x)} & 1-x \end{bmatrix} \quad \text{and} \quad \begin{bmatrix} x & 1-x \\ x & 1-x \end{bmatrix}, \tag{3.36}$$

respectively. Since, for $x \in \mathbb{C} \setminus \{0, 1\}$,

$$\begin{bmatrix} x & \sqrt{x(1-x)} \\ \sqrt{x(1-x)} & 1-x \end{bmatrix} = \begin{bmatrix} \sqrt[4]{\frac{x}{1-x}} & 0 \\ 0 & \sqrt[4]{\frac{1-x}{x}} \end{bmatrix} \begin{bmatrix} x & 1-x \\ x & 1-x \end{bmatrix} \begin{bmatrix} \sqrt[4]{\frac{1-x}{x}} & 0 \\ 0 & \sqrt[4]{\frac{x}{1-x}} \end{bmatrix}$$

and, moreover,

$$\begin{bmatrix} 1 & 0 \\ 0 & 0 \end{bmatrix} = \begin{bmatrix} \sqrt[4]{\frac{x}{1-x}} & 0 \\ 0 & \sqrt[4]{\frac{1-x}{x}} \end{bmatrix} \begin{bmatrix} 1 & 0 \\ 0 & 0 \end{bmatrix} \begin{bmatrix} \sqrt[4]{\frac{1-x}{x}} & 0 \\ 0 & \sqrt[4]{\frac{x}{1-x}} \end{bmatrix}$$

where $\sqrt[4]{\frac{x}{1-x}}$ is any number with $\left(\sqrt[4]{\frac{x}{1-x}}\right)^4 = \frac{x}{1-x}$, and $\sqrt[4]{\frac{1-x}{x}}$ is $\left(\sqrt[4]{\frac{x}{1-x}}\right)^{-1}$, both representations in (3.36) are equivalent.

Note also that the assertion of Theorem 3.2.4 is more general than that of Theorem 3.1.4 because it includes the inverse-closedness of $\mathscr{B}$ in $\mathscr{A}$. We can thus add a further point to Theorem 3.1.4:

(v) *an element $a \in \mathscr{B}$ is invertible in $\mathscr{A}$ if and only if the matrices $F_x(a)$ are invertible for all $x \in \sigma_{\mathscr{A}}(e - p - r + pr + rp) \setminus \{0, 1\}$ and if the numbers $G_m(a)$ are non-zero for all $m \in \sigma_{\mathscr{A}}(p + 2r) \cap \{0, 1, 2, 3\}$.*

Finally, notice that the invertibility assertion of the Halmos theorem (Theorem 3.1.1) follows from the general two projections theorem. But the latter theorem does not immediately imply a description of the C^*-algebras generated by two projections as an algebra of *continuous* functions.

3.2.8 The spectrum of an abstract SIO

Our next concern is to demonstrate how Theorem 3.2.4 can be used to compute the spectrum of singular integrals in the case that the spectrum of the operator

$$X = \sum_{i=1}^{N} (p_{2i-1} P p_{2i-1} + p_{2i} Q p_{2i})$$

is known. Let T be a non-empty proper subset of $\{1, 2, \ldots, 2N\}$ and set $p := \sum_{i \in T} p_i$ and $q := e - p$. Elements of the form $A := pPp + q \ (\in \mathcal{B})$ are called *abstract singular integral operators*. From Theorem 3.2.4 we conclude that the spectrum of A equals

$$\bigcup_{x \in \sigma_{\mathscr{F}}(X) \backslash \{0,1\}} \sigma(F_x(A)) \ \cup \bigcup_{m \in \sigma_{\mathscr{F}}(Y) \cap \{1,\ldots,2N\}} \sigma(G_m(A)),$$

where the choice of $\mathscr{F} \in \{\mathscr{A}, \mathscr{B}\}$ depends on whether we want to know the spectrum of A in $\mathscr{F} = \mathscr{A}$ or in $\mathscr{F} = \mathscr{B}$.

Let us first determine the spectrum of $F_x(A)$ for $x \in \sigma_{\mathscr{F}}(X) \setminus \{0,1\}$. Let $\lambda \in \mathbb{C}$ and set $D(x) := \det(F_x(A - \lambda e))$. Further, let t, t_o, and t_e refer to the number of the elements of the sets T, $T \cap \{1, 3, \ldots, 2N-1\}$, and $T \cap \{2, 4, \ldots, 2N\}$, respectively. Also put $v := t_o - t_e$. Changing the rows and columns of $F_x(A)$ in an appropriate way produces a matrix of the form

$$\begin{bmatrix} F_{11} & 0 \\ 0 & I \end{bmatrix} \tag{3.37}$$

where F_{11} is a $t \times t$ matrix and I is the $(2N - t) \times (2N - t)$ identity matrix. The determinant $D(x)$ of (3.37) is a polynomial of first degree in x (see the proof of Proposition 3.2.11), and

$$D(0) = (-\lambda)^{t_o} (1-\lambda)^{t_e} (1-\lambda)^{2N-t}, \quad D(1) = (1-\lambda)^{t_o} (-\lambda)^{t_e} (1-\lambda)^{2N-t},$$

the factors $(1-\lambda)^{2N-t}$ coming from the lower right corner in (3.37) and the other factors resulting from the upper left one. Thus,

$$D(x) = (1-\lambda)^{2N-t} [x(1-\lambda)^{t_o} (-\lambda)^{t_e} + (1-x)(-\lambda)^{t_o} (1-\lambda)^{t_e}].$$

Depending on whether $v > 0$, $v = 0$, or $v < 0$, this equals

$$D(x) = (1-\lambda)^{2N-t} (1-\lambda)^{t_e} (-\lambda)^{t_e} [x(1-\lambda)^v + (1-x)(-\lambda)^v],$$

$$D(x) = (1-\lambda)^{2N-t} (1-\lambda)^{t_o} (-\lambda)^{t_o},$$

or

$$D(x) = (1-\lambda)^{2N-t}(1-\lambda)^{t_o}(-\lambda)^{t_o}\left[x(-\lambda)^{|v|} + (1-x)(1-\lambda)^{|v|}\right],$$

respectively. Thus, if $v = 0$, then $\sigma(F_x(A)) = \{0, 1\}$. In the case $v > 0$, we have

$$x(1-\lambda)^v + (1-x)(-\lambda)^v = 0 \tag{3.38}$$

if and only if

$$\left(\frac{\lambda}{\lambda - 1}\right)^v = \frac{x}{x - 1} \tag{3.39}$$

(note that $x \neq 1$ by assumption and that (3.38) cannot vanish if $\lambda = 1$). Let $\zeta_0(x), \ldots, \zeta_{v-1}(x)$ denote the v roots of $x/(x-1)$. Then we infer from (3.39) that the spectrum of $F_x(A)$ equals

$$\{0, 1\} \cup \left\{\frac{\zeta_0(x)}{\zeta_0(x) - 1}, \ldots, \frac{\zeta_{v-1}(x)}{\zeta_{v-1}(x) - 1}\right\} \quad \text{if} \quad t_e > 0, \tag{3.40}$$

$$\{1\} \cup \left\{\frac{\zeta_0(x)}{\zeta_0(x) - 1}, \ldots, \frac{\zeta_{v-1}(x)}{\zeta_{v-1}(x) - 1}\right\} \quad \text{if} \quad t_e = 0. \tag{3.41}$$

In the case $v < 0$ we obtain analogously that $\sigma(F_x(A))$ is

$$\{0, 1\} \cup \left\{\frac{-1}{\zeta_0(x) - 1}, \ldots, \frac{-1}{\zeta_{|v|-1}(x) - 1}\right\} \quad \text{if} \quad t_o > 0, \tag{3.42}$$

$$\{1\} \cup \left\{\frac{-1}{\zeta_0(x) - 1}, \ldots, \frac{-1}{\zeta_{|v|-1}(x) - 1}\right\} \quad \text{if} \quad t_o = 0. \tag{3.43}$$

Finally, it is evident that $G_m(A) \in \{0, 1\}$ for all m, and it is also clear which value is actually assumed.

In the next chapter we are going to consider singular integral operators on admissible curves. In this context, the case where $\sigma_{\mathscr{F}}(X)$ is a circular arc running from 0 to 1 is of particular interest. For $0 < \gamma < 1$, set

$$\mathfrak{A}_\gamma := \left\{(1 + \coth((y + \mathbf{i}\gamma)\pi))/2 : -\infty < y < \infty\right\} \cup \{0, 1\}. \tag{3.44}$$

Suppose that $\sigma_{\mathscr{F}}(X) = \mathfrak{A}_\gamma$, and let $x \in \mathfrak{A}_\gamma \setminus \{0, 1\}$. Assume first that $v := t_o - t_e > 0$ and $t_e > 0$. Then $\sigma(F_x(A))$ is given by (3.40). If $x = (1 + \coth((y + \mathbf{i}\gamma)\pi))/2$, then a short calculation gives $x/(x-1) = e^{2z}e^{2\pi \mathbf{i}\gamma}$. Consequently, the v roots of $\zeta^v = x/(x-1)$ are

$$\zeta_k(x) = e^{2z/v}e^{2\pi \mathbf{i}(v+k)/v} \quad \text{where} \quad k = 0, \ldots, v-1.$$

Thus, if x traces out $\mathfrak{A}_\gamma \setminus \{0, 1\}$, then

$$\zeta_k(x)/(\zeta_k(x) - 1) = (1 + \coth((z/v + \mathbf{i}(\gamma + k)/v)\pi))/2$$

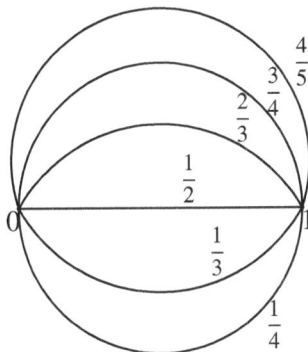

Fig. 3.1 The arc $\mathfrak{A}_\gamma$ for several values of γ. Note that for $\gamma = 1/2$ the "arc" is actually a line segment.

moves along the circular arc $\mathfrak{A}_{(\gamma+k)/\nu} \setminus \{0, 1\}$. In the case $\nu < 0$, we get similarly that if x ranges over $\mathfrak{A}_\gamma \setminus \{0, 1\}$, then $-1/(\zeta_k(x) - 1)$ runs along the circular arc $\mathfrak{A}_{(\gamma+k)/|\nu|} \setminus \{0, 1\}$. Taking into account that spectra are closed we obtain the following from (3.40)–(3.43):

Theorem 3.2.18. *Let $\mathscr{F}$ be $\mathscr{A}$ or $\mathscr{B}$. If $\sigma_{\mathscr{F}}(X)$ is the circular arc $\mathfrak{A}_\gamma$, then the spectrum of the abstract singular integral operator is*

$$\sigma_{\mathscr{F}}(pPp+q) = \begin{cases} \{0, 1\} & \text{if } \nu = 0, \\ \cup_{k=0}^{\nu-1}\mathfrak{A}_{(\gamma+k)/\nu} & \text{if } \nu > 0, \\ \cup_{k=0}^{|\nu|-1}\mathfrak{A}_{(\gamma+k)/|\nu|} & \text{if } \nu < 0. \end{cases}$$

3.2.9 Exercises

Exercise 3.2.1. Find explicit examples showing that all possibilities predicted by the N projections theorem regarding existence of the one-dimensional representations indeed occur.

Exercise 3.2.2. Let $\mathbb{A}$ be the disk algebra introduced in Example 1.2.23. Describe the smallest closed subalgebra of $\mathbb{A}^{2\times 2}$ which contains the following functions on the unit disk

$$P : t \mapsto \begin{bmatrix} t & t-1 \\ t & t-1 \end{bmatrix}, \quad p_1 : t \mapsto \begin{bmatrix} 1 & 0 \\ 0 & 0 \end{bmatrix} \quad \text{and} \quad p_2 : t \mapsto \begin{bmatrix} 0 & 0 \\ 0 & 1 \end{bmatrix}.$$

Exercise 3.2.3. In the previous exercise, substitute $\mathbb{A}$ by $C^1[0,1]$, the Banach algebra of all continuously differentiable functions defined on $[0,1]$, and describe the resulting algebra.

Exercise 3.2.4. Consider again $\mathbb{A}^{2 \times 2}$ with the functions P and p_1 as in Exercise 3.2.2 and with the flip

$$t \mapsto \begin{bmatrix} 0 & 1 \\ 1 & 0 \end{bmatrix}.$$

Are the conditions of the N projections theorem fulfilled?

3.3 Algebras generated by two idempotents and a flip which changes the orientation

The goal of the next two sections is to study the invertibility of elements in Banach algebras which are generated by two idempotents and one flip element. More precisely, let $\mathscr{F}$ be a complex Banach algebra with identity element e, and let p, r and j be elements of $\mathscr{F}$ satisfying

$$p^2 = p, \quad r^2 = r \quad \text{and} \quad j^2 = e. \tag{3.45}$$

The smallest closed subalgebra of $\mathscr{F}$ which contains the idempotents p and r and the flip j will be denoted by $\mathscr{A}$. Clearly, this definition implies that $e \in \mathscr{A}$. We consider two different sets of additional conditions which describe the interaction between the idempotents and the flip, each of which is inspired by a particular class of applications. Either we require

$$jpj = e - p \quad \text{and} \quad jrj = e - r, \tag{3.46}$$

or

$$jpj = p \quad \text{and} \quad jrj = e - r. \tag{3.47}$$

Example 3.3.1. Let $\mathscr{F} = \mathscr{L}\left(L^2(\mathbb{T})\right)$, and let χ stand for the characteristic function of the upper half circle $\{z \in \mathbb{C} : |z| = 1, \operatorname{Im} z > 0\}$. The operators

$$P: \sum_{k=-\infty}^{+\infty} a_k t^k \mapsto \sum_{k=0}^{+\infty} a_k t^k, \quad (Ru)(t) := \chi(t)u(t), \quad (Ju)(t) := \frac{1}{t}u(1/t)$$

satisfy conditions (3.45) and (3.46) in place of p, r and j, respectively.   □

Example 3.3.2. In the context of the previous example, substitute J by the operator $(J_1 u)(t) := u(-t)$. Then P, R and J_1 satisfy the conditions (3.45) and (3.47).   □

Note that the flip J in Example 3.3.1 changes the orientation of the unit circle, whereas the flip J_1 in Example 3.3.2 preserves the orientation. That is why one usually refers to a flip which satisfies (3.46) as *orientation changing* whereas flips which satisfy (3.47) are called *orientation preserving*.

Note also that the distinction between idempotents and flips is not essential. Indeed, $e - 2p$ is a flip for every idempotent p, and $(e + j)/2$ is an idempotent for every flip j. So we can also think of $\mathscr{A}$ as an algebra which is generated by three idempotents or, likewise, by three flips. It is the specific form of the conditions (3.45)–(3.47) and the actual applications in operator theory which let us choose the preferred terminology.

We will soon see that the two sets of conditions in (3.46) and (3.47) imply rather different structures for the corresponding algebras. So we divide the treatment of algebras generated by two idempotents and one flip into two sections and start in this section with algebras where the flip changes the orientation. Thus, we assume that conditions (3.45) and (3.46) are satisfied throughout this section.

We start with some elementary observations.

Proposition 3.3.3. *The subalgebra of $\mathscr{A}$ which is generated by p and j is isomorphic to $\mathbb{C}^{2\times 2}$, and the isomorphism can be chosen in such a way that it takes the elements*

$$e_{11} = p, \quad e_{12} = pj(e - p), \quad e_{21} = (e - p)jp, \quad e_{22} = e - p$$

into the matrices

$$E_{11} = \begin{bmatrix} 1 & 0 \\ 0 & 0 \end{bmatrix}, \quad E_{12} = \begin{bmatrix} 0 & 1 \\ 0 & 0 \end{bmatrix}, \quad E_{21} = \begin{bmatrix} 0 & 0 \\ 1 & 0 \end{bmatrix}, \quad E_{22} = \begin{bmatrix} 0 & 0 \\ 0 & 1 \end{bmatrix},$$

respectively.

Proof. Let $\mathscr{E} \subset \mathscr{A}$ stand for the linear space spanned by e_{11}, e_{12}, e_{21} and e_{22}. These four elements are linearly independent. Indeed, suppose

$$\alpha p + \beta pj(e - p) + \gamma(e - p)jp + \delta(e - p) = 0 \tag{3.48}$$

with some α, β, γ, $\delta \in \mathbb{C}$. Multiplying (3.48) by p or $e - p$ from the left- or right-hand sides yields

$$\alpha p = 0, \quad \beta pj(e - p) = 0, \quad \gamma(e - p)jp = 0, \quad \delta(e - p) = 0.$$

The assumption $p = 0$ together with the axiom $jpj = e - p$ would imply $0 = e$ which was excluded. Thus, $\alpha = 0$ and, analogously, $\delta = 0$. Similarly, if $pj(e - p)$ were 0, then $pj(e - p)j = p$ would be 0, too, which is impossible as we have just seen. Thus, $\beta = 0$ and, analogously, $\gamma = 0$.

So, $\mathscr{E}$ is a four-dimensional subspace of $\mathscr{A}$ which contains the elements $p = e_{11}$, $e = e_{11} + e_{22}$ and $j = e_{12} + e_{21}$. Moreover, $\mathscr{E}$ is an algebra, since

$$e_{ij} \cdot e_{kl} = \delta_{jk} e_{il} \tag{3.49}$$

(with δ_{jk} referring to the Kronecker delta), as one easily checks. Therefore, $\mathscr{E}$ is the smallest algebra, which contains p and j, and the algebra is already closed due to its finite dimensionality. The isomorphy of $\mathscr{E}$ to $\mathbb{C}^{2\times 2}$ (with e_{ij} corresponding to E_{ij}) is an immediate consequence of (3.49) and of the analogous relations holding for the matrices E_{ij}.   ∎

For every $i, j \in \{1, 2\}$, we define mappings

$$w_{ij} : \mathscr{F} \to \mathscr{F}, \quad a \mapsto e_{1i} a e_{j1} + e_{2i} a e_{j2} \tag{3.50}$$

and set

$$w : \mathscr{F} \to \mathscr{F}^{2\times 2}, \quad a \mapsto \begin{bmatrix} w_{11}(a) & w_{12}(a) \\ w_{21}(a) & w_{22}(a) \end{bmatrix}. \tag{3.51}$$

Evidently,

$$w(e) = \begin{bmatrix} e & 0 \\ 0 & e \end{bmatrix}, \quad w(p) = \begin{bmatrix} e & 0 \\ 0 & 0 \end{bmatrix}, \quad w(j) = \begin{bmatrix} 0 & e \\ e & 0 \end{bmatrix}. \tag{3.52}$$

Proposition 3.3.4. *Let $a \in \mathscr{F}$ be an element satisfying $jaj = e - a$. Then:*

(i) $w(a) = \begin{bmatrix} b & c \\ -c & e-b \end{bmatrix}$ *with* $b = pap + (e-p)(e-a)(e-p)$ *and* $c = (pa - ap)j$.

(ii) *The elements b and c commute if and only if* $p(a^2 - a) = (a^2 - a)p$.

Proof. (i) The axiom $jaj = e - a$ and the equalities (3.52) imply that $w(a) = \begin{bmatrix} b & c \\ -c & e-b \end{bmatrix}$ with some elements $b, c \in \mathscr{F}$. Further,

$$\begin{aligned} b &= w_{11}(a) = pap + (e-p)jaj(e-p) = pap + (e-p)(e-a)(e-p), \\ c &= w_{12}(a) = pa(e-p)j + (e-p)ja(e-p) \\ &= (pa(e-p) + (e-p)(e-a)p)j = (pa - ap)j. \end{aligned}$$

(ii) It is easy to check that

$$bc - cb = p(a^2 - a)jp + (e-p)(a^2 - a)j(e-p).$$

Thus, $bc - cb = 0$ if and only if

$$p(a^2 - a)jp = 0 \text{ and } (e-p)(a^2 - a)j(e-p) = 0.$$

Since $jpj = e - p$ and j is invertible, this is equivalent to

$$p(a^2 - a)(e-p) = (e-p)(a^2 - a)p = 0$$

which, in turn, is equivalent to $p(a^2 - a) = (a^2 - a)p$.   ∎

In particular, Proposition 3.3.4 applies to the element $a = r$; so let $b, c \in \mathscr{A}$ be defined by

$$\begin{aligned} b &:= prp + (e-p)(e-r)(e-p) = e - p - r + pr + rp, \tag{3.53} \\ c &:= (pr - rp)j. \tag{3.54} \end{aligned}$$

Then we have $bc = cb$ due to Proposition 3.3.4 (ii) and, moreover,

$$c^2 = b(b-e), \tag{3.55}$$

the latter fact being a consequence of $r^2 = r$ and

$$\begin{bmatrix} b & c \\ -c & e-b \end{bmatrix}^2 = \begin{bmatrix} b & c \\ -c & e-b \end{bmatrix} = w(r).$$

Now observe that finite products having the matrices $w(e)$, $w(p)$, $w(j)$ and $w(r)$ as factors are matrices with entries which are polynomials in b and c.

Let $\mathscr{C}$ denote the smallest closed subalgebra of $\mathscr{A}$ which contains the elements b, c and e. The following is a summary of Theorem 1.1.17 and Propositions 3.3.3–3.3.4.

Proposition 3.3.5. *Let $\mathscr{A}$ be a Banach algebra with identity e which is generated by elements p, r, j satisfying (3.45) and (3.46), and let $\mathscr{C}$ be the smallest closed subalgebra of $\mathscr{A}$ generated by e and the elements b, c, defined in (3.53) and (3.54). Then $\mathscr{C}$ is the center of $\mathscr{A}$, and the mapping w defined by (3.51) is a continuous isomorphism from $\mathscr{A}$ onto $\mathscr{C}^{2\times2}$. The action of w on the generating elements of $\mathscr{A}$ is given by (3.52) and Proposition 3.3.4 (i).*

Thus, an element $a \in \mathscr{A}$ is invertible in $\mathscr{A}$ if and only if the matrix $w(a)$ is invertible in $\mathscr{C}^{2\times2}$. Since $\mathscr{C}$ is commutative, an element of $\mathscr{C}^{2\times2}$ is invertible if and only if its determinant (which can be defined as is usual) is invertible in $\mathscr{C}$, and the latter invertibility is subject to commutative Gelfand theory. If we let $M_\mathscr{C}$ refer to the maximal ideal space of $\mathscr{C}$ and $\widehat{} : \mathscr{C} \to C(M_\mathscr{C})$ to the Gelfand transform, then we arrive at the following result.

Theorem 3.3.6. *An element $a \in \mathscr{A}$ is invertible in $\mathscr{A}$ if and only if the function*

$$M_\mathscr{C} \to \mathbb{C}^{2\times2}, \quad x \mapsto \begin{bmatrix} \widehat{w_{11}(a)}(x) & \widehat{w_{12}(a)}(x) \\ \widehat{w_{21}(a)}(x) & \widehat{w_{22}(a)}(x) \end{bmatrix}$$

is invertible at every point $x \in M_\mathscr{C}$.

This theorem provides us with the desired matrix-valued calculus for the algebra $\mathscr{A}$. The symbol can be realized as a 2×2 matrix function on the maximal ideal space $M_\mathscr{C}$ of $\mathscr{C}$, and it is continuous if $M_\mathscr{C}$ is endowed with its Gelfand topology.

Remark 3.3.7. The information $bc = cb$ and $c^2 = b(b-e)$ about b and c is all that can be derived from the axioms (3.45) and (3.46). Indeed, if b' and c' are elements of a Banach algebra $\mathscr{F}'$ with identity e' with $b'c' = c'b'$ and $c'^2 = b'(b'-e')$ then

$$e = \begin{bmatrix} e' & 0 \\ 0 & e' \end{bmatrix}, \quad p = \begin{bmatrix} e' & 0 \\ 0 & 0 \end{bmatrix}, \quad j = \begin{bmatrix} 0 & e' \\ e' & 0 \end{bmatrix}, \quad r = \begin{bmatrix} b' & c' \\ -c' & e'-b' \end{bmatrix}$$

are elements of $\mathscr{F}'^{2\times2}$ which satisfy (3.45) and (3.46). Thus, the problem of establishing a matrix-valued symbol calculus for an algebra $\mathscr{A}$ which is generated by elements p, r, j satisfying (3.45) and (3.46) is completely equivalent to the determination of the maximal ideal space of the commutative Banach algebra which is generated by e and by elements b, c satisfying $c^2 = b(b - e)$ and to the computation of the Gelfand transform, at least for the generating elements b and c. This will be done in the forthcoming subsections.   □

3.3.1 Properties of the maximal ideal space $M_{\mathscr{C}}$

We are going to establish some characterizing properties of the maximal ideal space $M_{\mathscr{C}}$ of $\mathscr{C}$. Since $\mathscr{C}$ is the center of $\mathscr{A}$, one has $\sigma_{\mathscr{C}}(d) = \sigma_{\mathscr{A}}(d)$ for every $d \in \mathscr{C}$.

Since $\mathscr{C}$ is both a commutative algebra and generated by its elements b and c, $M_{\mathscr{C}}$ is homeomorphic to the joint spectrum $\sigma_{\mathscr{C}}(b,c)$ of b and c in $\mathscr{C}$, see Exercise 2.1.5. Recall from Proposition 1.3.24 that the joint spectrum $\sigma_{\mathscr{C}}(b,c)$ coincides with the set of all pairs $(x,y) \in \mathbb{C} \times \mathbb{C}$ such that the smallest closed ideal of $\mathscr{C}$ which contains $b - xe$ and $c - ye$ is proper.

Throughout the following, we shall identify $M_{\mathscr{C}}$ and $\sigma_{\mathscr{C}}(b,c)$. Then we have

$$\widehat{\mathcal{P}(b,c)}(x,y) = \mathcal{P}(x,y) \tag{3.56}$$

for every polynomial $\mathcal{P}$ in two variables and every pair $(x,y) \in M_{\mathscr{C}}$. We have already remarked that if $(x,y) \in M_{\mathscr{C}}$, then $x \in \sigma_{\mathscr{C}}(b)$. Conversely, given a point $x \in \sigma_{\mathscr{C}}(b)$, there is (at least) one $y \in \sigma_{\mathscr{C}}(c)$ such that $(x,y) \in M_{\mathscr{C}}$. This is a simple consequence of (3.55) and of the fact that the range of the Gelfand transform of an element coincides with the spectrum of that element. Hence, $M_{\mathscr{C}} = \sigma_{\mathscr{C}}(b,c)$ can be decomposed into fibers over $\sigma_{\mathscr{C}}(b)$: Given $x \in \sigma_{\mathscr{C}}(b)$, set $M_{\mathscr{C}}^x := \{(x,y) \in M_{\mathscr{C}}\}$. Then none of the fibers $M_{\mathscr{C}}^x$ is empty, and $\cup_{x \in \sigma_{\mathscr{C}}(b)} M_{\mathscr{C}}^x = M_{\mathscr{C}}$.

Proposition 3.3.8. *Let $x \in \sigma_{\mathscr{C}}(b)$. Then the fiber $M_{\mathscr{C}}^x$ is either a singleton $\{(x,y)\}$ or a doubleton $\{(x,y),(x,-y)\}$. In both cases, $y \in \mathbb{C}$ is a number such that $y^2 = x(x-1)$.*

Proof. From (3.56) we conclude that the value of the Gelfand transform of $c^2 - b(b - e)$ at (x,y) is $y^2 - x(x - 1)$. Since $c^2 - b(b - e) = 0$, this implies that

$$y^2 = x(x - 1) \quad \text{for every } (x,y) \in M_{\mathscr{C}}. \tag{3.57}$$

Thus, given $x \in \sigma_{\mathscr{C}}(b)$, there are at most two numbers y such that $(x,y) \in M_{\mathscr{C}}$.   ■

We denote the sets

$$\{x \in \sigma_{\mathscr{C}}(b) : M_{\mathscr{C}}^x \text{ is a singleton}\} \quad \text{and} \quad \{x \in \sigma_{\mathscr{C}}(b) : M_{\mathscr{C}}^x \text{ is a doubleton}\}$$

by Σ_1 and Σ_2, respectively.

There is an alternate way to derive this fibration, namely via Allan's local principle. We shall apply this theorem with $\mathscr{C}$ as the larger algebra and $\mathrm{alg}\{b\}$ in place of the subalgebra of the center of $\mathscr{C}$. We have seen in Exercise 2.1.3 that the maximal ideal space of the (singly generated) Banach algebra $\mathrm{alg}\{b\}$ is homeomorphic to the spectrum of b in this algebra. Moreover, the only maximal ideals $x \in \sigma_{\mathrm{alg}\{b\}}(b)$ which generate proper ideals $\mathscr{I}_x$ in $\mathscr{C}$ are those in $\sigma_{\mathscr{C}}(b)$. Thus, Allan's local principle offers the possibility of localizing $\mathscr{C}$ over $\sigma_{\mathscr{C}}(b)$. Given $x \in \sigma_{\mathscr{C}}(b)$, we denote the related local algebra $\mathscr{C}/\mathscr{I}_x$ by $\mathscr{C}_x$ and the canonical homomorphism from $\mathscr{C}$ onto $\mathscr{C}_x$ by Φ_x.

Proposition 3.3.9. *Let $(x,y) \in \sigma_{\mathscr{C}}(b) \times \sigma_{\mathscr{C}}(c)$. Then $(x,y) \in M_{\mathscr{C}}$ if and only if $y \in \sigma_{\mathscr{C}_x}(\Phi_x(c))$.*

Proof. If $(x,y) \notin M_{\mathscr{C}}$, then the ideal of $\mathscr{C}$ which is generated by $b - xe$ and $c - ye$ is not proper; that is, there are elements $f, g \in \mathscr{C}$ such that

$$f(b - xe) + g(c - ye) = e.$$

Applying the local homomorphism Φ_x to this equality and taking into account that $\Phi_x(b) = x\Phi_x(e)$, we find

$$\Phi_x(g)\left(\Phi_x(c) - y\Phi_x(e)\right) = \Phi_x(e),$$

i.e., $y \notin \sigma_{\mathscr{C}_x}(\Phi_x(c))$. Let, conversely, $y \notin \sigma_{\mathscr{C}_x}(\Phi_x(c))$. Then there are elements $f \in \mathscr{C}$ and $j \in \mathscr{I}_x$ such that
$$f(c - ye) = e + j.$$
The set $\mathscr{C}(b - xe)$ is dense in $\mathscr{I}_x$; hence, there are elements $g, h \in \mathscr{C}$ with $\|h\| < 1$ and $j = -g(b - xe) + h$. Consequently,

$$(e + h)^{-1}g(b - xe) + (e + h)^{-1}f(c - ye) = e$$

which implies that $(x,y) \notin M_{\mathscr{C}}$.   ∎

In particular, if $x \in \Sigma_1$ (respectively $x \in \Sigma_2$), then $\sigma_{\mathscr{C}_x}(\Phi_x(c))$ consists of one (respectively two) points, and the maximal ideal space of $\mathscr{C}_x$ is a singleton (respectively a doubleton). We set $\Sigma_0 := \sigma_{\mathscr{C}}(b) \cap \{0,1\}$. Evidently, $\Sigma_0 \subseteq \Sigma_1$.

Theorem 3.3.10. *The following assertions hold:*

(i) *the set $\Sigma_2 \cup \Sigma_0$ is closed in $\sigma_{\mathscr{C}}(b)$ (hence, compact);*
(ii) *the set $\Sigma_1 \setminus \Sigma_0$ is open in $\sigma_{\mathscr{C}}(b)$, and the mapping which assigns to every $x \in \Sigma_1 \setminus \Sigma_0$ the second component of the (uniquely determined) pair $(x,y) \in M_{\mathscr{C}}$ is continuous on $\Sigma_1 \setminus \Sigma_0$.*

Proof. (i) Let x_0 belong to the closure of $\Sigma_2 \cup \Sigma_0$. What we have to show is that $x_0 \in \Sigma_2 \cup \Sigma_0$. If $x_0 \in \Sigma_0$, then we are done. Let $x_0 \notin \Sigma_0$. Then there is a sequence $(x_n) \subseteq \Sigma_2$ which converges to x_0. Our assumption $x_0 \notin \Sigma_0$ guarantees that $x_0^2 - x_0 \neq$

0, thus, we can find an open neighborhood U of $x_0^2 - x_0$ in $\mathbb{C}$ such that there exists a continuous branch, sqrt say, of the square root function on U. Hence,

$$\sigma\left(\Phi_{x_n}(c)\right) = \left\{\mathrm{sqrt}(x_n^2 - x_n), -\mathrm{sqrt}(x_n^2 - x_n)\right\}$$

for all sufficiently large n. From Proposition 2.2.3 (ii) we infer that

$$\lim_{n\to\infty} \sigma\left(\Phi_{x_n}(c)\right) = \left\{\mathrm{sqrt}(x_0^2 - x_0), -\mathrm{sqrt}(x_0^2 - x_0)\right\} \subseteq \sigma\left(\Phi_{x_0}(c)\right).$$

But $\mathrm{sqrt}(x_0^2 - x_0) \neq -\mathrm{sqrt}(x_0^2 - x_0)$ because $x_0 \notin \Sigma_0$, hence, $\sigma\left(\Phi_{x_0}(c)\right)$ consists of two points, i.e., $x_0 \in \Sigma_2$.

(ii) It is immediate from part (i) that $\Sigma_1 \setminus \Sigma_0$ is an open subset of $\sigma_\mathscr{C}(b)$. What remains to show is the continuity of the mapping

$$\Sigma_1 \setminus \Sigma_0 \to \sigma_\mathscr{C}(c), \quad x \mapsto (x,y) \mapsto y.$$

If $x_0 \in \Sigma_1 \setminus \Sigma_0$, then $x_0^2 - x_0 \neq 0$, and one can again choose a neighborhood U of $x_0^2 - x_0$ in $\mathbb{C}$ as well as a continuous branch sqrt of the square root function on U such that

$$y_0 = y(x_0) = \mathrm{sqrt}(x_0^2 - x_0).$$

Let V be an open neighborhood of x_0 in $\Sigma_1 \setminus \Sigma_0$ such that $x^2 - x \in U$ for all $x \in V$. Then, for every $x \in V$, either $y(x) = \mathrm{sqrt}(x^2 - x)$ or $y(x) = -\mathrm{sqrt}(x^2 - x)$. Assume there exists a sequence $(x_n) \subseteq V$ tending to x_0 such that $y(x_n) = -\mathrm{sqrt}(x_n^2 - x_n)$ for every n. Since

$$y(x_n) = -\mathrm{sqrt}(x_n^2 - x_n) \to -\mathrm{sqrt}(x_0^2 - x_0)$$

we conclude, again via Proposition 2.2.3 (ii), that

$$-y_0 = -\mathrm{sqrt}(x_0^2 - x_0) \in \sigma\left(\Phi_{x_0}(c)\right).$$

Since $y_0 \neq -y_0$ (a consequence of $x_0 \notin \Sigma_0$) and $y_0 \in \sigma\left(\Phi_{x_0}(c)\right)$ (by assumption), this contradicts the hypothesis $x_0 \in \Sigma_1$. Therefore,

$$y = y(x) = \mathrm{sqrt}(x^2 - x) \text{ for all } x \in V$$

which implies the continuity of the function $x \mapsto y$ at each point of $\Sigma_1 \setminus \Sigma_0$.   ∎

The preceding theorem characterizes the potential symbol algebras completely. Indeed, let Σ' be a compact subset of $\mathbb{C}$, set $\Sigma'_0 = \Sigma' \cap \{0,1\}$, and let Σ'_1, Σ'_2 be subsets of Σ' such that

$$\Sigma'_1 \cap \Sigma'_2 = \emptyset, \quad \Sigma'_1 \cup \Sigma'_2 = \Sigma', \quad \Sigma'_0 \subseteq \Sigma'_1, \quad \Sigma'_2 \cup \Sigma'_0 \text{ is compact}$$

and such that there is a continuous function $y' = y'(x)$ on $\Sigma'_1 \setminus \Sigma'_0$ satisfying

$$y'(x)^2 = x^2 - x \text{ for all } x \in \Sigma'_1 \setminus \Sigma'_0.$$

Let $M'_\mathscr{C}$ be the set

$$M'_\mathscr{C} := \{(x,y) : x \in \Sigma'_1 \setminus \Sigma'_0,\ y = y'(x)\} \cup \{(x,y) : x \in \Sigma'_2 \cup \Sigma'_0,\ y^2 = x^2 - x\},$$

and define functions e', p', j', r' from $M'_\mathscr{C}$ into $\mathbb{C}^{2\times 2}$ by

$$e'(x,y) = \begin{bmatrix} 1 & 0 \\ 0 & 1 \end{bmatrix}, \quad p'(x,y) = \begin{bmatrix} 1 & 0 \\ 0 & 0 \end{bmatrix}, \quad j'(x,y) = \begin{bmatrix} 0 & 1 \\ 1 & 0 \end{bmatrix}$$

and

$$r'(x,y) = \begin{bmatrix} x & y \\ -y & 1-x \end{bmatrix}.$$

Then e', p', j' and r' satisfy the axioms (3.45) and (3.46) in place of e, p, j and r, respectively. The indicator elements corresponding to e', p', j' and r' are

$$b'(x,y) = \begin{bmatrix} x & 0 \\ 0 & x \end{bmatrix} \quad \text{and} \quad c'(x,y) = \begin{bmatrix} y & 0 \\ 0 & y \end{bmatrix},$$

and for the corresponding maximal ideal space and its subsets one finds $M_\mathscr{C} = M'_\mathscr{C}$, $\Sigma_1 = \Sigma'_1$, and $\Sigma_2 = \Sigma'_2$.

3.3.2 Determination of the maximal ideal space $M_\mathscr{C}$

Now we are going to point out one way to identify the components Σ_0, Σ_1 and Σ_2 of $\sigma_\mathscr{C}(b)$ as well as the correct branch of the square root function for every connected component of $\Sigma_1 \setminus \Sigma_0$. In particular, we shall see that the knowledge of the (global) spectrum of only one element of $\mathscr{C}$ is sufficient for the desired identifications. This element is $b+c$. To get this, we need an elementary observation.

Proposition 3.3.11. *Let x_1, x_2, y_1, $y_2 \in \mathbb{C}$ satisfy the conditions*

$$y_1^2 = x_1^2 - x_1, \quad y_2^2 = x_2^2 - x_2,\ and\ x_1 + y_1 = x_2 + y_2.$$

Then $x_1 = x_2$ and $y_1 = y_2$.

Proof. Squaring the equality $x_1 - x_2 = y_2 - y_1$ we get

$$x_1^2 - 2x_1x_2 + x_2^2 = y_1^2 - 2y_1y_2 + y_2^2 = x_1^2 - x_1 - 2y_1y_2 + x_2^2 - x_2$$

so that

$$2x_1x_2 - x_1 - x_2 = 2y_1y_2.$$

Squaring once more, we obtain

$$x_1^2 + x_2^2 + 4x_1^2x_2^2 - 4x_1^2x_2 - 4x_1x_2^2 + 2x_1x_2 = 4y_1^2y_2^2$$
$$= 4(x_1^2 - x_1)(x_2^2 - x_2) = 4x_1^2x_2^2 - 4x_1^2x_2 - 4x_1x_2^2 + 4x_1x_2$$

so that

$$x_1^2 - 2x_1x_2 + x_2^2 = (x_1 - x_2)^2 = 0,$$

whence the assertion follows.   ∎

Now we can characterize the points in $M_{\mathscr{C}}$.

Proposition 3.3.12. *The mapping*

$$M_{\mathscr{C}} \to \mathbb{C}, \quad (x,y) \mapsto x+y \tag{3.58}$$

is a bijection between $M_{\mathscr{C}}$ and $\sigma_{\mathscr{C}}(b+c)$.

Proof. From (3.56) we infer that

$$\widehat{b+c}(x,y) = x+y. \tag{3.59}$$

Since the range of the Gelfand transform of an element coincides with the spectrum of that element, (3.59) implies that (3.58) is a mapping from $M_{\mathscr{C}}$ onto $\sigma_{\mathscr{C}}(b+c)$. It remains to verify the injectivity of (3.58). Let $z \in \sigma_{\mathscr{C}}(b+c)$. Then, as we have just seen, there is (at least) one point $(x,y) \in M_{\mathscr{C}}$ such that $z = x+y$. But $y^2 = x^2 - x$ for every pair $(x,y) \in M_{\mathscr{C}}$ by (3.57). Hence, the point (x,y) is unique by Proposition 3.3.11.   ∎

Actually, Proposition 3.3.11 yields a little bit more than we have employed in the proof of Proposition 3.3.12. Namely, every point $z \in \sigma_{\mathscr{C}}(b+c)$ can be uniquely written as $x+y$ where $y^2 = x^2 - x$. We conclude from Proposition 3.3.12 that then, necessarily, $x \in \sigma_{\mathscr{C}}(b)$, $y \in \sigma_{\mathscr{C}}(c)$, and $(x,y) \in M_{\mathscr{C}}$. Moreover, if we have the summands x and y at our disposal, we can ask whether or not $z' = x - y$ also belongs to $\sigma_{\mathscr{C}}(b+c)$. If no, then $x \in \Sigma_1$, if yes, then $x \in \Sigma_2$. Finally, if $x \in \Sigma_1 \setminus \Sigma_0$ then the associated y indicates which branch of the square root function has to be chosen in an open neighborhood of x. These observations allow one to determine $M_{\mathscr{C}}$, Σ_0, Σ_1, Σ_2 as well as the explicit form of the symbol calculus in practice.

We summarize these facts in the following theorem.

Theorem 3.3.13. *Let $\mathscr{A}$ be a Banach algebra generated by elements p, r and j satisfying (3.45) and (3.46), and let $\mathscr{C}$ be the smallest closed subalgebra of $\mathscr{A}$ generated by e and by the elements b and c defined by (3.53) and (3.54). Further, let $M_{\mathscr{C}}$ consist of all pairs $(x,y) \in \mathbb{C} \times \mathbb{C}$ where $x^2 - x = y^2$ and $x+y \in \sigma_{\mathscr{C}}(b+c)$.*

(i) *The set $M_{\mathscr{C}}$ (provided with the topology inherited from that of $\mathbb{C} \times \mathbb{C}$) is homeomorphic to the maximal ideal space of $\mathscr{C}$.*

(ii) *The mapping* smb *which assigns to e, p, j and r a matrix-valued function on $M_{\mathscr{C}}$ by*

$$(\mathrm{smb}\,e)(x,y) = \begin{bmatrix} 1 & 0 \\ 0 & 1 \end{bmatrix}, \quad (\mathrm{smb}\,p)(x,y) = \begin{bmatrix} 1 & 0 \\ 0 & 0 \end{bmatrix}$$

$$(\mathrm{smb}\,j)(x,y) = \begin{bmatrix} 0 & 1 \\ 1 & 0 \end{bmatrix}, \quad (\mathrm{smb}\,r)(x,y) = \begin{bmatrix} x & y \\ -y & 1-x \end{bmatrix}$$

extends to a continuous homomorphism from $\mathscr{A}$ into $C(M_{\mathscr{C}}, \mathbb{C}^{2\times 2})$.

(iii) *An element $a \in \mathscr{A}$ is invertible if and only if $(\mathrm{smb}\,a)(x,y)$ is invertible for every $(x,y) \in M_{\mathscr{C}}$.*

Example 3.3.14. Here we discuss C^*-algebras generated by two projections and one flip, and compare the results with [149]. Let $\mathscr{F}$ be a C^*-algebra with identity e, and let p, r, j be self-adjoint elements of $\mathscr{F}$ satisfying

$$p^2 = p, \quad r^2 = r, \quad j^2 = e, \quad jpj = e - p, \quad jqj = e - r. \tag{3.60}$$

Further we suppose

$$\sigma(prp) = [0,1] \quad \text{and} \quad \mathbf{i}\,pjrp \geq 0, \tag{3.61}$$

where $\mathbf{i}$ denotes the imaginary unit. Define $b_0 := prp$ and $c_0 := \mathbf{i}\,pjrp$. It is elementary to check that

$$b = b_0 + jb_0 j \quad \text{and} \quad -\mathbf{i}c = c_0 + jc_0 j. \tag{3.62}$$

Further, if a is an element of the algebra $p\mathscr{A}p$ (with identity element p), then jaj belongs to the algebra $(e-p)\mathscr{A}(e-p)$ (with identity $e-p$), and $a - \lambda p$ is invertible in $p\mathscr{A}p$ if and only if $jaj - \lambda(e-p)$ is invertible in $(e-p)\mathscr{A}(e-p)$. From this observation one easily gets

$$\sigma_{\mathscr{A}}(a) \subseteq \sigma_{\mathscr{A}}(a + jaj) \cup \{0\} \quad \text{for every } a \in p\mathscr{A}p.$$

This fact together with (3.62) implies that (3.61) is equivalent to

$$\sigma(b) = [0,1] \quad \text{and} \quad -\mathbf{i}c \geq 0.$$

Since $(-\mathbf{i}c)^2 = b - b^2$, this further implies that $-\mathbf{i}c$ is the (non-negative) square root of the non-negative element $b - b^2$, whence $c \in \mathrm{alg}\{e,b\}$. Consequently, $\sigma(b) = [0,1]$ coincides with its component Σ_1. If we let $\sqrt{\cdot}$ refer to the branch of the square root function on $[0,1]$ with $\sqrt{1} = 1$, then Theorem 3.3.13 applies to this context as follows.

Theorem 3.3.15 (Power). *Let $\mathscr{A}$ be a C^*-algebra generated by self-adjoint elements e, p, j, r satisfying (3.60) and (3.61). Then $\mathscr{A}$ is isometrically isomorphic to the algebra $C([0,1], \mathbb{C}^{2\times 2})$, and the isomorphism maps e, p, j, r to the functions $\hat{e}$, $\hat{p}$, $\hat{j}$, $\hat{r}$ given by*

$$\hat{e}(x) = \begin{bmatrix} 1 & 0 \\ 0 & 1 \end{bmatrix}, \qquad \hat{p}(x) = \begin{bmatrix} 1 & 0 \\ 0 & 0 \end{bmatrix},$$

$$\hat{j}(x) = \begin{bmatrix} 0 & 1 \\ 1 & 0 \end{bmatrix}, \; \hat{r}(x) = \begin{bmatrix} x & i\sqrt{x(1-x)} \\ -i\sqrt{x(1-x)} & 1-x \end{bmatrix}.$$

For a proof, observe that $M_{\mathscr{C}}$ can be identified with Σ_1 (because $\Sigma_2 = \emptyset$). Thus, only functions depending on one variable x appear. That $\mathscr{A}$ is isomorphic to the whole algebra $C\left([0,1], \mathbb{C}^{2 \times 2}\right)$ is a consequence of the Gelfand-Naimark theorem for commutative C^*-algebras, which implies that

$$\mathscr{C} = \text{alg}\{e, b\} \cong C\left(\sigma(b)\right) = C[0,1].$$

Note that the form of the symbol in Theorem 3.3.15 differs slightly from the one used by Power ([149, Lemma 7.1]). Power's choice follows from the one presented here by multiplying $\hat{e}$, $\hat{p}$, $\hat{j}$ and $\hat{r}$ by $\begin{bmatrix} 1 & 0 \\ 0 & i \end{bmatrix}$ and $\begin{bmatrix} 1 & 0 \\ 0 & -i \end{bmatrix}$ from the left- and right- hand side, respectively.   □

Example 3.3.16. Now we weaken the assumptions (3.61) in the previous example and require merely that

$$\sigma_{\mathscr{C}}(b) \setminus \Sigma_0 \text{ is connected and } \Sigma_2 = \emptyset. \tag{3.63}$$

Further, we will not consider C^*-algebras only, but Banach algebras again. Then we conclude from Theorem 3.3.10 that there is a *continuous* function $y = y(x)$ on $\sigma_{\mathscr{C}}(b) \setminus \Sigma_0$ which satisfies $y^2 = x^2 - x$ such that

$$(\text{smb}\, e)(x) = (\text{smb}\, e)\,(x, y(x)) = \begin{bmatrix} 1 & 0 \\ 0 & 1 \end{bmatrix},$$

$$(\text{smb}\, p)(x) = (\text{smb}\, p)\,(x, y(x)) = \begin{bmatrix} 1 & 0 \\ 0 & 0 \end{bmatrix},$$

$$(\text{smb}\, j)(x) = (\text{smb}\, j)\,(x, y(x)) = \begin{bmatrix} 0 & 1 \\ 1 & 0 \end{bmatrix}.$$

Further, $(\text{smb}\, r)(x) = (\text{smb}\, r)\,(x, y(x))$ is either

$$\begin{bmatrix} x & y(x) \\ -y(x) & 1-x \end{bmatrix} \quad \text{or} \quad \begin{bmatrix} x & -y(x) \\ y(x) & 1-x \end{bmatrix}$$

for every $x \in \sigma_{\mathscr{C}}(b) \setminus \Sigma_0$ (and, thus, for every $x \in \sigma_{\mathscr{C}}(b)$, since $y(0) = y(1) = 0$). Consequently, one can consider the symbol as a function which depends on only *one* complex variable, and to determine this function it is sufficient to determine the sign of y in the symbol of r for only *one* point $x \in \sigma_{\mathscr{C}}(b) \setminus \Sigma_0$; then this sign is the same for all points $x \in \sigma_{\mathscr{C}}(b)$.   □

3.3.3 An alternate description of $M_{\mathscr{C}}$

Let $\mathscr{A}_0$ refer to the smallest (not necessarily closed) subalgebra of $\mathscr{F}$ which contains p, r and j, define $\mathscr{C}_0$ as the smallest (not necessarily closed) subalgebra of $\mathscr{A}_0$ which contains e, b and c, and $\mathrm{alg}_0\{b\}$ as the non-closed algebra generated by e and b. As above we conclude that $\mathrm{alg}_0\{b\}$ is in the center of $\mathscr{A}_0$ and that $\mathscr{C}_0$ coincides with the center of $\mathscr{A}_0$. The following proposition offers an alternate way to derive Theorem 3.3.6 via Corollary 2.6.22.

Proposition 3.3.17. *The algebra $\mathscr{A}_0$ is a module of dimension at most four over $\mathscr{C}_0$ and a module of dimension at most eight over* $\mathrm{alg}_0\{b\}$.

Proof. The first assertion is an immediate consequence of Theorem 3.3.6 (choose p, pj, jp and $e-p$ as a basis of the module), but it can also be verified directly without effort. For the second assertion, define elements $e_1, \ldots, e_8$ by

$$\begin{aligned}
e_1 &= pr, & e_5 &= e_1 j = j e_4, \\
e_2 &= (e-p)r, & e_6 &= e_2 j = j e_3, \\
e_3 &= p(e-r), & e_7 &= e_3 j = j e_2, \\
e_4 &= (e-p)(e-r), & e_8 &= e_4 j = j e_1,
\end{aligned}$$

and let $\mathscr{D}$ stand for the set of all elements $\sum r_i(b)e_i$ with polynomials r_i in b. Evidently, $\mathscr{D}$ is a linear space. For a proof that $\mathscr{D}$ is an algebra it is sufficient to check the multiplication table

$\cdot$	e_1	e_2	e_3	e_4
e_1	be_1	$(1-b)e_1$	be_3	$-be_3$
e_2	be_2	$(1-b)e_2$	$(b-1)e_4$	$(1-b)e_4$
e_3	$(1-b)e_1$	$(b-1)e_1$	$(1-b)e_3$	be_3
e_4	$-be_2$	be_2	$(1-b)e_4$	be_4

as well as the identities

$$e_m e_{4+n} = e_m e_n j,$$
$$e_{4+m} e_n = e_m j e_n = e_m e_{5-n} j,$$
$$e_{4+m} e_{4+n} = e_m j e_n j = e_m e_{5-n}$$

for $m, n \in \{1, 2, 3, 4\}$. Thus, $\mathscr{D}$ is an algebra, and p, r and j belong to $\mathscr{D}$ because

$$p = e_1 + e_3, \quad r = e_1 + e_2 \quad \text{and} \quad j = e_5 + e_6 + e_7 + e_8.$$

Conversely, every algebra which contains p, r and j also includes $\mathscr{D}$, which proves the second assertion. ∎

Employing Allan's local principle, we can localize the algebra $\mathscr{A}$ over its central subalgebra $\mathrm{alg}\{b\}$, the maximal ideal space of which is homeomorphic to $\sigma_{\mathscr{A}}(b)$. We denote the local coset of $a \in \mathscr{A}$ at $x \in \sigma_{\mathscr{A}}(b)$ by $\Phi_x(a)$. Since $\mathscr{D}$ is dense in $\mathscr{A}$, the image $\Phi_x(\mathscr{D})$ of $\mathscr{D}$ under Φ_x is dense in $\Phi_x(\mathscr{A}) = \mathscr{A}/\mathscr{I}_x$. But

$$\Phi_x\left(\sum r_i(b)e_i\right) = \sum r_i(x)\Phi_x(e_i),$$

thus, $\Phi_x(\mathscr{D})$ is a complex linear space of dimension at most eight. In particular, this implies that $\Phi_x(\mathscr{D})$ is closed in $\mathscr{A}/\mathscr{I}_x$. Hence, $\mathscr{A}/\mathscr{I}_x = \Phi_x(\mathscr{D})$, and $\mathscr{A}/\mathscr{I}_x$ is a linear space of dimension at most eight. We are going to show that the dimension of $\mathscr{A}/\mathscr{I}_x$ is equal to eight.

Proposition 3.3.18. *The cosets $\Phi_x(e_1),\dots,\Phi_x(e_8)$ are linearly independent if and only if the cosets $\Phi_x(e_1)$ and $\Phi_x(e_7)$ are linearly independent.*

Proof. Suppose there are complex numbers α_i with $\sum|\alpha_i| \neq 0$ such that

$$\alpha_1\Phi_x(e_1) + \dots + \alpha_8\Phi_x(e_8) = 0.$$

Combining multiplying this equality by $\Phi_x(p)$ or $\Phi_x(e-p)$ from the left-hand side with $\Phi_x(r)$ or $\Phi_x(e-r)$ from the right-hand side, we obtain the four equations

$$\alpha_1\Phi_x(e_1) + \alpha_7\Phi_x(e_7) = \alpha_1\Phi_x(pr) + \alpha_7\Phi_x(pjr) = 0,$$
$$\alpha_2\Phi_x(e_2) + \alpha_8\Phi_x(e_8) = \alpha_2\Phi_x((e-p)r) + \alpha_8\Phi_x((e-p)jr) = 0,$$
$$\alpha_3\Phi_x(e_3) + \alpha_5\Phi_x(e_5) = \alpha_3\Phi_x(p(e-r)) + \alpha_5\Phi_x(pj(e-r)) = 0,$$
$$\alpha_4\Phi_x(e_4) + \alpha_6\Phi_x(e_6) = \alpha_4\Phi_x((e-p)(e-r)) + \alpha_6\Phi_x((e-p)j(e-r)) = 0.$$
$$\tag{3.64}$$

Thus, at least one of the pairs

$$(\Phi_x(e_1), \Phi_x(e_7)), \ (\Phi_x(e_2), \Phi_x(e_8)), \ (\Phi_x(e_3), \Phi_x(e_5)), \ (\Phi_x(e_4), \Phi_x(e_6)) \quad (3.65)$$

is linearly dependent. For definiteness, let $(\Phi_x(e_1), \Phi_x(e_7))$ be a pair with this property. Multiplying the equality (3.64) from the left by $\Phi_x(j)$ we get

$$\alpha_1\Phi_x(jpr) + \alpha_7\Phi_x(jpjr) = \alpha_1\Phi_x((e-p)jr) + \alpha_7\Phi_x((e-p)r) = 0,$$

whence

$$\alpha_7\Phi_x(e_2) + \alpha_1\Phi_x(e_8) = 0.$$

Analogously, (3.64) implies

$$\alpha_7\Phi_x(e_3) + \alpha_1\Phi_x(e_5) = 0 \quad \text{and} \quad \alpha_7\Phi_x(e_4) + \alpha_1\Phi_x(e_6) = 0.$$

Hence, all pairs (3.65) are linearly dependent. Repeating these arguments with another pair in place of $(\Phi_x(e_1), \Phi_x(e_7))$, we conclude that, whenever one of the pairs (3.65) is linearly dependent, then all pairs (3.65) have this property. Thus, if the $\Phi_x(e_i), i = 1,\dots,8$, are linearly dependent, then $\Phi_x(e_1)$ and $\Phi_x(e_7)$ are linearly dependent. The reverse conclusion is obvious. ∎

Proposition 3.3.19. *The dimension of $\mathscr{A}/\mathscr{I}_x$ is either four or eight.*

Proof. We have already remarked that the dimension of $\mathscr{A}/\mathscr{I}_x$ cannot be greater than eight. If it is less than eight, then the preceding proposition and its proof imply that the dimension is not greater than four. Suppose the dimension of $\mathscr{A}/\mathscr{I}_x$ is less than four. Then there are complex numbers α_i with $\sum |\alpha_i| \neq 0$ such that

$$\alpha_1 \Phi_x(pr) + \alpha_2 \Phi_x((e-p)r) + \alpha_3 \Phi_x(p(e-r)) + \alpha_4 \Phi_x((e-p)(e-r)) = 0.$$

Combining multiplication by $\Phi_x(p)$ or $\Phi_x(e-p)$ from the left, with $\Phi_x(r)$ or $\Phi_x(e-r)$ from the right-hand side, yields

$$\alpha_1 \Phi_x(pr) = \alpha_2 \Phi_x((e-p)r) = \alpha_3 \Phi_x(p(e-r)) = \alpha_4 \Phi_x((e-p)(e-r)) = 0.$$

Since $\sum |\alpha_i| \neq 0$, this shows that at least one of the cosets $\Phi_x(e_i)$ is zero. For definiteness, let $\Phi_x(e_1) = \Phi_x(pr) = 0$ (the other possibilities can be treated similarly). Then, for all $x \in \sigma_{\mathscr{A}}(b)$,

$$\begin{aligned} x\Phi_x(e) = \Phi_x(b) &= \Phi_x(prp + (e-p)(e-r)(e-p)) \\ &= \Phi_x(pr)\Phi_x(p) + \Phi_x((e-r)(e-p)) + \Phi_x(pr)\Phi_x(e-p) \\ &= \Phi_x((e-p)(e-r)). \end{aligned}$$

First, let $x \neq 0$. Then the latter identity shows that $\Phi_x(e-p)$ and $\Phi_x(e-r)$ are invertible from one side. Since a one-sided invertible idempotent is necessarily the identity, one gets $\Phi_x(p) = \Phi_x(r) = 0$. As in the proof of Proposition 3.3.18, this implies that $\Phi_x(e) = \Phi_x(0)$, i.e., $\mathscr{A}/\mathscr{I}_x = \{0\}$ or $I_x = \mathscr{A}$. This is impossible for $x \in \sigma_{\mathscr{A}}(b)$ as we have already remarked.

So let $x = 0$. We conclude from the linear dependence of $\Phi_x(pr)$ and $\Phi_x(pjr)$ that $\Phi_x(pjr) = 0$ and, consequently,

$$\Phi_x((e-p)r) = \Phi_x(j)\Phi_x(pjp) = 0, \quad \Phi_x(p(e-r)) = \Phi_x(pjr)\Phi_x(j) = 0.$$

Thus,

$$\Phi_x(e) = \Phi_x(pr + p(e-r) + (e-p)r + (e-p)(e-r)) = \Phi_x(0),$$

which is impossible again. Hence, cosets $\Phi_x(e_1), \ldots, \Phi_x(e_4)$ are linearly independent, and the dimension of $\mathscr{A}/\mathscr{I}_x$ is four.   ∎

Now we are in a position to derive the aimed local characterization of the components Σ_1 and Σ_2 of $\sigma_{\mathscr{A}}(b)$. Recall that $\sigma_{\mathscr{A}}(b) = \sigma_{\mathscr{C}}(b)$.

Theorem 3.3.20. *The following assertions are equivalent for $x \in \sigma_{\mathscr{A}}(b) \setminus \{0, 1\}$:*

(i.1) $\dim \mathscr{A}/\mathscr{I}_x = 8$;
(i.2) $\Phi_x(pr)$ *and* $\Phi_x(pjr)$ *are linearly independent;*
(i.3) $\Phi_x(b)$ *and* $\Phi_x(c)$ *are linearly independent;*
(i.4) $x \in \Sigma_2$.

The following assertions are equivalent for $x \in \sigma_{\mathscr{A}}(b)$:

(ii.1) dim $\mathscr{A}/\mathscr{I}_x = 4$;

(ii.2) $\Phi_x(pr)$ and $\Phi_x(pjr)$ are linearly dependent;

(ii.3) $\Phi_x(b)$ and $\Phi_x(c)$ are linearly dependent;

(ii.4) $x \in \Sigma_1$.

Proof. The equivalences (i.1)$\Leftrightarrow$(i.2) and (i.1)$\Leftrightarrow$(ii.2) are shown in Propositions 3.3.18 and 3.3.19.

(i.1), (i.2)$\Rightarrow$(i.4): If $\mathscr{A}/\mathscr{I}_x$ is an eight-dimensional vector space, then every coset $\Phi_x(a)$ (which can be thought of as a multiplication operator on $\mathscr{A}/\mathscr{I}_x$) corresponds to a certain complex 8×8 matrix. Straightforward computation yields

$$\Phi_x(e) \mapsto \mathrm{diag}(1,1,1,1,1,1,1,1),$$
$$\Phi_x(p) \mapsto \mathrm{diag}(1,0,1,0,1,0,1,0),$$
$$\Phi_x(j) \mapsto \begin{bmatrix} 0 & J \\ J & 0 \end{bmatrix} \text{ with } J = \begin{bmatrix} 0 & 0 & 0 & 1 \\ 0 & 0 & 1 & 0 \\ 0 & 1 & 0 & 0 \\ 1 & 0 & 0 & 0 \end{bmatrix},$$

and

$$\Phi_x(r) \mapsto \mathrm{diag}(Q_1, Q_2, Q_1, Q_2)$$

with

$$Q_1 = \begin{bmatrix} x & 1-x \\ x & 1-x \end{bmatrix}, \quad Q_2 = \begin{bmatrix} x & -x \\ x-1 & 1-x \end{bmatrix}.$$

Thus,

$$\Phi_x(c) = \Phi_x((pr-rp)j) \mapsto \begin{bmatrix} 0 & 0 & 0 & C_1 \\ 0 & 0 & C_2 & 0 \\ 0 & C_1 & 0 & 0 \\ C_2 & 0 & 0 & 0 \end{bmatrix}$$

with

$$C_1 = \begin{bmatrix} 1-x & 0 \\ 0 & -x \end{bmatrix}, \quad C_2 = \begin{bmatrix} -x & 0 \\ 0 & 1-x \end{bmatrix}.$$

The eigenvalues of the 8×8 matrix corresponding to $\Phi_x(c)$ are $y_{1/2} = \pm\sqrt{x(x-1)}$ as one easily checks. Since $x \notin \{0,1\}$, one has $y_1 \neq y_2$, i.e., this matrix has two different eigenvalues, whence $x \in \Sigma_2$.

(ii.1), (ii.2)$\Rightarrow$(ii.3)$\Rightarrow$(ii.4): If the $\Phi_x(e_i)$, $i = 1, \ldots, 8$, are linearly dependent then (as we have checked in Proposition 3.3.18 and its proof) there is an $\alpha \in \mathbb{C} \setminus \{0\}$ such that

$$\alpha \Phi_x(pr) = \Phi_x(pjr), \tag{3.66}$$
$$\alpha \Phi_x((e-p)(e-r)) = \Phi_x((e-p)j(e-r)). \tag{3.67}$$

Multiplying (3.66) (resp. (3.67)) by $\Phi_x(p)$ (resp. $\Phi_x(e-p)$) from the right-hand side and adding the equalities we find

$$\begin{aligned} \alpha\Phi_x(b) &= \Phi_x\left(prp+(e-p)(e-r)(e-p)\right) \\ &= \Phi_x\left(pjrp+(e-p)j(e-r)(e-p)\right) \\ &= \Phi_x\left((p(e-r)(e-p)+(e-p)rp)j\right) = \Phi_x(-c) \end{aligned}$$

whence (ii.3). Now (ii.4) is almost evident: the spectrum of $\Phi_x(b)$ is a singleton (namely $\{x\}$), hence $\sigma\left(\Phi_x(c)\right)$ is a singleton, too, i.e., $x \in \Sigma_1$.

(i.4)$\Rightarrow$(i.1): Assume (i.1) to be violated. Then, by Proposition 3.3.19, (ii.1) is valid, which implies (ii.4) as we have just checked. But (ii.4) contradicts (i.4). The implications (ii.4)$\Rightarrow$(ii.1) and (i.3)$\Rightarrow$(i.1) follow analogously. ∎

We have seen in Example 3.3.14 that it may happen that c belongs to the algebra generated by e and b. This is not only a C^*-effect but can be observed in many cases of practical importance (for example, we will see that Hankel operators with special piecewise constant generating functions belong to algebras generated by Toeplitz operators on l^p or $L^p(\mathbb{R}^+)$). In Section 4.5 one of those cases will be presented. On the other hand, if 0 or 1 are inner points of $\sigma_{\mathscr{C}}(b)$ in $\mathbb{C}$, then c cannot be an element of $\text{alg}\{e,b\}$. Indeed, in this case Σ_1 cannot coincide with $\sigma_{\mathscr{C}}(b)$ since there is no continuous branch of the square root function in the neighborhood of 0, thus Σ_2 is not empty, but then $\Phi_x(c)$ and $\Phi_x(b)$ are linearly independent for every $x \in \Sigma_2$ by the preceding theorem. It is still an open question whether c always belongs to $\text{alg}\{b\}$ whenever $\Phi_x(b)$ and $\Phi_x(c)$ are linearly dependent for every $x \in \sigma_{\mathscr{A}}(b)$.

3.4 Algebras generated by two projections and a flip which preserves the orientation

Now we turn our attention to algebras with a flip which preserves the orientation. Thus, we now let $\mathscr{A}$ be a complex Banach algebra with identity element e and with elements p, r and j satisfying conditions (3.45) and (3.47).

Lemma 3.4.1. *The non-closed algebra $\mathscr{A}^0$ generated by p, r and j is a module of dimension at most 16 over the center $\mathscr{C}$ of $\mathscr{A}^0$.*

Proof. Let $t = (p-r)^2$ and $w = e - 2t$. It is not difficult to check that t and w commute with p and r, and that w^2 is in $\mathscr{C}$. Set

$$z_1 = e, \quad z_2 = p, \quad z_3 = r, \quad z_4 = pr, \tag{3.68}$$

and let $\mathscr{D}$ stand for the set of all elements $\sum x_i(t)z_i$ where the x_i are polynomials in t. The multiplication table

	z_1	z_2	z_3	z_4
z_1	z_1	z_2	z_3	z_4
z_2	z_2	z_2	z_4	z_4
z_3	z_3	$z_2+z_3-tz_1-z_4$	z_3	$(e-t)z_3$
z_4	z_4	$(e-t)z_2$	z_4	$(e-t)z_4$

shows that $\mathscr{D}$ is an algebra. It is easy to see that the algebra $\mathscr{D}$ coincides with $\mathrm{alg}\{p,r\}$; thus, every element x of $\mathrm{alg}\{p,r\}$ can be written as a linear combination

$$x_1(t)z_1 + x_2(t)z_2 + x_3(t)z_3 + x_4(t)z_4. \tag{3.69}$$

Now let $a \in \mathscr{A}^0$. We write a as $a = x + yj$ with $x,y \in \mathrm{alg}\{p,r\}$, then we write x and y in the form (3.69) with coefficients x_i and y_i, and finally we write each of these coefficients in the form $x_i(t) = x_{i1}(t)(w^2) + x_{i2}(t)(w^2)w$. Thus,

$$
\begin{aligned}
a &= \sum_{i=1}^{4} x_i(t)z_i + \sum_{i=1}^{4} y_i(t)z_i j \\
&= \sum_{i=1}^{4} x_{i1}(t)(w^2)z_i + \sum_{i=1}^{4} x_{i2}(t)(w^2)wz_i + \sum_{i=1}^{4} y_{i1}(t)(w^2)z_i j + \sum_{i=1}^{4} y_{i2}(t)(w^2)wz_i j
\end{aligned}
$$

which shows the assertion. ∎

Together with Corollary 2.6.22, this lemma implies that there exists a matrix symbol of order ≤ 4 for the algebra $\mathscr{A}$. Let I_n stand for the $n \times n$ identity matrix and $\mathrm{tr}A$ for the trace of the matrix A. The next lemma excludes that $\mathscr{A}$ has representations of order 1 or 3.

Lemma 3.4.2. *Let $R,J \in \mathbb{C}^{n \times n}$, with $R^2 = R$, $J^2 = I_n$ and $JRJ = I_n - R$. Then n is even.*

Proof. From $JRJ + R = I$ we get $n = \mathrm{tr}I = 2\mathrm{tr}R$. Since R is an idempotent, the trace of R is an integer. Thus, n is even. ∎

In the next sections, we are going to examine the representations of the algebra $\mathscr{A}$ of dimension 4 and 2.

3.4.1 Four-dimensional representations of the algebra $\mathscr{A}$

For $x \in \mathbb{C} \setminus \{0, \frac{1}{2}, 1\}$, let $\Phi_x : \mathscr{A}^0 \to \mathbb{C}^{4 \times 4}$ be the homomorphism defined by

$$
\Phi_x(p) = \begin{bmatrix} 1 & 0 & 0 & 0 \\ 0 & 0 & 0 & 0 \\ 0 & 0 & 1 & 0 \\ 0 & 0 & 0 & 0 \end{bmatrix}, \qquad
\Phi_x(j) = \begin{bmatrix} 0 & 0 & 1 & 0 \\ 0 & 0 & 0 & 1 \\ 1 & 0 & 0 & 0 \\ 0 & 1 & 0 & 0 \end{bmatrix}, \tag{3.70}
$$

and

$$
\Phi_x(r) = \begin{bmatrix} x & \sqrt{x(1-x)} & 0 & 0 \\ \sqrt{x(1-x)} & 1-x & 0 & 0 \\ 0 & 0 & 1-x & -\sqrt{x(1-x)} \\ 0 & 0 & -\sqrt{x(1-x)} & x \end{bmatrix}. \tag{3.71}
$$

Proposition 3.4.3. *Let $x \in \mathbb{C} \setminus \{0, \frac{1}{2}, 1\}$. Then $\operatorname{Im} \Phi_x = \mathbb{C}^{4\times 4}$. Moreover, for each irreducible four-dimensional representation Φ of $\mathscr{A}$, there is an $x \in \mathbb{C} \setminus \{0, \frac{1}{2}, 1\}$ and a basis of the space of the representation such that $\Phi = \Phi_x$ in this basis.*

Proof. To prove the first assertion, write q for $e - p$ and set

$$
B := \Phi_x\left(\frac{1}{\sqrt{x(1-x)}}(prq + qrp)\right) = \begin{bmatrix} 0 & 1 & 0 & 0 \\ 1 & 0 & 0 & 0 \\ 0 & 0 & 0 & -1 \\ 0 & 0 & -1 & 0 \end{bmatrix},
$$

$$
C := \Phi_x\left(\frac{1}{2x-1}(prp + (x-1)p)\right) = \begin{bmatrix} 1 & 0 & 0 & 0 \\ 0 & 0 & 0 & 0 \\ 0 & 0 & 0 & 0 \\ 0 & 0 & 0 & 0 \end{bmatrix},
$$

and $J := \Phi_x(j)$. From

$$
\begin{bmatrix} x_{11} & x_{12} & 0 & 0 \\ x_{21} & x_{22} & 0 & 0 \\ 0 & 0 & 0 & 0 \\ 0 & 0 & 0 & 0 \end{bmatrix} = x_{11}C + x_{12}CB + x_{21}BC + x_{22}BCB
$$

we conclude that every 4×4 matrix with an arbitrary 2×2 block in its upper left corner and zero blocks at the other corners belongs to the range of Φ_x. The identities

$$
\begin{bmatrix} 0 & 0 \\ X & 0 \end{bmatrix} = J\begin{bmatrix} X & 0 \\ 0 & 0 \end{bmatrix}, \quad \begin{bmatrix} 0 & X \\ 0 & 0 \end{bmatrix} = \begin{bmatrix} X & 0 \\ 0 & 0 \end{bmatrix}J, \quad \text{and} \quad \begin{bmatrix} 0 & 0 \\ 0 & X \end{bmatrix} = J\begin{bmatrix} X & 0 \\ 0 & 0 \end{bmatrix}J
$$

allow one to move this 2×2 block into one of the other corners of the matrix. Hence, each 4×4 matrix with a 2×2 block in one of its corners and zero blocks at the other corners lies in the range of Φ_x. Since this range is a linear space, we conclude that $\operatorname{Im} \Phi_x = \mathbb{C}^{4\times 4}$.

To prove the second assertion of the proposition, let $s = 2p - e$ and $h = 2r - e$ denote the flips associated with the projections p and r, respectively. Then $\mathscr{A} = \operatorname{alg}\{s, h, j\}$, and conditions (3.45) and (3.47) are equivalent to

$$
s^2 = h^2 = j^2 = e, \quad jsj = s, \quad hjh = -j. \tag{3.72}
$$

Let Φ be an irreducible four-dimensional representation of the algebra $\mathscr{A}$. Since $(\Phi(j))^2 = (\Phi(h))^2 = I_4$ and $\Phi(h)\Phi(j)\Phi(h) = -\Phi(j)$, one has $\operatorname{tr} \Phi(j) = 0$ and

$\sigma(\Phi(j)) \subset \{-1,1\}$. Hence, in a suitable basis,

$$\Phi(j) = \begin{bmatrix} I_2 & 0 \\ 0 & -I_2 \end{bmatrix}. \tag{3.73}$$

Using the equalities (3.72) and (3.73) we then obtain

$$\Phi(h) = \begin{bmatrix} 0 & A \\ A^{-1} & 0 \end{bmatrix} \quad \text{and} \quad \Phi(s) = \begin{bmatrix} M_1 & 0 \\ 0 & M_2 \end{bmatrix} \tag{3.74}$$

with 2×2 blocks A and M_i where A is invertible and $M_i^2 = I_2$. It easy to check that the matrices M_1 and M_2 are diagonalizable. Let $S = \mathrm{diag}(S_1, S_2)$ be a block diagonal matrix which diagonalizes $\Phi(s)$. Then the matrices $S^{-1}\Phi(j)S$, $S^{-1}\Phi(h)S$ and $S^{-1}\Phi(s)S$ have still the same block structure as in (3.73) and (3.74). We can thus assume, without loss of generality, that the M_i are diagonal matrices with $M_i^2 = I_2$.

If $M_1 = \pm I_2$, then each of the matrices $\Phi(s)$, $\Phi(h)$ and $\Phi(j)$ is of the form

$$\begin{bmatrix} AD_{11}A^{-1} & AD_{12} \\ D_{21}A^{-1} & D_{22} \end{bmatrix} \tag{3.75}$$

with 2×2 diagonal matrices D_{ik}. One easily checks that the set of all matrices of the form (3.75) constitutes a proper subalgebra of $\mathbb{C}^{4\times4}$. Thus, $\mathrm{alg}\{\Phi(s), \Phi(h), \Phi(j)\} \neq \mathbb{C}^{4\times4}$, and the representation Φ is reducible, which contradicts the hypothesis that Φ is an irreducible four-dimensional representation. Consequently, $M_1 = \mathrm{diag}(1, -1)$, and in a similar way one gets $M_2 = \mathrm{diag}(1, -1)$.

Next we show that none of the entries a_{ik} of A is zero. Suppose that $a_{11} = 0$. Then each of the matrices $\Phi(s)$, $\Phi(h)$ and $\Phi(j)$ is of the form

$$\begin{bmatrix} * & 0 & 0 & * \\ * & * & * & * \\ * & * & * & * \\ * & 0 & 0 & * \end{bmatrix}. \tag{3.76}$$

The set of all matrices of this form is a proper subalgebra of $\mathbb{C}^{4\times4}$; hence Φ is reducible, a contradiction. Analogously, if the entry a_{12}, a_{21} or a_{22} is zero, then $\mathrm{alg}\{\Phi(s), \Phi(h), \Phi(j)\}$ is contained in a subalgebra of $\mathbb{C}^{4\times4}$ consisting of all matrices of the form

$$\begin{bmatrix} * & 0 & * & 0 \\ * & * & * & * \\ * & 0 & * & 0 \\ * & * & * & * \end{bmatrix}, \quad \begin{bmatrix} * & * & * & * \\ 0 & * & 0 & * \\ * & * & * & * \\ 0 & * & 0 & * \end{bmatrix} \quad \text{or} \quad \begin{bmatrix} * & * & * & * \\ 0 & * & * & 0 \\ 0 & * & * & 0 \\ * & * & * & * \end{bmatrix},$$

respectively, which is again impossible under our hypotheses. Thus, all entries of A are non-zero. But then there exists an invertible diagonal matrix T such that

$$T\Phi(h)T^{-1} = \begin{bmatrix} 0 & N \\ N & 0 \end{bmatrix} \quad \text{where} \quad N = \begin{bmatrix} \alpha & \beta \\ \beta & \alpha \end{bmatrix} \tag{3.77}$$

with complex numbers α and β such that $\alpha^2 + \beta^2 = 1$ and $\alpha\beta \neq 0$. To get the second assertion of the proposition, set $\alpha = 2x - 1$ and $\Phi_x(a) := H\Phi(a)H^{-1}$ where $\Phi(s)$, $\Phi(h)$ and $\Phi(j)$ are defined by (3.73) and (3.74), $M_1 = M_2 = \text{diag}(1, -1)$, and $H = KT$ with

$$K = \frac{1}{\sqrt{2}} \begin{bmatrix} I_2 & I_2 \\ I_2 & -I_2 \end{bmatrix}.$$

It is not difficult to check that then $\Phi_x(p)$, $\Phi_x(j)$ and $\Phi_x(r)$ are of the form (3.70) and (3.71), respectively. Note that the conditions $\alpha\beta \neq 0$ and $\alpha^2 + \beta^2 = 1$ imply that $x \notin \{0, \frac{1}{2}, 1\}$. ∎

Remark 3.4.4. The representation Φ_{1-x} is equivalent to Φ_x; precisely,

$$\Phi_{1-x}(a) = \Phi_x(js)\Phi(a)\Phi_x(js)^{-1}$$

for every $a \in \mathscr{A}$. Moreover, if the square root $\sqrt{x(1-x)}$ in the definition of Φ_x is substituted by $-\sqrt{x(1-x)}$, then we obtain a representation Φ_x^- which is again equivalent to Φ_x via $\Phi_x^-(a) = \Phi_x(s)\Phi(a)\Phi_x(s)^{-1}$. □

3.4.2 Two-dimensional representations of the algebra $\mathscr{A}$

For $k \in \{-2, -1, 1, 2\}$, let $\Psi_k : \mathscr{A}^0 \to \mathbb{C}^{2\times 2}$ be the homomorphism defined by

$$\Psi_{-2}(p) := \begin{bmatrix} 0 & 0 \\ 0 & 1 \end{bmatrix}, \ \Psi_{-1}(p) := \begin{bmatrix} 0 & 0 \\ 0 & 0 \end{bmatrix}, \ \Psi_1(p) := \begin{bmatrix} 1 & 0 \\ 0 & 1 \end{bmatrix}, \ \Psi_2(p) := \begin{bmatrix} 1 & 0 \\ 0 & 0 \end{bmatrix},$$

$$\Psi_k(r) := \frac{1}{2} \begin{bmatrix} 1 & 1 \\ 1 & 1 \end{bmatrix} \quad \text{and} \quad \Psi_k(j) := \begin{bmatrix} 1 & 0 \\ 0 & -1 \end{bmatrix} \tag{3.78}$$

The proof of the following proposition is straightforward and left as an exercise.

Proposition 3.4.5. *Let $k \in \{-2, -1, 1, 2\}$. Then $\text{Im } \Psi_k = \mathbb{C}^{2\times 2}$. Moreover, for each irreducible two-dimensional representation Ψ of $\mathscr{A}$, there is a $k \in \{-2, -1, 1, 2\}$ and a basis in the space of the representation such that $\Psi = \Psi_k$ in this basis.*

Remark 3.4.6. From Proposition 3.4.3 we infer that the representations Φ_x are reducible for $x \in \{0, \frac{1}{2}, 1\}$. Further, Φ_0 and Φ_1 are equivalent by Remark 3.4.4. The relations between $\Phi_{1/2}$, Φ_1 and Ψ_k for $k \in \{-2, -1, 1, 2\}$ are given by

$$\Phi_{1/2} = R_1^{-1} \begin{bmatrix} \Psi_1 & 0 \\ 0 & \Psi_{-1} \end{bmatrix} R_1 \quad \text{and} \quad \Phi_1 = R_2^{-1} \begin{bmatrix} \Psi_{-2} & 0 \\ 0 & \Psi_2 \end{bmatrix} R_2 \tag{3.79}$$

with

$$R_1 := \frac{1}{\sqrt{2}} \begin{bmatrix} 1 & 0 & 1 & 0 \\ 1 & 0 & -1 & 0 \\ 0 & 1 & 0 & 1 \\ 0 & -1 & 0 & 1 \end{bmatrix} \quad \text{and} \quad R_2 := \frac{1}{\sqrt{2}} \begin{bmatrix} 1 & 0 & 1 & 0 \\ 0 & 1 & 0 & -1 \\ 0 & 1 & 0 & 1 \\ 1 & 0 & -1 & 0 \end{bmatrix}.$$

□

3.4.3 Main results

We are able now to state the main results regarding algebras generated by two idempotents and an orientation preserving flip of order 2.

Theorem 3.4.7 (Krupnik, Spigel). *Let $\mathscr{A}$ be a Banach algebra generated by elements p, r and j which satisfy (3.45) and (3.47). Set $\sigma_{prp} := \sigma_{\mathscr{A}}(prp) \setminus \{0, \frac{1}{2}, 1\}$ and $\sigma := \sigma_{\mathscr{A}}(b) \cap \{-2, -1, 1, 2\}$ where $b = s + 2j(rs + sr) + sr - rs$ and $s = 2p - e$. Then:*

(i) *the algebra $\mathscr{A}$ has a matrix symbol of order 4;*
(ii) *for any $x \in \sigma_{prp}$ and $k \in \sigma$, the homomorphisms Φ_x defined in (3.70) and Ψ_k defined in (3.78) can be extended to the algebra $\mathscr{A}$;*
(iii) *the set of homomorphisms*

$$\{\Phi_x\}_{x \in \sigma_{prp}} \cup \{\Psi_k\}_{k \in \sigma} \tag{3.80}$$

is a matrix symbol of order 4 for $\mathscr{A}$.

Proof. By Lemma 3.4.1, the non-closed algebra $\mathscr{A}^0$ generated by p, r and j is a module of dimension at most 16 over its center. Hence, by Corollary 2.6.22, there is a matrix symbol of order $n \leq 4$ for $\mathscr{A}$, that is, there exists a set $\{v_\tau\}$ of representations of dimension $l(\tau) \leq 4$. By Lemma 3.4.2, these representations must be of dimension two or four. We described these representations in Propositions 3.4.3 and 3.4.5. In particular, we have seen that there are sets $\Sigma_1(p, r, j) \subset \mathbb{C} \setminus \{0, \frac{1}{2}, 1\}$ and $\Sigma_2(p, r, j) \subset \{-2, -1, 1, 2\}$ such that

$$\{\Phi_x\}_{x \in \Sigma_1} \cup \{\Psi_k\}_{k \in \Sigma_2} \tag{3.81}$$

is a matrix symbol of order 4 for $\mathscr{A}$.

It remains to identify the sets Σ_1 and Σ_2. The spectrum $\sigma_{\mathscr{A}}(prp)$ is related with the spectra of Φ_x and Ψ_k by

$$\sigma_{\mathscr{A}}(prp) = \bigcup_{x \in \Sigma_1} \sigma(\Phi_x(prp)) \cup \bigcup_{k \in \Sigma_2} \sigma(\Psi_k(prp))$$

$$= \{0\} \cup \Sigma_1 \cup \widehat{\Sigma}_1 \cup \bigcup_{k \in \Sigma_2} \sigma(\Psi_k(prp)),$$

where $\widehat{\Sigma}_1 := 1 - \Sigma_1$ and $\cup_{k \in \Sigma_2} \sigma(\Psi_k(prp)) \subset \{0, \frac{1}{2}, 1\}$. Since

$$\Sigma_1 \cup \widehat{\Sigma}_1 \cap \left\{0, \frac{1}{2}, 1\right\} = \emptyset,$$

we conclude that $\sigma_{\mathscr{A}}(prp) \setminus \{0, \frac{1}{2}, 1\} = \Sigma_1 \cup \widehat{\Sigma}_1$. But Φ_x and Φ_{1-x} are equivalent representations by Remark 3.4.4. Hence, Σ_1 can be taken as $\sigma_{\mathscr{A}}(prp) \setminus \{0, \frac{1}{2}, 1\}$. Concerning Σ_2, note that

$$\sigma_{\mathscr{A}}(b) = \bigcup_{x \in \Sigma_1} \sigma(\Phi_x(b)) \cup \bigcup_{k \in \Sigma_2} \sigma(\Psi_k(b)).$$

A straightforward computation yields that $\sigma(\Psi_k(b)) \cap \{-2, -1, 1, 2\} = \{k\}$ for $k = -2, -1, 1, 2$ and $\sigma(\Phi_x(b)) \cap \{-2, -1, 1, 2\} = \emptyset$ for $x \in \Sigma_1$, whence the equality $\Sigma_2 = \sigma_{\mathscr{A}}(b) \cap \{-2, -1, 1, 2\}$. ∎

Corollary 3.4.8. *Let $\mathscr{A}$ be a Banach algebra generated by elements p, r and j which satisfy (3.45) and (3.47), and assume that each of the numbers $\frac{1}{2}$ and 1 is either a regular point of prp or a non-isolated point of $\sigma_{\mathscr{A}}(prp)$. Let $a, b \in \mathrm{alg}\{p, r\}$. Then $a + bj$ is invertible if and only if $a - bj$ is invertible.*

Proof. Let 1 be a regular point of prp. Since $1 \in \sigma(\Psi_1(prp))$ and $1 \in \sigma(\Psi_{-1}(prp))$, this implies that $\pm 1 \notin \sigma$. Analogously, if $\frac{1}{2}$ is a regular point of prp, then $\pm \frac{1}{2} \notin \sigma$.

Now assume that 1 is a point of $\sigma_{\mathscr{A}}(prp)$ which is not isolated. Then there exists a sequence $\{\lambda_n\}$ in $\sigma_{\mathscr{A}}(prp)$ with $\lambda_n \neq 1$ for each n which tends to 1. Since

$$\sigma(\Phi_{\lambda_n}(qrq)) = \sigma(\Phi_{\lambda_n}(prp))$$

for $q = e - p$, we have $\lambda_n \in \sigma_{\mathscr{A}}(qrq)$, whence $1 \in \sigma_{\mathscr{A}}(qrq)$. It is easy to check that, for $x \in \Sigma_1$,

$$1 \notin \sigma(\Phi_x(prp)) \quad \text{and} \quad 1 \notin \sigma(\Phi_x(qrq)).$$

Moreover,

$$1 \notin \bigcup_{k \in \Sigma_2 \setminus \{1\}} \sigma(\Psi_k(prp)) \quad \text{and} \quad 1 \notin \bigcup_{k \in \Sigma_2 \setminus \{1\}} \sigma(\Psi_k(qrq)).$$

Thus $\pm 1 \in \sigma$, and one can replace the two irreducible representations $\Psi_{\pm 1}$ in (3.81) by the reducible representation Φ_1.

Finally, let $\frac{1}{2}$ be a point in $\sigma_{\mathscr{A}}(prp)$ which is not isolated. Let $\{\lambda_n\} \subset \sigma_{\mathscr{A}}(prp)$ be a sequence with $\lambda_n \neq \frac{1}{2}$ which tends to $\frac{1}{2}$. For every $a \in \mathscr{A}^0$, one has $\sigma(\Phi_{\lambda_n}(a)) \subset \sigma_{\mathscr{A}}(a)$ and, thus, $\sigma(\Phi_{\frac{1}{2}}(a)) \subset \sigma_{\mathscr{A}}(a)$. Put $a = prp + j$. It is then easy to check that

$$\left\{-\frac{1}{2}, \frac{3}{2}\right\} \subset \sigma\left(\Phi_{\frac{1}{2}}(prp + j)\right)$$

and

$$\left\{-\frac{1}{2},\frac{3}{2}\right\} \cap \sigma(\Phi_x(prp+j)) = \emptyset$$

for $x \neq \frac{1}{2}$. Also one has

$$\left\{-\frac{1}{2},\frac{3}{2}\right\} \cap \sigma(\Psi_{\pm 2}(prp+j)) = \emptyset$$

and

$$\sigma(\Psi_2(prp+j)) = \left\{0,\frac{3}{2}\right\}, \quad \sigma(\Psi_{-2}(prp+j)) = \left\{-\frac{1}{2},1\right\}.$$

We conclude that $\pm 2 \in \sigma$ and that the representations $\Psi_{\pm 2}$ can be replaced by the representation $\Phi_{1/2}$. Thus the set $\{\Psi_x\}_{x \in \sigma_{\mathscr{A}}(prp)}$ is a matrix symbol for $\mathscr{A}$. To get the assertion note that, for arbitrary elements $a,b \in \mathrm{alg}\{p,r\}$,

$$\det \Psi_x(a+bj) \neq 0 \Leftrightarrow \det \Psi_x(a-bj) \neq 0;$$

so $a+bj$ is invertible if and only if the same is true for $a-bj$. ■

Remark 3.4.9. By Theorem 3.1.4, the algebra $\mathrm{alg}\{p,r\}$ possesses a matrix symbol of order 2, say $\{\Upsilon_x\}_{x \in X}$. The equality

$$\begin{bmatrix} a+bj & 0 \\ 0 & a-bj \end{bmatrix} = \frac{1}{2} \begin{bmatrix} e & j \\ e & -j \end{bmatrix} \begin{bmatrix} a & b \\ jbj & jaj \end{bmatrix} \begin{bmatrix} e & e \\ j & -j \end{bmatrix} \tag{3.82}$$

then shows that the homomorphisms

$$\tilde{\Upsilon}_x : \mathscr{A} \to \mathbb{C}^{l(x) \times l(x)}, \quad \tilde{\Upsilon}_x(a+bj) := \begin{bmatrix} \Upsilon_x(a) & \Upsilon_x(b) \\ \Upsilon_x(jbj) & \Upsilon_x(jaj) \end{bmatrix}$$

constitute a matrix symbol of order 4 for $\mathscr{A}$. □

3.5 Coefficient algebras

One possible extension of the results for algebras generated by idempotents needed for applications concerns the multidimensional case. It happens that this case can be considered by the introduction of a *coefficient algebra*. We will show the proof in detail for the two projections theorem, and enunciate without proof the generalized results for other cases.

3.5.1 A general version of the two projections theorem

We start by extending the notion of a symbol mapping defined in Section 1.2.1. Let $\mathscr{A}$, $\mathscr{B}$, $\mathscr{C}$ and $\mathscr{D}$ be unital Banach algebras with $\mathscr{A} \subseteq \mathscr{B}$ and $\mathscr{C} \subseteq \mathscr{D}$, and let $W : \mathscr{B} \to \mathscr{D}$ be a bounded linear mapping. This mapping is called a *symbol map* for $\mathscr{A}$ if:

(i) $W(\mathscr{A}) \subseteq \mathscr{C}$;
(ii) $W(e) = e$;
(iii) $W(a)W(b) = W(ab)$ and $W(b)W(a) = W(ba)$ for all $a \in \mathscr{A}$ and $b \in \mathscr{B}$;
(iv) the invertibility of $W(a)$ in $\mathscr{D}$ implies the invertibility of a in $\mathscr{B}$ for all $a \in \mathscr{A}$.

The element $W(a)$ is then referred to as the symbol of a. If W is a symbol map and $a \in \mathscr{A}$ is invertible in $\mathscr{B}$ then, by (ii) and (iii), the symbol $W(a)$ of a is invertible in $\mathscr{D}$. Since the reverse implication holds by (iv), we have

$$\sigma_{\mathscr{B}}(a) = \sigma_{\mathscr{D}}(W(a)) \quad \text{for all } a \in \mathscr{A}. \tag{3.83}$$

We write $\mathrm{smb}(\mathscr{A}, \mathscr{B}; \mathscr{C}, \mathscr{D})$ for the set of all symbol maps in the sense of (i)–(iv) and abbreviate $\mathrm{smb}(\mathscr{A}, \mathscr{A}; \mathscr{C}, \mathscr{D})$ to $\mathrm{smb}(\mathscr{A}; \mathscr{C}, \mathscr{D})$ and $\mathrm{smb}(\mathscr{A}, \mathscr{A}; \mathscr{D}, \mathscr{D})$ to $\mathrm{smb}(\mathscr{A}; \mathscr{D})$. The elements of the latter set are exactly the bounded homomorphisms from $\mathscr{A}$ into $\mathscr{D}$ to which we referred earlier as *symbols*. The need to generalize the earlier notion of a symbol comes from the idea of embedding $\mathscr{A}$ into a larger algebra, $\mathscr{B}$, and to study the invertibility of elements of $\mathscr{A}$ in $\mathscr{B}$. Note that the subalgebra $\mathscr{A}$ of $\mathscr{B}$ is inverse-closed if and only if the embedding operator from $\mathscr{A}$ into $\mathscr{B}$ belongs to $\mathrm{smb}(\mathscr{A}; \mathscr{B})$. It is obvious that

$$\mathrm{smb}(\mathscr{A}, \mathscr{B}; \mathscr{C}, \mathscr{D}) \subseteq \mathrm{smb}(\mathscr{A}, \mathscr{B}; \mathscr{D}) \tag{3.84}$$

for all subalgebras $\mathscr{C}$ of $\mathscr{D}$ and that

$$\mathrm{smb}(\mathscr{A}, \mathscr{B}; \mathscr{C}) \subseteq \mathrm{smb}(\mathscr{A}, \mathscr{B}; \mathscr{C}, \mathscr{D}) \tag{3.85}$$

whenever $\mathscr{C}$ is inverse-closed in $\mathscr{D}$. The latter inclusion is also a consequence of the general implication

$$W_1 \in \mathrm{smb}(\mathscr{A}, \mathscr{B}; \mathscr{C}, \mathscr{D}), \ W_2 \in \mathrm{smb}(\mathscr{C}, \mathscr{D}; \mathscr{E}, \mathscr{F}) \Rightarrow W_2 \circ W_1 \in \mathrm{smb}(\mathscr{A}, \mathscr{B}; \mathscr{E}, \mathscr{F}). \tag{3.86}$$

Given Banach algebras $\mathscr{C}_1$ and $\mathscr{C}_2$, we denote by $\mathscr{C}_1 \times \mathscr{C}_2$ the Banach algebra of all ordered pairs (c_1, c_2) with $c_1 \in \mathscr{C}_1$, $c_2 \in \mathscr{C}_2$, provided with component-wise operations and the norm $\|(c_1, c_2)\| := \max\{\|c_1\|, \|c_2\|\}$. Likewise one can think of the pair (c_1, c_2) as the diagonal matrix $\mathrm{diag}\,(c_1, c_2)$.

Now let $\mathscr{A}$ be a unital Banach algebra with identity element e, let p and r be idempotents in $\mathscr{A}$, and let $\mathscr{G}$ be a unital subalgebra of $\mathscr{A}$ the elements of which commute with p and r. Let $\mathscr{B}$ denote the smallest closed subalgebra of $\mathscr{A}$ which contains the algebra $\mathscr{G}$ and the idempotents p and r. The elements of $\mathscr{G}$ will play the role of coefficients in what follows.

As in Section 3.1 one observes that the element[2] $\tilde{c} := (p - r)^2 = p + r - pr - rp$ commutes with both p and r and therefore with all elements of $\mathscr{B}$. Hence, the smallest closed subalgebra alg$\{\tilde{c}\}$ of $\mathscr{B}$ which contains $\tilde{c}$ and the identity element e is a central subalgebra of $\mathscr{B}$, a fact which offers the applicability of Allan's local principle in this context.

Before stating the general version of the two projections theorem we have to introduce some more notation. We let $\mathscr{D}$ stand for the commutant of the element $\tilde{c}$ in $\mathscr{A}$, that is

$$\mathscr{D} := \{d \in \mathscr{A} : \tilde{c}d = d\tilde{c}\}.$$

Obviously, $\mathscr{D}$ forms a closed subalgebra of $\mathscr{A}$, $\mathscr{D}$ is inverse-closed in $\mathscr{A}$, and $\mathscr{D}$ contains $\mathscr{B}$. Thus, the problem of investigating invertibility of elements of $\mathscr{B}$ in $\mathscr{A}$ can be reduced to studying invertibility of elements of $\mathscr{B}$ in $\mathscr{D}$.

Further, it is immediate from the definition of $\mathscr{D}$ that alg$\{\tilde{c}\}$ is a central subalgebra of $\mathscr{D}$. Since this subalgebra is singly generated, the maximal ideal space of alg$\{\tilde{c}\}$ can be identified with the spectrum of $\tilde{c}$ in alg$\{\tilde{c}\}$ by Exercise 2.1.3.

For every maximal ideal t of alg$\{\tilde{c}\}$, we let $\mathscr{I}_t$ stand for the smallest closed ideal of $\mathscr{D}$ which contains t, and we write $\mathscr{D}_t$ for the quotient algebra $\mathscr{D}/\mathscr{I}_t$, b_t for the coset $b + \mathscr{I}_t$ which contains the element $b \in \mathscr{D}$, and $\mathscr{B}_t$ and $\mathscr{G}_t$ for the images of $\mathscr{B}$ and $\mathscr{G}$ under the canonical homomorphism $b \mapsto b_t$. Note that $\mathscr{I}_t = \mathscr{D}$ whenever t belongs to the spectrum of $\tilde{c}$ in alg$\{\tilde{c}\}$ but not to the spectrum of $\tilde{c}$ in $\mathscr{D}$ by Proposition 2.2.9. It is thus sufficient to consider the above definitions for maximal ideals $t \in \sigma_{\mathscr{A}}(\tilde{c})$.

In order to avoid some technicalities regarding the one-dimensional representations of the algebra, let us suppose here and hereafter that

$$0, 1 \in \mathbb{C} \text{ are not isolated points of } \sigma_{\mathscr{A}}(\tilde{c}). \tag{3.87}$$

In the case that the intersection $\sigma_{\mathscr{A}}(\tilde{c}) \cap \{0, 1\}$ is not empty we further require that

$$\mathscr{B}_t \text{ is inverse-closed in } \mathscr{D}_t \text{ for } t \in \sigma_{\mathscr{A}}(\tilde{c}) \cap \{0, 1\}. \tag{3.88}$$

If $0 \in \sigma_{\mathscr{A}}(\tilde{c})$, then we let $\tilde{I}_0$ refer to the smallest closed ideal of $\mathscr{B}_0$ which contains the element $(p - r)_0$. Possibly, this ideal consists of the zero element only. Later on we shall see that $\tilde{I}_0$ is always proper. Further we let I_{00} and I_{01} stand for the smallest closed ideals of the quotient algebra $\mathscr{B}_0/\tilde{I}_0$ which contain the cosets $e_0 - \frac{p_0 + r_0}{2} + \tilde{I}_0$ and $\frac{p_0 + r_0}{2} + \tilde{I}_0$, respectively. We write $\mathscr{B}_{00}$ and $\mathscr{B}_{01}$ for the algebras $(\mathscr{B}_0/\tilde{I}_0)/I_{00}$ and $(\mathscr{B}_0/\tilde{I}_0)/I_{01}$, and we let a_{00} and a_{01} stand for the cosets $(a_0 + \tilde{I}_0) + I_{00}$ and $(a_0 + \tilde{I}_0) + I_{01}$, respectively.

Analogously, if $1 \in \sigma_{\mathscr{A}}(\tilde{c})$, then $\tilde{I}_1$ refers to the smallest closed ideal of $\mathscr{B}_1$ which contains $(e - p - r)_1$, and I_{10} and I_{11} denote the smallest closed ideals of $\mathscr{B}/\tilde{I}_1$ which contain the elements $\frac{e_1 - p_1 + r_1}{2} + \tilde{I}_1$ and $\frac{e_1 + p_1 - r_1}{2} + \tilde{I}_1$, respectively. The quotient algebras $(\mathscr{B}_1/\tilde{I}_1)/I_{10}$ and $(\mathscr{B}_1/\tilde{I}_1)/I_{11}$ will be abbreviated to $\mathscr{B}_{10}$ and $\mathscr{B}_{11}$, and the cosets $(a_1 + \tilde{I}_1) + I_{10}$ and $(a_1 + \tilde{I}_1) + I_{11}$ to a_{10} and a_{11}, respectively.

---

[2] This element is equal to the element $e - c$ of Section 3.1, and it plays a similar role. To use c or $\tilde{c} = e - c$ is a question of convenience depending on the context.

Finally, let $\mathscr{F}$ stand for the algebra of all bounded functions on $\sigma_{\mathscr{A}}(\tilde{c})$ which take a value in $\mathscr{D}_t^{2\times2}$ at the point $t \in \sigma_{\mathscr{A}}(\tilde{c}) \setminus \{0,1\}$ and a value in $\mathscr{B}_{t0} \times \mathscr{B}_{t1}$ at $t \in \sigma_{\mathscr{A}}(\tilde{c}) \cap \{0,1\}$.

Theorem 3.5.1. *Let the algebras $\mathscr{A}, \mathscr{B}, \mathscr{D}, \mathscr{G}$ and the elements p, r be as above, and suppose that the hypotheses (3.87) and (3.88) are fulfilled. Then there exists a symbol map*

$$\Phi \in \mathrm{smb}\,(\mathscr{B}, \mathscr{D}; \mathscr{F}),$$

and this map can be chosen such that it sends the elements $g \in \mathscr{G}$, p, and r to the functions

$$t \mapsto \begin{cases} \begin{bmatrix} 1 & 0 \\ 0 & 1 \end{bmatrix} g_t & \text{if } t \in \sigma_{\mathscr{A}}(\tilde{c}) \setminus \{0,1\}, \\[2mm] \begin{bmatrix} g_{t0} & 0 \\ 0 & g_{t1} \end{bmatrix} & \text{if } t \in \sigma_{\mathscr{A}}(\tilde{c}) \cap \{0,1\}, \end{cases}$$

$$t \mapsto \begin{cases} \begin{bmatrix} 1 & 0 \\ 0 & 0 \end{bmatrix} e_t & \text{if } t \in \sigma_{\mathscr{A}}(\tilde{c}) \setminus \{0,1\}, \\[2mm] \begin{bmatrix} e_{t0} & 0 \\ 0 & 0 \end{bmatrix} & \text{if } t \in \sigma_{\mathscr{A}}(\tilde{c}) \cap \{0,1\}, \end{cases}$$

$$t \mapsto \begin{cases} \begin{bmatrix} 1-t & \sqrt{t(1-t)} \\ \sqrt{t(1-t)} & t \end{bmatrix} e_t & \text{if } t \in \sigma_{\mathscr{A}}(\tilde{c}) \setminus \{0,1\}, \\[2mm] \begin{bmatrix} e_{t0} & 0 \\ 0 & 0 \end{bmatrix} & \text{if } t \in \sigma_{\mathscr{A}}(\tilde{c}) \cap \{0\}, \\[2mm] \begin{bmatrix} 0 & 0 \\ 0 & e_{t1} \end{bmatrix} & \text{if } t \in \sigma_{\mathscr{A}}(\tilde{c}) \cap \{1\}, \end{cases}$$

respectively. Herein, $\sqrt{t(1-t)}$ refers to an arbitrarily chosen complex number the square of which is $t(1-t)$.

Taking into account the inverse-closedness of $\mathscr{D}$ in $\mathscr{A}$ one can rephrase the assertion of Theorem 3.5.1 as follows: *An element $a \in \mathscr{B}$ is invertible in $\mathscr{A}$ if and only if its symbol $\Phi(a)$ is invertible in $\mathscr{F}$.*

Proof. We shall verify the existence of symbol maps

$$\Phi_t \in \mathrm{smb}\,(\mathscr{B}_t, \mathscr{D}_t; \mathscr{D}_t^{2\times2}) \quad \text{if} \quad t \in \sigma_{\mathscr{A}}(\tilde{c}) \setminus \{0,1\}$$

and

$$\Phi_t \in \mathrm{smb}\,(\mathscr{B}_t, \mathscr{B}_t; \mathscr{B}_{t0} \times \mathscr{B}_{t1}) \quad \text{if} \quad t \in \sigma_{\mathscr{A}}(\tilde{c}) \cap \{0,1\}$$

which send the cosets $g_t \in \mathscr{G}_t$, p_t and r_t to the matrices

$$\begin{bmatrix} 1 & 0 \\ 0 & 1 \end{bmatrix} g_t, \quad \begin{bmatrix} 1 & 0 \\ 0 & 0 \end{bmatrix} e_t, \quad \begin{bmatrix} 1-t & \sqrt{t(1-t)} \\ \sqrt{t(1-t)} & t \end{bmatrix} e_t \qquad (3.89)$$

if $t \in \sigma_{\mathscr{A}}(\tilde{c}) \setminus \{0,1\}$, to

$$\begin{bmatrix} g_{t0} & 0 \\ 0 & g_{t1} \end{bmatrix}, \begin{bmatrix} e_{t0} & 0 \\ 0 & 0 \end{bmatrix}, \begin{bmatrix} e_{t0} & 0 \\ 0 & 0 \end{bmatrix} \tag{3.90}$$

if $t \in \sigma_{\mathscr{A}}(\tilde{c}) \cap \{0\}$, and to

$$\begin{bmatrix} g_{t0} & 0 \\ 0 & g_{t1} \end{bmatrix}, \begin{bmatrix} e_{t0} & 0 \\ 0 & 0 \end{bmatrix}, \begin{bmatrix} 0 & 0 \\ 0 & e_{t1} \end{bmatrix} \tag{3.91}$$

if $t \in \sigma_{\mathscr{A}}(\tilde{c}) \cap \{1\}$. Once this is done, then Allan's local principle will imply that the mapping Φ which sends the element $b \in \mathscr{D}$ to the function

$$t \mapsto \Phi_t(b_t) \tag{3.92}$$

is the desired symbol map, and we are done.

Consider the elements $N := e - p - r$ and $S := p - r$. One easily checks that

$$S^2 + N^2 = e \quad \text{and} \quad SN + NS = 0. \tag{3.93}$$

We first suppose that $t \in \sigma_{\mathscr{A}}(S^2) \setminus \{0,1\}$. Then $te_t - S_t^2 = 0$, that is $S_t^2 = te_t$, and from (3.93) we conclude that $N_t^2 = (1-t)e_t$. Choose complex numbers α and β such that $\alpha^2 = t$ and $\beta^2 = 1 - t$, and set $s_t := \frac{1}{\alpha} S_t$ and $n_t := \frac{1}{\beta} N_t$. Then

$$s_t^2 = n_t^2 = e_t \quad \text{and} \quad s_t n_t + n_t s_t = 0. \tag{3.94}$$

Consider the mapping

$$\eta_t : \mathscr{D}_t \to \mathscr{D}_t^{2 \times 2}, \quad b_t \mapsto \frac{1}{2} \begin{bmatrix} e_t & e_t \\ n_t & -n_t \end{bmatrix} \begin{bmatrix} e_t & 0 \\ 0 & s_t \end{bmatrix} \begin{bmatrix} b_t & 0 \\ 0 & b_t \end{bmatrix} \begin{bmatrix} e_t & 0 \\ 0 & s_t \end{bmatrix} \begin{bmatrix} e_t & n_t \\ e_t & -n_t \end{bmatrix}. \tag{3.95}$$

The matrices standing on the left- and right-hand side of the diagonal matrix $\mathrm{diag}\,(b_t, b_t)$ in (3.95) are inverse to each other. Thus, η_t is a bounded algebra homomorphism, and we claim that η_t is even a symbol map, precisely

$$\eta_t \in \mathrm{smb}(\mathscr{D}_t; \mathscr{D}_t^{2 \times 2}).$$

Indeed, if $\eta_t(b_t)$ is invertible in $\mathscr{D}_t^{2 \times 2}$, then the diagonal matrix $\mathrm{diag}\,(b_t, b_t)$ is invertible in $\mathscr{D}_t^{2 \times 2}$ and, hence, b_t is invertible in $\mathscr{D}_t$. Let us emphasize that η_t maps $g_t \in \mathscr{G}_t$, s_t, and n_t to the matrices

$$\begin{bmatrix} g_t & 0 \\ 0 & g_t \end{bmatrix}, \begin{bmatrix} s_t & 0 \\ 0 & -s_t \end{bmatrix}, \begin{bmatrix} 0 & e_t \\ e_t & 0 \end{bmatrix} \tag{3.96}$$

respectively. For the next step, let p_+, p_-, and j abbreviate the matrices

$$\frac{1}{2} \begin{bmatrix} e_t + s_t & 0 \\ 0 & e_t + s_t \end{bmatrix}, \frac{1}{2} \begin{bmatrix} e_t - s_t & 0 \\ 0 & e_t - s_t \end{bmatrix}, \begin{bmatrix} 0 & e_t \\ e_t & 0 \end{bmatrix},$$

respectively, and define a mapping μ_t by

$$\mu_t : \mathscr{D}_t \to \mathscr{D}_t^{2\times2}, \quad b_t \mapsto p_+ \eta_t(b_t)p_+ + p_- j\eta_t(b_t)jp_-.$$

We are going to show that μ_t is a symbol map, more exactly,

$$\mu_t \in \mathrm{smb}(\mathscr{B}_t, \mathscr{D}_t; \mathscr{D}_t^{2\times2}). \tag{3.97}$$

It is evident that μ_t is a linear bounded operator from $\mathscr{D}_t$ into $\mathscr{D}_t^{2\times2}$. Taking into account that $p_+^2 = p_+$, $p_-^2 = p_-$, $j^2 = e_t$ and $p_+ + p_- = e_t$, it is also obvious that μ_t is unital. Let us verify the semi-multiplicativity condition (iii), that is, given $a_t \in \mathscr{B}_t$ and $b_t \in \mathscr{D}_t$ we have to show that

$$\mu_t(a_t)\mu_t(b_t) = \mu_t(a_t b_t) \quad \text{and} \quad \mu_t(b_t)\mu_t(a_t) = \mu_t(b_t a_t). \tag{3.98}$$

Explicitly written, the left-hand side of the first of these identities equals

$$(p_+ \eta_t(a_t)p_+ + p_- j\eta_t(a_t)jp_-)(p_+ \eta_t(b_t)p_+ + p_- j\eta_t(b_t)jp_-)$$
$$= p_+ \eta_t(a_t)p_+ \eta_t(b_t)p_+ + p_- j\eta_t(a_t)p_- \eta_t(b_t)jp_-. \tag{3.99}$$

Now use that p_+ and p_- commute with each element of $\eta_t(\mathscr{B}_t)$, which comes from the fact that the cosets $g_t \in \mathscr{G}_t$, s_t, and n_t span the whole algebra $\mathscr{B}_t$ and from the special form of the matrices (3.96). Thus, the element of (3.99) is equal to

$$p_+ \eta_t(a_t)\eta_t(b_t)p_+ + p_- j\eta_t(a_t)\eta_t(b_t)jp_-,$$

which coincides with $\mu_t(a_t b_t)$ by definition. The second identity in (3.98) follows analogously. It remains to show that invertibility of $\mu_t(a_t)$ in $\mathscr{D}_t^{2\times2}$ implies that of a_t in $\mathscr{D}_t$ for all $a_t \in \mathscr{B}_t$ or, equivalently since η_t is a symbol map, that of $\eta_t(a_t)$ in $\mathscr{D}_t^{2\times2}$. Let $m \in \mathscr{D}_t^{2\times2}$ be the inverse of $\mu_t(a_t)$. A straightforward computation gives

$$p_+ \mu_t(a_t)p_+ + p_- j\mu_t(a_t)jp_- = \eta_t(a_t)$$

for all $a_t \in \mathscr{B}_t$. Thus,

$$\eta_t(a_t)(p_+ mp_+ + p_- jmjp_-)$$
$$= (p_+ \mu_t(a_t)p_+ + p_- j\mu_t(a_t)jp_-)(p_+ mp_+ + p_- jmjp_-)$$
$$= p_+ \mu_t(a_t)mp_+ + p_- j\mu_t(a_t)mjp_-$$
$$= \mathrm{diag}(e_t, e_t),$$

which implies the invertibility of $\eta_t(a_t)$ in $\mathscr{D}_t^{2\times2}$ and

$$\eta_t(a_t)^{-1} = p_+ mp_+ + p_- jmjp_-.$$

This proves (3.97). To finish the proof in the case that $t \in \sigma_{\mathscr{A}}(S^2) \setminus \{0,1\}$ we choose complex numbers γ and δ such that $\gamma^2 = 1 + \alpha$ and $\delta^2 = 1 - \alpha$ and define a 2×2 matrix d by

$$d := \frac{1}{\sqrt{2}} \begin{bmatrix} \gamma & -\delta \\ -\delta & -\gamma \end{bmatrix} e_t.$$

Since $d^2 = \mathrm{diag}(e_t, e_t)$, it is obvious that the mapping

$$\Phi_t : \mathscr{D}_t \to \mathscr{D}_t^{2\times 2}, \quad b_t \mapsto d\mu_t(b_t)d$$

belongs to $\mathrm{smb}(\mathscr{B}_t, \mathscr{D}_t; \mathscr{D}_t^{2\times 2})$, and a direct computation yields that $\Phi_t(g_t)$, $\Phi_t(p_t)$ and $\Phi_t(r_t)$ are just the matrices (3.89).

Now let $t = 0 \in \sigma_{\mathscr{A}}(S^2)$. Then one can argue as above to get that $S_0^2 = 0$. Thus, S_0 and N_0 are subject to the relations

$$S_0^2 = 0, \quad N_0^2 = e_0, \quad S_0 N_0 + N_0 S_0 = 0. \tag{3.100}$$

If $a_0 \in \mathscr{B}_0$ is invertible in $\mathscr{D}_0$, then a_0 is also invertible in $\mathscr{B}_0$ by hypothesis (3.88). Our first claim is that the canonical homomorphism

$$\nu_0 : \mathscr{B}_0 \to \mathscr{B}_0/\tilde{I}_0, \quad a_0 \mapsto a_0 + \tilde{I}_0$$

is a symbol map, precisely

$$\nu_0 \in \mathrm{smb}(\mathscr{B}_0; \mathscr{B}_0/\tilde{I}_0). \tag{3.101}$$

For, it is sufficient to show that the ideal $\tilde{I}_0$ belongs to the radical of $\mathscr{B}_0$ or, since $\tilde{I}_0$ is generated by the element $S_0 = p_0 - r_0$, that S_0 belongs to that radical. Equivalently, by Proposition 1.3.3, we have to show that $e_0 + a_0 S_0$ is left invertible in $\mathscr{B}_0$ for all $a_0 \in \mathscr{B}_0$. But the invertibility of that element follows from

$$(e_0 - a_0 S_0)(e_0 + a_0 S_0) = e_0 - a_0 S_0 a_0 S_0 = e_0$$

(note that each element $a_0 \in \mathscr{B}_0$ can be approximated by elements of the form $g_0^1 e_0 + g_0^2 S_0 + g_0^3 N_0 + g_0^4 S_0 N_0$ with $g_0^i \in \mathscr{G}_0$ and then use (3.100) to obtain $a_0 S_0 a_0 S_0 = 0$).

Let p_+ and p_- stand for the elements $\nu_0(\frac{e_0 + N_0}{2})$ and $\nu_0(\frac{e_0 - N_0}{2})$ in $\mathscr{B}_0/\tilde{I}_0$. Since $N_0^2 = e_0$, p_+ and p_- are complementary idempotents. We claim that these idempotents are non-trivial, that is, neither p_+ nor p_- is the identity element. Indeed, suppose p_+ is the identity, hence invertible. Then $\frac{e_0 + N_0}{2}$ is an invertible element of $\mathscr{B}_0$ by (3.101). Since $\frac{e_0 + N_0}{2}$ is an idempotent, this implies that $\frac{e_0 + N_0}{2} = e_0$ or, equivalently, $N_0 = e_0$. Since one also has $S_0^2 = 0$, these two identities combine to give $p_0 = r_0 = 0$. Consequently, the element $e_0 - p_0$ is invertible in $\mathscr{B}_0$. Then Proposition 2.2.3 (i) ensures that $e_t - p_t$ is invertible in $\mathscr{D}_t$ for all $t \in \sigma_{\mathscr{A}}(S^2)$ which belong to a certain neighborhood of 0 (note that such points exist by hypothesis (3.87)). But, as we have already seen,

$$\Phi_t(e_t - p_t) = \begin{bmatrix} 0 & 0 \\ 0 & e_t \end{bmatrix}$$

which fails to be invertible. This contradiction shows that p_+ is not the identity. The assertion for p_- follows similarly.

Now a little thought shows that each element of $\mathscr{B}_0/\tilde{I}_0$ can be approximated by elements of the form

$$(g_0^+ + \tilde{I}_0)p_+ + (g_0^- + \tilde{I}_0)p_- \quad \text{with} \quad g_0^+, g_0^- \in \mathscr{G}_0. \tag{3.102}$$

Hence, the set of all complex linear combinations $\alpha p_+ + \beta p_-$ forms a central subalgebra of $\mathscr{B}_0/\tilde{I}_0$, and one can study the invertibility of the elements (3.102) by employing Allan's local principle again. Since the central subalgebra is singly generated by p_+, its maximal ideal space is homeomorphic to the spectrum of p_+, and the latter equals $\{0,1\}$ because of the non-triviality of p_+. The corresponding "local ideals" are just the ideals I_{00} and I_{01} introduced above. Hence, by Allan's local principle, an element $a_0 \in \mathscr{B}_0$ is invertible in $\mathscr{B}_0$ if and only if $(a_0 + \tilde{I}_0) + I_{00} =: a_{00}$ is invertible in $\mathscr{B}_{00}$ and $(a_0 + \tilde{I}_0) + I_{01} =: a_{01}$ is invertible in $\mathscr{B}_{01}$ or, equivalently, if the diagonal matrix

$$\mathrm{diag}\,(a_{00}, a_{01}) \tag{3.103}$$

is invertible in $\mathscr{B}_{00} \times \mathscr{B}_{01}$. The simple observation that the homomorphism which assigns to $a_0 \in \mathscr{B}_0$ the matrix (3.103) sends the elements $g_0 \in \mathscr{G}_0$, p_0 and r_0 to the matrices (3.90), respectively, completes the proof for $t = 0$. In the case $t = 1 \in \sigma_{\mathscr{A}}(S^2)$, the proof runs analogously. ∎

Let us complete this result by a few corollaries. The first observation is that an analog of Corollary 3.1.5 of Theorem 3.1.4 holds in the present setting as well. In particular, one has the following.

Corollary 3.5.2. *Let $\mathscr{D}_p$ and $\mathscr{D}_{e-p}$ be the Banach algebras $\{pbp : b \in \mathscr{D}\}$ and $\{(e-p)b(e-p) : b \in \mathscr{D}\}$, respectively. If $\{0,1\} \subset \sigma_{\mathscr{D}_p}(prp)$, then*

$$\sigma_{\mathscr{D}}(c) = \sigma_{\mathscr{D}_p}(prp) = \sigma_{\mathscr{D}_{e-p}}((e-p)(e-r)(e-p)).$$

The proof is the same as that of Corollary 3.1.5.

The next corollary concerns the repeated factorization at the points 0 and 1 which seems to make the whole story rather non-transparent. This impression will be moderated by showing that the quotient algebras by the ideals $\tilde{I}_0$ are always of a simple structure.

Corollary 3.5.3. *The algebra $\mathscr{B}_0$ is the direct sum of its subalgebra which is generated by $\mathscr{G}_0$ and N_0 and of the ideal $\tilde{I}_0$,*

$$\mathscr{B}_0 = \mathrm{alg}\{\mathscr{G}_0, N_0\} + \tilde{I}_0.$$

In particular, the quotient algebra $\mathscr{B}_0/\tilde{I}_0$ is isomorphic to $\mathrm{alg}\{\mathscr{G}_0, N_0\}$.

Proof. Now let p_+ and p_- stand for the idempotents $\frac{e_0+N_0}{2}$ and $\frac{e_0-N_0}{2}$, respectively, and define a mapping $\omega : \mathscr{B}_0 \to \mathscr{B}_0$ by

$$\omega(a_0) =: p_+ a_0 p_+ + p_- a_0 p_-.$$

Straightforward computation shows that ω is a homomorphism and, since p_+ and p_- are complementary idempotents, that $\omega(\omega(a_0)) = \omega(a_0)$ for all $a_0 \in \mathscr{B}_0$. The identity $\omega^2 = \omega$ implies that $\mathscr{B}_0$ decomposes into the direct sum

$$\mathscr{B}_0 = \mathrm{Im}\,\omega \dotplus \mathrm{Ker}\,\omega.$$

Noting that $\omega(g_0) = g_0$ for $g_0 \in \mathscr{G}_0$, $\omega(N_0) = N_0$, and $\omega(S_0) = 0$, we find that $\mathrm{Im}\,\omega = \mathrm{alg}(\mathscr{G}_0, N_0)$ and $\mathrm{Ker}\,\omega \supseteq \tilde{I}_0$. Now recall that each element of $\mathscr{B}_0$ can be approximated by elements of the form $g_0^1 e_0 + g_0^2 S_0 + g_0^3 N_0 + g_0^4 S_0 N_0$ with $g_0^i \in \mathscr{G}_0$. Since the elements of this form belong evidently to $\mathrm{alg}\{\mathscr{G}_0, N_0\} + \tilde{I}_0$, we conclude that $\mathrm{Ker}\,\omega \subseteq \tilde{I}_0$. ∎

There are situations where no further factorization at 0 and 1 is needed.

Corollary 3.5.4. *Let the hypotheses be as in Theorem* 3.5.1 *and suppose moreover that*

$$\mathscr{G}_t \text{ is simple for } t \in \sigma_{\mathscr{A}}(S^2) \cap \{0,1\}. \qquad (3.104)$$

Then there exists a symbol map $\Phi \in \mathrm{smb}(\mathscr{B}, \mathscr{G}; \mathscr{F})$, *where now* $\mathscr{F}$ *stands for the algebra of all functions on* $\sigma_{\mathscr{A}}(S^2)$ *which take a value in* $\mathscr{D}_t^{2\times 2}$ *at* $t \in \sigma_{\mathscr{A}}(S^2)$. *The map* Φ *sends the elements* $g \in \mathscr{G}$, p *and* r *to the functions*

$$t \mapsto \begin{bmatrix} 1 & 0 \\ 0 & 1 \end{bmatrix} g_t, \quad t \mapsto \begin{bmatrix} 1 & 0 \\ 0 & 0 \end{bmatrix} e_t \quad and \quad t \mapsto \begin{bmatrix} 1-t & \sqrt{t(1-t)} \\ \sqrt{t(1-t)} & t \end{bmatrix} e_t,$$

respectively.

Proof. The proof rests on the simple fact that if $\mathscr{C}_1$ and $\mathscr{C}_2$ are Banach algebras and $H : \mathscr{C}_1 \to \mathscr{C}_2$ is a bounded homomorphism then $\mathrm{Im}\,H$ is isomorphic to $\mathscr{C}_1/\mathrm{Ker}\,H$. If, moreover, $\mathscr{C}_1$ is simple then either $\mathrm{Ker}\,H = \{0\}$ or $\mathrm{Ker}\,H = \mathscr{C}_1$; hence, $\mathrm{Im}\,H$ is either isomorphic to $\mathscr{C}_1$, or $\mathrm{Im}\,H = \{0\}$. Thus, if $\mathscr{G}_0$ is simple, then $\mathscr{G}_{00}$ is either isomorphic to $\mathscr{G}_0$, or it consists of the zero element only. But the latter is impossible since p_+ and p_- generate proper ideals as we have already seen. Hence, $\mathscr{G}_{00}$ is isomorphic to $\mathscr{G}_0$, and applying this isomorphism to the symbol map quoted in Theorem 3.5.1 we obtain the assertion. ∎

In the case that $\mathscr{G}$ is simple, the same arguments show that (3.104) is satisfied. In particular, Corollary 3.5.4 holds if the coefficient algebra $\mathscr{G}$ is $\mathbb{C}^{k\times k}$, i.e., in the matrix case.

To avoid misunderstandings, let us mention a formal difference between Theorems 3.1.4 and 3.5.1. In Theorem 3.1.4, the central role is played by the element $prp + (e-p)(e-r)(e-p) = e - p - r + pr + rp$, in contrast to the element S^2 in Theorem 3.5.1. Since both elements are related by

$$prp + (e-p)(e-r)(e-p) = e - S^2,$$

we have

$$\sigma_{\mathscr{A}}(prp + (e-p)(e-r)(e-p)) = 1 - \sigma_{\mathscr{A}}(S^2)$$

which implies that the symbol maps of Theorems 3.1.4 and 3.5.1 can be transformed into each other by substituting $t \leftrightarrow 1 - t$. One possible reason to prefer to work with the element $prp + (e-p)(e-r)(e-p)$ is that the spectra of $prp + (e-p)(e-r)(e-p)$ and prp coincide under suitable conditions and that prp can quite often be identified with a local Toeplitz operator. And the spectra of local Toeplitz operators are known in many situations.

Finally we discuss the inverse-closedness of $\mathscr{B}$ in $\mathscr{D}$.

Corollary 3.5.5. *Let the hypotheses be as in Theorem 3.5.1 and suppose moreover that*

$$\sigma_{\mathscr{A}}(S^2) = \sigma_{\mathscr{B}}(S^2) \tag{3.105}$$

and

$$\mathscr{G}_t^{2\times 2} \text{ is inverse-closed in } \mathscr{D}_t^{2\times 2} \text{ for } t \in \sigma_{\mathscr{A}}(S^2) \setminus \{0,1\}. \tag{3.106}$$

Then $\mathscr{B}$ is inverse-closed in $\mathscr{D}$, and the homomorphism Φ from Theorem 3.5.1 is a symbol map in $\mathrm{smb}(\mathscr{B}; \mathscr{F})$.

Proof. Let Φ refer to the symbol map established in Theorem 3.5.1, and write Φ' for the symbol map which results from the same theorem by choosing $\mathscr{A} := \mathscr{F}$. Both mappings coincide on $\mathscr{B}$ because of (3.105), and it is easy to see that $\Phi_t(a)$ and $\Phi_t'(a)$ belong to $\mathscr{G}_t^{2\times 2}$ for all $t \in \sigma_{\mathscr{A}}(S^2) \setminus \{0,1\}$ and all $a \in \mathscr{B}$. Thus, if $a \in \mathscr{B}$ is invertible in $\mathscr{D}$, then $\Phi_t(a)$ is invertible in $\mathscr{D}_t^{2\times 2}$ and, by (3.106), also in $\mathscr{G}_t^{2\times 2}$. The latter implies the invertibility of $\Phi_t'(a)$ in $\mathscr{G}_t^{2\times 2}$ and, consequently, that of a in $\mathscr{B}$. ∎

We conclude by an inverse-closedness result which is based on Corollary 1.2.32. Note that, if $\mathscr{G}$ is isomorphic to $\mathbb{C}^{k\times k}$, then $\mathscr{G}$ is simple, and the elements of $\mathscr{G}$ and $\mathscr{G}^{2\times 2}$ have thin (actually, discrete) spectra.

Corollary 3.5.6. *Let $\mathscr{A}, \mathscr{B}, \mathscr{D}, \mathscr{G}$ and p, r be as above, suppose (3.87) and (3.105) to be fulfilled, and let $\mathscr{G}$ be isomorphic to $\mathbb{C}^{k\times k}$. Then there exists a symbol map $\Phi \in \mathrm{smb}(\mathscr{B}, \mathscr{D}; \mathscr{F})$ where now $\mathscr{F}$ stands for the algebra of all functions on $\sigma_{\mathscr{A}}(c)$ with values in $\mathbb{C}^{2k\times 2k}$. The map Φ sends the elements $g \in \mathscr{G} \cong \mathbb{C}^{k\times k}$, p and r to the matrix functions*

$$t \mapsto \begin{bmatrix} g & 0 \\ 0 & g \end{bmatrix}, \quad t \mapsto \begin{bmatrix} e & 0 \\ 0 & 0 \end{bmatrix} \quad \text{and} \quad t \mapsto \begin{bmatrix} (1-t)e & \sqrt{t(1-t)}e \\ \sqrt{t(1-t)}e & te \end{bmatrix}$$

where e stands for the $k \times k$ unit matrix. Moreover, $\mathscr{B}$ is inverse-closed in $\mathscr{D}$, and Φ belongs to $\mathrm{smb}(\mathscr{B}; \mathscr{F})$.

3.5.2 N projections theorem

Again let $\mathscr{A}$ be a Banach algebra with identity e, let $\{p_i\}_{i=1}^{2N}$ be a partition of unity into projections and P be an idempotent in $\mathscr{A}$ such that the axioms (3.5) and (3.6) hold. The smallest closed subalgebra of $\mathscr{A}$ containing the partition $\{p_i\}$ as well as the element P will be denoted by $\mathscr{B}$ again. Suppose $\mathscr{G}$ is a simple and closed subalgebra of $\mathscr{A}$ containing e and having the property that

$$p_i g = g p_i \quad \text{and} \quad gP = Pg \quad \text{for all} \quad i = 1, \ldots, 2N \quad \text{and} \quad g \in \mathscr{G}.$$

The algebra $\mathscr{G}$ is the *coefficient algebra*. It is possible to derive a version of Theorem 3.2.4 which provides us with an invertibility symbol for the smallest closed subalgebra $\mathscr{C}$ of $\mathscr{A}$ which contains the partition $\{p_i\}$, the idempotent P and the algebra $\mathscr{G}$. Here is the formulation of such a version.

Theorem 3.5.7. *Let $\mathscr{C}$ be as above and let $\mathscr{G}$ be simple.*

(i) *If $x \in \sigma_{\mathscr{B}}(X) \setminus \{0, 1\}$, then the mapping*

$$F_x : \{P, p_1, \ldots, p_{2N}\} \cup \mathscr{G} \to \mathscr{G}^{2N \times 2N}$$

given by

$$F_x(p_i) = \operatorname{diag}(0, \ldots, 0, I, 0, \ldots, 0),$$

with the I standing at the ith place,

$$F_x(P) = \operatorname{diag}(I, -I, I, -I, \ldots, I, -I) \times$$

$$\begin{bmatrix} x & x-1 & x-1 & x-1 & \cdots & x-1 & x-1 \\ x & x-1 & x-1 & x-1 & \cdots & x-1 & x-1 \\ x & x & x & x-1 & \cdots & x-1 & x-1 \\ x & x & x & x-1 & \cdots & x-1 & x-1 \\ \vdots & \vdots & \vdots & \vdots & \ddots & \vdots & \vdots \\ x & x & x & x & \cdots & x & x-1 \\ x & x & x & x & \cdots & x & x-1 \end{bmatrix},$$

and

$$F_x(g) = \operatorname{diag}(g, g, \ldots, g),$$

extends to a continuous algebra homomorphism from $\mathscr{C}$ onto $\mathscr{G}^{2N \times 2N}$.

(ii) *If $m \in \sigma_{\mathscr{B}}(Y) \cap \{1, \ldots, 4N\}$, then the mapping*

$$G_m : \{P, p_1, \ldots, p_{2N}\} \cup \mathscr{G} \to \mathscr{G}$$

given by

$$G_{4m}(p_i) := \begin{cases} I & \text{if } i = 2m, \\ 0 & \text{if } i \neq 2m, \end{cases} \qquad G_{4m}(P) = 0,$$

$$G_{4m-1}(p_i) := \begin{cases} I & \text{if } i = 2m, \\ 0 & \text{if } i \neq 2m, \end{cases} \qquad G_{4m-1}(P) = I,$$

$$G_{4m-2}(p_i) := \begin{cases} I & \text{if } i = 2m-1, \\ 0 & \text{if } i \neq 2m-1, \end{cases} \qquad G_{4m-2}(P) = I,$$

$$G_{4m-3}(p_i) := \begin{cases} I & \text{if } i = 2m-1, \\ 0 & \text{if } i \neq 2m-1, \end{cases} \qquad G_{4m-3}(P) = 0,$$

where $m = 1, \ldots, N$, and by $G_m(g) = g$ extends to a continuous algebra homomorphism from $\mathscr{C}$ onto $\mathscr{G}$.

(iii) An element $C \in \mathscr{C}$ is invertible in $\mathscr{C}$ if and only if the matrices $F_x(C)$ are
invertible for all $x \in \sigma_{\mathscr{B}}(X) \setminus \{0, 1\}$ and the elements $G_m(C)$ are invertible for
all $m \in \sigma_{\mathscr{B}}(Y) \cap \{1, \ldots, 4N\}$.

(iv) An element $C \in \mathscr{C}$ is invertible in $\mathscr{A}$ if and only if the matrices $F_x(C)$ are
invertible for all $x \in \sigma_{\mathscr{A}}(X) \setminus \{0, 1\}$ and the elements $G_m(C)$ are invertible for
all $m \in \sigma_{\mathscr{A}}(Y) \cap \{1, \ldots, 4N\}$.

Observe that the conditions of the theorem are satisfied if, for example, $\mathscr{G}$ is the
algebra $\mathbb{C}^{n \times n}$ which yields just the matrix version of Theorem 3.2.4. It is possible to
consider also the case where $\mathscr{G}$ is not simple. In [55] the case for the two projections
theorem with a coefficient algebra was resolved, and the techniques used therein are
applicable to the N projections case.

3.5.3 Two projections and a flip

Let $\mathscr{F}$ be a Banach algebra with identity element e. Let p, r, j be elements of $\mathscr{F}$
satisfying the axioms (3.45) and (3.46). Let $\mathscr{G}$ be a unital subalgebra of $\mathscr{F}$ the elements of which commute with p, r and j. Denote the smallest closed subalgebra of
$\mathscr{F}$ containing $\mathscr{G}$, p, r and j by $\mathscr{A}_g$. We are interested in criteria for the invertibility
of elements from $\mathscr{A}_g$ in $\mathscr{F}$. We embed it into a larger subalgebra of $\mathscr{F}$. Let $\mathscr{H}$
stand for the set

$$\{f \in \mathscr{F} \ : \ fb = bf \text{ and } fc = cf\}, \tag{3.107}$$

where b and c are defined by (3.53) and (3.54). $\mathscr{H}$ forms a closed subalgebra of $\mathscr{F}$.
Moreover, $\mathscr{H}$ is inverse-closed in $\mathscr{F}$ and contains $\mathscr{A}_g$. Thus, the problem has been
reduced to the study of the invertibility of elements from $\mathscr{A}_g$ in $\mathscr{H}$. Now extend the
mapping w from (3.51) to the algebra $\mathscr{H}$,

$$w : \mathscr{H} \to \mathscr{H}^{2 \times 2}, \quad a \mapsto \begin{bmatrix} w_{11}(a) & w_{12}(a) \\ w_{21}(a) & w_{22}(a) \end{bmatrix}. \tag{3.108}$$

The images of the generating elements of $\mathscr{A}_\mathscr{G}$ are given by (3.52), Proposition 3.3.4 (i), and

$$w(g) = \begin{bmatrix} g & 0 \\ 0 & g \end{bmatrix} \quad \text{for } g \in \mathscr{G}. \tag{3.109}$$

Let $\mathscr{C} = \mathrm{alg}\{b,c\}$ be defined as before and let $\mathscr{C}_\mathscr{G}$ be the smallest closed subalgebra of $\mathscr{A}_\mathscr{G}$ which contains $\mathscr{C}$ and $\mathscr{G}$. The proposition below is the analog of Proposition 3.3.4.

Proposition 3.5.8. *The following assertions hold:*

(i) *The set $\mathscr{D} := w_{11}(\mathscr{H})$ is a closed subalgebra of $\mathscr{H}$ which contains e.*
(ii) *The mapping w from (3.108) is a continuous isomorphism from $\mathscr{H}$ onto $\mathscr{D}^{2\times 2}$. The image of w with respect to $\mathscr{A}_\mathscr{G}$ is $\mathscr{C}_\mathscr{G}^{2\times 2}$.*
(iii) *$\mathscr{C}$ belongs to the center of $\mathscr{H}$.*

Thus, an element $a \in \mathscr{A}_\mathscr{G}$ is invertible in $\mathscr{H}$ if and only if the matrix $w(a)$ is invertible in $\mathscr{D}^{2\times 2}$. As a consequence of part (iii) of the preceding proposition, Allan's local principle applies to $\mathscr{H}$ with $\mathscr{C}$ as the central subalgebra, and to $\mathscr{D}^{2\times 2}$ with

$$\left\{ \begin{bmatrix} c & 0 \\ 0 & c \end{bmatrix} : c \in \mathscr{C} \right\}$$

as the central subalgebra. The maximal ideal space of $\mathscr{C}$ is $M_\mathscr{C}$. We know the structure of this set from Sections 3.3.1 and 3.3.2. Let $M_\mathscr{C}^0$ stand for the collection of those $(x,y) \in M_\mathscr{C}$ for which $\mathscr{I}_{(x,y)} \neq \mathscr{A}$. Given $(x,y) \in M_\mathscr{C}^0$, we denote the related local algebra $\mathscr{H}/\mathscr{I}_{(x,y)}$ (resp. $\mathscr{D}^{2\times 2}/\mathscr{I}_{(x,y)}^{2\times 2}$) by $\mathscr{H}_{(x,y)}$ (resp. $\mathscr{D}_{(x,y)}^{2\times 2}$) and the canonical homomorphism from $\mathscr{H}$ onto $\mathscr{H}_{(x,y)}$ by $\Phi_{(x,y)}$. Set $\mathscr{G}_{(x,y)} = \Phi_{(x,y)}(\mathscr{G})$. Then Allan's local principle states the following.

Proposition 3.5.9. *An element $a \in \mathscr{A}_\mathscr{G}$ is invertible in $\mathscr{H}$ if and only if the function*

$$M_\mathscr{C}^0 \to \mathscr{G}_{(x,y)}^{2\times 2}, \quad (x,y) \mapsto \begin{bmatrix} \Phi_{(x,y)}(w_{11}(a)) & \Phi_{(x,y)}(w_{12}(a)) \\ \Phi_{(x,y)}(w_{21}(a)) & \Phi_{(x,y)}(w_{22}(a)) \end{bmatrix}$$

is invertible in $\mathscr{D}_{(x,y)}^{2\times 2}$ at every point $(x,y) \in M_\mathscr{C}^0$.

The functions which correspond to the elements $g \in \mathscr{G}$, p, j and r, are given by

$$(x,y) \mapsto \begin{bmatrix} \Phi_{(x,y)}(g) & 0 \\ 0 & \Phi_{(x,y)}(g) \end{bmatrix},$$

$$(x,y) \mapsto \begin{bmatrix} \Phi_{(x,y)}(e) & 0 \\ 0 & 0 \end{bmatrix},$$

$$(x,y) \mapsto \begin{bmatrix} 0 & \Phi_{(x,y)}(e) \\ \Phi_{(x,y)}(e) & 0 \end{bmatrix},$$

$$(x,y) \mapsto \begin{bmatrix} x\Phi_{(x,y)}(e) & y\Phi_{(x,y)}(e) \\ -y\Phi_{(x,y)}(e) & (1-x)\Phi_{(x,y)}(e) \end{bmatrix}.$$

Before applying this result to a simple coefficient algebra we shall give some characterization of the set $M_{\mathscr{C}}^0$. The local principle states that

$$\sigma_{\mathscr{D}}(b+c) = \bigcup_{(x,y)\in M_{\mathscr{C}}^0} \sigma_{\mathscr{D}_{(x,y)}}\big(\Phi_{(x,y)}(b+c)\big) = \bigcup_{(x,y)\in M_{\mathscr{C}}^0} \{x+y\},$$

and, hence, the mapping (3.58) from Proposition 3.3.12 sends $M_{\mathscr{C}}^0$ onto $\sigma_{\mathscr{D}}(b+c)$. But this means nothing except that the restriction of (3.58) onto $M_{\mathscr{C}}^0$,

$$M_{\mathscr{C}}^0 \to \mathbb{C}, \quad (x,y) \mapsto x+y \tag{3.110}$$

is a bijection between $M_{\mathscr{C}}^0$ and $\sigma_{\mathscr{D}}(b+c)$.

Proposition 3.5.10. $\sigma_{\mathscr{D}}(b+c) = \sigma_{\mathscr{H}}(b+c) = \sigma_{\mathscr{F}}(b+c)$.

Proof. The second equality follows from the inverse-closedness of $\mathscr{H}$ in $\mathscr{F}$. The first one is a consequence of Proposition 3.5.8 (ii) if we take into consideration that $w(b) = \mathrm{diag}(b,b)$ and $w(c) = \mathrm{diag}(c,c)$. ∎

We can specialize Propositions 3.5.9 and 3.5.10 by choosing $\mathscr{A}_{\mathscr{G}}$ itself as the larger algebra (in place of $\mathscr{F}$). Comparing the invertibility criteria in both cases leads to:

Corollary 3.5.11. *The algebra $\mathscr{A}_{\mathscr{G}}$ is inverse-closed in $\mathscr{F}$ if and only if the following two conditions are fulfilled:*

$$\sigma_{\mathscr{F}}(b+c) = \sigma_{\mathscr{C}_{\mathscr{G}}}(b+c), \tag{3.111}$$

$$\mathscr{G}_{(x,y)}^{2\times 2} \text{ is inverse-closed in } \mathscr{D}_{(x,y)}^{2\times 2}. \tag{3.112}$$

In what follows we suppose $\mathscr{G}$ to be simple, i.e., $\mathscr{G}$ possesses trivial ideals only. Then condition (3.112) proves to be fulfilled. Combining Propositions 3.5.9, 3.5.10 and Corollary 3.5.11 yields:

Theorem 3.5.12. *Let $\mathscr{F}$, $\mathscr{G}$, $\mathscr{A}_{\mathscr{G}}$, $\mathscr{H}$, $\mathscr{C}$ and p, r, j be defined as above. Let $M_{\mathscr{C}}$ be the maximal ideal space of $\mathscr{C}$. Further, let $M_{\mathscr{C}}^0$ consist of all pairs $(x,y) \in \mathbb{C}\times\mathbb{C}$ where $x^2 - x = y^2$ and $x+y \in \sigma_{\mathscr{F}}(b+c)$.*

(i) *The set $M_{\mathscr{C}}^0$ coincides with the collection of those $(x,y) \in M_{\mathscr{C}}$ for which $\mathscr{I}_{(x,y)} \neq \mathscr{H}$.*

(ii) *The mapping* smb *which assigns to $g \in \mathscr{G}$, p, j and r a matrix-valued function on $M_{\mathscr{C}}^0$ by*

$$(\mathrm{smb}\, g)(x,y) = \begin{bmatrix} g & 0 \\ 0 & g \end{bmatrix}, \ (\mathrm{smb}\, p)(x,y) = \begin{bmatrix} e & 0 \\ 0 & 0 \end{bmatrix},$$

$$(\mathrm{smb}\, j)(x,y) = \begin{bmatrix} 0 & e \\ e & 0 \end{bmatrix}, \ (\mathrm{smb}\, r)(x,y) = \begin{bmatrix} xe & ye \\ -ye & (1-x)e \end{bmatrix}$$

extends to a continuous homomorphism from $\mathscr{A}_{\mathscr{G}}$ into $C(M_{\mathscr{C}}^0, \mathscr{G}^{2\times 2})$.

(iii) *An element $a \in \mathscr{A}_{\mathscr{G}}$ is invertible in $\mathscr{F}$ if and only if $(\text{smb}\,a)(x,y)$ is invertible for every $(x,y) \in M^0_{\mathscr{C}}$.*

(iv) *If, in addition, condition (3.111) holds, then $M^0_{\mathscr{C}} = M_{\mathscr{C}}$, and $\mathscr{A}_{\mathscr{G}}$ is inverse-closed in $\mathscr{F}$.*

Let us point out the case that $\mathscr{G}$ is isomorphic to $\mathbb{C}^{k \times k}$. Then the preceding theorem associates with every element $a \in \mathscr{A}_{\mathscr{G}}$ a matrix symbol of order $2k$ which answers the invertibility of a in $\mathscr{F}$. There is a somewhat different way to derive this result. Since $\mathbb{C}^{k \times k}$ is a full matrix algebra in $\mathscr{A}_{\mathscr{G}}$ we can define a homomorphism $\hat{w}$ from $\mathscr{H}$ into $\mathscr{H}^{2k \times 2k}$ which maps $\mathscr{A}_{\mathscr{G}}$ onto $\mathbb{C}^{2k \times 2k}$. Now Allan's local principle applies and leads to an analogous result.

3.6 Notes and comments

Algebras generated by two projections appear in many places and have thus attracted a lot of attention. Already in the late 1960's, Halmos [83] and Pedersen [136] studied the C^*-algebra generated by two self-adjoint projections p and r under the assumption that $\sigma(prp) = [0,1]$. Besides this spectral condition, Halmos also supposed that the projections are in general position, that is, $X \cap Y = \{0\}$ whenever $X, Y \in \{\text{Ker } p, \text{Im } p, \text{Ker } r, \text{Im } r\}$ and $X \neq Y$. These conditions are fulfilled, for example, if p is the operator of multiplication by the characteristic function of the upper semi-circle $\mathbb{T}_+ := \{z \in \mathbb{T} : \text{Im } z > 0\}$ acting on $L^2(\mathbb{T})$, and r is the operator $(I + S_{\mathbb{T}})/2$ where $S_{\mathbb{T}}$ stands for the operator of singular integration on $L^2(\mathbb{T})$. Theorem 3.1.1 appeared with full proof in [196] also. Independently, this result was derived in [149]. Note that the case when $\sigma(prp) = [0,1]$ is of particular importance, since it distinguishes the universal C^*-algebra generated by two self-adjoint projections.

While Halmos' paper from 1969 is certainly the most influential in this field, the theory of two Hilbert space projections has a much longer history, including Krein, Krasnoselski, Milman [106], Dixmier [38] and Davis [35], to mention some of the main contributions. For a detailed overview on this history as well as for a survey on applications of the two projections theorem (mainly in the fields of linear algebra and Hilbert space theory), we refer to the recent "gentle guide" by Böttcher and Spitkovsky [23].

As far as we know, Banach algebras generated by two idempotents were first studied in the 1988 paper [167] by two of the authors. In that paper, the spectrum of prp is supposed to be connected and to include the points 0 and 1. The approach of [167] is based on Krupnik's theory of Banach PI-algebras. In its final form, Theorem 3.1.4 first appeared in [75]. The proof presented above is in the spirit of [167].

In general, algebras generated by three idempotents are of an involved structure, and there is no hope for a general classification. In order to say something substantial about them, one has to impose strong restrictions on the generating elements of the algebra, which often come from modeling specific applications.

Banach algebras with generators which fulfill conditions (3.45) and (3.46) are of particular interest. They occur as local models in the theory of Toeplitz and Hankel operators. Theorem 3.3.15, which is the first result in the direction and treats the C^*-case, belongs to Power [149]. The Banach algebra generated by two idempotents and orientation reversing flip was considered by Finck and two of the authors in [56], where Theorem 3.3.13 is derived. A forerunner of that paper is [167] where we showed that this algebra satisfies the standard polynomial F_4. Banach algebras generated by elements which satisfy (3.45) and (3.47) (i.e., the case of an orientation preserving flip) were completely studied by Krupnik and Spigel in [110]. The N projections theorem 3.2.4 appeared in [12] and is the result of joint efforts of Böttcher, Gohberg, Yu. Karlovich, Krupnik, Spitkovsky, and two of the authors. The results of Section 3.5 are mainly based on [55].

Let us finally mention that [112] and [152] present further interesting examples of finitely generated Banach algebras, which turn out to be PI-algebras. See also [132] for an overview of finitely generated (not necessarily normed) algebras.

Part II
Case Studies

Chapter 4
Singular integral operators

Algebras generated by singular integral operators and convolution operators are one of the numerous instances where local principles and projection theorems have been successfully applied. And conversely, it was mainly the study of these algebras which stimulated the development of local principles and projection theorems as a tool in operator theory. We will now give a thorough exposition of algebras generated by singular integral operators on Lebesgue spaces over admissible curves. Note that we have already encountered singular integral operators on the unit circle in the previous chapters.

4.1 Curves and algebras

4.1.1 Admissible curves

Let $\mathbb{C}$ be the complex plane and Ω be a (possibly unbounded) subset of $\mathbb{C}$. The set of complex-valued continuous functions defined on Ω is denoted by $C(\Omega)$. Let $\lambda \in \,]0,1[$. A continuous function f is said to be *Hölder continuous* in Ω with exponent λ if there is a constant c such that

$$|f(t) - f(s)| \leq c|t - s|^{\lambda}$$

for all $s, t \in \Omega$. The set of Hölder continuous functions on Ω will be denoted by $C^{\lambda}(\Omega)$.

A *Lyapunov arc* , Γ, is an oriented, bounded curve in the complex plane which is homeomorphic to the closed interval $[0, 1]$ and which fulfills the *Lyapunov condition*, i.e., the tangent to Γ exists at each point $t \in \Gamma$, and the smallest angle $\theta(t)$ between the tangent and the real axis, measured from the latter counterclockwise, is a Hölder continuous function on Γ.

Let $\gamma : [0, 1] \to \Gamma$ be a homeomorphism describing the Lyapunov arc Γ. At the endpoints $\gamma(0)$ and $\gamma(1)$ of Γ, we define *one-sided tangents* as the half lines

S. Roch et al., *Non-commutative Gelfand Theories*, Universitext,
DOI 10.1007/978-0-85729-183-7_4, © Springer-Verlag London Limited 2011

$\{\gamma(0) + t\gamma'(0) : t > 0\}$ and $\{\gamma(1) + t\gamma'(1) : t < 0\}$, respectively, where γ' represents the derivative of γ.

Let $\Gamma_1, \ldots, \Gamma_m$ be Lyapunov arcs and assume that, for each pair (i, j) with $i \neq j$, the intersection $\Gamma_i \cap \Gamma_j$ is either empty, or it consists of exactly one point z, which is an endpoint of both Γ_i and Γ_j and which has the property that the one-sided tangents of Γ_i and Γ_j at z do *not* coincide. In this case we set $n_j(z) := 1$ if Γ_j is directed away from z and $n_j(z) := -1$ if Γ_j is directed towards z. A (bounded) *admissible curve* is the union of a finite number of Lyapunov arcs which are subject to these conditions.

Fig. 4.1 An admissible curve...   ...and a non-admissible one.

If an oriented admissible curve Γ divides the plane into two (not necessarily connected) parts, each of them having Γ as its boundary and being located at one side of Γ, then we say that Γ is *closed*.

Let k be a positive integer. A point $z \in \Gamma$ is said to have *order k* if, for each sufficiently small closed neighborhood $\mathcal{U}$ of z, the set $\Gamma \cap \mathcal{U}$ is the union of k Lyapunov arcs $\Gamma_1, \ldots, \Gamma_k$ each of them having z as one of its endpoints and having no other points besides z in common. The curve Γ is called *simple* if all its points have an order less than three.

An unbounded curve Γ is called *admissible* if there is a point $z_0 \in \mathbb{C} \setminus \Gamma$ such that the image of Γ under the mapping $z \mapsto (z - z_0)^{-1}$ is a bounded admissible curve. Of course, if this holds for one point z_0, then it holds for every point $z_0 \in \mathbb{C} \setminus \Gamma$.

It is sometimes useful to consider an unbounded (hence, non-compact) curve Γ as a subset of the one-point compactification $\dot{\mathbb{C}} = \mathbb{C} \cup \{\infty\}$ of the complex plane $\mathbb{C}$, with a basis of neighborhoods of the point $\infty \in \dot{\mathbb{C}}$ given by the sets $\Omega_\varepsilon = \{z : |z| > 1/\varepsilon\}$ where $\varepsilon > 0$. We then denote by $\dot{\Gamma}$ the smallest compact set in $\dot{\mathbb{C}}$ containing the unbounded admissible curve Γ. Note that $\dot{\Gamma} = \Gamma$ for every bounded admissible curve.

Remark 4.1.1. Admissible curves include curves like arcs, circles, polygons, or lines. On the other hand, curves with cusps or spirals are not admissible in our sense. The reader who is interested in singular integral operators on L^p-spaces over general (Carleson) curves with general (Muckenhoupt) weights is referred to the ground-breaking and prize-winning monograph [14]. We restrict our attention to the admittedly simple case of admissible curves since our main emphasis is on illustrating the use of local principles (which also play a major role in [14]) and since admissible curves behave locally as star-shaped unions of half-axes. The latter fact

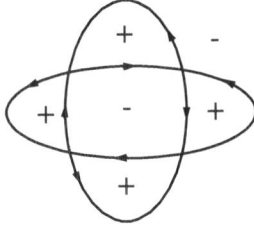 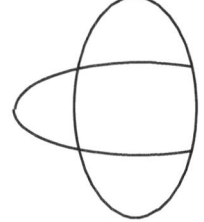

Fig. 4.2 A closed curve...   ...and a non-closed one.

allows one to use Mellin techniques to identify the arising local algebras in a convenient way, as we shall see below.   □

4.1.2 Lebesgue spaces and multiplication operators

Let Γ be an admissible curve in $\mathbb{C}$. By a *weight* on Γ we will always mean a *power* (or *Khvedelidze*) weight, which is a function on Γ of the form

$$w(t) := \prod_{i=0}^{n} |t - t_i|^{\alpha_i} \tag{4.1}$$

with $t_0 \in \mathbb{C} \setminus \Gamma$, $t_1, \ldots, t_n \in \Gamma$, and $\alpha_0, \ldots, \alpha_n \in \mathbb{R}$.

Let $1 < p < \infty$, and let w be a weight on Γ. By $L^p(\Gamma, w)$ (or simply $L^p(\Gamma)$ if $w \equiv 1$) we denote the weighted Lebesgue space of all measurable functions u on Γ such that

$$\int_{\Gamma} |u(t)|^p w^p(t) |dt| < \infty$$

with norm

$$\|u\|_{\Gamma, p, w} := \left(\int_{\Gamma} |u(t)|^p w^p(t) |dt| \right)^{\frac{1}{p}}.$$

The dual space of $L^p(\Gamma, w)$ can be identified with $L^q(\Gamma, w^{-1})$ where $1/p + 1/q = 1$ in the usual way. Further, we write $L^\infty(\Gamma)$ for the space of all essentially bounded measurable functions on Γ, i.e., for the space of all measurable functions $a : \Gamma \to \mathbb{C}$ for which there exists a non-negative constant N such that the set $\{z \in \Gamma : |a(z)| > N\}$ has Lebesgue measure zero. The norm in $L^\infty(\Gamma)$ is the essential supremum norm, i.e., the smallest N with this property.

A well-known fact from measure theory states that the set of all continuous functions on Γ with compact support is dense in $L^p(\Gamma, w)$ if $-1/p < \alpha_i$ for $i = 1, \ldots, n$. The following is a consequence of that fact.

Proposition 4.1.2. *If $-1/p < \alpha_i$ for $i = 1, \ldots, n$, then $L^2(\Gamma) \cap L^p(\Gamma, w)$ is dense in $L^p(\Gamma, w)$.*

Every function $f \in L^\infty(\Gamma)$ induces an operator fI of multiplication via

$$(fIu)(t) := f(t)u(t). \tag{4.2}$$

which is obviously bounded on $L^p(\Gamma, w)$. Note that[1]

$$\|fI\| = \|f\|_\infty. \tag{4.3}$$

The algebra $L^\infty(\Gamma)$ can thus be considered as a (commutative) subalgebra of $\mathscr{L}(L^p(\Gamma, w))$.

In Section 2.2.6 we have already met piecewise continuous functions on the unit circle. More generally, a function is called *piecewise continuous* on the admissible curve Γ if, at each point $z_0 \in \Gamma$ of order k, it possesses finite limits as $z \to z_0$ along each Lyapunov arc having z_0 as its endpoint. We denote the space of all functions which are continuous at each point of $\dot{\Gamma}$ by $C(\dot{\Gamma})$, and we write $PC(\dot{\Gamma})$ for the space of all piecewise continuous functions on $\dot{\Gamma}$. The spaces $C(\dot{\Gamma})$ and $PC(\dot{\Gamma})$, equipped with pointwise operations and the norm of $L^\infty(\Gamma)$, are Banach algebras. The set of all piecewise constant functions is dense in $PC(\dot{\Gamma})$ (see, for example, [43, Lemma 2.9]).

4.1.3 SIOs on admissible curves

Let Γ be an admissible curve and w a weight on Γ. The *singular integral operator* (or SIO for short) S_Γ on Γ is defined as the Cauchy principal value integral

$$\lim_{\varepsilon \to 0} \frac{1}{\pi \mathbf{i}} \int_{\Gamma \setminus \Gamma_{t,\varepsilon}} \frac{u(s)}{s - t} \, ds, \qquad t \in \Gamma, \tag{4.4}$$

with $\Gamma_{t,\varepsilon}$ referring to the part of Γ within an ε-radius ball centered at the point t. For $u \in L^p(\Gamma, w)$, the singular integral exists almost everywhere on Γ.

A basic result whose proof can be found in several textbooks on singular integral operators (for instance, [73, Chapter 1] or [120, Chapter 2]) is the boundedness of the singular integral operator under some restrictions on the weight. We will use the result in the following form, which is sufficient for our purposes. For a general result (stating that S_Γ is bounded on $L^p(\Gamma, w)$ if and only if Γ is a Carleson curve and w is a Muckenhoupt weight), see again [14].

If the curve Γ is bounded, we denote by $\mathcal{A}_p(\Gamma)$ the set of all weights of the form (4.1) with $0 < 1/p + \alpha_i < 1$ for $i = 1, \ldots, n$. In case Γ is unbounded, we let $\mathcal{A}_p(\Gamma)$ stand for the set of all weights with $0 < 1/p + \alpha_i < 1$ for $i = 1, \ldots, n$ and $0 < 1/p + \alpha_0 + \sum_{i=1}^n \alpha_i < 1$.

---

[1] When the curve Γ is evident, we simply write $\| \cdot \|_\infty$ for the norm in $L^\infty(\Gamma)$.

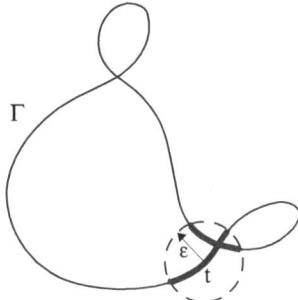

Fig. 4.3 The set $\Gamma_{t,\varepsilon}$, for an admissible curve Γ.

Theorem 4.1.3. *Let* $1 < p < \infty$, Γ *an admissible curve, and* $w \in \mathcal{A}_p(\Gamma)$. *Then the singular integral operator* S_Γ *is bounded on* $L^p(\Gamma, w)$.

In what follows we will work exclusively on spaces $L^p(\Gamma, w)$ where p, Γ and w are as in the previous theorem.

Let $\mathcal{A} := \mathcal{A}(C(\Gamma), S_\Gamma, w)$ denote the smallest closed subalgebra of $\mathcal{L}(L^p(\Gamma, w))$ which contains the singular integral operator S_Γ and all operators of multiplication by functions in $C(\dot{\Gamma})$, and abbreviate by $\mathcal{K} := \mathcal{K}(L^p(\Gamma, w))$ the ideal of the compact operators on $L^p(\Gamma, w)$.

The following commutator result provides the basis for localization in algebras generated by singular integral operators.

Theorem 4.1.4. *Let* $w \in \mathcal{A}_p(\Gamma)$ *and* $f \in C(\dot{\Gamma})$. *Then the operator* $f S_\Gamma - S_\Gamma f I$ *is compact on* $L^p(\Gamma, w)$.

Proof. First suppose that Γ is a bounded admissible curve and that f is a rational function without poles on Γ. Then

$$\big((S_\Gamma f - f S_\Gamma)u\big)(t) = \frac{1}{\pi \mathbf{i}} \int_\Gamma \frac{f(s) - f(t)}{s - t} u(s)\,ds. \tag{4.5}$$

Since f is rational, the function $(s,t) \mapsto \frac{f(s)-f(t)}{s-t}$ is continuous. Thus, $S_\Gamma f I - f S_\Gamma$ is an integral operator with continuous kernel on a compact set and, hence, compact (see, for instance, [93]). Now let $f \in C(\Gamma)$. The set of the rational functions without poles on Γ is dense in $C(\Gamma)$ in the L^∞ norm by the Stone-Weierstrass Theorem (see, for instance, [26, Section V.8]). Thus, there is a sequence $(f_n)_{n \in \mathbb{N}}$ of rational functions such that $\|f - f_n\|_\infty \to 0$ as $n \to \infty$. Since

$$\|(S_\Gamma f I - f S_\Gamma) - (S_\Gamma f_n I - f_n S_\Gamma)\| \le 2\|f - f_n\|_\infty \|S_\Gamma\|, \tag{4.6}$$

the operator $S_\Gamma f I - f S_\Gamma$ is the uniform limit of a sequence of compact operators and, thus, compact. This proves the assertion for bounded curves.

Now let Γ be an unbounded admissible curve. By definition, there is a $z_0 \in \mathbb{C} \setminus \Gamma$ such that the image of Γ under the mapping $z \mapsto (z - z_0)^{-1}$ is a bounded admissible curve Γ'. For $\lambda \in \Gamma'$, let $w'(\lambda) := w((\lambda - z_0)^{-1})(\lambda - z_0)^{-2/p}$ and consider the operator

$$B : L^p(\Gamma, w) \rightarrow L^p(\Gamma', w'), \quad (Bu)(\lambda) := u\left(\frac{1}{\lambda - z_0}\right) \qquad (4.7)$$

which is an isometry between these spaces. We have to show that the operator $K : L^p(\Gamma', w') \rightarrow L^p(\Gamma', w')$ given by

$$K := B(S_\Gamma fI - fS_\Gamma)B^{-1}$$

is compact. The argument is the same as above, since the kernel

$$k(\mu, \lambda) := \frac{f\left(\frac{1}{\mu - z_0}\right) - f\left(\frac{1}{\lambda - z_0}\right)}{\frac{1}{\mu - z_0} - \frac{1}{\lambda - z_0}} \frac{1}{(\mu - z_0)^2}$$

of K is continuous on $\Gamma' \times \Gamma'$ if f is rational without poles on Γ.   ∎

Theorem 4.1.5. *The algebra $\mathscr{A}(C(\Gamma), S_\Gamma, w)$ contains the ideal $\mathscr{K}$.*

Proof. First let Γ be bounded. Employing the density of the set of continuous functions in $L^p(\Gamma, w)$ and in its dual space, it is not difficult to verify that the set of all operators of the form $a(\phi S_\Gamma - S_\Gamma \phi I)bI$ with $a, b \in C(\Gamma)$ and $\phi(t) := t$ is dense in the set of all operators in $\mathscr{L}(L^p(\Gamma, w))$ with range dimension equal to 1. Since the rank one operators span a dense subset of the compact operators on $L^p(\Gamma, w)$ (see [26, VI, Section 3, Exercises 19, 20]), the assertion follows.

Now let Γ be unbounded and K be a compact operator in $\mathscr{L}(L^p(\Gamma, w))$. Define a sequence of functions $\chi_n \in C(\Gamma)$ which have compact support and for which $\chi_n(z) = 1$ for $|z| < n$. The sequence of multiplication operators $\chi_n I$ tends strongly to the identity operator on $L^p(\Gamma, w)$, and the sequence of adjoint operators $(\chi_n I)^*$ tends strongly to the identity operator on $(L^p(\Gamma, w))^*$. By Lemma 1.4.7, we thus have $\|\chi_n K \chi_n I - K\| \rightarrow 0$, and it remains to show that the operators $\chi_n K \chi_n I$ belong to $\mathscr{A}(C(\Gamma), S_\Gamma, w)$ for every n. This follows from the first part of this proof, since $\chi_n K \chi_n I$ can be considered as an operator on a bounded curve.   ∎

4.1.4 SIOs with continuous coefficients on closed curves

We have already observed that $S_\Gamma^2 = I$ in the case that Γ is the unit circle. This fact holds for general closed curves. A proof of the following theorem can be found in [120, Chapter II, Theorem 3.1] for the case of bounded closed admissible curves. The proof for unbounded closed admissible curves follows from the result for bounded curves, as in Theorem 4.1.4, by employing the mapping $z \mapsto (z - z_0)^{-1}$ with $z_0 \notin \Gamma$.

Theorem 4.1.6. *Let $1 < p < \infty$, Γ an admissible curve, and $w \in A_p(\Gamma)$. If the curve Γ is closed, then $S_\Gamma^2 = I$.*

Let Γ be a closed curve. We consider the algebra $\mathscr{A}(C(\Gamma), S_\Gamma, w)$ generated by the singular integral operator and all operators of multiplication by continuous functions. From Theorems 4.1.4 and 4.1.5 we infer that the subalgebra $\mathscr{A}^{\mathscr{K}} := \mathscr{A}/\mathscr{K}$ of the Calkin algebra is commutative and, thus, subject to Gelfand's representation theorem. We will briefly indicate how Gelfand theory works in this setting and start with identifying the maximal ideal space of $\mathscr{A}^{\mathscr{K}}$. Since we do not refer to these results in what follows, and since these results will follow without effort from the results for SIOs with piecewise continuous coefficients, we will not give all details here.

Since Γ is closed, we have $S_\Gamma^2 = I$ by Theorem 4.1.6. Thus, the operators $P_\Gamma := (I + S_\Gamma)/2$ and $Q_\Gamma := (I - S_\Gamma)/2$ are projections[2], and one can easily check that every coset $A + \mathscr{K} \in \mathscr{A}^{\mathscr{K}}$ can be written in the form

$$A + \mathscr{K} = f P_\Gamma + g Q_\Gamma + \mathscr{K} \text{ with } f, g \in C(\dot{\Gamma}).$$

The following can be checked in a similar way to Proposition 2.1.10.

Proposition 4.1.7. *Every proper ideal of $\mathscr{A}^{\mathscr{K}}$ is contained in an ideal of the form*

$$\mathscr{I}_{P, X_0} := \{ f P_\Gamma + g Q_\Gamma + \mathscr{K} : f(X_0) = 0 \}$$

or

$$\mathscr{I}_{Q, X_0} := \{ f P_\Gamma + g Q_\Gamma + \mathscr{K} : g(X_0) = 0 \}$$

with a certain subset X_0 of $\dot{\Gamma}$.

This proposition implies that the maximal ideals of $\mathscr{A}^{\mathscr{K}}$ are of the form

$$\mathscr{I}_{P, x_0} := \{ f P_\Gamma + g Q_\Gamma + \mathscr{K} : f(x_0) = 0 \}$$

or

$$\mathscr{I}_{Q, x_0} := \{ f P_\Gamma + g Q_\Gamma + \mathscr{K} : g(x_0) = 0 \}$$

with some point $x_0 \in \dot{\Gamma}$. These ideals are maximal and by Theorem 1.3.5 closed. Thus, the maximal ideal space of $\mathscr{A}^{\mathscr{K}}$ is homeomorphic to $\dot{\Gamma} \times \{0, 1\}$, and the Gelfand transform of the coset $A + \mathscr{K} = f P_\Gamma + g Q_\Gamma + \mathscr{K}$ is

$$(\widehat{A + \mathscr{K}})(x, n) = \begin{cases} f(x) & \text{if } n = 0, \\ g(x) & \text{if } n = 1. \end{cases}$$

In particular, the coset $A + \mathscr{K}$ is invertible if and only if $f(x) \neq 0$ and $g(x) \neq 0$ for all $x \in \dot{\Gamma}$. It is also easy to see that the radical of $\mathscr{A}^{\mathscr{K}}$ is $\{0\}$.

The description of $\mathscr{A}^{\mathscr{K}}$ just obtained allows us to derive conclusions about the Fredholm property and the essential spectrum of operators in $\mathscr{A}$. The issue of in-

---

[2] Remember Exercise 1.2.18.

vertibility is more delicate, and local principles cannot help, since invertibility of singular integral operators is a non-local property. Fortunately, in the case at hand, there is an effectively checkable (non-local) criterion for the invertibility of a Fredholm singular integral operator. For that we need the index (or winding number) of a function around the origin $0 \in \mathbb{C}$. The *index* of $a : \Gamma \to \mathbb{C}$ is defined as

$$\operatorname{ind} a = \frac{1}{2\pi}[\arg a(x)]_\Gamma$$

where $[\cdot]_\Gamma$ measures the increment of the expression in brackets taken as the result of a circuit around Γ in the counter-clockwise direction. Then one has the following result, a proof of which can be found in [73, Chapter 3] or [120, Chapter 3].

Theorem 4.1.8. *Let $f, g \in C(\dot{\Gamma})$. Then the operator $fP_\Gamma + gQ_\Gamma$ is Fredholm on $L^p(\Gamma, w)$ if and only if $f(x)g(x) \neq 0$ for all $x \in \dot{\Gamma}$. If this condition is satisfied, then this operator is invertible (resp. invertible from the left, invertible from the right) if the integer* ind f/g *is zero (resp. positive, negative). The same assertion holds for the operator $P_\Gamma fI + Q_\Gamma gI$.*

4.2 Singular integral operators on homogeneous curves

We now turn our attention to singular integral operators on the real line and, more generally, on homogeneous curves. Homogeneous curves are star-shaped unions of rotated half-axes. Their special geometry permits the use of techniques like the Fourier transform and homogenization. The importance of these curves lies in the fact that they will serve as local models of general admissible curves.

4.2.1 SIOs on the real line

Let Γ be a subinterval of the real axis $\mathbb{R}$, and let a weight w on Γ be defined by

$$w(t) = |t^2 + 1|^{\alpha_0/2} \prod_{i=1}^{n} |t - t_i|^{\alpha_i}. \tag{4.8}$$

Note that w is not of the form (4.1), but it is equivalent to a weight $\tilde{w}$ of this form in the sense that $w/\tilde{w}$ is bounded below and above by positive constants. We consider $L^p(\Gamma, w)$ as a closed subspace of $L^p(\mathbb{R}, w)$ in the natural way. In particular, we identify the identity operator on $L^p(\Gamma, w)$ with the operator $\chi_\Gamma I$ of multiplication by the characteristic function of the interval Γ, acting on $L^p(\mathbb{R}, w)$. More generally, a linear bounded operator on $L^p(\Gamma, w)$ is identified with the operator $\chi_\Gamma A \chi_\Gamma I$ acting on $L^p(\mathbb{R}, w)$. These identifications will be used without further comment.

An important property of the singular integral operator on the real line becomes visible via the Fourier transform. The *Fourier transform F* is defined on the

Schwartz space of rapidly decreasing infinitely differentiable functions u by

$$(Fu)(y) = \int_{-\infty}^{+\infty} e^{-2\pi iyx} u(x)\, dx, \; y \in \mathbb{R}. \tag{4.9}$$

The Fourier transform is invertible on the Schwartz space, and its inverse F^{-1} is given by

$$(F^{-1}v)(x) = \int_{-\infty}^{+\infty} e^{2\pi ixy} v(y)\, dy, \; x \in \mathbb{R}. \tag{4.10}$$

The operators F and F^{-1} can be extended continuously to bounded and unitary operators on the Hilbert space $L^2(\mathbb{R})$ (this is Plancherel's theorem, see [26, Theorem 6.17] or [193, Theorem 48]). It is also known that the Fourier transform extends continuously to a bounded operator from $L^p(\mathbb{R})$ into $L^q(\mathbb{R})$ when $1 < p \le 2$ and $q := p/(p-1)$ (see, for instance, [193, Theorem 74]).

For $1 < p < \infty$ and w as above, let $\mathcal{M}_{p,w}$ denote the set of all *Fourier multipliers*, i.e., the set of all functions $a \in L^\infty(\mathbb{R})$ with the following property: if $u \in L^2(\mathbb{R}) \cap L^p(\mathbb{R}, w)$, then $F^{-1}aFu \in L^p(\mathbb{R}, w)$, and there is a constant $c_{p,w}$ independent of u such that $\|F^{-1}aFu\|_{\Gamma,p,w} \le c_{p,w}\|u\|_{\Gamma,p,w}$. If $a \in \mathcal{M}_{p,w}$, then the operator $F^{-1}aF : L^2(\mathbb{R}) \cap L^p(\mathbb{R}, w) \to L^p(\mathbb{R}, w)$ extends continuously to a bounded operator on $L^p(\mathbb{R}, w)$. This extension is called the *operator of (Fourier) convolution by* a and will be denoted by $W^0(a)$. The function a is also called the *generating function*[3] of the operator $W^0(a)$.

The set $\mathcal{M}_{p,w}$ (written as $\mathcal{M}_p$ if $w \equiv 1$) of all multipliers forms a Banach algebra when equipped with the usual operations and the norm

$$\|a\|_{\mathcal{M}_{p,w}} := \|W^0(a)\|_{\mathscr{L}(L^p(\mathbb{R},w))}. \tag{4.11}$$

Let $\overline{\mathbb{R}}$ stand for the two-point compactification of $\mathbb{R}$ by $\pm\infty$ with bases of neighborhoods of $\pm\infty$ given by the sets $V_\varepsilon^\pm = \{t : \pm t > 1/\varepsilon\}$.

To describe some further properties of the multiplier classes, we need the total variation $V(a)$ of a function a on an interval $\Gamma = [c, d] \subseteq \overline{\mathbb{R}}$, which is defined as

$$V(a) := \sup\left(\sum_{k=1}^{n} |a(t_k) - a(t_{k-1})|\right) \tag{4.12}$$

where the supremum is taken over all partitions $c \le t_0 < t_1 < \ldots < t_n \le d$ of the interval Γ. Functions with finite total variation are bounded and measurable. The set of all bounded functions on Γ with finite total variation will be denoted by $BV(\Gamma)$. This set is a Banach space under the norm

$$\|a\|_{BV} := \|a\|_\infty + V(a).$$

---

[3] Sometimes a is also called the *symbol* or *presymbol* of $W^0(a)$.

Remark 4.2.1. Note that the definition of total variation can be extended in a natural way to curves homeomorphic to an interval.   □

Proposition 4.2.2. *Let* $1 < p < \infty$, $1/p + 1/q = 1$, *and* $w \in A_p(\mathbb{R})$.

(i) *If* $w(t) = w(-t)$ *for all* $t \in \mathbb{R}$, *then* $\mathcal{M}_{p,w} = \mathcal{M}_{q,w} \subseteq \mathcal{M}_2 = L^\infty(\mathbb{R})$ *and*

$$\|a\|_{\mathcal{M}_{p,w}} = \|a\|_{\mathcal{M}_{q,w}} \geq \|a\|_{\mathcal{M}_2} = \|a\|_\infty.$$

(ii) *The set* $BV(\overline{\mathbb{R}})$ *is contained in* $\mathcal{M}_{p,w}$, *and the* Stechkin *inequality*

$$\|a\|_{\mathcal{M}_{p,w}} \leq \|S_\mathbb{R}\|_{p,w} (\|a\|_\infty + V(a)) \tag{4.13}$$

holds for $a \in BV(\overline{\mathbb{R}})$.

The proofs of the above results can easily be derived from Proposition 2.4 and Theorem 2.11 in [43].

In general it is hard to decide whether a given function is a multiplier. Stechkin's inequality provides us with an easy-to-check sufficient condition for a function to have this property. There are also no general results concerning the invertibility of multipliers. Some partial results exist however, mainly for spaces without weights, as we will see.

The *convolution* $k * u$ of functions k and u on the real line is the function on $\mathbb{R}$ defined by

$$(k * u)(t) := \int_{-\infty}^{+\infty} k(t - s)u(s)\,ds$$

whenever this makes sense (which happens, for example, if $k, u \in L^1(\mathbb{R})$). In the latter case (see for instance [193, Section 2.1]) the relation

$$F(k * u) = (Fk)(Fu), \tag{4.14}$$

also known as the convolution theorem, is valid.

How this relation can be useful is shown in the next two examples. If $k \in L^1(\mathbb{R})$, then the operator acting on $L^1(\mathbb{R})$ defined by

$$(Wu)(t) := \int_{-\infty}^{+\infty} k(t - s)u(s)\,ds \tag{4.15}$$

is just a convolution operator $W^0(a)$ with $a = Fk$. But Equation (4.14) is valid even for some functions $k \notin L^1(\mathbb{R})$. Let sgn denote the *sign* function on $\mathbb{R}$, which takes the value 1 on $[0, \infty[$ and the value -1 otherwise, and for $\xi \in \mathbb{R}$ define sgn_ξ by $\text{sgn}_\xi(x) := \text{sgn}(x - \xi)$. It turns out (see [43, Lemma 1.35 and Section 2]) that the operator

$$(W_\xi u)(t) := \frac{1}{\pi \mathbf{i}} \int_{-\infty}^{+\infty} \frac{e^{\mathbf{i}\xi(s-t)}}{s - t} u(s)\,ds, \quad t \in \mathbb{R}, \tag{4.16}$$

coincides with the convolution operator with $\mathrm{sgn}_\xi = F\left(\frac{1}{\pi i}\frac{e^{i\xi t}}{t}\right)$ as its generating function, that is,

$$W_\xi \equiv W^0(\mathrm{sgn}_\xi) \tag{4.17}$$

and, in particular,

$$S_\mathbb{R} \equiv W^0(\mathrm{sgn}). \tag{4.18}$$

From these equations we deduce an important property of the operators W_ξ (which is expected at least in the case of the singular integral operator since $\dot{\mathbb{R}}$, the one-point compactification of the real line, is a closed curve): It is its own inverse. In fact $W_\xi^2 = W^0(\mathrm{sgn}_\xi)W^0(\mathrm{sgn}_\xi) = W^0(1) = I$.

Remark 4.2.3. The identity (4.18) indicates that the theory of singular integral operators on the real line can be considered as a part of the theory of convolution operators. We would like to emphasize that the reverse is also true, that is, knowledge on singular integral operators helps to understand large classes of convolution operators. This relation will become evident when studying convolution operators by local principles. □

For $s, t \in \mathbb{R}$, consider the following kinds of shift operators, both with norm 1,

$$U_s : L^p(\mathbb{R}, w) \mapsto L^p(\mathbb{R}, w), \quad (U_s u)(x) = e^{-2\pi i x s} u(x) \tag{4.19}$$

and

$$V_t : L^p(\mathbb{R}, w) \mapsto L^p(\mathbb{R}, w_t), \quad (V_t u)(x) = u(x - t) \tag{4.20}$$

with $w_t(x) = w(x - t)$. Clearly, $U_s^{-1} = U_{-s}$ and $V_t^{-1} = V_{-t}$.

The following lemma can be proved by writing the operators explicitly and making an obvious substitution of variables. For each multiplier a, define a multiplier $V_s a$ by $(V_s a)(x) := a(x - s)$.

Lemma 4.2.4. *Let $s \in \mathbb{R}$ and $a \in M_{p,w}$. Then*

$$U_{-s}W^0(a)U_s = W^0(V_s a) \quad \text{and} \quad V_s W^0(a)V_{-s} = W^0(a).$$

Moreover, on $L^2(\mathbb{R})$, one has

$$FU_s = V_{-s}F \quad \text{and} \quad FV_s = U_s F.$$

The proof of the next result is left as an exercise.

Lemma 4.2.5. *The operators V_t tend weakly to zero as $t \to \pm\infty$.*

An operator $A \in \mathscr{L}(L^p(\mathbb{R}))$ with $V_s A V_{-s} = A$ for all $s \in \mathbb{R}$ is called *translation invariant*[4]. By Lemma 4.2.4, convolution operators are translation invariant. The following proposition states that the converse is also true (see [87, Section 1.1]).

---

[4] Sometimes the expression *shift invariant* is also used, but as there are different types of shifts, it can be misleading.

Proposition 4.2.6 (Hörmander). *Let $A \in \mathscr{L}(L^p(\mathbb{R}))$ be translation invariant. Then there exists a multiplier $a \in \mathfrak{M}_p$ such that $A = W^0(a)$.*

Proposition 4.2.7. *The only compact convolution operator on $L^p(\mathbb{R})$ is the zero operator.*

Proof. Suppose $W^0(a)$ is compact for some multiplier a. Since $W^0(a)$ is translation invariant, one has

$$W^0(a) = V_s W^0(a) V_{-s},$$

and the right-hand side of this equality tends strongly to zero as $s \to \infty$ by Lemma 1.4.6 because the operators V_{-s} tend weakly to zero and the operators V_s are uniformly bounded with respect to s. ■

Proposition 4.2.8. *Let $a \in \mathfrak{M}_p$. The operator $W^0(a) \in \mathscr{L}(L^p(\mathbb{R}))$ is invertible if and only if it is a Fredholm operator.*

Proof. Let $W^0(a)$ be a Fredholm operator. Then there is a bounded operator B and a compact operator K on $L^p(\mathbb{R})$ such that $BW^0(a) = I + K$. Applying the shift operators to both sides of this equality we obtain

$$V_s B V_{-s} V_s W^0(a) V_{-s} = I + V_s K V_{-s} \iff B_s W^0(a) = I + V_s K V_{-s}$$

with $B_s := V_s B V_{-s}$. Since K is compact, and by Lemma 4.2.5, the right-hand side of this equation tends strongly to the identity. Hence, $B_s W^0(a) u \to u$ for all $u \in L^p(\mathbb{R}, w)$, which implies that the kernel of $W^0(a)$ is trivial. Since $W^0(a)$ is a Fredholm operator by assumption, this operator is invertible from the left. The invertibility from the right can be shown by passing to the adjoint operator. ■

A function $a \in L^\infty(\mathbb{R})$ is said to be *bounded away from zero* if there is an $\varepsilon > 0$ such that the measure of the set $\{t \in \mathbb{R} : |a(t)| \le \varepsilon\}$ is zero.

Proposition 4.2.9. *If $a \in \mathfrak{M}_p$ and $W^0(a)$ is invertible on $L^p(\mathbb{R})$, then a is bounded away from zero.*

Proof. Since the operator $W^0(a)$ is translation invariant, its inverse is translation invariant, too, and hence of the form $W^0(a)^{-1} = W^0(b)$ with $b \in \mathfrak{M}_p \subset L^\infty(\mathbb{R})$ by Proposition 4.2.6. Then 1 $ab = 1$, hence a is invertible in $L^\infty(\mathbb{R})$, whence the assertion. ■

4.2.2 SIOs on the half axis

Let $\mathbb{R}^+ := [0, +\infty[$. Define the operator $E_{p,w} : L^p(\mathbb{R}^+, w) \to L^p(\mathbb{R})$ by

$$(E_{p,w}v)(t) := 2\pi e^{\frac{2\pi t}{p}} w(e^{2\pi t}) v(e^{2\pi t}). \tag{4.21}$$

This operator is invertible, and its inverse $E_{p,w}^{-1} : L^p(\mathbb{R}) \to L^p(\mathbb{R}^+, w)$ is given by

$$(E_{p,w}^{-1}u)(x) = \frac{1}{2\pi} \frac{1}{x^{1/p}w(x)} u\left(\frac{1}{2\pi}\ln(x)\right). \tag{4.22}$$

An easy computation yields that

$$\|E_{p,w}v\|_{L^p(\mathbb{R})} = 2\pi^{1-1/p}\|v\|_{L^p(\mathbb{R}^+, w)}. \tag{4.23}$$

Let $a \in BV(\overline{\mathbb{R}})$. Then the operator

$$M^0(a) := E_{p,w}^{-1}W^0(a)E_{p,w} \tag{4.24}$$

is bounded on $L^p(\mathbb{R}^+, w)$, and (4.23) together with the Stechkin inequality for $W^0(a)$ imply the estimate

$$\|M^0(a)\|_{p,w} \le \|S_{\mathbb{R}}\|_{\mathscr{L}(L^p(\mathbb{R}))}(\|a\|_\infty + V(a)) \tag{4.25}$$

which is also referred to as the Stechkin inequality. From (4.3) and (4.23) we further conclude

$$\|M^0(a)\|_2 = \|a\|_\infty. \tag{4.26}$$

The operator $M^0(a)$ is called the *Mellin convolution* with generating function a. The operator $M_{p,w} := FE_{p,w}$ is called the *Mellin transform* which is explicitly given by

$$(M_{p,w}v)(y) = \int_0^{+\infty} t^{1/p-1-\mathbf{i}y}w(t)v(t)\,dt, \tag{4.27}$$

with $t^{x+\mathbf{i}y} := t^x e^{\mathbf{i}y\log t} = t^x(\cos(y\log t) + \mathbf{i}\sin(y\log t))$ for $x, y \in \mathbb{R}$ and $t > 0$.

As in the case of Fourier convolutions, one might call functions $a \in \mathcal{M}_{p,w}$ *Mellin multipliers* and define a corresponding *multiplier norm* by

$$\|a\|_{M^0,p,w} := \|M^0(a)\|_{\mathscr{L}(L^p(\mathbb{R}^+, w))} \tag{4.28}$$

under which the set $\mathcal{M}_{p,w}$, with the usual operations, becomes a Banach algebra. But as a consequence of (4.23), the classes of Fourier and Mellin multipliers coincide, and also the norms $\|a\|_{\mathcal{M}_{p,w}}$ and $\|a\|_{M^0,p,w}$ are the same. So we shall simply speak of multipliers instead of Fourier or Mellin multipliers. Note, however, that the operator $M^0(a)$ depends on the weight w in general.

Write $C_{p,w}(\overline{\mathbb{R}})$ for the closure in $\mathcal{M}_{p,w}$ of the set of all functions with finite total variation which are continuous on $\overline{\mathbb{R}}$. Let a_1 denote the function $t \mapsto 1$, and let a_2 be any other function in $BV(\overline{\mathbb{R}})$ which is continuous and one-to-one on $\overline{\mathbb{R}}$ and for which $a_2(\overline{\mathbb{R}})$ is a smooth curve which does not contain the origin. By (4.25), the functions a_1 and a_2 belong to $\mathcal{M}_{p,w}$. The proof of the following proposition appeared for the first time in [189].

Proposition 4.2.10. *The algebra $C_{p,w}(\mathbb{R})$ coincides with the smallest closed subalgebra of $\mathcal{M}_{p,w}$ which contains a_1 and a_2.*

Proof. Let $\mathscr{A}$ denote the Banach algebra generated by a_1 and a_2 in $\mathcal{M}_{p,w}$. By definition, $\mathscr{A} \subseteq C_{p,w}(\mathbb{R})$. To get the reverse inclusion, put $\Gamma := a_2(\mathbb{R})$ and consider the operator $Q : C(\Gamma) \to C(\mathbb{R})$ by

$$Q : C(\Gamma) \to C(\mathbb{R}), \qquad (Qf)(t) = f(a_2(t)). \tag{4.29}$$

Clearly, Q is an isometric isomorphism which does not change the total variation of f (see Remark 4.2.1). Our first claim is the inclusion $Q(C^1(\Gamma)) \subseteq \mathscr{A}$ where $C^1(\Gamma)$ stands for the Banach space of all continuously differentiable functions on the curve Γ. Let $f \in Q(C^1(\Gamma))$ and $g = Q^{-1}f$. Approximate g by a sequence $\{g_n\}_{n\in\mathbb{N}}$ of polynomials in the norm of $C^1(\Gamma)$. Then the sequence $\{g_n\}$ approximates g in the norm of $BV(\Gamma)$ and, hence, the sequence $\{f_n\}$ with $f_n = Qg_n$ approximates f in the norms of both $C(\mathbb{R})$ and $BV(\mathbb{R})$. The inequality (4.25) shows that then $\|f - f_n\|_{\mathcal{M}_{p,w}} \to 0$, and since $f_n \in \mathscr{A}$ we get our claim.

Next we will verify that $C(\mathbb{R}) \cap BV(\mathbb{R}) \subseteq \mathscr{A}$. Given $f \in C(\mathbb{R}) \cap BV(\mathbb{R})$, there exists a sequence $\{f_n\}_{n\in\mathbb{N}}$ of functions $f_n \in Q(C^1(\Gamma))$ which approximates f in the norm of $C(\mathbb{R})$ and which is uniformly bounded in the norm of $BV(\mathbb{R})$. By what has already been shown, $f_n \in \mathscr{A}$. The Stein-Weiss interpolation theorem (Theorem 4.8.1 in the appendix of this chapter) implies

$$\|f_n - f\|_{M^0,p,w} \le \|f_n - f\|_{M^0,2,1}^{1-\theta} \|f_n - f\|_{M^0,q,w^{1/\theta}}^{\theta}, \tag{4.30}$$

for a suitably chosen θ such that $|1 - 2/p| < \theta < 1$, and $q = \frac{2p\theta}{2+p(\theta-1)}$. By the Stechkin inequality (4.25), the sequence $\|f_n - f\|_{M^0,q,w^{1/\theta}}$ is uniformly bounded, and the sequence (f_n) converges to f in the norm of $\mathcal{M}_{2,1}^0$ ($= L^\infty(\mathbb{R})$ by (4.26)). Hence, $\|f_n - f\|_{M^0,p,w} \to 0$, which implies $C(\mathbb{R}) \cap BV(\mathbb{R}) \subseteq \mathscr{A}$ and thus, $C_{p,w}(\mathbb{R}) \subseteq \mathscr{A}$. ∎

Let $\alpha \in \mathbb{R}$, and specialize the weight as $w_\alpha(t) := t^\alpha$. Similarly to the Fourier transform, one has the following formal relation on $L^p(\mathbb{R}^+, w_\alpha)$:

$$M_{p,w_\alpha} \left(\int_0^{+\infty} k\left(\frac{t}{s}\right) u(s) s^{-1} \, ds \right) = (M_{p,w_\alpha}k)(M_{p,w_\alpha}u). \tag{4.31}$$

Thus, if the given integrals are well defined, the operator M defined by

$$(Mu)(x) := \int_0^{+\infty} k\left(\frac{t}{s}\right) u(s) s^{-1} \, ds \tag{4.32}$$

can also be understood as the Mellin convolution $M^0(a)$ with generating function $a := M_{p,w_\alpha}k$. The following result gives a concrete example of an operator of this type.

For $z_0 \in \mathbb{C}$ with $0 < \Re(z_0) < 1$, define the function

$$s_{z_0}(y) := \coth\left((y + iz_0)\pi\right). \tag{4.33}$$

Proposition 4.2.11. *The function s_{z_0} has finite total variation and is, hence, a multiplier. The associated Mellin convolution operator is given by*

$$\left(M^0(s_{z_0})u\right)(t) = \frac{1}{\pi i} \int_0^{+\infty} \left(\frac{s}{t}\right)^\sigma \frac{u(s)}{s-t}\, ds, \quad t \in \mathbb{R}^+, \tag{4.34}$$

where $\sigma := 1/p + \alpha - z_0$.

Proof. From

$$\frac{1}{\pi i} \int_0^{+\infty} \left(\frac{s}{t}\right)^\sigma \frac{u(s)}{s-t}\, ds = \frac{1}{\pi i} \int_0^{+\infty} \left(\frac{s}{t}\right)^\sigma \frac{u(s)}{1 - \frac{t}{s}} s^{-1} ds \tag{4.35}$$

it becomes clear that we have to find the Mellin transform of the function

$$k(t) = \frac{1}{\pi i} \frac{t^{-\sigma}}{1-t}.$$

For $0 < \Re(\upsilon) < 1$, Formulas 3.238.1 and 3.238.2 in [79] give

$$\frac{1}{2\pi i} \int_{-\infty}^{+\infty} \frac{|t|^{\upsilon-1}}{1-t}\, dt = -\frac{i}{2} \cot\left(\frac{\pi}{2}\upsilon\right)$$

and

$$\frac{1}{2\pi i} \int_{-\infty}^{+\infty} \frac{|t|^{\upsilon-1}}{1-t} \operatorname{sgn}(t)\, dt = \frac{i}{2} \tan\left(\frac{\pi}{2}\upsilon\right).$$

Adding these two identities and substituting $\upsilon := 1/p + \alpha - \sigma - iy$ we get

$$\left(M_{p,w_\alpha} k\right)(y) = \frac{1}{\pi i} \int_0^{+\infty} t^{1/p + \alpha - 1 - iy} \frac{t^{-\sigma}}{1-t}\, dt$$

$$= \frac{i}{2}\left(\tan\left((1/p + \alpha - \sigma - iy)\pi/2\right) - \cot\left((1/p + \alpha - \sigma - iy)\pi/2\right)\right)$$

$$= \coth\left((y + i(1/p + \alpha - \sigma))\pi\right). \tag{4.36}$$

Now the assertion follows from (4.31). ∎

Let $\tau > 0$. We enlarge our arsenal of shift operators by the multiplicative shifts Z_τ, acting on $L^p(\mathbb{R}^+, w_\alpha)$ by

$$(Z_\tau f)(x) := \tau^{-1/p - \alpha} f(x/\tau). \tag{4.37}$$

Evidently, $Z_\tau^{-1} = Z_{\tau^{-1}}$ and $\|Z_\tau\| = 1$.

Lemma 4.2.12. *The operators $Z_\tau^{\pm 1}$ converge weakly to zero as $\tau \to \infty$.*

Proof. The dual of the space $L^p(\mathbb{R}^+, w_\alpha)$ is the space $L^q(\mathbb{R}^+, w_{-\alpha})$ with $1/q := 1 - 1/p$. Let $u := \chi_{[a,b]}$ and $v := \chi_{[c,d]}$ be characteristic functions of intervals in $\mathbb{R}^+$. Then

$$\langle v, Z_\tau u \rangle = \int_c^d \tau^{-\alpha - 1/p} \chi_{[a\tau, b\tau]}(x)\, dx \leq \tau^{-\alpha - 1/p}(d - c) \to 0$$

as $\tau \to \infty$ since $-\alpha - 1/p < 0$. This implies that $\langle \tilde{v}, Z_\tau \tilde{u} \rangle \to 0$ for arbitrary piecewise constant functions $\tilde{u}$ and $\tilde{v}$ in $L^p(\mathbb{R}^+, w_\alpha)$ with a finite number of jumps. As these functions are dense in $L^p(\mathbb{R}^+, w_\alpha)$, the assertion for Z_τ follows from Lemma 1.4.1. For Z_τ^{-1} and characteristic functions u and v as above one has

$$\langle v, Z_\tau^{-1} u \rangle = \int_c^d \tau^{\alpha + 1/p} \chi_{[\frac{a}{\tau}, \frac{b}{\tau}]}(x)\, dx \leq \tau^{\alpha + 1/p}\left(\frac{b}{\tau} - \frac{a}{\tau}\right) \to 0.$$

Now one can argue as above.   ∎

Lemma 4.2.13. *Mellin convolution operators commute with multiplicative shift operators.*

Proof. We have to show that $Z_\tau^{-1} M^0(a) Z_\tau = M^0(a)$ for all $\tau > 0$. This equality follows immediately from the definition of $M^0(a)$ in (4.24) and from the fact that $E_{p,w_\alpha} Z_\tau = V_s E_{p,w_\alpha}$ with $s = \frac{1}{2\pi} \log(\tau)$.   ∎

Operators which commute with multiplicative shifts are also called *homogeneous operators*. The proof of the following result is left as an exercise.

Proposition 4.2.14. *The operator $M^0(a) \in \mathcal{L}(L^p(\mathbb{R}^+, w_\alpha))$ is invertible if and only if it is Fredholm.*

We return for a moment to the operator in Proposition 4.2.11. The range of the generating function s_{z_0} is a circular arc which joins -1 and 1 and passes through the point $-\mathbf{i}\cot(\pi\mathfrak{R}(z_0))$ (see Figure 4.4). By Proposition 4.2.14, both the spectrum and the essential spectrum of $M^0(s_{z_0})$ are equal to the closure of the range of s_{z_0}. Note that putting $\sigma = 0$ in formula (4.34) gives the singular integral operator on $\mathbb{R}^+$. From the discussion above and the Stechkin inequality (4.25) we thus conclude immediately that $S_{\mathbb{R}^+}$ is bounded on $L^p(\mathbb{R}^+, t^\alpha)$ for $-\frac{1}{p} < \alpha < \frac{1}{q}$. Note that the latter condition for α is equivalent to membership of the weight t^α in $A_p(\mathbb{R}^+)$.

We consider a few more concrete examples of Mellin convolutions. For $0 < \upsilon := 1/p + \alpha < 1$ and $0 < \mathfrak{R}(\beta) < 2\pi$, let

$$h_\beta(y) := \frac{e^{(y + \mathbf{i}\upsilon)(\pi - \beta)}}{\sinh((y + \mathbf{i}\upsilon)\pi)}. \tag{4.38}$$

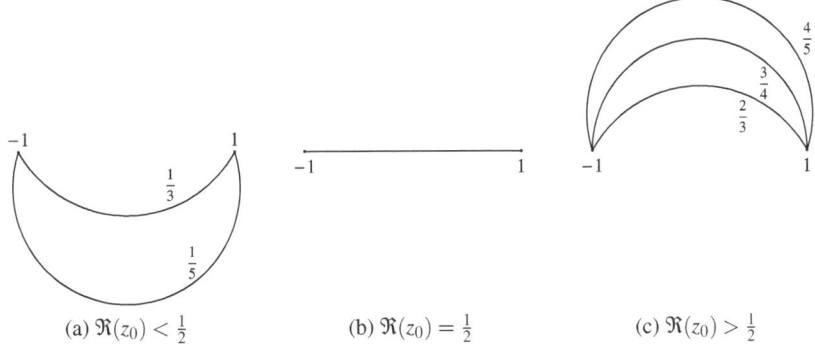

(a) $\Re(z_0) < \frac{1}{2}$ (b) $\Re(z_0) = \frac{1}{2}$ (c) $\Re(z_0) > \frac{1}{2}$

Fig. 4.4 The spectrum of $M^0(s_{z_0})$ for several values of $\Re(z_0)$.

Proposition 4.2.15. *Under the above conditions for α and β, the function h_β is a Mellin multiplier on $L^p(\mathbb{R}^+, t^\alpha)$, and the corresponding Mellin convolution operator on this space is given by*

$$\left(M^0(h_\beta)u\right)(t) = \frac{1}{\pi i}\int_0^{+\infty}\frac{u(s)}{s - e^{i\beta}t}\,ds. \tag{4.39}$$

Moreover,

$$S_{\mathbb{R}^+}^2 - I = M^0(h_\beta)M^0(h_{2\pi-\beta}). \tag{4.40}$$

Proof. The multiplier property comes from the Stechkin inequality. To prove (4.39), we have to calculate the Mellin transform of the function

$$k(t) = \frac{1}{\pi i}\frac{1}{s - e^{i\beta}t},$$

which leads to the evaluation of the integral

$$\left(M_{p,w_\alpha}k\right)(y) = \frac{1}{\pi i}\int_0^{+\infty}\frac{t^{1/p+\alpha-1-iy}}{1 - e^{i\beta}t}\,dt. \tag{4.41}$$

This integral coincides with $h_\beta(y)$ by Formula 3.194.4 in [79]. Finally, since we are working with Mellin convolutions, it is sufficient to verify the identity (4.40) on the level of generating functions. The corresponding equality

$$s_{1/p+\alpha}^2 - 1 = h_\beta h_{2\pi-\beta},$$

can be verified straightforwardly.   ∎

Operators of the form (4.39) are also called *Hankel operators*. For real β, the spectrum of $M^0(h_\beta)$ can easily be obtained from the spectrum of $M^0(h_\pi)$ since

$$|h_\beta(y)| = e^{v(\pi-\beta)}|h_\pi(y)| \quad \text{and} \quad \arg(h_\beta(y)) = v(\pi-\beta) + \arg(h_\pi(y)).$$

The spectrum of $M^0(h_\beta)$ is shown in Figure 4.5 for several values of v and β. Note that it always includes the point 0.

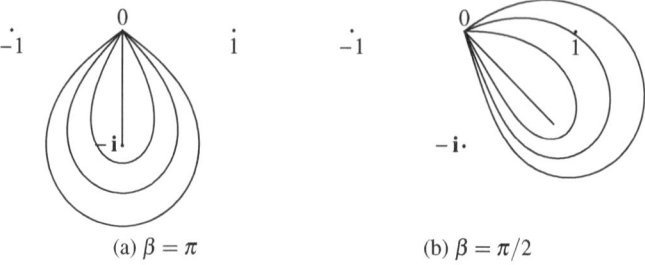

(a) $\beta = \pi$ (b) $\beta = \pi/2$

Fig. 4.5 The spectrum of $M^0(h_\beta)$ for $v = \frac{1}{2}$ (gives the straight line) and $v = \frac{2}{3}, \frac{3}{4}, \frac{4}{5}$.

4.2.3 The algebra of SIOs on $\mathbb{R}^+$

In this section, we are going to examine the smallest closed subalgebra $\mathscr{E}_{p,\alpha}$ of $\mathscr{L}(L^p(\mathbb{R}^+, t^\alpha))$ which contains the identity operator I and the singular integral operator $S_{\mathbb{R}^+}$. Let S_σ and H_β abbreviate the operators $M^0(s_{1/p+\alpha-\sigma})$ and $M^0(h_\beta)$, respectively (compare propositions 4.2.11 and 4.2.15).

Proposition 4.2.16. *Let* $0 < 1/p + \alpha < 1$, $0 < \mathfrak{R}(1/p + \alpha - \sigma) < 1$, *and* $0 < \mathfrak{R}(\beta) < 2\pi$. *Then*

(i) $S_\sigma \in \mathscr{E}_{p,\alpha}$ *and* $H_\beta \in \mathscr{E}_{p,\alpha}$;
(ii) *the operators* I *and* S_σ *span a dense subalgebra of* $\mathscr{E}_{p,\alpha}$;
(iii) *the algebras* $\mathscr{E}_{p,\alpha}$ *and* $\mathscr{E}_{p,0}$ *are isometrically isomorphic by the mapping* $A \mapsto t^\alpha A t^{-\alpha} I$, *and the image of the operator* $S_0 \equiv S_{\mathbb{R}^+}$ *under this mapping is* $S_{-\alpha}$;
(iv) *the algebra* $\mathscr{E}_{p,\alpha}$ *is inverse-closed in* $\mathscr{L}(L^p(\mathbb{R}^+, t^\alpha))$.

Proof. By Proposition 4.2.10, the mapping $a \mapsto M^0(a)$ is an isometric isomorphism from $C_{p,t^\alpha}(\overline{\mathbb{R}})$ onto $\mathscr{E}_{p,\alpha}$. This fact proves (i) and (ii). To prove (iii), note that the mapping $A \mapsto t^\alpha A t^{-\alpha} I$ is an isometric isomorphism from $\mathscr{L}(L^p(\mathbb{R}^+, t^\alpha))$ onto $\mathscr{L}(L^p(\mathbb{R}^+, 0))$. By (ii), and since $t^\alpha S_{\mathbb{R}^+} t^{-\alpha} I = S_{-\alpha}$, the image of the algebra $\mathscr{E}_{p,\alpha}$ under this mapping is contained in $\mathscr{E}_{p,0}$. Conversely, if $S_{\mathbb{R}^+}$ is the singular integral operator in $\mathscr{E}_{p,0}$ then, again by (i) and (ii), the operator $t^{-\alpha} S_{\mathbb{R}^+} t^\alpha I \in \mathscr{E}_{p,\alpha}$ spans, together with the identity, the whole algebra $\mathscr{E}_{p,\alpha}$. Thus, the isomorphism maps $\mathscr{E}_{p,\alpha}$ onto $\mathscr{E}_{p,0}$. The inverse-closedness (assertion (iv)) follows from Corollary 1.2.32, applied to the dense subalgebra of $\mathscr{E}_{p,\alpha}$ generated by $S_{\mathbb{R}^+}$. Note that the operators in this algebra have a thin spectrum which coincides with the closure of the range of the generating Mellin symbol.   ∎

Let $\mathcal{N}_{p,\alpha}$ denote the smallest closed ideal of the Banach algebra $\mathscr{E}_{p,\alpha}$ which contains the operator H_π.

Proposition 4.2.17. *Let $0 < 1/p + \alpha < 1$. Then*

(i) *for $M^0(a) \in \mathscr{E}_{p,\alpha}$, one has $M^0(a) \in \mathcal{N}_{p,\alpha}$ if and only if $a(\pm\infty) = 0$;*

(ii) *for $0 < 1/p + \alpha - \Re(\sigma) < 1$, the algebra $\mathscr{E}_{p,\alpha}$ decomposes into the direct sum*

$$\mathscr{E}_{p,\alpha} = \mathbb{C}I \dotplus \mathbb{C}S_\sigma \dotplus \mathcal{N}_{p,\alpha};$$

(iii) *for $0 < \Re(\beta) < 2\pi$, the smallest closed ideal of $\mathscr{E}_{p,\alpha}$ which contains the operator H_β coincides with $\mathcal{N}_{p,\alpha}$.*

Proof. Let a be a polynomial in $s_{1/p+\alpha}$ with vanishing absolute term. Since $a(\pm\infty) = 0$, there is a polynomial a_1 such that $a(s_{1/p+\alpha}) = (1 - s_{1/p+\alpha}^2)a_1(s_{1/p+\alpha})$. By Proposition 4.2.15, $1 - s_{1/p+\alpha}^2 = h_\pi^2$, whence

$$M^0(a) = -H_\pi^2 M^0(a_1) \subseteq H_\pi \mathscr{E}_{p,\alpha} \subseteq \mathcal{N}_{p,\alpha}.$$

Since the set of all polynomials in $s_{1/p+\alpha}$ with vanishing absolute term is dense in the set of all functions in $C_{p,t^\alpha}(\overline{\mathbb{R}})$ which vanish at $\pm\infty$ by Proposition 4.2.10, we conclude that $M^0(a) \in \mathcal{N}_{p,\alpha}$. Conversely, let $M^0(a) \in \mathcal{N}_{p,\alpha}$. Then a can be written as $a = a_1 h_\beta$ with $a_1 \in C_{p,t^\alpha}(\overline{\mathbb{R}})$. This implies $a(\pm\infty) = 0$, and assertion (i) is proved.

For a proof of (ii), let $M^0(a) \in \mathscr{E}_{p,\alpha}$. Write a as

$$a = \frac{a(+\infty) + a(-\infty)}{2} + \frac{a(+\infty) - a(-\infty)}{2} s_{1/p+\alpha} + a_0$$

with $a_0(\pm\infty) = 0$. Then

$$M^0(a) = \frac{a(+\infty) + a(-\infty)}{2} I + \frac{a(+\infty) - a(-\infty)}{2} S_{\mathbb{R}^+} + M^0(a_0)$$

with $M^0(a_0) \in \mathcal{N}_{p,\alpha}$ by (i).

To prove assertion (iii), let $\mathcal{N}$ denote the smallest closed ideal of $\mathscr{E}_{p,\alpha}$ which contains the operator H_β. From $H_\beta \in \mathcal{N}_{p,\alpha}$ by (i), one gets immediately that $\mathcal{N} \subseteq \mathcal{N}_{p,\alpha}$. The reverse inclusion follows from the equality $1 - s_{1/p+\alpha}^2 = -h_\beta h_{2\pi-\beta}$ by arguments similar to those in the proof of (i).   ∎

Here is another example of an integral operator which can be described as a Mellin convolution.

Example 4.2.18. Let $\sigma, \beta \in \mathbb{C}$ and $n \in \mathbb{Z}^+$, and set $\upsilon := 1/p + \alpha - \sigma$. If $0 < \Re(\upsilon) < n+1$ and $0 < \Re(\beta) < 2\pi$ then the function

$$b(y) := (-1)^n \binom{\upsilon - \mathbf{i}y - 1}{n} \frac{e^{(\pi-\beta)(y-\mathbf{i}\pi)}}{\sinh((y+\mathbf{i}\upsilon)\pi)} \tag{4.42}$$

is a Mellin multiplier on $L^p(\mathbb{R}^+, t^\alpha)$, and the corresponding Mellin convolution operator is given by

$$(M^0(b)f)(t) = \frac{1}{\pi\mathbf{i}} \int_0^{+\infty} \frac{s^{n+\sigma} t^{-\sigma}}{(s - e^{\mathbf{i}\beta} t)^{n+1}} f(s)\, ds. \tag{4.43}$$

Moreover, the operator $M^0(b)$ belongs to the algebra $\mathscr{E}_{p,\alpha}$ by Proposition 4.2.17. The relation between (4.43) and (4.42) follows from Formula 3.194.4 in [79].   □

4.2.4 Algebras of SIOs on homogeneous curves

Homogeneous curves are unions of rotated half axes. They will serve us as local models of admissible curves. More precisely, a homogeneous curve is a union $\Gamma := \cup_{j=1}^k e^{\mathbf{i}\beta_j}\mathbb{R}^+$ with real numbers $0 \le \beta_1 < \beta_2 < \ldots < \beta_k < 2\pi$. We suppose that each half axis $e^{\mathbf{i}\beta_j}\mathbb{R}^+$ is equipped with an orientation and put $\nu_j := 1$ if $e^{\mathbf{i}\beta_j}\mathbb{R}^+$ is directed away from 0 and $\nu_j := -1$ if $e^{\mathbf{i}\beta_j}\mathbb{R}^+$ is directed towards 0. Further, we write χ_j for the characteristic function of $e^{\mathbf{i}\beta_j}\mathbb{R}^+$ and let $\chi := \{\chi_1, \ldots, \chi_k\}$. Note that by Theorem 4.1.3, the singular integral operator S_Γ is bounded on $L^p(\Gamma, |t|^\alpha)$ if $1 < p < \infty$ and $0 < 1/p + \alpha < 1$.

   Our goal is to study some algebras of operators on $L^p(\Gamma, |t|^\alpha)$ generated by the singular integral operator S_Γ and by operators of multiplication by piecewise continuous functions. Given a set A of functions in $L^\infty(\Gamma)$, let $\mathscr{A}(A, S_\Gamma, \alpha)$ denote the smallest closed subalgebra of $\mathscr{L}(L^p(\Gamma, |t|^\alpha))$ which contains the operator S_Γ and all operators of multiplication by functions in A.

   One can think of operators in $\mathscr{A}(A, S_\Gamma, \alpha)$ as a matrix of operators on the semi-axis. For we let $L_k^p(\mathbb{R}^+, t^\alpha)$ denote the Banach space of all vectors $g := (g_1, \ldots, g_k)^T$ with entries $g_j \in L^p(\mathbb{R}^+, t^\alpha)$, endowed with the norm

$$\|g\| := \left(\sum_{j=1}^k \|g_j\|_{\mathbb{R}^+, p, t^\alpha}^p\right)^{1/p}, \tag{4.44}$$

and consider the mapping

$$\eta : L^p(\Gamma, |t|^\alpha) \to L_k^p(\mathbb{R}^+, t^\alpha), \qquad (\eta f)(t) := \begin{pmatrix} f(e^{\mathbf{i}\beta_1} t) \\ \vdots \\ f(e^{\mathbf{i}\beta_k} t) \end{pmatrix}. \tag{4.45}$$

Clearly, η is a bijective isometry, and the mapping $A \mapsto \eta^{-1}A\eta$ is an isometric isomorphism between the algebras $\mathscr{L}(L^p(\Gamma, |t|^\alpha))$ and $\mathscr{L}(L_k^p(\mathbb{R}^+, t^\alpha))$. For $1 < p < \infty$ and $0 < 1/p + \alpha < 1$, let $\mathscr{E}_{p,\alpha}^k \subseteq \mathscr{L}(L_k^p(\mathbb{R}^+, t^\alpha))$ denote the Banach algebra of all matrices $[A_{ij}]_{i,j=1}^k$ with entries $A_{ij} \in \mathscr{E}_{p,\alpha}$, and write $\mathscr{E}_{p,\alpha}^{k,\mathcal{N}}$ for the subset of $\mathscr{E}_{p,\alpha}^k$ of all matrices $[A_{ij}]_{i,j=1}^k$ with $A_{ij} \in \mathcal{N}_{p,\alpha}$ whenever $i \neq j$. Since $\mathcal{N}_{p,\alpha}$ is a two-sided ideal of $\mathscr{E}_{p,\alpha}$, the set $\mathscr{E}_{p,\alpha}^{k,\mathcal{N}}$ is a closed subalgebra of $\mathscr{E}_{p,\alpha}^k$.

Proposition 4.2.19. *The mapping $A \mapsto \eta A \eta^{-1}$ is an isometric isomorphism from the algebra $\mathscr{A}(\chi, S_\Gamma, \alpha)$ onto $\mathscr{E}_{p,\alpha}^{k,\mathcal{N}}$. Thereby,*

$$\eta S_\Gamma \eta^{-1} = \begin{bmatrix} S_{\mathbb{R}^+} & H_{2\pi+\beta_1-\beta_2} & \cdots & H_{2\pi+\beta_1-\beta_k} \\ H_{\beta_2-\beta_1} & S_{\mathbb{R}^+} & \cdots & H_{2\pi+\beta_2-\beta_k} \\ \vdots & \vdots & \ddots & \vdots \\ H_{\beta_k-\beta_1} & H_{\beta_k-\beta_2} & \cdots & S_{\mathbb{R}^+} \end{bmatrix} \operatorname{diag}_{1 \le j \le k}(v_j) \qquad (4.46)$$

and

$$\eta \chi_i \eta^{-1} = \operatorname{diag}(0, \ldots, 0, I, 0, \ldots, 0), \qquad (4.47)$$

the I standing at the ith position. Thus, the mapping $A \mapsto \eta A \eta^{-1}$ provides an invertibility symbol for the algebra $\mathscr{A}(\chi, S_\Gamma, \alpha)$.

Proof. A straightforward computation of the entries of the operator $[A_{ij}]_{i,j=1}^k := \eta S_\Gamma \eta^{-1}$ yields

$$A_{ij} = \begin{cases} v_j S_{\mathbb{R}^+} & \text{if } i = j, \\ v_j M^0(h_{\beta_i-\beta_j}) & \text{if } i > j, \\ v_j M^0(h_{2\pi+\beta_i-\beta_j}) & \text{if } i < j. \end{cases} \qquad (4.48)$$

Thus, $\eta S_\Gamma \eta^{-1} \in \mathscr{E}_{p,\alpha}^{k,\mathcal{N}}$. Since $\eta \chi_j \eta^{-1} = \operatorname{diag}(0, \ldots, 0, I, 0, \ldots, 0)$ with the identity standing at the jth position we get in fact that $\eta \mathscr{A}(\chi, S_\Gamma, \alpha)\eta^{-1} \subseteq \mathscr{E}_{p,\alpha}^{k,\mathcal{N}}$. It remains to verify that the mapping $A \mapsto \eta A \eta^{-1}$ maps $\mathscr{A}(\chi, S_\Gamma, \alpha)$ onto $\mathscr{E}_{p,\alpha}^{k,\mathcal{N}}$.

Let $B = [B_{ij}]_{i,j=1}^k \in \mathscr{E}_{p,\alpha}^{k,\mathcal{N}}$ and put $A^{ij} := [\delta_{mi}\delta_{nj}B_{ij}]_{m,n=1}^k$ for each pair (i,j) of subscripts, with δ referring to the Kronecker symbol again. Since $B = \sum_{i,j=1}^k A^{ij}$, it suffices to prove that $A^{ij} \in \eta \mathscr{A}(\chi, S_\Gamma, \alpha)\eta^{-1}$ for $1 \le i, j \le k$. First let $i = j$. Then

$$\eta \chi_j \eta^{-1} = [\delta_{mj}\delta_{nj}I]_{m,n=1}^k \quad \text{and} \quad v_j \eta \chi_j S_\Gamma \chi_j \eta^{-1} = [\delta_{mj}\delta_{nj}S_{\mathbb{R}^+}]_{m,n=1}^k.$$

By definition, these operators generate the algebra $[\delta_{mj}\delta_{nj}\mathscr{E}_{p,\alpha}]_{m,n=1}^k$. If $i > j$ then, by Proposition 4.2.17 (iii), one can approximate B_{ij} by sums of the form $\sum_l C_l M^0(h_{\beta_i-\beta_j})D_l$ with operators $C_l, D_l \in \mathscr{E}_{p,\alpha}$. Put

$$\tilde{C}_l = [\delta_{mi}\delta_{ni}C_l]_{m,n=1}^k \quad \text{and} \quad \tilde{D}_l = [\delta_{mj}\delta_{nj}D_l]_{m,n=1}^k.$$

By what has been just proved, both $\tilde{C}_l$ and $\tilde{D}_l$ belong to $\eta \mathscr{A}(\chi, S_\Gamma, \alpha)\eta^{-1}$. Since

$$v_j \eta \chi_i S_\Gamma \chi_j \eta^{-1} = [\delta_{mi} \delta_{nj} M^0 (h_{\beta_i - \beta_j})]_{m,n=1}^k$$

by (4.48), the operator A^{ij} can be approximated by sums of the form

$$\sum_l \widetilde{C}_l v_j \eta \chi_i S_\Gamma \chi_j \eta^{-1} \widetilde{D}_l \in [\delta_{mi} \delta_{nj} \mathcal{N}_{p,\alpha}]_{m,n=1}^k$$

as desired. For $i < j$, the proof is analogous.   ∎

Interlude. We digress for a moment to describe another setting where the conditions of the N projections theorem 3.2.4 are satisfied. Let $\Gamma := \cup_{j=1}^{2n} e^{i\beta_j} \mathbb{R}^+$ with angles $0 \leq \beta_1 < \beta_2 < \ldots < \beta_{2n} < 2\pi$ and with the special orientation $v_j = (-1)^j$ for $1 \leq j \leq 2n$. Let p_j stand for the operator of multiplication by the characteristic function χ_j of $e^{i\beta_j} \mathbb{R}^+$, and set $P := (I + S_\Gamma)/2$ and $Q := I - P$. Then, evidently, condition (3.4) holds. To show condition (3.5), i.e., the equality $P(p_{2j-1} + p_{2j})P = (p_{2j-1} + p_{2j})P$, one can use the representations of the operators S_Γ and p_j derived in the previous proposition. Thus, all computations can be done on the level of generating functions of special Mellin convolution operators. Using the (already mentioned) identity $S_{\mathbb{R}^+}^2 - I = H_\beta H_{2\pi - \beta}$ as well as the identities

$$H_{\alpha+\beta}(I + S_{\mathbb{R}^+}) = H_\alpha H_\beta \quad \text{and} \quad H_{\alpha+\beta}(I - S_{\mathbb{R}^+}) = -H_\alpha H_{2\pi+\beta} = -H_{2\pi+\alpha} H_\beta,$$

it is then straightforward to obtain (3.5) and (3.6). Finally, summing the identities $P(p_{2j-1} + p_{2j})P = (p_{2j-1} + p_{2j})P$ for j between 1 and n we get $P^2 = P$, which is (3.3). Note that the latter equality is equivalent to $S_\Gamma^2 = I$ (which we already know since Γ is a closed curve under the above assumptions).

Note also that for any homogeneous curve Γ, there is a closed homogeneous curve $\widetilde{\Gamma}$ (which satisfies the conditions for the orientation imposed above) such that $\Gamma \subseteq \widetilde{\Gamma}$ and the orientation of Γ coincides with the orientation inherited from $\widetilde{\Gamma}$. Then operators in $\mathscr{A}(\chi, S_\Gamma, \alpha)$ can be studied via operators in the corresponding algebra $\mathscr{A}(\widetilde{\chi}, S_{\widetilde{\Gamma}}, \alpha)$. For instance, the operator S_Γ is invertible if and only if the operator $\chi_\Gamma S_{\widetilde{\Gamma}} \chi_\Gamma + (1 - \chi_\Gamma) I_{\widetilde{\Gamma}}$ is invertible. This observation can be helpful for operators on Banach spaces where a result like Proposition 4.2.19 is not available.   ∎

Corollary 4.2.20. *The algebra $\mathscr{A}(\chi, S_\Gamma, \alpha)$ is inverse-closed in $\mathscr{L}(L^p(\Gamma, |t|^\alpha))$.*

Proof. By the preceding proposition, it is sufficient to show that $\mathscr{E}_{p,\alpha}^{k,\mathcal{N}}$ is an inverse-closed subalgebra of $\mathscr{L}(L_k^p(\mathbb{R}^+, t^\alpha))$. This will follow once we have shown that $\mathscr{E}_{p,\alpha}^{k,\mathcal{N}}$ is inverse-closed in $\mathscr{E}_{p,\alpha}^k$ and $\mathscr{E}_{p,\alpha}^k$ is inverse-closed in $\mathscr{L}(L_k^p(\mathbb{R}^+, t^\alpha))$. The first assertion follows immediately from the explicit formula of the inverse matrix using cofactors. The second assertion is equivalent to the inverse-closedness of $\mathscr{E}_{p,\alpha}$ in $\mathscr{L}(L^p(\mathbb{R}^+, t^\alpha))$ by Proposition 1.2.35 (note that the entries of the matrices in

(4.46) and (4.47) commute). The inverse-closedness of $\mathscr{E}_{p,\alpha}$ in $\mathscr{L}\left(L^p(\mathbb{R}^+, t^\alpha)\right)$ was proved in Proposition 4.2.16 (iv). ∎

Another remarkable consequence of Proposition 4.2.19 is that the algebra $\mathscr{A}(\chi, S_\Gamma, \alpha)$ depends neither on the values of the angles β_j nor on the orientation of the axes $e^{i\beta_j}\mathbb{R}^+$. Moreover, by Proposition 4.2.16 (iii), it does not depend either on the weight $|t|^\alpha$. For a precise statement, let Γ_k denote the curve

$$\Gamma_k = \cup_{j=0}^{k-1} e^{\frac{2\pi j i}{k}} \mathbb{R}^+ \qquad (4.49)$$

with each axis being directed away from 0.

Corollary 4.2.21. *The algebras* $\mathscr{A}(\chi, S_\Gamma, \alpha)$ *and* $\mathscr{A}(\chi, S_{\Gamma_k}, 0)$ *are isometrically isomorphic.*

Proof. By Proposition 4.2.19, the algebras $\mathscr{A}(\chi, S_\Gamma, \alpha)$ and $\mathscr{E}_{p,\alpha}^{k,\mathscr{N}}$ are isometrically isomorphic, as well as the algebras $\mathscr{A}(\chi, \Gamma, 0)$ and $\mathscr{E}_{p,0}^{k,\mathscr{N}}$. That the algebras $\mathscr{E}_{p,\alpha}^{k,\mathscr{N}}$ and $\mathscr{E}_{p,0}^{k,\mathscr{N}}$ are isometrically isomorphic follows from Proposition 4.2.16 (iii) in combination with Proposition 4.2.17 (i) (recall that the symbol of the operator $t^{-\alpha}H_\beta t^\alpha I$ is given by (4.42)). ∎

So far we have studied operators with piecewise constant coefficients. Operators with piecewise continuous coefficients can be reduced to this special class by means of a *homogenization* technique which we are going to introduce now. We will need the multiplicative shift operators Z_τ previously defined on $\mathbb{R}^+$ by (4.37), the definition of which can be naturally extended to general homogeneous curves. Let $A \in \mathscr{L}\left(L^p(\Gamma, |t|^\alpha)\right)$ be an operator such that the strong limit s-lim$_{\tau\to 0} Z_\tau^{-1} A Z_\tau$ (respectively the strong limit s-lim$_{\tau\to\infty} Z_\tau^{-1} A Z_\tau$) exists. We denote this strong limit by

$$H_0(A) := \operatorname*{s\text{-}lim}_{\tau\to 0} Z_\tau^{-1} A Z_\tau \quad \text{resp.} \quad H_\infty(A) := \operatorname*{s\text{-}lim}_{\tau\to\infty} Z_\tau^{-1} A Z_\tau. \qquad (4.50)$$

It is easy to see that the set of all operators A with this property forms a Banach algebra and that H_0 (resp. H_∞) acts continuously and homomorphically on this algebra. Moreover,

$$\|H_0(A)\| \le \|A\| \quad \text{resp.} \quad \|H_\infty(A)\| \le \|A\|. \qquad (4.51)$$

For $a \in PC(\dot{\Gamma})$, we denote the one-sided limits of a at $\{0, \infty\}$ by

$$a_j(0) := \lim_{\substack{t\to 0 \\ t\in e^{i\beta_j}\mathbb{R}^+}} a(t), \quad a_j(\infty) := \lim_{\substack{t\to\infty \\ t\in e^{i\beta_j}\mathbb{R}^+}} a(t). \qquad (4.52)$$

Proposition 4.2.22. *The strong limits* (4.50) *exist for every operator A which belongs to* $\mathscr{A}(PC(\dot{\Gamma}), S_\Gamma, \alpha)$. *In particular,*

(i) $H_0(S_\Gamma) = H_\infty(S_\Gamma) = S_\Gamma$;

(ii) *for $a \in PC(\dot{\Gamma})$,*

$$H_0(aI) = \sum_{j=1}^{k} a_j(0)\chi_j I, \quad H_\infty(aI) = \sum_{j=1}^{k} a_j(\infty)\chi_j I; \tag{4.53}$$

(iii) *if K is compact then $H_0(K) = H_\infty(K) = 0$;*
(iv) *H_0 and H_∞ are homomorphisms from $\mathscr{A}(PC(\dot{\Gamma}), S_\Gamma, \alpha)$ onto $\mathscr{A}(\chi, S_\Gamma, \alpha)$.*

Proof. Since H_0 and H_∞ are continuous homomorphisms, it suffices to check assertions (i) – (iii). Assertion (i) follows by straightforward computation. The proof of (ii) will be illustrated only for the curve $\Gamma = \mathbb{R}^+$ from which the general case easily follows. Let $a \in PC(\mathbb{R}^+)$ and denote by $\chi_+ I$ the operator of multiplication by the characteristic function of $\mathbb{R}^+$. Then $(Z_\tau^{-1} a Z_\tau u)(x) = a(\tau x)u(x)$, whence

$$\|(a(+\infty)\chi_+ - Z_\tau^{-1} a Z_\tau)u\|^p = \int_0^{+\infty} |(a(+\infty) - a(\tau x))u(x)|^p x^{\alpha p} \, dx.$$

Given $\varepsilon > 0$, choose $x_\varepsilon \in \mathbb{R}^+$ such that $|a(+\infty) - a(x)|^p \|u\|^p < \frac{\varepsilon^p}{2}$ for $x \geq x_\varepsilon$. For some $\tau > 1$, we write the above integral as the sum

$$\int_0^{\frac{x_\varepsilon}{\tau}} (|a(+\infty) - a(\tau x)| \, |u(x)|)^p x^{\alpha p} \, dx + \int_{\frac{x_\varepsilon}{\tau}}^{+\infty} (|a(+\infty) - a(\tau x)| \, |u(x)|)^p x^{\alpha p} \, dx$$

which is not greater than

$$\max_{0 < x < x_\varepsilon} |a(+\infty) - a(x)|^p \int_0^{\frac{x_\varepsilon}{\tau}} |u(x)|^p x^{\alpha p} \, dx + \frac{\varepsilon^p}{2}.$$

For sufficiently large τ, the first term of this sum becomes as small as desired. Thus, $H_\infty(aI) = a(\infty)\chi_+ I$. Analogously, $H_0(aI) = a(0)\chi_+ I$.

Assertion (iii) follows from Lemma 1.4.7, since the operators Z_τ^{-1} are uniformly bounded and $Z_\tau \rightharpoonup 0$ weakly by Lemma 4.2.12. ∎

4.2.5 Algebras of SIOs and piecewise continuous functions on the real line

The goal of this section is to illustrate how the tools developed so far (local principles, projection theorems, homogenization) apply to the study of the Fredholm property of singular integral operators with piecewise continuous coefficients. We shall do this here in a context which allows us to avoid technicalities: we work on the real line and in spaces without weight. The general setting is the subject of the following section.

Consider the smallest closed subalgebra $\mathscr{A}$ of $\mathscr{L}(L^p(\mathbb{R}))$ which contains the singular integral operator $S_{\mathbb{R}}$ and all operators of multiplication by piecewise continuous functions, thus $\mathscr{A} = \mathscr{A}(PC(\dot{\mathbb{R}}), S_{\mathbb{R}})$. Note that the one-point compactification $\dot{\mathbb{R}}$ is a closed curve in the sense of Section 4.1.1. The object of our interest is the subalgebra $\mathscr{A}^{\mathscr{K}}$ of the Calkin algebra.

Theorem 4.1.4 implies that $C(\dot{\mathbb{R}})I + \mathscr{K} = \{fI + \mathscr{K} : f \in C(\dot{\mathbb{R}})\}$ is a central subalgebra of $\mathscr{A}^{\mathscr{K}}$. This algebra is isometrically isomorphic to the algebra $C(\dot{\mathbb{R}})$ by Proposition 1.4.11. Thus, its maximal ideal space is homeomorphic to $\dot{\mathbb{R}}$, with the maximal ideal $\{(fI) + \mathscr{J} : f \in C(\dot{\mathbb{R}}), f(x) = 0\}$ corresponding to $x \in \dot{\mathbb{R}}$ (compare with Section 1.4.3).

Let $\mathscr{I}_x$ denote the smallest closed ideal of $\mathscr{A}^{\mathscr{K}}$ which contains the maximal ideal x of $C(\dot{\mathbb{R}})I + \mathscr{K}$. Allan's local principle transfers the invertibility problem in $\mathscr{A}^{\mathscr{K}}$ to a family of invertibility problems, one in each of the local algebras $\mathscr{A}_x^{\mathscr{K}} := \mathscr{A}^{\mathscr{K}}/\mathscr{I}_x$. Let $\Phi_x^{\mathscr{K}}$ denote the canonical homomorphism from $\mathscr{A}$ to $\mathscr{A}_x^{\mathscr{K}}$. The following lemma describes the local representatives of operators of multiplication. For $x \in \mathbb{R}$, let χ_x denote the characteristic function of the interval $[x, +\infty[$, and set $\chi_\infty := 1 - \chi_0$. Further, write $a(x^\pm)$ for the one-sided limits of the function $a \in PC(\dot{\mathbb{R}})$ at $x \in \mathbb{R}$. For $x = \infty$, we set $a(\infty^-) = a(+\infty)$ and $a(\infty^+) = a(-\infty)$.

Lemma 4.2.23. *Let $a \in PC(\dot{\mathbb{R}})$ and $x \in \dot{\mathbb{R}}$.*

(i) *If a is continuous at x and $a(x) = 0$, then $\Phi_x^{\mathscr{K}}(aI) = 0$;*
(ii) *$\Phi_x^{\mathscr{K}}(aI) = \Phi_x^{\mathscr{K}}(a(x^-)(1-\chi_x)I + a(x^+)\chi_x I)$.*

Proof. (i) First let $x \neq \infty$. For $\varepsilon > 0$, choose a continuous function $f_\varepsilon : \dot{\mathbb{R}} \to [0, 1]$ with $f_\varepsilon(x) = 1$ the support of which is contained in the interval $[x - \varepsilon, x + \varepsilon]$. It easy to see that $\Phi_x^{\mathscr{K}}(f_\varepsilon I)$ is the identity element of the local algebra at x. Thus,

$$\|\Phi_x^{\mathscr{K}}(aI)\| = \|\Phi_x^{\mathscr{K}}(aI)\Phi_x^{\mathscr{K}}(f_\varepsilon I)\| = \|\Phi_x^{\mathscr{K}}(af_\varepsilon I)\| \leq \|af_\varepsilon\|_{L^\infty}.$$

The norm on the right-hand side can be as small as desired by choosing ε small enough. For $x = \infty$ the proof is similar, now with the support of f_ε contained in $\{y \in \mathbb{R} : |y| > 1/\varepsilon\}$.

(ii) Write a as

$$a = a(x^-)(1 - \chi_x) + a(x^+)\chi_x + a'.$$

Since a' is continuous at x and $a'(x) = 0$, the assertion follows from part (i). ∎

The following is an immediate consequence.

Proposition 4.2.24. *Every local algebra $\mathscr{A}_x^{\mathscr{K}}$ is unital and generated by two idempotents, namely $\Phi_x^{\mathscr{K}}(\chi_x I)$ and $\Phi_x^{\mathscr{K}}(P_{\mathbb{R}})$.*

Thus, the local algebras are subject to the two projections theorem, with $p = \Phi_x^{\mathscr{K}}(\chi_x I)$ and $q = \Phi_x^{\mathscr{K}}(P_{\mathbb{R}})$. The application of that theorem requires some knowledge on the spectrum of

$$pqp = \Phi_x^{\mathscr{H}}(\chi_x P_{\mathbb{R}} \chi_x I). \tag{4.54}$$

A convenient way to determine this spectrum is the use of homogenization techniques. Let V_t refer to the (additive) shifts defined in Section 4.2.1, and write χ_+ for χ_0 and χ_- for $1 - \chi_0$. We consider the set of all operators A for which the strong limit $H_\infty(A)$, defined in (4.50), and all strong limits

$$H_x(A) := \operatorname*{s-lim}_{\tau \to +\infty} Z_\tau V_{-x} A V_x Z_\tau^{-1}$$

with $x \in \mathbb{R}$ exist. This set is an algebra, and every H_x is a homomorphism on this algebra.

Proposition 4.2.25. *The strong limits* $H_x(A)$ *exist for all* $A \in \mathscr{A}$ *and* $x \in \dot{\mathbb{R}}$. *In particular:*

   (i) $H_x(S_{\mathbb{R}}) = S_{\mathbb{R}}$;
   (ii) $H_x(aI) = a(x^-)\chi_- I + a(x^+)\chi_+ I$ *for* $a \in PC(\dot{\mathbb{R}})$;
   (iii) $H_x(K) = 0$ *for* K *compact.*

Proof. The case $x = \infty$ is proved in Proposition 4.2.22. Let $x \neq \infty$. Then the proof of (i) is a straightforward calculation using the definition of the singular integral operator. For assertion (ii), note that

$$\begin{aligned}
(Z_\tau V_{-x} a V_x Z_\tau^{-1} u)(t) &= (a V_x Z_\tau^{-1} u)(t/\tau + x) \\
&= a(t/\tau + x)(V_x Z_\tau^{-1} u)(t/\tau + x) \\
&= a(t/\tau + x)u(t)
\end{aligned}$$

and pass to the limit as $\tau \to \infty$. Assertion (iii) follows from Lemma 1.4.6, since the operators $Z_\tau^{\pm}$ are uniformly bounded and tend weakly to zero by Lemma 4.2.12. ∎

Proposition 4.2.25 implies that each H_t maps the algebra $\mathscr{A}$ to the Banach algebra $\operatorname{alg}\{S_{\mathbb{R}}, \chi_+\}$ generated by $S_{\mathbb{R}}$ and $\chi_+ I$. By assertion (iv) of the same proposition, the quotient homomorphisms (denoted by the same symbol) $H_t : \mathscr{A}^{\mathscr{H}} \to \operatorname{alg}\{S_{\mathbb{R}}, \chi_+\}$ are well defined. Moreover, due to (ii), the homomorphism H_x maps the local ideal at x to 0 and is, thus, also well defined on the quotient algebra $\mathscr{A}_x^{\mathscr{H}}$. So finally we have a family of homomorphisms $H_x : \mathscr{A}_x^{\mathscr{H}} \to \operatorname{alg}\{S_{\mathbb{R}}, \chi_+\}$.

In order to show that these homomorphisms are in fact isomorphisms, consider the homomorphism $W_x' : \operatorname{alg}\{S_{\mathbb{R}}, \chi_+\} \to \mathscr{A}_x^{\mathscr{H}}$ defined by

$$W_x'(A) := \Phi_x^{\mathscr{H}}(V_x A V_{-x}).$$

Using Propositions 4.2.24 and 4.2.25 one can easily check to see that every homomorphism W_x' is onto and that W_x' is the inverse of H_x. Thus, H_x is indeed an (even isometric) isomorphism between the local algebra $\mathscr{A}_x^{\mathscr{H}}$ and the operator algebra $\operatorname{alg}\{S_{\mathbb{R}}, \chi_+\}$.

This fact can be employed to determine the spectrum of the indicator element (4.54) needed for the application of the two-projections theorem. Indeed, this spectrum coincides with the spectrum of the singular integral operator $\chi_+ P_\mathbb{R} \chi_+ I = (I + S_{\mathbb{R}^+})/2$ which is a circular arc, namely the closure of the range of the function

$$\mathbb{R} \to \mathbb{C}, \qquad y \mapsto (1 + \coth((y + \mathbf{i}/p)\pi))/2$$

(see Proposition 4.2.11 and Figure 4.4). Now one can use the two projections theorem to associate with every local coset $\Phi_x^{\mathscr{K}}(A)$ a 2×2-matrix function defined on this arc with the property that the coset is invertible if and only if the associated matrix function is invertible.

As already mentioned, the same result follows from homogenization directly, without invoking the two projections theorem explicitly. This is a consequence of Proposition 4.2.19 and of the fact that the entries in the (2×2 matrix) operators in $\mathscr{E}_{p,\alpha}^{k,\mathcal{N}}$ are Mellin convolutions.

Since, by Allan's local principle, the family $\{H_x\}_{x \in \mathbb{R}}$ of homomorphisms is sufficient for $\mathscr{A}^{\mathscr{K}}$, we can summarize as follows.

Theorem 4.2.26. *Let $A \in \mathscr{A} := \mathscr{A}(PC(\dot{\mathbb{R}}), S_\mathbb{R})$. Then the coset of A modulo compact operators is invertible in $\mathscr{A}^{\mathscr{K}}$ if and only if the operators $H_x(A)$ are invertible for all $x \in \mathbb{R}$. The invertibility of these operators can effectively be checked via Proposition 4.2.19.*

The preceding theorem in combination with Corollary 1.2.32 easily implies that the algebra $\mathscr{A}^{\mathscr{K}}$ is inverse-closed in the Calkin algebra. Thus, Theorem 4.2.26 is actually a criterion for the Fredholm property of singular integral operators.

Theorem 4.2.27. *Let $\mathscr{A} := \mathscr{A}(PC(\dot{\mathbb{R}}), S_\mathbb{R})$. The algebra $\mathscr{A}^{\mathscr{K}}$ is inverse-closed in the Calkin algebra on $L^p(\mathbb{R})$, and the algebra $\mathscr{A}$ is inverse-closed in $\mathscr{L}(L^p(\mathbb{R}))$. Thus, an operator $A \in \mathscr{A}$ is Fredholm if and only if the operators $H_x(A)$ are invertible for all $x \in \dot{\mathbb{R}}$.*

One should mention that in the case at hand, homogenization indeed yields *much more* than the two projections theorem: It provides us with an isometric isomorphism between each local algebra and an algebra of operators. Since the operator algebra is basically an algebra of convolutions, this implies, for example, that the local algebras have a trivial radical. Questions of this kind cannot be answered by the two projections theorem.

4.3 Algebras of singular integral operators on admissible curves

4.3.1 Algebras of SIOs and piecewise continuous functions on admissible curves

Let Γ be an admissible curve and $w \in \mathcal{A}_p(\Gamma)$ a power weight of the form (4.1). We consider the smallest closed subalgebra $\mathscr{A}(PC(\dot{\Gamma}), S_\Gamma, w)$ of $\mathscr{L}(L^p(\Gamma, w))$ which contains the operator S_Γ and all operators of multiplication by a piecewise continuous function.

In what follows, we need some additional notation. Let z be a finite point of Γ of order $k(z)$. There is a closed neighborhood $\mathscr{U}$ of z such that $\mathscr{U} \cap \Gamma$ is a union $\cup_{j=1}^{k(z)} \Gamma_j$ of Lyapunov arcs Γ_j with $\cap_{j=1}^{k(z)} \Gamma_j = \{z\}$. We denote the angle between the tangents of Γ_1 and Γ_j at z by $\beta_j \equiv \beta_j(z)$. Without loss of generality we can assume that $0 = \beta_1 < \ldots < \beta_{k(z)} < 2\pi$. If Γ is unbounded and $z = \infty$, then the angles β_j are defined as the angles at the point 0 of the image of Γ under the conformal mapping $z \mapsto (z - z_0)^{-1}$ with respect to some point $z_0 \notin \Gamma$.

Given the weight w, define a function

$$\alpha : \Gamma \to \mathbb{R}, \quad z \mapsto \begin{cases} \alpha_i & \text{if } z = t_i, i \in \{1, \ldots, n\}, \\ \sum_{i=0}^{n} \alpha_i & \text{if } z = \infty, \\ 0 & \text{if } z \notin \{t_1, \ldots, t_n, \infty\} \end{cases} \tag{4.55}$$

and call $w_z(t) := |t|^{\alpha(z)}$ with $t \in \mathbb{C}$ the local weight function at the point z. Finally, abbreviate the Calkin algebra $\mathscr{A}(PC(\dot{\Gamma}), S_\Gamma, w)/\mathscr{K}$ to $\mathscr{A}^{\mathscr{K}}(\Gamma, w)$ and write Φ for the canonical homomorphism from $\mathscr{A}(PC(\dot{\Gamma}), S_\Gamma, w)$ onto $\mathscr{A}^{\mathscr{K}}(\Gamma, w)$.

We will employ localization in order to study the algebra $\mathscr{A}(PC(\dot{\Gamma}), S_\Gamma, w)$. First we have to identify a suitable central subalgebra of $\mathscr{A}^{\mathscr{K}}(\Gamma, w)$. The following is an immediate consequence of Theorem 4.1.4.

Proposition 4.3.1. *The cosets* $fI + \mathscr{K}$ *with* $f \in C(\dot{\Gamma})$ *belong to the center of* $\mathscr{A}^{\mathscr{K}}(\Gamma, w)$.

We can use Allan's local principle to localize $\mathscr{A}^{\mathscr{K}}(\Gamma, w)$ over the central subalgebra $\Phi(C(\dot{\Gamma}))$ generated by the set of cosets $\{(fI) + \mathscr{K} : f \in C(\dot{\Gamma})\}$. This algebra is isomorphic to the algebra $C(\dot{\Gamma})$, and its maximal ideal space is homeomorphic to $\dot{\Gamma}$, where the maximal ideal corresponding to $x \in \dot{\Gamma}$ is $\{(fI) + \mathscr{K} : f \in C(\dot{\mathbb{R}}), f(x) = 0\}$. Moreover, the pair consisting of the algebra $\mathscr{A}^{\mathscr{K}}(\Gamma, w)$ and its subalgebra $\Phi(C(\dot{\Gamma}))$ is a faithful localizing pair by Theorem 2.3.5.

For $z \in \dot{\Gamma} = M(\Phi(C(\dot{\Gamma})))$, we denote the corresponding local ideal of $\mathscr{A}^{\mathscr{K}}(\Gamma, w)$ by $\mathscr{I}_z$, the quotient algebra $\mathscr{A}^{\mathscr{K}}(\Gamma, w)/\mathscr{I}_z$ by $\mathscr{A}_z^{\mathscr{K}}(\Gamma, w)$, and the canonical homomorphism from $\mathscr{A}^{\mathscr{K}}(\Gamma, w)$ onto this quotient algebra by Φ_z. To each $z \in \dot{\Gamma}$, we associate the suitable curve $\Gamma_z := \cup_{j=1}^{k(z)} e^{i\beta_j} \mathbb{R}^+$ where $e^{i\beta_j} \mathbb{R}^+$ is endowed with the same orientation as Γ_j (to z or away from z).

The following proposition establishes an isomorphism between local algebras on $\dot{\Gamma}$ and local algebras on the homogeneous curves Γ_z. Note that, in this setting, "localization" does not only imply a localization (freeze in) of the coefficients of the operator, but also of the weight function and even of the underlying curve. For every measurable set M, let χ_M denote its characteristic function.

Proposition 4.3.2. *Let* $z \in \dot{\Gamma}$. *Then the local algebras* $\mathscr{A}_z^{\mathscr{K}}(\Gamma, w)$ *and* $\mathscr{A}_0^{\mathscr{K}}(\Gamma_z, w_z)$ *are isometrically isomorphic. The isomorphism can be chosen so that* $\Phi_z(S_\Gamma)$ *and* $\Phi_z(\chi_{\Gamma_j} I)$ *are carried over into* $\Phi_0(S_{\Gamma_z})$ *and* $\Phi_0(\chi_{e^{i\beta_j} \mathbb{R}^+} I)$, *respectively.*

Proof. First let $z \in \Gamma$. Choose a (sufficiently small) neighborhood $\mathscr{U}$ of z which contains none of the weight points $t_0, t_1, \ldots, t_n$ besides, possibly, the point z itself, and for which $\mathscr{U} \cap \Gamma$ can be written as a union $\cup_{j=1}^{k(z)} \Gamma_j$ of Lyapunov arcs Γ_j such that $\cap_{j=1}^{k(z)} \Gamma_j = \{z\}$. For each function a and each set M, let $a_{|M}$ denote the restriction of a onto M. We will prove the assertion by verifying that the following algebras are isometrically isomorphic to each other:

- $\mathscr{A}_1 := \mathscr{A}_z^{\mathscr{K}}(\Gamma, w);$
- $\mathscr{A}_2 := \mathscr{A}_z^{\mathscr{K}}(\Gamma \cap \mathscr{U}, w_{|\Gamma \cap \mathscr{U}});$
- $\mathscr{A}_3 := \mathscr{A}_z^{\mathscr{K}}(\Gamma \cap \mathscr{U}, w_{z|\Gamma \cap \mathscr{U}});$
- $\mathscr{A}_4 := \mathscr{A}_0^{\mathscr{K}}(\Gamma_z \cap \mathscr{U}', w_{z|\Gamma_z \cap \mathscr{U}'});$
- $\mathscr{A}_5 := \mathscr{A}_0^{\mathscr{K}}(\Gamma_z, w_z).$

Step 1 $\mathscr{A}_1 \cong \mathscr{A}_2$: Choose a function $f \in C(\dot{\Gamma})$ with $f(z) = 1$ and supp $f \subseteq \mathscr{U}$. Then the desired isomorphy follows immediately from

$$\Phi_z(S_\Gamma) = \Phi_z(fI)\Phi_z(S_\Gamma)\Phi_z(fI) = \Phi_z(fS_\Gamma fI) = \Phi_z(fS_{\Gamma \cap \mathscr{U}} fI) = \Phi_z(S_{\Gamma \cap \mathscr{U}}) \tag{4.56}$$

and from

$$\Phi_z(aI) = \Phi_z(fI)\Phi_z(aI)\Phi_z(fI) = \Phi_z(fafI) = \Phi_z(fa_{|\Gamma \cap \mathscr{U}} fI) = \Phi_z(a_{|\Gamma \cap \mathscr{U}} I) \tag{4.57}$$

for each function $a \in PC(\dot{\Gamma})$.

Step 2 $\mathscr{A}_2 \cong \mathscr{A}_3$: For $t \in \Gamma \cap \mathscr{U}$, set $\vartheta(t) := w(t)w_z^{-1}(t)$. The function ϑ is continuous on $\Gamma \cap \mathscr{U}$ and does not degenerate there. Hence, the Banach spaces $L^p(\Gamma \cap \mathscr{U}, w_{|\Gamma \cap \mathscr{U}})$ and $L^p(\Gamma \cap \mathscr{U}, w_{z|\Gamma \cap \mathscr{U}})$ coincide, and the mapping

$$T : L^p(\Gamma \cap \mathscr{U}, w_{|\Gamma \cap \mathscr{U}}) \to L^p(\Gamma \cap \mathscr{U}, w_{z|\Gamma \cap \mathscr{U}}), \quad f \mapsto \vartheta f$$

is an isometry. Thus, the mapping

$$\mathscr{L}\big(L^p(\Gamma \cap \mathscr{U}, w_{|\Gamma \cap \mathscr{U}})\big) \to \mathscr{L}\big(L^p(\Gamma \cap \mathscr{U}, w_{z|\Gamma \cap \mathscr{U}})\big) : \quad A \mapsto TAT^{-1}$$

is an isometry, too. Since $TAT^{-1} - A$ is a compact operator for each $A \in \mathscr{A}(\Gamma \cap \mathscr{U}, w_{|\Gamma \cap \mathscr{U}})$ by Theorem 4.1.4, this mapping determines an isometric isomorphism between the quotient algebras $\mathscr{A}^{\mathscr{K}}(\Gamma \cap \mathscr{U}, w_{|\Gamma \cap \mathscr{U}})$ and $\mathscr{A}^{\mathscr{K}}(\Gamma \cap \mathscr{U}, w_{z|\Gamma \cap \mathscr{U}})$ under the action of which the algebra $\Phi(C(\dot{\mathbb{R}}))$ remains invariant. The conclusion is, that the local algebras at $z \in \Gamma$ are isometrically isomorphic, too.

Step 3 $\mathscr{A}_3 \cong \mathscr{A}_4$: Let $\mathscr{U}'$ be a neighborhood of 0 and $\upsilon : \Gamma_z \cap \mathscr{U}' \to \Gamma \cap \mathscr{U}$ a homeomorphism subject to the following conditions:

(i) the restriction of υ onto $e^{i\beta_j}\mathbb{R}^+ \cap \mathscr{U}'$ has a Hölder continuous derivative, and the image of this restriction is $\Gamma_j \cap \mathscr{U}$;

(ii) the function υ' defined at inner points $t \in e^{i\beta_j}\mathbb{R}^+ \cap \mathscr{U}'$ by

$$v'(t) = \frac{1}{t(s)} \left. \frac{d(t(s))}{ds} \right|_{s=te^{-i\beta_j}}$$

can be extended to a continuous function on $\Gamma_z \cap \mathcal{U}'$;

(iii) $v(0) = z$, $v'(0) = 1$.

It is not hard to see that a neighborhood $\mathcal{U}'$ and a function v with these properties always exist. The homeomorphism v determines an isomorphism Υ from $L^p(\Gamma \cap \mathcal{U}, w_{z|_{\Gamma \cap \mathcal{U}}})$ onto $L^p(\Gamma_z \cap \mathcal{U}, w_{z|_{\Gamma_z \cap \mathcal{U}}})$ by $(\Upsilon f)(t) := f(v(t))$. Our first claim is that the algebras $\mathscr{A}_3$ and $\mathscr{A}_4$ are algebraically isomorphic. For brevity, put $A^v := \Upsilon A \Upsilon^{-1}$ for each operator $A \in \mathscr{L}(L^p(\Gamma \cap \mathcal{U}, w_{z|_{\Gamma \cap \mathcal{U}}}))$. If we knew that

$$S^v_{\Gamma \cap \mathcal{U}} = S_{\Gamma \cap \mathcal{U}'} + K \quad \text{with } K \text{ compact}, \tag{4.58}$$

then we could get our claim as follows: The operator K^v is compact for each compact K, and if a is piecewise continuous then $(aI)^v$ is the operator of multiplication by the piecewise continuous function a^v. Consequently, the mapping $A \mapsto A^v$ settles an isomorphism between the quotient algebras $\mathscr{A}^{\mathscr{K}}(\Gamma \cap \mathcal{U}, w_{z|_{\Gamma \cap \mathcal{U}}})$ and $\mathscr{A}^{\mathscr{K}}(\Gamma_z \cap \mathcal{U}', w_{z|_{\Gamma_z \cap \mathcal{U}'}})$. Moreover, this isomorphism implies an isomorphism between the corresponding local algebras at the point z since the algebra $\Upsilon C(\Gamma \cap \mathcal{U})\Upsilon^{-1}$ coincides with $C(\Gamma_z \cap \mathcal{U})$ and since $f^v(0) = f(z)$ for each continuous f. So we are left with verifying (4.58). Let χ_j denote the characteristic function of the arc Γ_j, and write $\tilde{\chi}_j$ for the characteristic function of $e^{i\beta_j}\mathbb{R}^+ \cap \mathcal{U}'$. The equality

$$
\begin{aligned}
&S^v_{\Gamma \cap \mathcal{U}} - S_{\Gamma \cap \mathcal{U}'} \\
&= \sum_{i,j=1}^{k} \tilde{\chi}_i (S^v_{\Gamma \cap \mathcal{U}} - S_{\Gamma \cap \mathcal{U}'}) \tilde{\chi}_j I \\
&= \sum_{i,j=1}^{k} \left((\chi_i S_{\Gamma \cap \mathcal{U}} \chi_j I)^v - \tilde{\chi}_i S_{\Gamma \cap \mathcal{U}'} \tilde{\chi}_j I \right) \\
&= \sum_{\substack{i,j=1 \\ i \le j}}^{k} \left(S^v_{\Gamma_i \cup \Gamma_j} - S_{(e^{i\beta_i}\mathbb{R}^+ \cup e^{i\beta_j}\mathbb{R}^+) \cap \mathcal{U}'} \right) - (k-1) \sum_{j=1}^{k} \left(S^v_{\Gamma_j} - S_{e^{i\beta_j}\mathbb{R}^+} \right)
\end{aligned} \tag{4.59}
$$

with $k \equiv k(z)$ shows that it suffices to consider the operators in the first and in the second sum, which correspond to $k = 2$ and $k = 1$, respectively. We need the following lemma.

Lemma 4.3.3. *Let Γ_i, Γ_j, $\mathcal{U}'$, z, v, β_i, β_j be as above and assume that Γ_i is directed to z and Γ_j away from z. Then the operator*

$$S^v_{\Gamma_i \cup \Gamma_j} - S_{(e^{i\beta_i}\mathbb{R}^+ \cup e^{i\beta_j}\mathbb{R}^+) \cap \mathcal{U}'} \tag{4.60}$$

is compact.

Proof. Writing the operator (4.60) applied to a function u as an integral we obtain

$$\frac{1}{\pi i} \int_{(e^{i\beta_i}\mathbb{R}^+ \cup e^{i\beta_j}\mathbb{R}^+) \cap \mathscr{U}'} \frac{k(s,t)}{s-t} u(s)\, ds \tag{4.61}$$

with

$$k(s,t) = \frac{v'(s)(s-t)}{v(s) - v(t)} - 1. \tag{4.62}$$

We shall verify that

$$k(s,t) = O(|s-t|^\alpha) \quad \text{for some } \alpha \in (0,1), \tag{4.63}$$

which implies that the integral (4.61) has a weakly singular kernel and is, hence, compact (see, for instance, [93, Section X.2]). First, suppose s and t are in the same segment, say in $e^{i\beta_i}\mathbb{R}^+ \cap \mathscr{U}'$. Without loss of generality, assume $\beta_j = 0$. Then, by condition (i) for v,

$$v'(s) - v'(t) = O(|s-t|^\alpha)$$

with some $\alpha \in (0,1)$, whence

$$v(s) - v(t) - v'(t)(s-t) = O(|s-t|^{\alpha+1}).$$

Substituting these two equalities into (4.62) we immediately get (4.63). Now let $s \in e^{i\beta_i}\mathbb{R}^+$ and $t \in e^{i\beta_j}\mathbb{R}^+$ and assume again $\beta_i = 0$. Then

$$v(s) = v(0) + v'(0)s + O(|s|^{1+\alpha}) = s + O(|s-t|^{1+\alpha}),$$
$$v(t) = t + O(|s-t|^{1+\alpha}),$$

and

$$v'(s) - v'(0) = v'(s) - 1 = O(|s-t|^\alpha).$$

The above gives

$$k(s,t) = \frac{(1 + O(|s-t|^\alpha))(s-t) - (s-t) + O(|s-t|^{1+\alpha})}{(s-t) + O(|s-t|)^{1+\alpha}}$$

and this is just (4.63). The assertion is proved. ∎

Resuming the proof of Proposition 4.3.2 we will now show that, for the special cases $k=1$ and $k=2$, Lemma 4.3.3 implies (4.58). Indeed, if $k=1$, then we multiply (4.60) from both sides by $\tilde{\chi}_j I$ to obtain the compactness of $S^v_{\Gamma_j} - S_{e^{i\beta_j}\mathbb{R}^+}$. If $k=2$ and Γ_i and Γ_j are directed as in Lemma 4.3.3 then there is nothing to prove. Suppose they are not. For instance, let both Γ_i and Γ_j be directed away from z. In this case consider a new curve $\tilde{\Gamma}$ which is obtained from $\Gamma_i \cup \Gamma_j$ by changing the orientation of Γ_j. Define analogously the curve $\tilde{\Gamma}_z$. Then, obviously,

$$S_{\Gamma_i \cup \Gamma_j} = -S_{\tilde{\Gamma}}\chi_i I + S_{\tilde{\Gamma}}\chi_j I.$$

Thus, there exist compact operators K_1 and K_2 such that

$$S_{\Gamma_i \cup \Gamma_j}^{\upsilon} = (-S_{\tilde{\Gamma}}\chi_i I + S_{\tilde{\Gamma}}\chi_j I)^{\upsilon} = -S_{\tilde{\Gamma}}^{\upsilon}\tilde{\chi}_i I + S_{\tilde{\Gamma}}^{\upsilon}\tilde{\chi}_j I$$
$$= -S_{\tilde{\Gamma}_z}\tilde{\chi}_i I + S_{\tilde{\Gamma}_z}\tilde{\chi}_j I + K_1 = S_{(e^{i\beta_i}\mathbb{R}_+ \cup e^{i\beta_j}\mathbb{R}_+) \cap \mathcal{U}'} + K_2, \qquad (4.64)$$

which gives (4.58) and, consequently, our claim.

To finish the third step we have to check whether Υ induces an isometry between the corresponding local algebras. Let $f \in L^p(\Gamma \cap \mathcal{U}, w_{z|_{\Gamma \cap \mathcal{U}}})$. Then

$$\|\Upsilon f\|^p = \int_{\Gamma_z \cap \mathcal{U}'} |f(\upsilon(t))|^p w_z^p(t) \, |dt|$$
$$= \int_{\Gamma \cap \mathcal{U}} |f(s)|^p w_z^p(\upsilon^{-1}(s)) \left| \frac{d\upsilon^{-1}(s)}{ds} \right| \, |ds|$$
$$\leq \sup_{s \in \Gamma \cap \mathcal{U}} \frac{w_z^p(\upsilon^{-1}(s))}{w_z^p(s)} \left| \frac{d\upsilon^{-1}(s)}{ds} \right| \, \|f\|^p. \qquad (4.65)$$

By hypotheses (i)–(iii), the function

$$h(s) := \frac{w_z^p(\upsilon^{-1}(s))}{w_z^p(s)} \left| \frac{d\upsilon^{-1}(s)}{ds} \right| \qquad (4.66)$$

is continuous on $\Gamma \cap \mathcal{U}$, and $h(z) = 1$. Hence, given $\varepsilon > 0$, we find a sufficiently small neighborhood $\mathcal{V}$ of z such that $h(s) \leq 1 + \varepsilon$ for $s \in \mathcal{V}$. Analogously, for all $u \in L^p(\Gamma_z \cap \mathcal{U}', w_{z|_{\Gamma_z \cap \mathcal{U}'}})$ and for all continuous functions g with $g(0) = 1$ and with sufficiently small support, we have

$$\|\upsilon^{-1}(gu)\| \leq \|gu\|(1 + \varepsilon)^{1/p}.$$

Now let $A \in \mathscr{A}(\Gamma, w)$, let K be a compact operator, and choose f as in Step 1 and g as above. Since

$$\|\Phi_0(A^{\upsilon})\| = \|\Phi_0(\Upsilon(A + K)\Upsilon^{-1})\| = \|\Phi_0(\Upsilon f(A + K)\Upsilon^{-1}gI)\|$$

we obtain

$$\|\Phi_0(A^{\upsilon})\| \leq \|\Upsilon f(A + K)\Upsilon^{-1}gI\|$$
$$= \sup_{\|u\|=1} \|\Upsilon f(A + K)\Upsilon^{-1}gu\|$$
$$\leq \sup_{\|u\|=1} \|f(A + K)\Upsilon^{-1}gu\|(1 + \varepsilon)^{1/p}$$
$$\leq \sup_{\|u\|=1} \|f(A + K)\|\|\Upsilon^{-1}gu\|(1 + \varepsilon)^{1/p}$$
$$\leq \|g_1(A + K)\|(1 + \varepsilon)^{2/p}.$$

Since

$$\|\Phi_0(A)\| = \inf\{\|\Phi(gA)\| : g \in C(\dot{\Gamma}) \text{ with } g(0) = 1\},$$

by Proposition 2.2.4, we conclude that

$$\|\Phi_0(A^\upsilon)\| \leq \|\Phi_z(A)\|(1+\varepsilon)^{2/p}.$$

Letting ε go to zero, we finally obtain $\|\Phi_0(A^\upsilon)\| \leq \|\Phi_z(A)\|$. The reverse inequality can be verified analogously.

Step 4 $\mathscr{A}_4 \cong \mathscr{A}_5$: This proof follows the same lines as the proof of the first step.

Finally we have to examine the case when $z = \infty$. Via the mapping $z \mapsto (z - z_0)^{-1}$ with $z_0 \notin \Gamma \cap \Gamma_z$ and the transformation defined in (4.7), we trace back this case to the situation when $z = 0$, which has already been treated. This finishes the proof of Proposition 4.3.2. ∎

For $\Gamma_z = \cup_{j=1}^{k(z)} e^{\mathrm{i}\beta_j} \mathbb{R}^+$ as above, let χ_z denote the set of characteristic functions of the half lines $e^{\mathrm{i}\beta_j} \mathbb{R}^+$, $j = 1, \ldots, k(z)$.

Proposition 4.3.4. *The local algebras $\mathscr{A}_0^{\mathscr{K}}(\Gamma_z, w_z)$ and $\mathscr{A}_\infty^{\mathscr{K}}(\Gamma_z, w_z)$ are both isomorphic to the algebra $\mathscr{A}(\chi_z, S_{\Gamma_z}, \alpha(z))$ defined in Section 4.2.4.*

Proof. Let $A \in \mathscr{A}(\Gamma_z, w_z)$. Then the strong limit $\mathsf{H}_0(A)$ exists, and it depends only on the coset $\Phi(A)$ by Proposition 4.2.22 (iii). Moreover, by Proposition 4.2.22 (ii), $\mathsf{H}_0(fI) = 0$ for each continuous function f on Γ_z with $f(0) = 0$. This shows that the operator $\mathsf{H}_0(A)$ only depends on the coset $\Phi_0(A)$. We denote the resulting quotient mapping $\Phi_0(A) \mapsto \mathsf{H}_0(A)$ by H_0 again. Clearly, the new H_0 is a homomorphism from $\mathscr{A}_0^{\mathscr{K}}(\Gamma_z, w_z)$ onto the algebra $\mathscr{A}(\chi_z, S_{\Gamma_z}, \alpha(z))$. We are going to show that H_0 is even an isometry, i.e., that

$$\|\Phi_0(A)\| = \|\mathsf{H}_0(A)\| \tag{4.67}$$

for $A \in \mathscr{A}(\Gamma_z, w_z)$. Taking into account the values of the norms of the operators $Z_\tau^\pm$, we conclude via the Banach-Steinhaus theorem that $\|\mathsf{H}_0(A)\| \leq \|A + K\|$ for all operators K with $\Phi_0(K) = 0$. Hence,

$$\|\mathsf{H}_0(A)\| \leq \inf_K \|A + K\| = \|\Phi_0(A)\|.$$

On the other hand, it is easy to see that $\Phi_0(\mathsf{H}_0(A)) = \Phi_0(A)$. So we have $\|\Phi_0(A)\| \leq \|\mathsf{H}_0(A)\|$, which gives equality (4.67). The proof for the algebra $\mathscr{A}_\infty^{\mathscr{K}}(\Gamma_z, w_z)$ is analogous. ∎

We are now in a position to formulate and prove the main theorem of this section.

Theorem 4.3.5. *Let Γ be an admissible curve and $w \in \mathcal{A}_p(\Gamma)$ a power weight of the form* (4.1). *Then:*

(i) *for each $z \in \dot{\Gamma}$, there is a homomorphism $A \mapsto \mathrm{smb}(A, z)$ from the algebra $\mathcal{A}(PC(\dot{\Gamma}), S_\Gamma, w)$ onto $\mathcal{E}_{p,\alpha(z)}^{k(z),\mathcal{N}}$. In particular,*

$$\mathrm{smb}(S_\Gamma, z) = \begin{bmatrix} S_{\mathbb{R}^+} & H_{2\pi+\beta_1-\beta_2} & \cdots & H_{2\pi+\beta_1-\beta_k} \\ H_{\beta_2-\beta_1} & S_{\mathbb{R}^+} & \cdots & H_{2\pi+\beta_2-\beta_k} \\ \vdots & \vdots & \ddots & \vdots \\ H_{\beta_k-\beta_1} & H_{\beta_k-\beta_2} & \cdots & S_{\mathbb{R}^+} \end{bmatrix} \mathrm{diag}_{1 \le j \le k}(v_j) \quad (4.68)$$

where $k := k(z)$ and $v_j := v_j(z)$, and

$$\mathrm{smb}(aI, z) = \mathrm{diag}_{1 \le j \le k(z)}(a_j(z)I) \quad (4.69)$$

for $a \in PC(\dot{\Gamma})$, with

$$a_j(z) := \lim_{\substack{t \to z \\ t \in \dot{\Gamma}_j}} a(t); \quad (4.70)$$

(ii) *the coset of an operator $A \in \mathcal{A}(PC(\dot{\Gamma}), S_\Gamma, w)$ modulo compact operators is invertible in $\mathcal{A}^{\mathcal{K}}(PC(\dot{\Gamma}), S_\Gamma, w)$ if and only if the operators $\mathrm{smb}(A, z)$ are invertible for all $z \in \dot{\Gamma}$;*

(iii) *the algebra $\mathcal{A}^{\mathcal{K}}(PC(\dot{\Gamma}), S_\Gamma, w)$ is inverse-closed in the Calkin algebra of $L^p(\Gamma, w)$, and the algebra $\mathcal{A}(PC(\dot{\Gamma}), S_\Gamma, w)$ is inverse-closed in $\mathscr{L}(L^p(\Gamma, w))$; thus, an operator $A \in \mathcal{A}(PC(\dot{\Gamma}), S_\Gamma, w)$ is Fredholm if and only if $\mathrm{smb}(A, z)$ is invertible for all $z \in \dot{\Gamma}$, and A possesses a regularizer in $\mathcal{A}(PC(\dot{\Gamma}), S_\Gamma, w)$ in this case;*

(iv) *for $A \in \mathcal{A}(PC(\dot{\Gamma}), S_\Gamma, w)$, one has $\|\Phi(A)\| = \sup_{z \in \dot{\Gamma}} \|\mathrm{smb}(A, z)\|$;*

(v) *the radicals of $\mathcal{A}^{\mathcal{K}}(PC(\dot{\Gamma}), S_\Gamma, w)$ and $\mathcal{A}^{\mathcal{K}}(C(\dot{\Gamma}), S_\Gamma, w)$ are trivial. In particular, if $A \in \mathcal{A}(PC(\dot{\Gamma}), S_\Gamma, w)$ and $\mathrm{smb}(A, z) = 0$ for all $z \in \dot{\Gamma}$, then A is compact;*

(vi) *the collection $\mathscr{F}(\Gamma, w)$ of all functions $\{B_z\}_{z \in \dot{\Gamma}}$ such that $B_z \in \mathcal{E}_{p,\alpha(z)}^{k(z),\mathcal{N}}$ and $\{B_z\}$ is upper semi-continuous with respect to $\mathcal{A}^{\mathcal{K}}(PC(\dot{\Gamma}), S_\Gamma, w)$, forms a Banach algebra under the norm $\|\{B_z\}\| := \sup_z \|B_z\|$, and the mapping which associates with $\Phi(A)$ the function $z \mapsto \{\mathrm{smb}(A, z)\}$ is an isometric isomorphism from the algebra $\mathcal{A}^{\mathcal{K}}(PC(\dot{\Gamma}), S_\Gamma, w)$ onto $\mathscr{F}(\Gamma, w)$.*

Proof. (i) For $z \in \dot{\Gamma}$ and $A \in \mathcal{A}(PC(\dot{\Gamma}), S_\Gamma, w)$, put $\mathrm{smb}(A, z) := (\eta H_0 \Upsilon \Phi_z)(A)$. The particular form of $\mathrm{smb}(S_\Gamma, w)$ and $\mathrm{smb}(a, z)$ results from Propositions 4.2.22, 4.3.2 and 4.3.4. The surjectivity of the mapping

$$\mathcal{A}(PC(\dot{\Gamma}), S_\Gamma, w) \to \mathcal{E}_{p,\alpha(z)}^{k(z),\mathcal{N}}, \quad A \mapsto \mathrm{smb}(A, z)$$

was established in Proposition 4.2.19.

(ii) By Allan's local principle, the coset of an operator $A \in \mathcal{A}(PC(\dot{\Gamma}), S_\Gamma, w)$ modulo compact operators is invertible in $\mathscr{A}^{\mathscr{K}}(PC(\dot{\Gamma}), S_\Gamma, w)$ if and only if $\Phi_z(A)$ is invertible in the corresponding local algebra $\mathscr{A}_z^{\mathscr{K}}(PC(\dot{\Gamma}), S_\Gamma, w)$ for each $z \in \dot{\Gamma}$. By Propositions 4.2.22, 4.3.2 and 4.3.4, this is equivalent to the invertibility of $\mathrm{smb}(A, z)$ in $\mathscr{E}_{p,\alpha(z)}^{k(z),\mathscr{N}}$. It remains to show that an operator $B \in \mathscr{E}_{p,\alpha}^{k,\mathscr{N}}$ is invertible in $\mathscr{E}_{p,\alpha}^{k,\mathscr{N}}$ if and only if it is invertible in $\mathscr{L}(L_k^p(\mathbb{R}^+, |t|^\alpha))$. This fact follows from Corollary 1.2.32. Here is another argument: Write $B \in \mathscr{E}_{p,\alpha}^{k,\mathscr{N}}$ as a matrix operator $[B_{ij}]_{i,j=1}^k$ with entries $B_{ij} \in \mathscr{E}_{p,\alpha}$. Since $\mathscr{E}_{p,\alpha}$ is a commutative algebra, the operator B is invertible if and only if $\det[B_{ij}]$ is invertible, and

$$B^{-1} = \frac{1}{\det[B_{ij}]}[C_{ij}]$$

with a certain operator $[C_{ij}] \in \mathscr{E}_{p,\alpha}^{k,\mathscr{N}}$ again. Moreover, since $\det[B_{ij}]$ is a Mellin multiplier, say $M^0(a)$ for some $a \in C_{p,t^\alpha}(\mathbb{R})$, the determinant $\det[B_{ij}]$ is invertible if and only if $a(z) \neq 0$ for all $z \in \mathbb{R}$. But the maximal ideal space of $\mathscr{E}_{p,\alpha} = C_{p,t^\alpha}(\mathbb{R})$ is homeomorphic to $\overline{\mathbb{R}}$ by Proposition 4.2.10, and so $a^{-1} \in C_{p,t^\alpha}(\mathbb{R})$, too. Thus, $(\det[B_{ij}])^{-1} = M^0(a^{-1}) \in \mathscr{E}_{p,\alpha}$.

(iii) The inverse-closedness of the algebra $\mathscr{A}^{\mathscr{K}}(PC(\dot{\Gamma}), S_\Gamma, w)$ in the Calkin algebra of $L^p(\Gamma, w)$ follows again via Corollary 1.2.32.

(iv) By Theorem 2.3.5, the algebra $\mathscr{A}^{\mathscr{K}}(PC(\dot{\Gamma}), S_\Gamma, w)$ forms, together with its subalgebra $\Phi(C(\dot{\Gamma}))$, a faithful localizing pair. Thus, the assertion follows immediately from Theorem 2.3.3 in combination with the fact that the mapping $\Phi_z(A) \mapsto \mathrm{smb}(A, z)$ is an isometry by Propositions 4.2.22, 4.3.2 and 4.3.4.

(v) Let $\Phi(A)$ belong to the radical of $\mathscr{A}^{\mathscr{K}}(PC(\dot{\Gamma}), S_\Gamma, w)$. If $\Phi(A)$ belongs to the part $\cap_{z \in \Gamma} \mathscr{I}_z$ of this radical then, by (iii), we have $\|\Phi(A)\| = 0$. Hence, A is compact. Otherwise, there is a $z \in \dot{\Gamma}$ such that $\Phi_z(A) \neq 0$. This implies that the coset $\Phi_z(A)$ is a non-zero element of the radical of the local algebra $\mathscr{A}_z^{\mathscr{K}}(PC(\dot{\Gamma}), S_\Gamma, w)$. But this is impossible since $\mathscr{A}_z^{\mathscr{K}}(PC(\dot{\Gamma}), S_\Gamma, w)$ is isomorphic to the algebra $\mathscr{E}_{p,\alpha}^{k,\mathscr{N}}$ of multipliers, which is semi-simple.

Assertion (vi) is an immediate consequence of Theorem 2.3.5 and of what we have already proved. ∎

A surprising consequence of Theorem 4.3.5, Proposition 4.2.19 and Corollary 4.2.21 is that the algebra $\mathscr{A}^{\mathscr{K}}(PC(\dot{\Gamma}), S_\Gamma, w)$ only depends on p and on the topological properties of Γ, but neither on the metric properties of Γ nor on the weight.

Corollary 4.3.6. *Let Γ_1 and Γ_2 be admissible curves such that Γ_1 and Γ_2 are homeomorphic, and let w_1 and w_2 denote weight functions of the form* (4.1) *such that $w_1, w_2 \in A_p(\Gamma)$. Then the quotient algebras*

$$\mathscr{A}^{\mathscr{K}}(PC(\dot{\Gamma}_i), S_{\Gamma_i}, w_i) \subset \mathscr{L}(L^p(\Gamma_i, w_i))/\mathscr{K}(L^p(\Gamma_i, w_i))$$

with $i = 1, 2$ are isometrically isomorphic to each other.

4.3.2 *Essential spectrum of the SIO*

We will now apply the results of the preceding section to determine the essential spectrum of the singular integral operator S_Γ on an admissible curve Γ. Allan's local principle reduces this problem to the determination of all local spectra $\sigma(\Phi_z(S_\Gamma))$ where $z \in \Gamma$ if Γ is a bounded curve and where $z \in \dot{\Gamma}$ if Γ is unbounded (with notation as in the previous section). As we know from the proof of Proposition 4.3.2,

$$\sigma_{\mathscr{A}_z^{\mathscr{K}}(\Gamma,w)}(\Phi_z(S_\Gamma)) = \sigma_{\mathscr{A}(\chi_z, S_{\Gamma_z}, \alpha(z))}(S_{\Gamma_z}).$$

Thus, it remains to compute the spectrum of $S_\Gamma \in \mathscr{L}(L^p(\Gamma, |t|^\alpha))$ when $\Gamma = \cup_{j=1}^N \Gamma_j$ with $\Gamma_j = e^{i\beta_j}\mathbb{R}^+$. We will do this in the general context of Section 3.2.8.

Suppose that t_o of the rays Γ_j are directed away from 0 and t_e of these rays are directed towards 0. Thus, $N = t_o + t_e$. We call $v := v(z) := t_o - t_e$ the *valency* of the point z. Further, we write p_j for the operator of multiplication by the characteristic function of Γ_j. Then $p_1, p_2, \ldots, p_N$ are idempotents whose sum is the identity and which satisfy $p_i p_j = \delta_{ij} p_i$. In general, $P_\Gamma := (I + S_\Gamma)/2$ is not an idempotent, so that Theorem 3.2.18 is not applicable immediately. We therefore embed Γ into a larger homogeneous curve Γ^E, where now $S_{\Gamma^E}^2 = I$. For, let

$$\Gamma^E := \cup_{j=1}^{2N} \Gamma_j^E \quad \text{where} \quad \Gamma_j^E = e^{i\gamma_j}\mathbb{R}^+$$

with $0 \leq \gamma_1 < \gamma_2 < \ldots < \gamma_{2N} < 2\pi$ and with Γ_j^E being oriented away from zero if j is even and towards zero if j is odd. The angles γ_j are chosen in such a way that all β_k occur among the γ_j and that Γ_k and Γ_j^E possess the same orientation if $\beta_k = \gamma_j$. The subset of $\{1, 2, \ldots, 2N\}$ which corresponds to the rays of the original curve Γ will be denoted by T, and we abbreviate the operator of multiplication by the characteristic function of Γ_j^E by p_j^E.

The result of this construction is that the operators p_j^E with $j = 1, \ldots, 2N$ and $P_{\Gamma^E} := (I + S_{\Gamma^E})/2$ are idempotents which satisfy the axioms (3.3)–(3.6). Thus, we may use the results of Section 3.2.8 to compute the spectrum of the singular integral operator

$$A := \left(\sum_{j \in T} p_j^E \right) P_{\Gamma^E} \left(\sum_{j \in T} p_j^E \right) + \sum_{j \notin T} p_j^E. \tag{4.71}$$

In order to apply Theorem 3.2.18, we need the spectrum of the operator

$$X^E := \sum_{i=1}^N (p_{2i-1}^E P_{\Gamma^E} p_{2i-1}^E + p_{2i}^E Q_{\Gamma^E} p_{2i}^E)$$

where $Q_{\Gamma^E} := I - P_{\Gamma^E}$.

By Proposition 3.2.6, the spectrum of X^E coincides with the spectrum of the singular projection

$$p_1^E X^E p_1^E = p_1^E P_{\Gamma^E} p_1^E = P_{\Gamma_1^E}$$

which is the circular arc $\mathfrak{A}_{\alpha+1/p}$. More precisely, we know that the spectrum of $P_{\Gamma_1^E}$ in $\mathscr{L}(L^p(\Gamma_1^E, |t|^\alpha))$ coincides with the arc $\mathfrak{A}_{\alpha+1/p}$. But since the complement in $\mathbb{C}$ of this arc is connected, Theorem 1.2.30 implies that the spectrum of $P_{\Gamma_1^E}$ equals $\mathfrak{A}_{\alpha+1/p}$ in every unital closed subalgebra of $\mathscr{L}(L^p(\Gamma_1^E, |t|^\alpha))$ which contains this operator. Since 0 and 1 belong to the circular arc, this also justifies the application of Proposition 3.2.6. Thus, specializing Theorem 3.2.18 to the present context yields the following.

Theorem 4.3.7. *The spectrum of the singular integral operator* (4.71) *on the space* $L^p(\Gamma^E, |t|^\alpha)$ *equals*

$$\{0, 1\} \quad for \quad v = 0, \tag{4.72}$$

$$\bigcup_{k=0}^{v-1} \mathfrak{A}_{(\alpha+1/p+k)/v} \quad for \quad v > 0, \tag{4.73}$$

$$\bigcup_{k=0}^{|v|-1} \mathfrak{A}_{(\alpha+1/p+k)/|v|} \quad for \quad v < 0. \tag{4.74}$$

Now it becomes clear that the spectrum of $P_\Gamma = (I + S_\Gamma)/2$ on $L^p(\Gamma, |t|^\alpha)$ is given by (4.72)–(4.74). Since $\sigma(S_\Gamma) = 2\sigma(P_\Gamma) - 1$, this settles the problem of computing the spectrum of the operator S_Γ of singular integration on $L^p(\Gamma, |t|^\alpha)$ for homogeneous curves Γ and, via the local principle, also the problem of determining the essential spectrum of S_Γ for general admissible curves Γ and L^p-spaces with power weights.

4.3.3 Essential norms

Assertion (iii) of Theorem 4.3.5 offers a way to compute essential norms of singular integral operators in terms of norms of associated homogeneous operators. To illustrate this, we consider a simple closed piecewise Lyapunov curve Γ, denote the angle of Γ at a point $z \in \Gamma$ by $\beta(z)$, put $\Gamma_z := \mathbb{R}^+ \cup e^{\mathrm{i}\beta(z)}\mathbb{R}^+$, and let χ_+ and χ_- refer to the characteristic functions of $\mathbb{R}^+$ and $e^{\mathrm{i}\beta(z)}\mathbb{R}^+$, respectively. Write $\alpha(z)$ for the exponent of the local weight function and, given $a \in PC(\Gamma)$, denote the one-sided limits of a at z by $a_\pm(z)$. Further, we use the notation $w_\alpha(t) := |t|^\alpha$.

Corollary 4.3.8. *Let* $z \in \Gamma$ *with* $0 < 1/p + \alpha(z) < 1$ *and* $a, b \in PC(\Gamma)$. *Then*

$$\|\Phi(aI + bS_\Gamma)\| =$$
$$\sup_{z\in\Gamma} \left\| (a_+(z)\chi_+ + a_-(z)\chi_-)I + (b_+(z)\chi_+ + b_-(z)\chi_-)S_{\Gamma_z} \right\|_{\mathscr{L}(L^p(\Gamma_z, w_{\alpha(z)}))}. \tag{4.75}$$

For the case of continuous coefficients this result was established in [5]. For general p, the determination of the norms of singular integral operators with piecewise constant coefficients (as the ones occurring on the right-hand side of (4.75)) is by no means evident.

The determination of these norms is much easier if $p = 2$. Then $L^2(\mathbb{R}, |t|^\alpha)$ becomes a Hilbert space with respect to the weighted inner product

$$\langle f,g\rangle_\alpha := \int_{\mathbb{R}} |t|^{2\alpha} f(t)\overline{g(t)}\, dt. \tag{4.76}$$

Consequently, the norm of the operator $cI + dS_{\mathbb{R}}$ with constant coefficients is equal to the square root of the spectral radius of $(cI + dS_{\mathbb{R}})(cI + dS_{\mathbb{R}})^*$ where "*" represents the adjoint with respect to the inner product (4.76). In order to keep the formulas readable we agree to write $\cos(x\pi)$ as $\cos x\pi$ (with evident modification for the other trigonometric functions) in this section.

Theorem 4.3.9. *Let* $c,d \in \mathbb{C}$ *and* $\upsilon := 1/2 + \alpha$ *with* $0 < \upsilon < 1$. *Then*

$$\|cI + dS_{\mathbb{R}}\|_{\mathscr{L}(L^2(\mathbb{R}, w_\alpha))} = \frac{1}{\sin \upsilon\pi}\left[|c|^2(1 - \cos^2 \upsilon\pi) + |d|^2(1 + \cos^2 \upsilon\pi) + \right.$$
$$\left. + \sqrt{4|d|^4\cos^2 \upsilon\pi + ((c\overline{d} + \overline{c}d)^2 - (c\overline{d} - \overline{c}d)^2\cos^2 \upsilon\pi)\sin^2 \upsilon\pi} \right]^{1/2}.$$

Proof. It easy to see that the formula holds for $d = 0$. So we can assume without loss of generality that $d = 1$. By Proposition 4.2.19, the norm of $cI + S_{\mathbb{R}}$ is equal to the norm of the matrix operator

$$\begin{bmatrix} cI + S_{\mathbb{R}^+} & -H_\pi \\ H_\pi & cI - S_{\mathbb{R}^+} \end{bmatrix},$$

considered on the Hilbert space $L_2^2(\mathbb{R}^+, w_\alpha)$. Thus,

$$\|cI + S_{\mathbb{R}}\|^2 = \left\| \begin{bmatrix} cI + S_{\mathbb{R}^+} & -H_\pi \\ H_\pi & cI - S_{\mathbb{R}^+} \end{bmatrix} \begin{bmatrix} \overline{c}I + S_{\mathbb{R}^+}^* & H_\pi^* \\ -H_\pi^* & \overline{c}I - S_{\mathbb{R}^+}^* \end{bmatrix} \right\|.$$

We write the operator on the right-hand side of this equality as $[B_{ij}]_{i,j=1}^2$. Since this operator is self-adjoint by construction, its norm is equal to its spectral radius. We determine the spectral radius via the Mellin transform. The Mellin symbols of the B_{ij} are given by

$$\left(M_{2,w_\alpha} B_{11} M_{2,w_\alpha}^{-1} \right)(z)$$
$$= \left(M_{2,w_\alpha}(S_{\mathbb{R}^+} S_{\mathbb{R}^+}^* + \overline{c}S_{\mathbb{R}^+} + cS_{\mathbb{R}^+}^* + |c|^2 + H_\pi H_\pi^*) M_{2,w_\alpha}^{-1} \right)(z)$$
$$= \frac{\cosh^2 z\pi + \cos^2 \upsilon\pi}{\cosh^2 z\pi - \cos^2 \upsilon\pi} + |c|^2 + \overline{c}\,\frac{\cosh(z+i\upsilon)\pi \sinh(z-i\upsilon)\pi}{\cosh^2 z\pi - \cos^2 \upsilon\pi}$$
$$+ \frac{\cosh(z-i\upsilon)\pi \sinh(z+i\upsilon)\pi}{\cosh^2 z\pi - \cos^2 \upsilon\pi};$$

$$\left(M_{2,w_\alpha}B_{12}M_{2,w_\alpha}^{-1}\right)(z) = \left(M_{2,w_\alpha}(S_{\mathbb{R}^+}H_\pi^* + H_\pi S_{\mathbb{R}^+}^* + cH_\pi^* + \overline{c}H_\pi)M_{2,w_\alpha}^{-1}\right)(z)$$

$$= \frac{\cosh(z+i\upsilon)\pi + \cosh(z-i\upsilon)\pi}{\cosh^2 z\pi - \cos^2 \upsilon\pi}$$

$$+ \frac{c\sinh(z+i\upsilon)\pi - \overline{c}\sinh(z-i\upsilon)\pi}{\cosh^2 z\pi - \cos^2 \upsilon\pi};$$

and

$$M_{2,w_\alpha}B_{21}M_{2,w_\alpha}^{-1} = \overline{M_{2,w_\alpha}B_{12}M_{2,w_\alpha}^{-1}};$$

$$M_{2,w_\alpha}B_{22}M_{2,w_\alpha}^{-1} = M_{2,w_\alpha}(S_{\mathbb{R}^+}S_{\mathbb{R}^+}^* - \overline{c}S_{\mathbb{R}^+} - cS_{\mathbb{R}^+}^* + |c|^2 + H_\pi H_\pi^*)M_{2,w_\alpha}^{-1}.$$

Hence, after multiplication by the common divisor $\cosh^2 z\pi - \cos^2 \upsilon\pi$, the eigenvalue equation becomes

$$(\cosh^2 z\pi - \cos^2 \upsilon\pi)\lambda^2$$
$$- 2\lambda(\cosh^2 z\pi - \cos^2 \upsilon\pi)(\cosh^2 z\pi + \cos^2 \upsilon\pi + |c|^2(\cosh^2 z\pi - \cos^2 \upsilon\pi))$$
$$+ (\cosh^2 z\pi + \cos^2 \upsilon\pi + |c|^2(\cosh^2 z\pi - \cos^2 \upsilon\pi))^2$$
$$- (\overline{c}\cosh(z+i\upsilon)\pi\sinh(z-i\upsilon)\pi + c\cosh(z-i\upsilon)\pi\sinh(z+i\upsilon)\pi)^2$$
$$- (\cosh(z+i\upsilon)\pi + \cosh(z-i\upsilon)\pi)^2 + (c\sinh(z+i\upsilon)\pi - \overline{c}\sinh(z-i\upsilon)\pi)^2 = 0.$$

A straightforward calculation yields that the largest eigenvalue $\lambda_{max}(z)$ is given by

$$\lambda_{max}(x) = |c|^2 + \frac{1+x}{1-x} + \frac{2}{1-x}\sqrt{x + (\Re(c)^2 + \Im(c)^2 x)(1-x)} \qquad (4.77)$$

where we have written $x := \cos^2 \upsilon\pi / \cosh^2 z\pi$.

In order to maximize $\lambda_{max}(z)$ with respect to $z \in \mathbb{R}$, we have to maximize $\lambda_{max}(x)$ with respect to $x \in]0, \cos^2 \upsilon\pi] \subseteq]0,1[$. For the function λ_{max} to attain its maximum at an inner point $x \in]0,1[$, it is necessary that the derivative $\lambda'_{max}(x)$ is zero. This happens if and only if

$$(x-1)^2 [(\Re(c)^2 + \Im(c)^2) + 1 + 2(\Re(c)^2 - \Im(c)^2)] = 0.$$

If the term in square brackets is zero, then λ'_{max} is identically zero and $\lambda_{max}(x)$ takes its maximum at every point of $]0,1[$. So let $x = 1$. The function λ_{max} has a pole at $x = 1$. So it is monotonically decreasing or increasing on $[0,1[$. Since

$$\lambda_{max}(0) = 1 + 2|\Re(c)| > 0$$

and

$$\lambda'_{max}(0) = 1 + \frac{1 + \Im(c)^2 - \Re(c)^2}{|\Re(c)|} + 1 + 2|\Re(c)| \geq 2 + \frac{\Im(c)^2}{|\Re(c)|} + |\Re(c)| > 0,$$

the function λ_{max} is increasing. Consequently, $\lambda_{max}(x)$ becomes maximal for maximal x, that is, for $x = \cos^2 \upsilon\pi$, whence $z = 0$. Now set $z = 0$ in (4.77) to get the assertion. ∎

Example 4.3.10. Setting $c = 0$ and $d = 1$ in Theorem 4.3.9 we get

$$\|S_{\mathbb{R}}\|_{\mathscr{L}(L^2(\mathbb{R}, w_\alpha))} = \|\Phi(S_{\mathbb{R}})\| = \frac{1 + |\cos \upsilon\pi|}{\sin \upsilon\pi} = \begin{cases} \cot \frac{\upsilon\pi}{2} & \text{if } 0 < \upsilon \leq \frac{1}{2}, \\ \tan \frac{\upsilon\pi}{2} & \text{if } \frac{1}{2} < \upsilon \leq 1. \end{cases}$$

□

Example 4.3.11. Setting $c = 1/2$ and $d = 1/2$ or $c = 1/2$ and $d = -1/2$ in Theorem 4.3.9 we obtain

$$\|P_{\mathbb{R}}\|_{\mathscr{L}(L^2(\mathbb{R}, w_\alpha))} = \|I - P_{\mathbb{R}}\|_{\mathscr{L}(L^2(\mathbb{R}, w_\alpha))} = \|\Phi(P_{\mathbb{R}})\| = \|\Phi(I - P_{\mathbb{R}})\| = \frac{1}{\sin \upsilon\pi}.$$

□

On curves with corners, the picture is more involved.

Theorem 4.3.12. *The norm of the singular integral operator S_Γ on the angle $\Gamma =: \mathbb{R}^+ \cup e^{i\beta}\mathbb{R}^+$ is*

$$\|S_\Gamma\|_{\mathscr{L}(L^2(\Gamma, w_\alpha))} = \|\Phi(S_\Gamma)\| = \left(\frac{1 + \sqrt{m}}{1 - \sqrt{m}}\right)^{1/2}$$

where

$$m = \sup_{y \in \mathbb{R}} \frac{\cos^2 \upsilon\pi + \sinh^2 y(\beta - \pi)}{\cosh^2 y\pi + \sinh^2 y(\beta - \pi)}. \tag{4.78}$$

In particular, if $\alpha = 0$ and $\beta \neq \pi$, then this supremum is attained at the point y_0 which is the only positive solution of the equation

$$\frac{\cosh \beta y}{\beta} = \frac{\cosh(2\pi - \beta)y}{2\pi - \beta}. \tag{4.79}$$

Proof. As in the proof of the preceding theorem,

$$\|S_\Gamma\| = \left\| \begin{bmatrix} S_{\mathbb{R}^+} & -H_\beta \\ H_{2\pi-\beta} & -S_{\mathbb{R}^+} \end{bmatrix} \begin{bmatrix} S_{\mathbb{R}^+}^* & H_{2\pi-\beta}^* \\ -H_\beta^* & -S_{\mathbb{R}^+}^* \end{bmatrix} \right\|$$

$$= \left\| \begin{bmatrix} S_{\mathbb{R}^+}S_{\mathbb{R}^+}^* + H_\beta H_\beta^* & S_{\mathbb{R}^+}H_{2\pi-\beta}^* + H_\beta S_{\mathbb{R}^+}^* \\ H_{2\pi-\beta}S_{\mathbb{R}^+}^* + S_{\mathbb{R}^+}H_\beta^* & S_{\mathbb{R}^+}S_{\mathbb{R}^+}^* + H_{2\pi-\beta}H_{2\pi-\beta}^* \end{bmatrix} \right\|.$$

We write the latter matrix as $[C_{ij}]_{i,j=1}^2$. For the Mellin symbols of the entries C_{ij} we find

$$(M_{2,w_\alpha} C_{11} M_{2,w_\alpha}^{-1})(y) = \frac{\sinh^2 y\pi + \cos^2 \upsilon\pi + e^{2y(\pi-\beta)}}{\cosh^2 y\pi - \cos \upsilon\pi};$$

$$(M_{2,w_\alpha} C_{12} M_{2,w_\alpha}^{-1})(y) = \frac{\cosh(y+i\upsilon)\pi e^{(\beta-\pi)(y-i\upsilon)} + e^{(\pi-\beta)(y+i\upsilon)}\cosh(y-i\upsilon)\pi}{\cosh^2 \pi y - \cos^2 \upsilon\pi};$$

$$(M_{2,w_\alpha} C_{21} M_{2,w_\alpha}^{-1})(y) = \overline{(M_{2,w_\alpha}(S_{\mathbb{R}^+} H_{2\pi-\beta} + H_\beta S_{\mathbb{R}^+}^*) M_{2,w_\alpha}^{-1})(y)};$$

$$(M_{2,w_\alpha} C_{22} M_{2,w_\alpha}^{-1})(y) = \frac{\sinh^2 \pi y + \cos^2 \upsilon\pi + e^{2y(\beta-\pi)}}{\cosh^2 \pi y - \cos^2 \upsilon\pi}.$$

A straightforward calculation leads to the eigenvalue equation

$$\lambda^2 - 2\lambda \frac{\cosh^2 y\pi - \cos^2 \upsilon y + 2\sinh^2 y(\pi-\beta)}{\cosh^2 y\pi - \cos^2 \upsilon\pi} + 1 = 0$$

which implies that the largest eigenvalue λ is given by

$$\sup_{y \in \mathbb{R}} \frac{\left(\sqrt{\cosh^2 y\pi + \sinh^2 y(\pi-\beta)} + \sqrt{\cos^2 \upsilon\pi + \sinh^2 y(\pi-\beta)}\right)^2}{\cosh^2 y\pi - \cos^2 \upsilon\pi}$$

$$= \sup_{y \in \mathbb{R}} \frac{\sqrt{\cosh^2 y\pi + \sinh^2 y(\pi-\beta)} + \sqrt{\cos^2 \upsilon\pi + \sinh^2 y(\pi-\beta)}}{\sqrt{\cosh^2 y\pi + \sinh^2 y(\pi-\beta)} - \sqrt{\cos^2 \upsilon\pi + \sinh^2 y(\pi-\beta)}} = \frac{1+\sqrt{m}}{1-\sqrt{m}}.$$

Now let $\upsilon = 1/2$. For the function

$$f(y) := \frac{\sinh^2 y(\pi-\beta)}{\cosh^2 y\pi + \sinh^2 y(\pi-\beta)} = \frac{1}{\frac{\cosh^2 y\pi}{\sinh^2 y(\pi-\beta)} + 1}$$

to become maximal it is necessary that the function

$$g(y) := \frac{\sinh y(\pi-\beta)}{\cosh y\pi}, \quad y \geq 0,$$

becomes maximal. The equation $g'(y) = 0$ is equivalent to (4.79). ∎

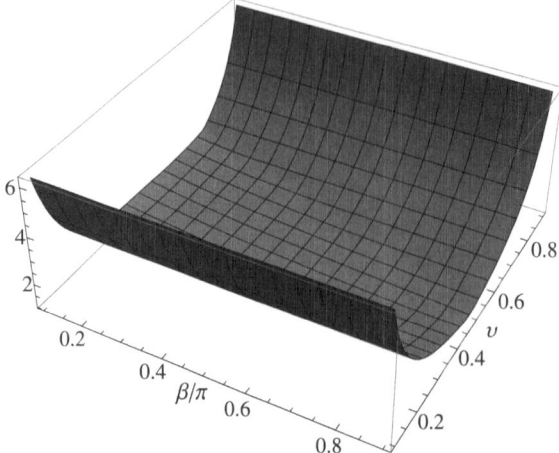

Fig. 4.6 The norm of S_Γ on the angle $\Gamma = \mathbb{R}^+ \cup e^{i\beta}\mathbb{R}^+$, as a function of β and υ. A cut-section of this figure for $\upsilon = 1/2$ is given below.

In [130], Nyaga took the trouble to compute the norms of the singular integral operator S_Γ on $\Gamma = \mathbb{R}^+ \cup e^{i\beta}\mathbb{R}^+$ for some special values of β. Here is his result.

$$\|S_\Gamma\|_{\mathscr{L}(L^2(\Gamma))} = \|\Phi(S_\Gamma)\| = \begin{cases} \frac{1+\sqrt{5}}{2} & \text{if } \beta = \frac{\pi}{3}, \\ \sqrt{2} & \text{if } \beta = \frac{\pi}{2}, \\ \frac{\sqrt{3+4\sqrt{3}+2\sqrt{6(\sqrt{3}-1)}}}{3} & \text{if } \beta = \frac{2\pi}{3}. \end{cases}$$

Compare also Figure 4.7 which shows the norms of S_Γ calculated numerically using computer software.

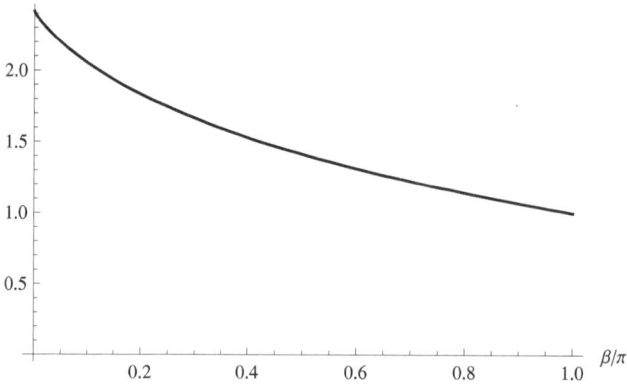

Fig. 4.7 The norm of S_Γ on the angle $\Gamma = \mathbb{R}^+ \cup e^{i\beta}\mathbb{R}^+$, for $\upsilon = 1/2$, as a function of β.

Another simple consequence of Theorem 4.3.12 which we shall need later on is the following.

Corollary 4.3.13. *For S_Γ in $\mathscr{L}\left(L^2(\Gamma, w_\alpha)\right)$, the essential norm $\|\Phi(S_\Gamma)\|$ is greater than or equal to 1. This norm is equal to 1 if and only if $\alpha = 0$ and $\beta = \pi$.*

Proof. Evidently, the norm cannot be smaller than 1. It is equal to 1 if and only if $m = 0$. But $m = 0$ holds if and only if $\alpha = 0$ and $\beta = \pi$. ∎

The corresponding result for the singular integral operator on the half axis reads as follows.

Theorem 4.3.14. *Let $\upsilon = 1/2 + \alpha$. Then*

$$\|S_{\mathbb{R}^+}\|_{\mathscr{L}(L^2(\mathbb{R}^+, w_\alpha))} = \|\Phi(S_{\mathbb{R}^+})\| = \begin{cases} 1 & \text{if } \upsilon \in \left[\frac{1}{4}, \frac{3}{4}\right], \\ |\cot \upsilon \pi| & \text{if } \upsilon \in \left]0, \frac{1}{4}\right[\cup \left]\frac{3}{4}, 1\right[. \end{cases}$$

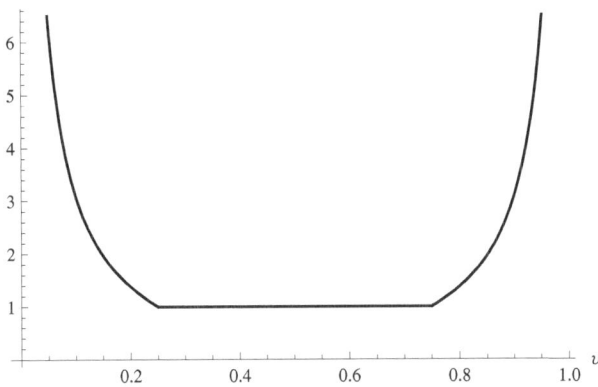

Fig. 4.8 The norm of $S_{\mathbb{R}^+}$, as a function of υ.

Proof. As above,

$$\|S_{\mathbb{R}^+}\|^2 = \sup_{y \in \mathbb{R}} \left| \frac{\cosh^2(y + i\upsilon)\pi}{\sinh^2(y + i\upsilon)\pi} \right| = \sup_{y \in \mathbb{R}} \frac{\cosh^2 y\pi - \sin^2 \upsilon \pi}{\cosh^2 y\pi - \cos^2 \upsilon \pi}$$

$$= 1 + \sup_{y \in \mathbb{R}} \frac{2\cos^2 \upsilon \pi - 1}{\cosh^2 y\pi - \cos^2 \upsilon \pi}.$$

If $2\cos^2 \upsilon \pi - 1 > 0$ or, equivalently, if $\upsilon \in \left]0, \frac{1}{4}\right[\cup \left]\frac{3}{4}, 1\right[$, then

$$\|S_{\mathbb{R}^+}\| = \sqrt{\frac{\cos^2 \upsilon \pi}{1 - \cos^2 \upsilon \pi}} = |\cot \upsilon \pi|.$$

If $2\cos^2 \upsilon \pi - 1 \le 0$ or, equivalently, if $\upsilon \in \left[\frac{1}{4}, \frac{3}{4}\right]$, then the supremum is reached for maximal $\cosh^2 y\pi$, that is, when $y \to +\infty$. In this case, $\|S_{\mathbb{R}^+}\| = 1$. ∎

4.4 Singular integral operators with Carleman shift changing orientation

Throughout this section, let Γ denote a simple, closed, bounded, admissible curve with $t : [0, L] \to \Gamma$ as its parameter representation, the arc length. A *Carleman shift* on Γ is a homeomorphism $\mu : \Gamma \to \Gamma$ such that μ^m is the identical mapping for some integer $m > 1$. A Carleman shift either changes or preserves the orientation of Γ. Singular equations with a Carleman shift preserving the orientation can be reduced to a system of singular integral equations without shift by a simple and standard procedure, see [95, 104, 114] for these and other facts. We shall therefore only treat shifts which do not preserve the orientation.

We shall consider shifts μ which satisfy the following conditions:

(i) μ changes the orientation of Γ;
(ii) the mapping $J_\mu : f \mapsto f \circ \mu$ defines a bounded linear operator on $L^p(\Gamma, w)$;
(iii) the derivative $\mu'(t) := \frac{d\mu(t(s))}{ds} \frac{1}{t'(s)}$ is piecewise Hölder continuous, and $\mu'(t) \ne 0$ for all $t \in \Gamma$.

Condition (i) implies that μ has exactly two fixed points, say t_0 and t_1, and that $\mu^2 = I$. Indeed, consider the functions t and $\mu \circ t$. Because of the change of orientation, there exist points $s_0, s_1 \in \Gamma$ such that $t(s_0) = \mu(t(s_0)) =: t_0$ and $t(s_1) = \mu(t(s_1)) =: t_1$. To get that $\mu^2 = I$, just note that the Carleman shift μ^2 preserves the orientation and has fixed points. So it must be the identity.

Let Γ_+ and Γ_- refer to the arcs joining t_0 with t_1 and t_1 with t_0, respectively, i.e., Γ_+ consists of all points of Γ which lie after t_0 and before t_1 with respect to the given orientation of Γ. Condition (ii) implies that, for each $z_0 \in \Gamma_+ \setminus \{t_0, t_1\}$, the weight function w is of the form

$$w(z) = |z - z_0|^{\alpha(z_0)} |\mu(z) - \mu(z_0)|^{\alpha(z_0)} w_1(z) \tag{4.80}$$

with w_1 being a function which is continuous at z_0 and such that $w_1(z_0) \ne 0$.

We let $\mathscr{B}(\Gamma, w, \mu)$ refer to the smallest closed subalgebra of $\mathscr{L}(L^p(\Gamma, w))$ which contains the operators S_Γ and J_μ and all operators of multiplication by functions in $PC(\Gamma)$. By Theorem 4.1.5, $\mathscr{K}(L^p(\Gamma, w)) \subseteq \mathscr{B}(\Gamma, w, \mu)$. So we can consider the quotient algebra $\mathscr{B}^{\mathscr{K}}(\Gamma, w, \mu) := \mathscr{B}(\Gamma, w, \mu) / \mathscr{K}(L^p(\Gamma, w))$. The canonical homomorphism from $\mathscr{B}(\Gamma, w, \mu)$ onto this quotient algebra will be denoted by Φ.

Given $z \in \Gamma_+$, write $\alpha(z)$ for the local exponent defined in (4.55), and let $\beta(z)$ stand for the angle of Γ at z as defined above. Further, for $a \in PC(\Gamma)$ we set

$$a_{\pm}(z) := \lim_{\substack{t \to z \\ t \gtrless z}} a(t). \qquad (4.81)$$

Finally, let $\overset{\circ 4}{\mathscr{E}}_{p,\alpha}$ stand for the subset of $\mathscr{E}^4_{p,\alpha}$ which consists of all matrix operators $[A_{ij}]^4_{i,j=1}$ of the form

$$\begin{bmatrix} * & \circ & \circ & * \\ \circ & * & * & \circ \\ \circ & * & * & \circ \\ * & \circ & \circ & * \end{bmatrix}$$

where $*$ symbolizes elements from $\mathscr{E}_{p,\alpha}$ and $\circ$ refers to elements from $\mathscr{N}_{p,\alpha}$. One can check, as in the proof of Theorem 4.3.5, that $\overset{\circ 4}{\mathscr{E}}_{p,\alpha}$ is a closed subalgebra of $\mathscr{E}^4_{p,\alpha}$ and that $\overset{\circ 4}{\mathscr{E}}_{p,\alpha}$ is inverse-closed in $\mathscr{L}(L^p_4(\mathbb{R}^+, t^\alpha))$.

Our goal is an analog of Theorem 4.3.5 for singular integral operators with shifts. We prepare this theorem with some auxiliary results. Let $C_\mu(\Gamma)$ denote the Banach algebra of all continuous functions f on Γ for which $f(t) = f(\mu(t))$ for all $t \in \Gamma$. From Theorem 4.1.4 and from

$$J_\mu f J_\mu = f \quad \text{for} \quad f \in C_\mu(\Gamma), \qquad (4.82)$$

we deduce that the algebra $\Phi(C_\mu(\Gamma))$ lies in the center of $\mathscr{B}^{\mathscr{K}}(\Gamma, w, \mu)$, a fact which allows us to use Allan's local principle. Moreover, one can easily show that the algebra $\mathscr{B}^{\mathscr{K}}(\Gamma, w, \mu)$ and its subalgebra $\Phi(C_\mu(\Gamma))$ form a faithful localizing pair.

We localize the algebra $\mathscr{B}^{\mathscr{K}}(\Gamma, w, \mu)$ over the maximal ideal space of $\Phi(C_\mu(\Gamma))$, which is homeomorphic to Γ_+. The localization at the fixed points $t = t_0$ and $t = t_1$ runs parallel to the case without shift. But at non-fixed points $t \in \Gamma_+ \setminus \{t_0, t_1\}$, we observe that every function $f \in C_\mu(\Gamma)$ which vanishes at t must vanish at $\mu(t)$, too. In that sense, we have a "double-point localization" at the non-fixed points, which causes some differences between the treatment of the local algebras at fixed points and at non-fixed points.

Given a point $z \in \Gamma_+$, we denote the local algebra at z by $\mathscr{B}^{\mathscr{K}}_z(\Gamma, w)$ and the canonical homomorphism from $\mathscr{B}(\Gamma, w, \mu)$ onto $\mathscr{B}^{\mathscr{K}}_z(\Gamma, w)$ by Φ_z. If z is a fixed point of μ, then the local weight w_z and the curve Γ_z are defined as in Section 4.3. In that case, Γ_z consists of two half axes having the same angle $\beta(z)$ between as the tangents of Γ at z. Since Γ is closed and oriented, one half axis must be oriented to 0 and the other away from 0. On Γ_z we consider the Carleman shift μ_z which sends a point $t \in \Gamma_z, t \neq 0$, to the point $\mu_z(t)$ which is located at the other half axis and has the same distance to 0 as t. Thus, μ_z has 0 and ∞ as its fixed points.

Proposition 4.4.1. *Let z be a fixed point of the Carleman shift μ. Then the local algebras $\mathscr{B}^{\mathscr{K}}_z(\Gamma, w, \mu)$ and $\mathscr{B}^{\mathscr{K}}_0(\Gamma_z, w_z, \mu_z)$ are isometrically isomorphic. The isomorphism can be chosen as in Proposition 4.3.2 where, additionally, $\Phi_z(J_\mu)$ will be carried over into $\Phi_0(J_{\mu_z})$.*

Proof. The proof proceeds as that of Proposition 4.3.2. Let $\mathcal{U}$ be a sufficiently small neighborhood of z such that $\mu(\mathcal{U} \cap \Gamma) = \mathcal{U} \cap \Gamma$. Then the local algebras

- $\mathcal{B}_1 := \mathcal{B}_z^{\mathcal{K}}(\Gamma, w, \mu)$;
- $\mathcal{B}_2 := \mathcal{B}_z^{\mathcal{K}}(\Gamma \cap \mathcal{U}, w|_{\Gamma \cap \mathcal{U}}, \mu|_{\Gamma \cap \mathcal{U}})$;
- $\mathcal{B}_3 := \mathcal{B}_z^{\mathcal{K}}(\Gamma \cap \mathcal{U}, w_z|_{\Gamma \cap \mathcal{U}}, \mu|_{\Gamma \cap \mathcal{U}})$;
- $\mathcal{B}_4 := \mathcal{B}_0^{\mathcal{K}}(\Gamma_z \cap \mathcal{U}', w_z|_{\Gamma_z \cap \mathcal{U}'}, \mu_z|_{\Gamma_z \cap \mathcal{U}'})$;
- $\mathcal{B}_5 := \mathcal{B}_0^{\mathcal{K}}(\Gamma_z, w_z, \mu_z)$

are isometrically isomorphic to each other. Indeed, the proofs of the first and last step are the same as in the proof of Proposition 4.3.2. In the second step some care is in order since the operator $TJ_\mu T^{-1} - J_\mu$ does not need to be compact. But since

$$TJ_\mu T^{-1} = \frac{\vartheta}{\mu \circ \vartheta} J_\mu$$

with the continuous function $\vartheta(\mu \circ \vartheta)^{-1}$ taking the value 1 at z, we get the isomorphy between $\mathcal{B}_2$ and $\mathcal{B}_3$ in the shift case, too. For a proof of $\mathcal{B}_3 \cong \mathcal{B}_4$ (= the third step), we have to choose a homeomorphism υ which fulfills conditions (i)–(iii) in the proof of Proposition 4.3.2 and which, moreover, satisfies the condition

(iv) $\mu\upsilon = \upsilon\mu$.

To get such a function, define υ on one of the half axes of Γ_z so that (i)–(iii) hold. Then set $\upsilon(t) := (\mu\upsilon\mu_z)(t)$ for all points t belonging to the other half axis. This function has the desired properties. The remainder of the proof runs as that of Proposition 4.3.2.   ∎

In order to represent the local algebra $\mathcal{B}_0^{\mathcal{K}}(\Gamma_z, w_z, \mu_z)$ as an algebra of homogeneous operators, define the mapping H_0 as in Section 4.2.4 and let $\mathcal{D}(\Gamma_z, w_z)$ refer to the smallest closed subalgebra of $\mathcal{L}(L^p(\Gamma_z, w_z))$ which contains the operators S_{Γ_z}, J_{μ_z} and the operators of multiplication by the characteristic functions of the two half axes which constitute Γ_z. Since $\mathsf{H}_0(J_{\mu_z}) = J_{\mu_z}$, the following can be proved in the same way as Proposition 4.3.4.

Proposition 4.4.2. *The mapping H_0 is an isometric isomorphism between the local algebra $\mathcal{B}_0^{\mathcal{K}}(\Gamma_z, w_z, \mu_z)$ and the algebra $\mathcal{D}(\Gamma_z, w_z)$.*

Now we turn our attention to the localization at non-fixed points of μ. Let $z \in \Gamma_+$ be a non-fixed point. Set $\tilde{z} := \mu(z)$, and denote the angle, with z (resp. with $\tilde{z}$) as its vertex and with the tangents of Γ at z (resp. at $\tilde{z}$) as its legs, by Γ_z' (resp. by $\Gamma_{\tilde{z}}'$). Choose real numbers φ and $\tilde{\varphi}$ such that the curves

$$\Gamma_z := e^{\mathrm{i}\varphi}(\Gamma_z' - z) + z \quad \text{and} \quad \Gamma_{\tilde{z}} := e^{\mathrm{i}\tilde{\varphi}}(\Gamma_{\tilde{z}}' - \tilde{z}) + \tilde{z},$$

obtained from Γ'_z and $\Gamma'_{\tilde{z}}$ by rotation, are disjoint. Further, let $\beta_1, \ldots, \beta_4 \in [0, 2\pi[$ be defined by (see Figure 4.9)

$$\Gamma_z = (e^{\mathrm{i}\beta_1}\mathbb{R}^+ \cup e^{\mathrm{i}\beta_2}\mathbb{R}^+) + z,$$
$$\Gamma_{\tilde{z}} = (e^{\mathrm{i}\beta_3}\mathbb{R}^+ \cup e^{\mathrm{i}\beta_4}\mathbb{R}^+) + \tilde{z}.$$

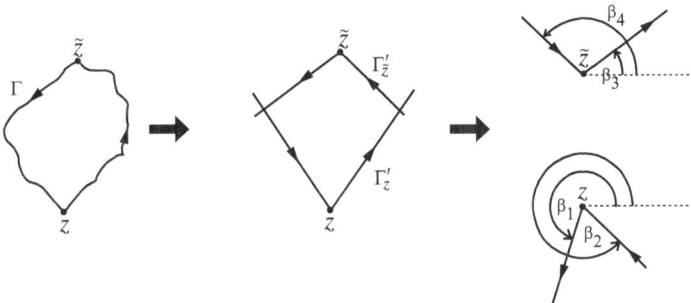

Fig. 4.9 The construction of Γ_z and $\Gamma_{\tilde{z}}$ from the curve Γ.

We define the Carleman shift $\mu_z : \Gamma_z \cup \Gamma_{\tilde{z}} \to \Gamma_z \cup \Gamma_{\tilde{z}}$ by

$$\mu_z(z + e^{\mathrm{i}\beta_j}t) = \tilde{z} + e^{\mathrm{i}\beta_{5-j}}t \quad \text{if } j = 1, 2 \tag{4.83}$$

and

$$\mu_z(\tilde{z} + e^{\mathrm{i}\beta_j}t) = z + e^{\mathrm{i}\beta_{5-j}}t \quad \text{if } j = 3, 4 \tag{4.84}$$

where $t \in \mathbb{R}^+$ and define the operator J_{μ_z} by $(J_{\mu_z}u)(t) := u(\mu_z(t))$ as above. Finally, denote the characteristic functions of the angles Γ_z and $\Gamma_{\tilde{z}}$ by χ_z and $\chi_{\tilde{z}}$ respectively, and put $w_z(t) = |t - z|^{\alpha(z)}$ and $w_{\tilde{z}}(t) = |t - \tilde{z}|^{\alpha(\tilde{z})}$.

Proposition 4.4.3. *Let z be a non-fixed point of the Carleman shift μ. Then the local algebras $\mathscr{B}_z^{\mathscr{K}}(\Gamma, w, \mu)$ and $\mathscr{B}_z^{\mathscr{K}}(\Gamma_z \cup \Gamma_{\tilde{z}}, \chi_z w_z + \chi_{\tilde{z}} w_{\tilde{z}}, \mu_z)$ are isometrically isomorphic. The isomorphism maps $\Phi_z^{\mathscr{K}}(S_\Gamma)$ to $\Phi_z^{\mathscr{K}}(S_{\Gamma_z \cup \Gamma_{\tilde{z}}})$, $\Phi_z^{\mathscr{K}}(a)$ with $a \in PC(\Gamma)$ to $\Phi_z^{\mathscr{K}}(a')$ with a certain $a' \in PC(\Gamma_z \cup \Gamma_{\tilde{z}})$, and $\Phi_z^{\mathscr{K}}(\mu)$ to $\Phi_z^{\mathscr{K}}(\mu_z)$.*

Proof. Let χ'_z and $\chi'_{\tilde{z}}$ denote the characteristic functions of Γ'_z and $\Gamma'_{\tilde{z}}$, respectively, and define numbers $\beta'_1, \ldots, \beta'_4 \in [0, 2\pi[$ by

$$\Gamma'_z = (e^{\mathrm{i}\beta'_1}\mathbb{R}^+ \cup e^{\mathrm{i}\beta'_2}\mathbb{R}^+) + z,$$
$$\Gamma'_{\tilde{z}} = (e^{\mathrm{i}\beta'_3}\mathbb{R}^+ \cup e^{\mathrm{i}\beta'_4}\mathbb{R}^+) + \tilde{z},$$

where we assume that $\beta'_j + \varphi = \beta_j \mod 2\pi$ for $j = 1, 2$ and $\beta'_j + \tilde{\varphi} = \beta_j \mod 2\pi$ for $j = 3, 4$. Further, we let μ'_z denote the Carleman shift defined on $\Gamma'_z \cup \Gamma'_{\tilde{z}}$ in a neighborhood of z and $\tilde{z}$ by

$$\mu_z'(z + e^{i\beta_j'}t) = \bar{z} + e^{i\beta_{5-j}'}t \quad \text{if } j = 1, 2 \tag{4.85}$$

and

$$\mu_z'(\bar{z} + e^{i\beta_j'}t) = z + e^{i\beta_{5-j}'}t \quad \text{if } j = 3, 4 \tag{4.86}$$

where $t \in \mathbb{R}^+$.

Now take a sufficiently small neighborhood $\mathscr{U}_0$ of z (resp. $\mathscr{U}_1$ of z) such that $\Gamma \cap \mathscr{U}_0$ (resp. $(\Gamma_z' \cup \Gamma_{\bar{z}}') \cap \mathscr{U}_1$) is invariant under action of μ (resp. μ_z') and such that $\Gamma \cap \mathscr{U}_0$ (resp. $(\Gamma_z' \cup \Gamma_{\bar{z}}') \cap \mathscr{U}_1$) is not connected but consists of two connected components. Let Γ^j (resp. Γ'^j) be the component of $\Gamma \cap \mathscr{U}_0$ (resp. $(\Gamma_z' \cup \Gamma_{\bar{z}}') \cap \mathscr{U}_1$) which contains the point z if $j = 0$ and the point $\bar{z}$ if $j = 1$, and let χ_j (resp. χ_j') refer to the characteristic function of Γ^j (resp. $\Gamma^{j'}$) with $j = 1, 2$. The same arguments as in the proofs of Propositions 4.3.2 and 4.4.1 show that the algebras

$$\mathscr{D}_1 := \mathscr{B}_z^{\mathscr{K}}(\Gamma, w, \mu) \tag{4.87}$$

and

$$\mathscr{D}_2 := \mathscr{B}_z^{\mathscr{K}}\left(\Gamma \cap \mathscr{U}_0, (\chi_0 w_z + \chi_1 w_{\bar{z}})|_{\Gamma \cap \mathscr{U}_0}, \mu|_{\Gamma \cap \mathscr{U}_0}\right) \tag{4.88}$$

are isometrically isomorphic. In the next step we show that the algebras $\mathscr{D}_2$ and

$$\mathscr{D}_3 := \mathscr{B}_z^{\mathscr{K}}\left((\Gamma_z' \cup \Gamma_{\bar{z}}') \cap \mathscr{U}_1, (\chi_0' w_z + \chi_1' w_{\bar{z}})|_{(\Gamma_z' \cup \Gamma_{\bar{z}}') \cap \mathscr{U}_1}, \mu'|_{(\Gamma_z' \cup \Gamma_{\bar{z}}') \cap \mathscr{U}_1}\right) \tag{4.89}$$

are isometrically isomorphic. Let $\upsilon_0 : \Gamma_z' \to \Gamma^0$ be a homeomorphism which is subject to the conditions:

(i) the restriction of υ_0 onto $e^{i\beta_j}\mathbb{R}^+ \cap \mathscr{U}'$ has a Hölder continuous derivative, and the image of this restriction is $\Gamma_j \cap \mathscr{U}$;

(ii) the function υ' defined at inner points t of $e^{i\beta_j}\mathbb{R}^+ \cap \mathscr{U}'$ by

$$\upsilon'(t) := \frac{1}{t(s)} \frac{d(t(s))}{ds}\Bigg|_{s=te^{-i\beta_j}}$$

can be extended to a continuous function on $\Gamma_z \cap \mathscr{U}'$;

(iii) $\upsilon_0(z) = z$, $\upsilon_0'(z) = 1$;

and put

$$\upsilon_1(t) := (\mu \upsilon_0 \mu_z)(t) \quad \text{for } t \in \Gamma_{\bar{z}}'. \tag{4.90}$$

Then υ_1 is a homeomorphism from $\Gamma_{\bar{z}}'$ onto Γ_1 which also fulfills (i)–(iii). Set

$$\upsilon(t) := \upsilon_j(t) \tag{4.91}$$

where $j = 0$ if $t \in \Gamma_z'$ and $j = 1$ if $t \in \Gamma_{\bar{z}}'$. Finally, for $u \in L^p((\Gamma_0 \cup \Gamma_1) \cap \mathscr{U}, \chi_0 w_z + \chi_1 w_{\bar{z}})$, define

$$(\Upsilon u)(t) := u(\upsilon(t)). \tag{4.92}$$

We claim that the mapping $A \mapsto \Upsilon A \Upsilon^{-1}$ is an isomorphism between the algebras

$$\mathscr{B}\left(\Gamma\cap\mathscr{U}_0,(\chi_0 w_z+\chi_1 w_{\bar z})_{|_{\Gamma\cap\mathscr{U}_0}},\mu_{|_{\Gamma\cap\mathscr{U}_0}}\right)$$

and

$$\mathscr{B}\left((\Gamma_z'\cup\Gamma_{\bar z}')\cap\mathscr{U}_1,(\chi_z' w_z+\chi_{\bar z}' w_{\bar z})_{|_{(\Gamma_z'\cup\Gamma_{\bar z}')\cap\mathscr{U}_1}},\mu_{|_{(\Gamma_z'\cup\Gamma_{\bar z}')\cap\mathscr{U}_1}}'\right).$$

The claim will follow once we have shown that $\Upsilon A\Upsilon^{-1}$ belongs to the latter algebra whenever A is in the first one. If $A = J_\mu$ then, by the definition of υ_1, $\Upsilon J_\mu\Upsilon^{-1} = J_{\mu'}$. If A is the operator of multiplication by the piecewise continuous function a, then $\Upsilon A\Upsilon^{-1}$ is the operator of multiplication by $a\circ\upsilon$. Finally, for A being a singular integral operator,

$$\Upsilon S_{\Gamma\cap\mathscr{U}}\Upsilon^{-1} = \Upsilon(\chi_0 S_{\Gamma\cap\mathscr{U}}\chi_0+\chi_1 S_{\Gamma\cap\mathscr{U}}\chi_0+\chi_0 S_{\Gamma\cap\mathscr{U}}\chi_1+\chi_1 S_{\Gamma\cap\mathscr{U}}\chi_1)\Upsilon^{-1}$$
$$= \Upsilon\chi_0 S_{\Gamma\cap\mathscr{U}}\chi_0\Upsilon^{-1}+\Upsilon\chi_1 S_{\Gamma\cap\mathscr{U}}\chi_1\Upsilon^{-1}+K_1$$

where K_1 is compact, since the operators $\chi_1 S_{\Gamma\cap\mathscr{U}}\chi_0$ and $\chi_0 S_{\Gamma\cap\mathscr{U}}\chi_1$ are compact (their kernels are bounded since Γ_0 and Γ_1 are disjoint). Repeating the arguments from the proof of Proposition 4.3.2 (third step) we obtain

$$\Upsilon S_{\Gamma\cap\mathscr{U}}\Upsilon^{-1} = \Upsilon\chi_0 S_{\Gamma\cap\mathscr{U}}\chi_0\Upsilon^{-1}+\Upsilon\chi_1 S_{\Gamma\cap\mathscr{U}}\chi_1\Upsilon^{-1}+K_1$$
$$= \chi_z' S_{\Gamma_z'\cup\Gamma_{\bar z}'\cap\mathscr{U}_1}\chi_z'+\chi_{\bar z}' S_{\Gamma_z'\cup\Gamma_{\bar z}'\cap\mathscr{U}_1}\chi_{\bar z}'+K_2$$
$$= S_{\Gamma_z'\cup\Gamma_{\bar z}'\cap\mathscr{U}_1}+K_3$$

with K_2 and K_3 compact. To finish the proof of the isomorphy $\mathscr{D}_2\cong\mathscr{D}_3$ notice that, for $f\in C_\mu(\Gamma\cap\mathscr{U})$, $\Upsilon f\Upsilon^{-1}$ is the operator of multiplication by a function in $C_{\mu'}((\Gamma_z'\cup\Gamma_{\bar z}')\cap\mathscr{U}_1)$ and that $\Upsilon f\Upsilon^{-1}(z) = f(z) = f(\bar z)$. Thus, the mapping $A\mapsto \Upsilon A\Upsilon^{-1}$ induces an isomorphism between the local algebras at $z\in\Gamma_+$. The same reasoning as in the proof of Proposition 4.3.2 shows that this local isomorphism is an isometry.

Finally we have to establish that the algebras $\mathscr{D}_3$ and

$$\mathscr{D}_4 := \mathscr{B}_z^{\mathscr{K}}\left(\Gamma_z\cup\Gamma_{\bar z},\chi_z w_z+\chi_{\bar z} w_{\bar z},\mu_z\right) \tag{4.93}$$

are isometrically isomorphic. For $t\in(\Gamma_z'\cup\Gamma_{\bar z}')\cap\mathscr{U}_1$, put

$$\zeta(t) := \begin{cases} e^{\mathrm{i}\psi}(t-z)+z & \text{if } t\in\Gamma_z', \\ e^{\mathrm{i}\tilde\psi}(t-z)+z & \text{if } t\in\Gamma_{\bar z}', \end{cases} \tag{4.94}$$

and let $\mathscr{U}_2$ be an open subset of $\mathbb{C}$ such that $\zeta((\Gamma_z'\cup\Gamma_{\bar z}')\cap\mathscr{U}_1) = (\Gamma_z\cup\Gamma_{\bar z})\cap\mathscr{U}_2$. As above, the homeomorphism ζ induces an isometric isomorphism from $L^p((\Gamma_z'\cup\Gamma_{\bar z}')\cap\mathscr{U}_1,\chi_z' w_z+\chi_{\bar z}' w_{\bar z|_{\mathscr{U}_1}})$ onto $L^p((\Gamma_z\cup\Gamma_{\bar z})\cap\mathscr{U}_2,\chi_z w_z+\chi_{\bar z} w_{\bar z|_{\mathscr{U}_2}})$ which we denote by ζ again. For this isomorphism, it is easy to verify the identities

$$\zeta S_{(\Gamma_z'\cup\Gamma_{\bar z}')\cap\mathscr{U}_1}\zeta^{-1} = S_{(\Gamma_z\cup\Gamma_{\bar z})\cap\mathscr{U}_2}+\text{ compact}$$

and

$$\zeta J_{\mu'_z}\zeta^{-1} = J_{\mu_z}, \qquad \zeta \chi'_z\zeta^{-1} = \chi_z, \qquad \zeta \chi'_{\tilde{z}}\zeta^{-1} = \chi_{\tilde{z}}.$$

These identities imply in a standard way that the algebras $\mathscr{D}_3$ and

$$\mathscr{D}'_4 := \mathscr{B}^{\mathscr{K}}_z \left((\Gamma_z \cup \Gamma_{\tilde{z}}) \cap \mathscr{U}_2, (\chi_z w_z + \chi_{\tilde{z}} w_{\tilde{z}})_{|\mathscr{U}_2}, \mu_z|_{\mathscr{U}_2} \right) \tag{4.95}$$

are isometrically isomorphic. Since, evidently, $\mathscr{D}'_4 \cong \mathscr{D}_4$, the proof is finished. ∎

Let χ_j $(j = 1,\dots,4)$ denote the characteristic functions of the sets $z + e^{\mathrm{i}\beta_j}\mathbb{R}^+$ if $j = 1,2$ and $\tilde{z} + e^{\mathrm{i}\beta_j}\mathbb{R}^+$ if $j = 3,4$, and let $\mathscr{D}_z$ refer to the smallest closed subalgebra of the Banach algebra $\mathscr{L}\left(L^p(\Gamma_z \cup \Gamma_{\tilde{z}}, \chi_z w_z + \chi_{\tilde{z}} w_{\tilde{z}}) \right)$ which contains the operators $(\chi_1 + \chi_2)S_{\Gamma_z \cup \Gamma_{\tilde{z}}}(\chi_1 + \chi_2) + (\chi_3 + \chi_4)S_{\Gamma_z \cup \Gamma_{\tilde{z}}}(\chi_3 + \chi_4)$, J_{μ_z}, and the four operators of multiplication by the χ_j.

Proposition 4.4.4. *The algebras* $\mathscr{B}^{\mathscr{K}}_z(\Gamma_z \cup \Gamma_{\tilde{z}}, \chi_z w_z + \chi_{\tilde{z}} w_{\tilde{z}}, \mu_z)$ *and* $\mathscr{D}_z$ *are isometrically isomorphic.*

Proof. This proof is a slight modification of the proof of Proposition 4.3.4. For $\tau \in \mathbb{R}^+$, consider the operator Z'_τ given by

$$(Z'_\tau u)(s) = \begin{cases} u(\frac{s-z}{\tau+z}) & \text{if } s \in \Gamma_z, \\ u(\frac{s-\tilde{z}}{\tau+\tilde{z}}) & \text{if } s \in \Gamma_{\tilde{z}}, \end{cases} \tag{4.96}$$

which maps the Banach space $L^p(\Gamma_z \cup \Gamma_{\tilde{z}}, \chi_z w_z + \chi_{\tilde{z}} w_{\tilde{z}})$ continuously onto itself. In analogy with Section 4.2.4, define the operator H'_0 via the strong limit

$$H'_0(A) = \text{s-}\lim_{\tau \to 0} Z'^{-1}_\tau A Z'_\tau, \tag{4.97}$$

provided this limit exists for A. In particular, one has

$$H'_0(S_{\Gamma_z \cup \Gamma_{\tilde{z}}}) = (\chi_1 + \chi_2)S_{\Gamma_z \cup \Gamma_{\tilde{z}}}(\chi_1 + \chi_2) + (\chi_3 + \chi_4)S_{\Gamma_z \cup \Gamma_{\tilde{z}}}(\chi_3 + \chi_4), \tag{4.98}$$

$$H'_0(J_{\mu_z}) = J_{\mu_z} \tag{4.99}$$

and, for $a \in PC(\Gamma_z \cup \Gamma_{\tilde{z}})$,

$$H'_0(aI) = a_1\chi_1 + \dots + a_4\chi_4 \tag{4.100}$$

where a_j denotes the one-sided limit of a as $t \to z$ (respectively $t \to \tilde{z}$) along the curve on which χ_j does not vanish. We will only prove (4.98). The two other identities can be verified as in Proposition 4.2.22. Write

$$S_{\Gamma_z \cup \Gamma_{\tilde{z}}} = (\chi_1 + \chi_2)S_{\Gamma_z \cup \Gamma_{\tilde{z}}}(\chi_1 + \chi_2) + (\chi_1 + \chi_2)S_{\Gamma_z \cup \Gamma_{\tilde{z}}}(\chi_3 + \chi_4)$$
$$+ (\chi_3 + \chi_4)S_{\Gamma_z \cup \Gamma_{\tilde{z}}}(\chi_1 + \chi_2) + (\chi_3 + \chi_4)S_{\Gamma_z \cup \Gamma_{\tilde{z}}}(\chi_3 + \chi_4).$$

For the first item on the right-hand side of this equality, we find at $s \in \Gamma_z$,

$$(Z_\tau'^{-1}(\chi_1+\chi_2)S_{\Gamma_z\cup\bar\Gamma_z}(\chi_1+\chi_2)Z_\tau'u)(s)$$

$$= \frac{1}{\pi\mathbf{i}}\int_{\Gamma_z}\frac{u((t-z)/(\tau+z))}{t-(s-z)\tau-z}\,dt = \frac{1}{\pi\mathbf{i}}\int_{\Gamma_z}\frac{u(y)}{y-s}\,dy$$

$$= ((\chi_1+\chi_2)S_{\Gamma_z\cup\bar\Gamma_z}(\chi_1+\chi_2)u)(s).$$

For the second item, we have $Z_\tau'^{-1}(\chi_1+\chi_2)S_{\Gamma_z\cup\bar\Gamma_z}(\chi_3+\chi_4)Z_\tau' \to 0$ strongly as $\tau \to 0$, since the operator $(\chi_1+\chi_2)S_{\Gamma_z\cup\bar\Gamma_z}(\chi_3+\chi_4)$ is compact. Similar arguments apply to the third and fourth items, whence (4.98). The remainder of the proof (which, in particular, includes the norm equality) runs as in Proposition 4.3.4. ∎

The following is the main result of this section.

Theorem 4.4.5.

(i) *For each $z \in \Gamma_+$, there is a homomorphism $A \mapsto \mathrm{smb}(A,z)$ from $\mathscr{B}(\Gamma,w,\mu)$ onto the algebra $\mathscr{E}^2_{p,\alpha(z)}$ if z is a fixed point of μ and onto the algebra $\mathscr{E}^4_{p,\alpha(z)}$ in the case z is a non-fixed point of μ. In particular, if z is a fixed point of μ, then*

$$\mathrm{smb}(S_\Gamma,z) = \begin{bmatrix} S_{\mathbb{R}^+} & -H_{2\pi-\beta(z)} \\ H_{\beta(z)} & -S_{\mathbb{R}^+} \end{bmatrix}, \tag{4.101}$$

$$\mathrm{smb}(J_\mu,z) = \begin{bmatrix} 0 & I \\ I & 0 \end{bmatrix}, \tag{4.102}$$

$$\mathrm{smb}(aI,z) = \begin{bmatrix} a_+(z)I & 0 \\ 0 & a_-(z)I \end{bmatrix}, \tag{4.103}$$

whereas if z is a non-fixed point,

$$\mathrm{smb}(S_\Gamma,z) = \begin{bmatrix} S_{\mathbb{R}^+} & -H_{2\pi-\beta(z)} & 0 & 0 \\ H_{\beta(z)} & -S_{\mathbb{R}^+} & 0 & 0 \\ 0 & 0 & S_{\mathbb{R}^+} & -H_{2\pi-\beta(\mu(z))} \\ 0 & 0 & H_{\beta(\mu(z))} & -S_{\mathbb{R}^+} \end{bmatrix}, \tag{4.104}$$

$$\mathrm{smb}(J_\mu,z) = \begin{bmatrix} 0 & 0 & 0 & I \\ 0 & 0 & I & 0 \\ 0 & I & 0 & 0 \\ I & 0 & 0 & 0 \end{bmatrix} \quad and \tag{4.105}$$

$$\mathrm{smb}(aI,z) = \begin{bmatrix} a_+(z)I & 0 & 0 & 0 \\ 0 & a_-(z)I & 0 & 0 \\ 0 & 0 & a_+(\mu(z))I & 0 \\ 0 & 0 & 0 & a_-(\mu(z))I \end{bmatrix}. \tag{4.106}$$

(ii) *The algebra $\mathscr{B}^{\mathscr{K}}(\Gamma,w,\mu)$ is inverse-closed in the Calkin algebra. An operator $A \in \mathscr{B}(\Gamma,w,\mu)$ is Fredholm if and only if $\mathrm{smb}(A,z)$ is invertible for all $z \in \Gamma_+$.*

(iii) $\|\Phi(A)\| = \sup_{z\in\Gamma_+} \|\mathrm{smb}(A,z)\|$ *for any* $A \in \mathscr{B}(\Gamma,w,\mu)$.

(iv) *The radical of* $\mathscr{B}^{\mathscr{K}}(\Gamma,w,\mu)$ *is trivial.*

(v) *The set* $\mathscr{F}(\Gamma,w)$ *of all functions* $(B_z)_{z\in\Gamma_+}$ *such that*

$$B_z \in \begin{cases} \mathscr{E}^2_{p,\alpha(z)} & \text{if } z \text{ is a fixed point of } \mu, \\ \mathring{\mathscr{E}}^4_{p,\alpha(z)} & \text{if } z \text{ is a non-fixed point of } \mu, \end{cases}$$

and $(B_z)_{z\in\Gamma_+}$ *is upper semi-continuous with respect to* $\mathscr{B}^{\mathscr{K}}(\Gamma,w,\mu)$ *forms a Banach algebra under the supremum norm, and the mapping which associates with* $\Phi(A)$ *the function* $z \mapsto (\mathrm{smb}(A,z))$ *is an isometric isomorphism from* $\mathscr{B}^{\mathscr{K}}(\Gamma,w)$ *onto* $\mathscr{F}(\Gamma,w)$.

Proof. (i) As in Section 4.2.4, one defines an isometric isomorphism η such that the mapping $\Theta : A \mapsto \eta A\eta^{-1}$ maps $\mathscr{D}(\Gamma_z,w_z)$ to $\mathscr{E}^2_{p,\alpha(z)}$ if z is a fixed point of the Carleman shift and it maps $\mathscr{D}_z$ to $\mathring{\mathscr{E}}^4_{p,\alpha(z)}$ in the case z is a non-fixed point. It is easy to see that Θ is onto. Now put $\mathrm{smb}(A,z) := (\Theta H_0 \Upsilon \Phi_z^{\mathscr{K}})(A)$ for fixed points z and $\mathrm{smb}(A,z) := (\Theta H_0' \Upsilon \Phi_z^{\mathscr{K}})(A)$ for non-fixed points z. The special form of the operators $\mathrm{smb}(S_\Gamma,z)$, $\mathrm{smb}(aI,z)$ and $\mathrm{smb}(J_{\mu_z},z)$ results from Propositions 4.4.1–4.4.4 in the same way as in Theorem 4.3.5.

(ii) One easily checks that if an operator $A \in \mathscr{E}^2_{p,\alpha(z)}$ (resp. $A \in \mathring{\mathscr{E}}^4_{p,\alpha(z)}$) is invertible, then its inverse belongs to $\mathscr{E}^2_{p,\alpha(z)}$ (resp. $\mathring{\mathscr{E}}^4_{p,\alpha(z)}$), again. The remainder of the proof is identical to that of Theorem 4.3.5. ∎

As before, let $\beta(z)$ denote the angle at $z \in \Gamma$.

Corollary 4.4.6. *Let* Γ *be a piecewise Lyapunov curve and* J_μ *be a Carleman shift changing the orientation of* Γ. *Then the operator* $S_\Gamma + J_\mu S_\Gamma J_\mu$ *is compact if and only if* $\beta(z) + \beta(\mu(z)) = 2\pi$ *for all* $z \in \Gamma$.

Proof. By Theorem 4.4.5, the symbol of $S_\Gamma + J_\mu S_\Gamma J_\mu$ at the fixed point $z \in \Gamma$ is the matrix

$$\begin{bmatrix} 0 & H_{\beta(z)} - H_{2\pi-\beta(z)} \\ H_{\beta(z)} - H_{2\pi-\beta(z)} & 0 \end{bmatrix},$$

whereas the symbol of this operator at a non-fixed point z is

$$\begin{bmatrix} 0 & H_{\beta(\mu(z))} - H_{2\pi-\beta(z)} & 0 & 0 \\ H_{\beta(z)} - H_{2\pi-\beta(\mu(z))} & 0 & 0 & 0 \\ 0 & 0 & 0 & H_{\beta(z)} - H_{2\pi-\beta(\mu(z))} \\ 0 & 0 & H_{\beta(\mu(z))} - H_{2\pi-\beta(z)} & 0 \end{bmatrix}.$$

In both cases, this symbol is the zero matrix if and only if $\beta(z) + \beta(\mu(z)) = 2\pi$. ∎

This corollary implies, in particular, that for curves having complementary angles at z and $\mu(z)$ the operator $J_\mu S_\Gamma J_\mu$ belongs to the algebra $\mathscr{A}(PC(\Gamma),S_\Gamma,w)$. This fact remains valid under more general conditions.

Corollary 4.4.7. *Let* Γ *be a piecewise Lyapunov curve and* J_μ *be a Carleman shift changing the orientation of* Γ. *Then the operator* $J_\mu S_\Gamma J_\mu$ *belongs to the algebra* $\mathscr{A}(PC(\Gamma), S_\Gamma, w)$.

Proof. Let $\widetilde{\mathscr{A}}(PC(\Gamma), S_\Gamma, w)$ denote the smallest closed subalgebra of $\mathscr{L}(L^p(\Gamma, w))$ which contains the algebra $\mathscr{A}(PC(\Gamma), S_\Gamma, w)$ and the operator $J_\mu S_\Gamma J_\mu$, and let $\widetilde{\mathscr{A}}^{\mathscr{K}}(PC(\Gamma), S_\Gamma, w)$ denote the image of $\widetilde{\mathscr{A}}(PC(\Gamma), S_\Gamma, w)$ in the Calkin algebra with canonical homomorphism Φ. Since the algebra $\Phi(C_\mu(\Gamma)I)$ belongs to the center of $\widetilde{\mathscr{A}}^{\mathscr{K}}(PC(\Gamma), S_\Gamma, w)$, we can identify the sets $\mathscr{A}$, $\mathscr{B}$ and $\mathscr{C}$ in Theorem 2.3.4 as the algebras $\widetilde{\mathscr{A}}^{\mathscr{K}}(PC(\Gamma), S_\Gamma, w)$, $\Phi(C_\mu(\Gamma)I)$ and $\mathscr{A}(PC(\Gamma), S_\Gamma, w)$, respectively. To apply this theorem, we must show that for each $z \in \Gamma_+ = M(\mathscr{B})$, there is an operator $A_z \in \mathscr{A}(PC(\Gamma), S_\Gamma, w)$ such that $\Phi_z(J_\mu S_\Gamma J_\mu - A_z) = 0$, where Φ_z refers to the canonical homomorphism from $\widetilde{\mathscr{A}}^{\mathscr{K}}(PC(\Gamma), S_\Gamma, w)$ onto the local algebra at the point z.

Let z be a fixed point of μ. The same reasoning as that in the proof of Theorem 4.4.5 shows that local algebra $\widetilde{\mathscr{A}}_z^{\mathscr{K}}(PC(\Gamma), S_\Gamma, w)$ is isometrically isomorphic to $\mathscr{E}_{p,\alpha(z)}^{2,\mathscr{N}}$. Moreover, the same isomorphism maps the local algebra $\mathscr{A}_z^{\mathscr{K}}(PC(\Gamma), S_\Gamma, w)$ (which is a closed subalgebra of $\widetilde{\mathscr{A}}_z^{\mathscr{K}}(PC(\Gamma), S_\Gamma, w)$) also onto $\mathscr{E}_{p,\alpha(z)}^{2,\mathscr{N}}$. Hence, the algebra $\widetilde{\mathscr{A}}_z^{\mathscr{K}}(PC(\Gamma), S_\Gamma, w)$ and its subalgebra coincide, whence the inclusion $\Phi_z(J_\mu S_\Gamma J_\mu) \in \mathscr{A}_z^{\mathscr{K}}(PC(\Gamma), S_\Gamma, w)$.

Similarly, if z be a non-fixed point, then both the algebra $\widetilde{\mathscr{A}}^{\mathscr{K}}(PC(\Gamma), S_\Gamma, w)$ and its subalgebra $\mathscr{A}_z^{\mathscr{K}}(PC(\Gamma), S_\Gamma, w)$ are isometrically isomorphic (again under the same isomorphism) to the closed subalgebra of $\mathscr{E}_{p,\alpha(z)}^4$ consisting of all matrices

$$\begin{bmatrix} A & 0 \\ 0 & B \end{bmatrix} \quad \text{with } A, B \in \mathscr{E}_{p,\alpha(z)}^{2,\mathscr{N}}.$$

Thus, in each case, there is an operator $A_z \in \mathscr{A}(PC(\Gamma), S_\Gamma, w)$ such that $J_\mu S_\Gamma J_\mu$ and A_z have the same symbol. Since this local symbol is one-to-one, the assertion follows. ∎

4.5 Toeplitz and Hankel operators

In this section, we are going to introduce the Hardy spaces H_w^p and the Toeplitz and Hankel operators acting on them. In particular, we will study the smallest closed subalgebra of $\mathscr{L}(H_w^p)$ containing all Toeplitz and Hankel operators with piecewise continuous generating functions. The use of Theorem 4.4.5 would provide us with a matrix symbol of order 4. In this case we shall see that there exists in fact a matrix symbol of order 2.

Throughout this section, let $1 < p < \infty$. We start with the space $L^p(\mathbb{T}, w)$, where $\mathbb{T}$ represents the complex unit circle oriented counter-clockwise, as usual. We consider the flip J,

$$(Ju)(t) := \frac{1}{t} u(1/t),$$

associated with the Carleman shift J_μ, $u(t) \mapsto u(1/t)$, the fixed points of which are -1 and $+1$. The weight w is assumed to fulfill the requirements of the last section such that both S and J be bounded linear operators on $L^p(\mathbb{T}, w)$. We denote by H_w^p the image space of $P := (I + S)/2$. The projection P is called the *Riesz projection*, and H_w^p is the *Hardy space*. The classical Hardy spaces H^p can be defined exactly in this way. For $n \in \mathbb{Z}$, let $e_n : \mathbb{T} \to \mathbb{C}$ be given by $e_n(t) = t^n$. The set $(e_n)_{n \in \mathbb{Z}^+}$ forms a basis of H_w^p, to which we refer as the canonical basis. One can easily see this by using the boundedness of P. Indeed, for the operators

$$V := Pe_1 P : H_w^p \to H_w^p \quad \text{and} \quad V^{(-1)} := Pe_{-1} P : H_w^p \to H_w^p$$

and for $n \geq 1$, one has

$$V^n = (Pe_1 P)^n = e_n P \quad \text{and} \quad (V^{(-1)})^n = (Pe_{-1} P)^n = Pe_{-n},$$

and the operators $P_n := I - V^n(V^{(-1)})^n$ are projections onto $\mathrm{span}\{e_0, e_1, ..., e_n\}$. Moreover, the sequence $(P_n)_{n \in \mathbb{Z}^+}$ is uniformly bounded. Since $\mathrm{span}\{e_n\}_{n \in \mathbb{Z}^+}$ is dense in H_w^p, the Banach-Steinhaus theorem (Theorem 1.4.2) ensures that (P_n) tends strongly to the identity operator in H_w^p which is equivalent to saying that $\{e_n\}_{n \in \mathbb{Z}^+}$ is a basis in H_w^p.

Let $a \in L^\infty(\mathbb{T})$. The operator $T(a)$ defined by

$$T(a) : H_w^p \to H_w^p, \quad u \mapsto PaPu,$$

is obviously bounded, and

$$\|T(a)\|_{\mathscr{L}(H_w^p)} \leq c_{p,w} \|a\|_\infty,$$

where $c_{p,w}$ is the norm of the Riesz projection P. This operator is called the *Toeplitz operator* on H_w^p with generating function a. The associated *Hankel operator* $H(a)$ is defined by

$$H(a) : H_w^p \to H_w^p, \quad u \mapsto PaQJu = PaJPu,$$

with $Q := I - P$. We further let $H(\tilde{a})$ denote the operator

$$H(\tilde{a}) : H_w^p \to H_w^p, \quad u \mapsto JQaPu = PJaPu.$$

This notation is justified since $H(\tilde{a}) = P\tilde{a}JP$ and $JaJ = \tilde{a}I$ with $\tilde{a}(t) = a(1/t)$. It is now readily verified that the matrix representations of $T(a)$ and $H(a)$ with respect to the canonical basis are exactly

$$[a_{j-k}]_{j,k=0}^\infty \quad \text{and} \quad [a_{j+k+1}]_{j,k=0}^\infty,$$

respectively, where $(a_j)_{j \in \mathbb{Z}}$ is the sequence of Fourier coefficients of a. Note that $H(a)$ is a compact operator if $a \in C(\mathbb{T})$. The next proposition provides some useful identities for Toeplitz and Hankel operators of products of functions.

Proposition 4.5.1. *Let* $a, b \in L^\infty(\mathbb{T})$. *Then*

$$T(ab) = T(a)T(b) + H(a)H(\tilde{b}),$$
$$H(ab) = T(a)H(b) + H(a)T(\tilde{b}).$$

Proof. Indeed, for the Toeplitz operator $T(ab)$ one has

$$T(ab) = PabP = Pa(P + QJJQ)bP$$
$$= PaPbP + PaQJJQbP = T(a)T(b) + H(a)H(\tilde{b}).$$

The proof of the second identity is similar.   ∎

Let $\mathscr{T}(PC)$ and $\mathscr{TH}(PC)$ denote the smallest closed subalgebras of $\mathscr{L}(H_w^p)$ containing all Toeplitz operators with piecewise continuous generating function and all Toeplitz and Hankel operators with piecewise continuous generating function, respectively.

Clearly, operators acting on H_w^p can be identified with operators acting on $L^p(\mathbb{T}, w)$ that are equal to zero on Im $(I - P)$. Thereby the spectrum or the essential spectrum of the continuation may differ only by the point zero from that of the original operator. Hence, $\mathscr{T}(PC)$ and $\mathscr{TH}(PC)$ can be identified with subalgebras of the algebra $\mathscr{B} := \mathscr{B}(\mathbb{T}, w, \mu)$ introduced in Section 4.4. The identity element in both algebras is the projection P, and this has to be taken into account when determining the essential spectrum of the elements belonging to these algebras.

Let us denote the images of the algebra $\mathscr{T}(PC)$ and $\mathscr{TH}(PC)$ under the symbol map smb given in Theorem 4.4.5 and related to the point $z \in \mathbb{T}_+ := \{z \in \mathbb{T} : \Im z \geq 0\}$ by $\mathscr{A}_z$ and $\mathscr{B}_z$, respectively. Suppose first that $z \in \{-1, 1\}$, i.e., z is a fixed point of μ. Notice that in this case we have $\text{smb}(J, z) = z \, \text{smb}(J_\mu, z)$. We will show that $\mathscr{B}_z$ is singly generated. Take, for definiteness, $z = 1$. It is easy to see that the operators

$$p := \frac{1}{2} \begin{bmatrix} I + S_{\mathbb{R}^+} & -H_\pi \\ H_\pi & I - S_{\mathbb{R}^+} \end{bmatrix}, \quad r := \begin{bmatrix} I & 0 \\ 0 & 0 \end{bmatrix} \quad \text{and} \quad j := \begin{bmatrix} 0 & I \\ I & 0 \end{bmatrix} \tag{4.107}$$

generate the algebra $\mathscr{E}_{p,\alpha(1)}^2$ due to the relations (4.101)–(4.103) and (3.45), (3.46). By Proposition 3.4.3, the center $\mathscr{C}$ of $\mathscr{E}_{p,\alpha(1)}^2$ is generated by the elements

$$d := prp + (e - p)(e - r)(e - p) = rpr + (e - r)(e - p)(e - r)$$

and

$$c := (pr - rp)j = prjp - (e - p)rj(e - p).$$

An easy computation yields

$$d = \frac{1}{2} \begin{bmatrix} I + S_{\mathbb{R}^+} & 0 \\ 0 & I + S_{\mathbb{R}^+} \end{bmatrix}, \tag{4.108}$$

$$c = \frac{1}{2} \begin{bmatrix} H_\pi & 0 \\ 0 & H_\pi \end{bmatrix}, \tag{4.109}$$

and Proposition 4.2.16 shows that $\mathscr{C}$ is singly generated by the element d, the spectrum of which is the range of the function

$$v_{p,\alpha}(y) := \frac{1}{2}\left(1 + \coth\left((y + \mathbf{i}(1/p + \alpha))\pi\right)\right), \quad y \in \mathbb{R}, \tag{4.110}$$

for $\alpha = \alpha(1)$ and corresponds to the arc $\mathfrak{A}_{1/p+\alpha(1)}$ defined in (3.44). According to Theorem 4.4.5 we get for $a, b \in PC(\mathbb{T})$

$$\mathrm{smb}(T(a), 1) = a(1^+)prp + a(1^-)p(e - r)p$$

and

$$\mathrm{smb}(H(b), 1) = b(1^+)prjp - b(1^-)prjp$$
$$= (b(1^+) - b(1^-))prjp.$$

Let $\varepsilon > 0$. Since $\mathscr{C}$ is singly generated by d, there is a polynomial q such that $\|q(d) - c\| < \varepsilon$. Hence,

$$\|q(prp) - prjp\| = \|p(q(d) - c)p\| \le \|p\|^2 \varepsilon$$

and these arguments show that $\mathscr{B}_1$ is singly generated by prp and that $\mathscr{A}_1 = \mathscr{B}_1$.

It remains to determine the spectrum of prp. Applying Proposition 1.2.6 we get

$$\sigma(prp) \setminus \{0\} = \sigma(rpr) \setminus \{0\},$$

and from (4.108) (note that $rdr = rpr$) we conclude

$$\sigma(prp) = \mathfrak{A}_{1/p+\alpha(1)}.$$

Using $j(prjp)j = -(e - p)rj(e - p)$ we see that $\sigma(prjp) \setminus \{0\} = \sigma(c) \setminus \{0\}$ and by (4.109)

$$\sigma(prjp) = \left\{\frac{1}{2}h_\pi(y) : y \in \mathbb{R}\right\},$$

where

$$h_\pi(y) = \frac{1}{\sinh\left((y + \mathbf{i}(1/p + \alpha(1)))\pi\right)}, \quad y \in \overline{\mathbb{R}}.$$

As we know (Exercise 2.1.3), the maximal ideal space of $\mathscr{B}_1$ is homeomorphic to $\sigma(prp)$, and the Gelfand transforms of prp and $prjp$ coincide with the functions $v_{p,\alpha(1)}$ and $\frac{1}{2}h_\pi$, respectively. Putting all these things together we get that for $a, b \in PC(\mathbb{T})$, the element $\mathrm{smb}(T(a) + H(b), 1)$ is invertible in $\mathscr{B}_1$ if and only if

$$a(1^+)v_{p,\alpha(1)}(y) + a(1^-)(1 - v_{p,\alpha(1)}(y)) + \frac{b(1^+) - b(1^-)}{2}h_\pi(y) \neq 0$$

for all $y \in \mathbb{R}$. Identical considerations hold true for the point -1, that is, for the local algebra $\mathscr{B}_{-1}$.

Now let $z \in \mathbb{T}_+ \setminus \{-1, 1\} =: \overset{\circ}{\mathbb{T}}_+$. Consider the circular arc with endpoints z and $\bar{z}$ which contains the point -1. Denote its characteristic function by χ_z. Due to Proposition 4.5.1,

$$\left(T(\chi_z) + H(\chi_z)\right)^2 = T(\chi_z) + H(\chi_z).$$

Consequently, the algebra $\mathscr{B}_z$ contains the idempotent

$$p_z := \mathrm{smb}\left(T(\chi_z) + H(\chi_z), z\right).$$

Take $f_0 \in C(\mathbb{T})$ such that $f_0(z) = 1$ and $f_0(\bar{z}) = 0$. Then an easy computation shows that $r_z := \mathrm{smb}(T(f_0), z)$ is also an idempotent in $\mathscr{B}_z$. We are going to prove that $\mathscr{B}_z$ is generated by p_z and r_z, and that

$$\begin{aligned}
r_z \mathrm{smb}\left(T(\chi_z), z\right)r_z &= r_z p_z r_z, \\
r_z \mathrm{smb}\left(T(1 - \chi_z), z\right)r_z &= r_z(e - p_z)r_z, \\
(e - r_z)\mathrm{smb}\left(T(\chi_z), z\right)(e - r_z) &= (e - r_z)p_z(e - r_z), \\
(e - r_z)\mathrm{smb}\left(T(1 - \chi_z), z\right)(e - r_z) &= (e - r_z)(e - p_z)(e - r_z),
\end{aligned} \tag{4.111}$$

where e denotes the unit element in $\mathscr{B}_z$. We will only prove the first equality in (4.111); the remaining equalities can be proved in an analogous manner. The first equality in (4.111) will follow as soon as we have checked that

$$r_z \mathrm{smb}\left(H(\chi_z), z\right)r_z = 0.$$

Using that $J\chi_z J = \tilde{\chi}_z I = \chi_z I$, an easy computation gives

$$T(f_0)H(\chi_z)T(f_0) = H(f_0\chi_z\tilde{f}_0) + H(f_0)T(\tilde{\chi}_z) + T(f_0\chi_z)H(\tilde{f}_0).$$

Since $f_0\chi_z\tilde{f}_0$, f_0 and $\tilde{f}_0$ are continuous functions, the symbol of this operator at the point z equals zero. Now observe that, for any $a \in PC(\mathbb{T})$,

$$\begin{aligned}
\mathrm{smb}\left(T(a), z\right) = {}& a(z^+)r_z\mathrm{smb}\left(T(\chi_z), z\right)r_z + a(z^-)r_z\mathrm{smb}\left(T(1 - \chi_z), z\right)r_z \\
& + a(\bar{z}^+)(e - r_z)\mathrm{smb}\left(T(1 - \chi_z), z\right)(e - r_z) \\
& + a(\bar{z}^-)(e - r_z)\mathrm{smb}\left(T(\chi_z), z\right)(e - r_z).
\end{aligned}$$

Using (4.111) we get

$$\begin{aligned}
\mathrm{smb}\left(T(a), z\right) = {}& a(z^+)r_z p_z r_z + a(z^-)r_z(e - p_z)r_z \\
& + a(\bar{z}^+)(e - r_z)(e - p_z)(e - r_z) + a(\bar{z}^-)(e - r_z)p_z(e - r_z), \quad (4.112)
\end{aligned}$$

and the above implies that $\mathrm{smb}(T(a),z)$ belongs to the algebra generated by p_z and r_z. Analogously one proves that, for any $b \in PC(\mathbb{T})$,

$$\mathrm{smb}(H(b),z) = b(z^+)r_zp_z(e-r_z) + b(z^-)r_z(e-p_z)(e-r_z)$$
$$+ b(\bar{z}^-)(e-r_z)p_zr_z + b(\bar{z}^+)(e-r_z)(e-p_z)r_z. \quad (4.113)$$

Hence, $\mathrm{smb}(H(b),z)$ is also in the algebra generated by p_z and r_z.

Now it is clear that $\mathscr{B}_z$ is generated by the idempotents p_z and r_z, and Theorem 3.1.4 applies. We must thus determine the spectrum of the element

$$c_z := p_zr_zp_z + (e-p_z)(e-r_z)(e-p_z).$$

We have $\sigma(p_zr_zp_z) \setminus \{0\} = \sigma(r_zp_zr_z) \setminus \{0\}$. The first equality in (4.111) shows that

$$\sigma(r_zp_zr_z) = \sigma\left(r_z\mathrm{smb}(T(\chi_z),z)r_z\right) = \sigma(rpr) = \sigma\left(\frac{1}{2}\begin{bmatrix} I+S_{\mathbb{R}^+} & 0 \\ 0 & 0 \end{bmatrix}\right),$$

where r and p are defined in (4.107). Thus, the spectrum of $r_zp_zr_z$ equals the range of the function given by (4.110). Because the point zero is not isolated in this spectrum, we also get that $\sigma(p_zr_zp_z) = \sigma(r_zp_zr_z)$. Now it is easy to conclude that the spectra of $p_zr_zp_z$ in $p_z\mathscr{B}_zp_z$ and in $\mathscr{B}_z$ coincide. Since also the point 1 is not isolated in the spectrum, we get by Corollary 3.1.5

$$\sigma(c) = \mathfrak{A}_{1/p+\alpha} \quad (4.114)$$

where $\alpha = \alpha(z)$ refers to the local exponent at z of the weight function in the remainder of this subsection. Now we use Theorem 3.1.4 and assign to the elements e, r_z, p_z the matrix functions

$$\begin{bmatrix} 1 & 0 \\ 0 & 1 \end{bmatrix}, \quad \begin{bmatrix} 1 & 0 \\ 0 & 0 \end{bmatrix}, \quad \begin{bmatrix} v_{p,\alpha} & \sqrt{v_{p,\alpha}(1-v_{p,\alpha})} \\ \sqrt{v_{p,\alpha}(1-v_{p,\alpha})} & 1-v_{p,\alpha} \end{bmatrix},$$

respectively, where $v_{p,\alpha}$ is the function (4.110). Using (4.112) and (4.113) it is easily seen that $\mathrm{smb}(T(a)+H(b),z)$ is mapped to

$$\begin{bmatrix} a(z^+)v_{p,\alpha}+a(z^-)(1-v_{p,\alpha}) & (b(z^+)-b(z^-))\sqrt{v_{p,\alpha}(1-v_{p,\alpha})} \\ (b(\bar{z}^-)-b(\bar{z}^+))\sqrt{v_{p,\alpha}(1-v_{p,\alpha})} & a(\bar{z}^+)v_{p,\alpha}+a(\bar{z}^-)(1-v_{p,\alpha}) \end{bmatrix}. \quad (4.115)$$

Clearly, $\mathrm{smb}(T(a)+H(b),z)$ is invertible if and only if the matrix function (4.115) is invertible for all $y \in \mathbb{R}$. The equality $v_{p,\alpha}(1-v_{p,\alpha}) = -\frac{1}{4}h_\pi^2$ leads to a simplification of (4.115), and we are able now to formulate the main result of this section.

Theorem 4.5.2. *The operator $T(a)+H(b) \in \mathscr{TH}(PC)$ is Fredholm on H_w^p if and only if the functions*

(i) $a(z^+)v_{p,\alpha}+a(z^-)(1-v_{p,\alpha})+z\frac{b(z^+)-b(z^-)}{2}h_\pi, \qquad z \in \{-1,1\},$

(ii) $\left[\begin{matrix} a(z^+)v_{p,\alpha}+a(z^-)(1-v_{p,\alpha}) & \frac{b(z^+)-b(z^-)}{2i}h_\pi \\ \frac{b(\bar{z}^-)-b(\bar{z}^+)}{2i}h_\pi & a(\bar{z}^+)v_{p,\alpha}+a(\bar{z}^-)(1-v_{p,\alpha}) \end{matrix} \right]$, $z\in\overset{\circ}{\mathbb{T}}_+$,

are invertible.

Remark 4.5.3. Obviously, the algebras $\mathscr{A}_z$ are commutative for $z\in\mathbb{T}_+$. Theorem 4.5.2 yields that $T(a)$ is Fredholm on H_w^p if and only if the functions

$$a(z^+)v_{p,\alpha}+a(z^-)(1-v_{p,\alpha})$$

are invertible for all $z\in\mathbb{T}$. $\square$

4.6 Singular integral operators with conjugation

The double layer potential operator is a prominent example of an operator which is constituted of a singular integral operator and the operator of complex conjugation (see also later in this section). This is one reason to consider the smallest closed algebra which is generated by the (linear) operators belonging to $\mathscr{A}(PC(\Gamma),S_\Gamma,w)$ and by the (antilinear) operator C of complex conjugation,

$$(Cf)(t):=\overline{f(t)}. \tag{4.116}$$

Proposition 4.6.1. *Let Γ be an admissible curve and let w be the weight* (4.1). *Then the linear operator $CS_\Gamma C$ belongs to the algebra $\mathscr{A}(PC(\dot{\Gamma}),S_\Gamma,w)$, and its symbol is given by*

$$\mathrm{smb}(CS_\Gamma C,z)=-\left[\begin{matrix} S_{\mathbb{R}^+} & H_{\beta_2-\beta_1} & \cdots & H_{\beta_k-\beta_1} \\ H_{2\pi+\beta_1-\beta_2} & S_{\mathbb{R}^+} & \cdots & H_{\beta_k-\beta_2} \\ \vdots & \vdots & \ddots & \vdots \\ H_{2\pi+\beta_1-\beta_k} & H_{2\pi+\beta_2-\beta_k} & \cdots & S_{\mathbb{R}^+} \end{matrix} \right]\mathrm{diag}_{1\le j\le k}(v_j(z)) \tag{4.117}$$

where $k=k(z)$ and $0\le\beta_1<\beta_2<\ldots<\beta_k<2\pi$ are the angles at $z\in\dot{\Gamma}$ as in Section 4.3.

Proof. Write $\widetilde{\mathscr{A}}(PC(\dot{\Gamma}),S_\Gamma,w)$ for the smallest closed subalgebra of $\mathscr{L}(L^p(\Gamma,w))$ which contains the algebra $\mathscr{A}(PC(\dot{\Gamma}),S_\Gamma,w)$ and the operator $CS_\Gamma C$. Further, let $\widetilde{\mathscr{A}}^{\mathscr{K}}(PC(\dot{\Gamma}),S_\Gamma,w)$ stand for the image of $\widetilde{\mathscr{A}}(PC(\dot{\Gamma}),S_\Gamma,w)$ in the respective Calkin algebra. Since $\Phi(C(\dot{\Gamma})I)$ lies in the center of $\widetilde{\mathscr{A}}^{\mathscr{K}}(PC(\dot{\Gamma}),S_\Gamma,w)$, we can replace the sets $\mathscr{A}$, $\mathscr{B}$ and $\mathscr{C}$ in Theorem 2.3.4 by the algebras $\widetilde{\mathscr{A}}^{\mathscr{K}}(PC(\dot{\Gamma}),S_\Gamma,w)$,

$\Phi(C(\dot{\Gamma})I)$ and $\mathscr{A}(PC(\dot{\Gamma}), S_\Gamma, w)$, respectively. For $z \in \dot{\Gamma} = M(\mathscr{B})$, let Φ_z denote the canonical homomorphism from $\widetilde{\mathscr{A}}^{\mathscr{K}}(PC(\dot{\Gamma}), S_\Gamma, w)$ onto the corresponding local algebra $\widetilde{\mathscr{A}}_z^{\mathscr{K}}(PC(\dot{\Gamma}), S_\Gamma, w)$ at the point z.

As in the proof of Proposition 4.3.2, the local algebra $\widetilde{\mathscr{A}}_z^{\mathscr{K}}(PC(\dot{\Gamma}), S_\Gamma, w)$ is isometrically isomorphic to the closed subalgebra of $\mathscr{L}(L^p(\Gamma, w))$ (with the local curve Γ_z and the local weight w_z being defined as in that proposition) which is generated by the operators $CS_\Gamma C$, S_{Γ_z}, and by the operators of multiplication by the characteristic functions of the half-axes $e^{i\beta_j}\mathbb{R}^+$ $(j = 1, \ldots, k(z))$. This isomorphism maps the subalgebra $\mathscr{A}_z^{\mathscr{K}}(PC(\dot{\Gamma}), S_\Gamma, w)$ of $\widetilde{\mathscr{A}}_z^{\mathscr{K}}(PC(\dot{\Gamma}), S_\Gamma, w)$ onto the algebra $\mathscr{A}(\chi, S_{\Gamma_z}, \alpha(z))$. It thus suffices to verify that $CS_{\Gamma_z}C \in \mathscr{A}(\chi, S_{\Gamma_z}, \alpha(z))$.

Let η denote the mapping (4.45). Since the mapping $A \mapsto \eta A \eta^{-1}$ is an isomorphism between $\mathscr{A}(\chi, S_{\Gamma_z}, \alpha(z))$ and $\mathscr{E}_{p,\alpha(z)}^{k(z),\mathscr{N}}$, it remains to show that $\eta CS_{\Gamma_z}C\eta^{-1}$ belongs to $\mathscr{E}_{p,\alpha(z)}^{k(z),\mathscr{N}}$.

Write $\eta CS_{\Gamma_z}C\eta^{-1}$ as $[B_{ij}]_{i,j=1}^{k(z)}$ with $B_{ij} \in \mathscr{L}(L^p(\mathbb{R}^+, w))$, and assume for simplicity that $v_j(z) = 1$ for all j. Then, for $u \in L^p(\mathbb{R}^+, w)$,

$$(B_{ij}u)(t) = -\frac{1}{\pi i}\overline{\int_{e^{i\beta_j}\mathbb{R}^+} \frac{\overline{u(e^{-i\beta_j}s)}}{s - e^{i\beta_i}t}\, ds} = -\frac{1}{\pi i}\overline{\int_{\mathbb{R}^+} \frac{\overline{u(s)}e^{i\beta_j}}{e^{i\beta_j}s - e^{i\beta_i}t}\, ds}$$

$$= -\frac{1}{\pi i}\int_{\mathbb{R}^+} \frac{u(s)}{s - e^{i(\beta_j - \beta_i)}t}\, ds = \begin{cases} -(H_{\beta_j - \beta_i}u)(t) & \text{if } i < j, \\ -(S_{\mathbb{R}^+}u)(t) & \text{if } i = j, \\ -(H_{2\pi + \beta_j - \beta_i}u)(t) & \text{if } i > j. \end{cases}$$

Consequently, $\eta CS_{\Gamma_z}C\eta^{-1} \in \mathscr{E}_{p,\alpha(z)}^{k(z),\mathscr{N}}$, and the proof is finished.   ■

Let $\tilde{L}^p(\Gamma, w)$ denote the *real* Banach space $L^p(\Gamma, w)$ (i.e., $\tilde{L}^p(\Gamma, w)$ is considered a linear space over $\mathbb{R}$, but it still consists of complex-valued functions), and let $\mathscr{C}(\Gamma, w)$ stand for the smallest closed *real* subalgebra of $\mathscr{L}(\tilde{L}^p(\Gamma, w))$ which encloses the operator C (which is now a linear operator on $\tilde{L}^p(\Gamma, w)$) and all operators from $\mathscr{A}(PC(\dot{\Gamma}), S_\Gamma, w)$. Further we denote by $\mathscr{C}^{\mathscr{K}}(\Gamma, w)$ the image of $\mathscr{C}(\Gamma, w)$ in the respective Calkin algebra.

Theorem 4.6.2. *Let Γ be an admissible curve.*

(i) *For each $z \in \dot{\Gamma}$, there exists a homomorphism $A \mapsto \text{smb}(A, z)$ from $\mathscr{C}(\Gamma, w)$ onto the closed subalgebra of $\mathscr{E}_{p,\alpha(z)}^{2k(z),\mathscr{N}}$ consisting of all matrices $[C_{ij}]_{i,j=1}^2$ with $C_{ij} \in \mathscr{E}_{p,\alpha(z)}^{k(z),\mathscr{N}}$. In particular, $\text{smb}(S_\Gamma, z) = [A_{ij}]_{i,j=1}^2$ with*

$$A_{11} = - \begin{bmatrix} S_{\mathbb{R}^+} & H_{2\pi+\beta_1-\beta_2} & \cdots & H_{2\pi+\beta_1-\beta_k} \\ H_{\beta_2-\beta_1} & S_{\mathbb{R}^+} & \cdots & H_{2\pi+\beta_2-\beta_k} \\ \vdots & \vdots & \ddots & \vdots \\ H_{\beta_k-\beta_1} & H_{\beta_k-\beta_2} & \cdots & S_{\mathbb{R}^+} \end{bmatrix} \operatorname{diag}_{1 \le j \le k}(v_j(z))$$

$$A_{22} = - \begin{bmatrix} S_{\mathbb{R}^+} & H_{\beta_2-\beta_1} & \cdots & H_{\beta_k-\beta_1} \\ H_{2\pi+\beta_1-\beta_2} & S_{\mathbb{R}^+} & \cdots & H_{\beta_k-\beta_2} \\ \vdots & \vdots & \ddots & \vdots \\ H_{2\pi+\beta_1-\beta_k} & H_{2\pi+\beta_2-\beta_k} & \cdots & S_{\mathbb{R}^+} \end{bmatrix} \operatorname{diag}_{1 \le j \le k}(v_j(z))$$

(where again $k = k(z)$) and $A_{21} = A_{12} = 0$,

$$\operatorname{smb}(C,z) = \begin{bmatrix} 0 \cdots 0 & I \cdots 0 \\ \vdots & \vdots & \vdots & \ddots & \vdots \\ 0 \cdots 0 & 0 \cdots & I \\ I \cdots 0 & 0 \cdots & 0 \\ \vdots & \ddots & \vdots & \vdots \\ 0 \cdots I & 0 \cdots 0 \end{bmatrix}$$

and, if a is piecewise continuous,

$$\operatorname{smb}(aI,z) = \operatorname{diag}(a_1(z), \ldots, a_{k(z)}(z), \overline{a_1(z)}, \ldots, \overline{a_{k(z)}(z)}).$$

(ii) *The algebra $\mathscr{C}^{\mathscr{K}}(\Gamma, w)$ is inverse-closed in the Calkin algebra. An operator $A \in \mathscr{C}(\Gamma, w)$ is Fredholm if and only if $\operatorname{smb}(A, z)$ is invertible for all $z \in \dot{\Gamma}$.*

(iii) *Set $\|\Phi(A)\| := \sup_{z \in \dot{\Gamma}} \|\operatorname{smb}(A, z)\|$. Then $\|\Phi(A)\|$ is an equivalent norm for $\mathscr{C}^{\mathscr{K}}(\Gamma, w)$.*

(iv) *The radical of $\mathscr{C}^{\mathscr{K}}(\Gamma, w)$ is trivial.*

Proof. Every operator $A \in \mathscr{C}(\Gamma, w)$ can be written as $A = X + YC$ with operators $X, Y \in \mathscr{A}(PC(\dot{\Gamma}), S_\Gamma, w)$. Indeed, simply put $X := \frac{1}{2}(A - iAi)$ and $Y := -\frac{i}{2}(iA - Ai)C$. For X and Y to be linear with respect to $\mathbb{C}$ it is necessary and sufficient that $-iXi = X$ and $-iYi = Y$, which is easily verified.

Approximate X and Y by $\mathbb{C}$-linear operators $B_n \in \mathscr{C}(\Gamma, w)$ which are finite sums of products $B_n = \sum_i \Pi_j B_{ij}$ where either $B_{ij} \in \mathscr{A}(PC(\dot{\Gamma}), S_\Gamma, w)$ or $B_{ij} = C$. It is easy to see that this is always possible. Clearly, for the operators B_n to be linear it is necessary that the operator C of conjugation occurs in each product $\Pi_j B_{ij}$ in an even number. From Proposition 4.6.1 we conclude then that $B_n \in \mathscr{A}(PC(\dot{\Gamma}), S_\Gamma, w)$ and, hence, X and Y both belong to $\mathscr{A}(PC(\dot{\Gamma}), S_\Gamma, w)$.

Now let $A = X + YC \in \mathscr{C}(\Gamma, w)$. Evidently, the operator $iAi = X - YC$ is Fredholm if and only if the operator A is Fredholm. Hence, the operator

$$\begin{bmatrix} A & 0 \\ 0 & iAi \end{bmatrix} = \begin{bmatrix} X + YC & 0 \\ 0 & X - YC \end{bmatrix} \in \mathscr{L}(\tilde{L}^p(\Gamma, w))$$

is Fredholm if and only if the operator A is Fredholm. It remains to apply the simple identity

$$\begin{bmatrix} X+YC & 0 \\ 0 & X-YC \end{bmatrix} = \frac{1}{2} \begin{bmatrix} I & C \\ I & -C \end{bmatrix} \begin{bmatrix} X & Y \\ CYC & CXC \end{bmatrix} \begin{bmatrix} I & I \\ C & -C \end{bmatrix} \tag{4.118}$$

in which the matrices $\begin{bmatrix} I & C \\ I & -C \end{bmatrix}$ and $\begin{bmatrix} I & I \\ C & -C \end{bmatrix}$ are invertible and where the operator $\begin{bmatrix} X & Y \\ CYC & CXC \end{bmatrix}$ belongs to $[\mathscr{A}(PC(\dot{\Gamma}), S_\Gamma, w)]_{2 \times 2}$ due to Proposition 4.6.1. Then all assertions of the theorem follow from equation (4.117) and Theorem 4.3.5. In particular, the norms $\|\Phi(A)\|$ and $\left\|\Phi\left(\begin{bmatrix} X & Y \\ CYC & CXC \end{bmatrix}\right)\right\|$ are equivalent by (4.118). ∎

The operator $V_\Gamma := (S_\Gamma + C S_\Gamma C)/2$ is known as the *double layer potential operator*. On a simple closed curve Γ, V_Γ acts via

$$(V_\Gamma u)(t) = \frac{1}{\pi} \int_\Gamma u(\tau) \frac{d}{dn_\tau} \log|t - \tau| ds_\tau, \quad t \in \Gamma,$$

where n_τ is the inner normal to Γ at $\tau \in \Gamma$ and ds_τ refers to the arc length differential (see Muskhelishvili's classical monograph [125]).

Corollary 4.6.3. *Let the curve Γ be admissible. Then the operator $S_\Gamma + C S_\Gamma C$ is compact if and only if Γ is a simple Lyapunov curve (i.e., without intersections and corners).*

Proof. By (4.117) and Theorem 4.3.5, $\text{smb}(S_\Gamma + C S_\Gamma C, z)$ is equal to the matrix

$$\begin{bmatrix} 0 & H_{2\pi+\beta_1-\beta_2} - H_{\beta_2-\beta_1} & \cdots & H_{2\pi+\beta_1-\beta_k} - H_{\beta_k-\beta_1} \\ H_{\beta_2-\beta_1} - H_{2\pi+\beta_1-\beta_2} & 0 & \cdots & H_{2\pi+\beta_2-\beta_k} - H_{\beta_k-\beta_2} \\ \vdots & \vdots & & \vdots \\ H_{\beta_k-\beta_1} - H_{2\pi+\beta_1-\beta_k} & H_{\beta_k-\beta_2} - H_{2\pi+\beta_2-\beta_k} & \cdots & 0 \end{bmatrix} \times D \tag{4.119}$$

where $D = \text{diag}_{1 \leq j \leq k}(v_j(z))$ and $k = k(z)$ again. This matrix is zero if and only if

$$\beta_i - \beta_j = 2\pi + \beta_j - \beta_i \quad \text{for all} \quad i, j = 1, \dots, k, \ i \neq j,$$

which is equivalent to $\beta_i = \pi + \beta_j$. This is only possible if $k = 2$ and Γ is smooth at z, or if $k = 1$. ∎

Finally, and only for completeness, we mention that (unlike the case without conjugation) the Fredholmness of an operator $A \in \mathscr{C}(\Gamma, w)$ essentially depends on the values of the angles of the curve Γ. This follows from the preceding corollary, and it follows also from a remarkable fact discovered by Nyaga [131]. He considered the operator $A = I + \sqrt{2} S_\Gamma + C$ on $L^2(\Gamma)$. If Γ is a circle then A is Fredholm whereas A fails to be Fredholm if Γ is the boundary of a rectangle.

4.7 C^*-algebras generated by singular integral operators

In this section we ask under which conditions the algebras $\mathscr{A}(PC(\dot{\Gamma}),S_\Gamma,w)$ and $\mathscr{A}(C(\dot{\Gamma}),S_\Gamma,w)$ carry the structure of a C^*-algebra. Let Γ be an admissible curve, and let the weight function w be given by (4.1). With respect to the bilinear form

$$\langle u,v\rangle_1 := \int_\Gamma u(t)\overline{v(t)}\,|dt|, \tag{4.120}$$

the Banach dual space of $L^2(\Gamma,w)$ can be identified with $L^2(\Gamma,w^{-1})$. We denote the adjoint of an operator A on $L^2(\Gamma,w)$ with respect to $\langle\cdot,\cdot\rangle_1$ by A^*. The Banach space $L^2(\Gamma,w)$ can also be considered as a Hilbert space, on defining a weighted inner product by

$$\langle u,v\rangle_w := \langle wu,wv\rangle_1. \tag{4.121}$$

We denote the adjoint of an operator $A \in \mathscr{L}(L^2(\Gamma,w))$ with respect to the new scalar product $\langle\cdot,\cdot\rangle_w$ by $A^\dagger$. To get the relation between A^* and $A^\dagger$ one can write

$$\begin{aligned}
\langle u,Av\rangle_w &= \langle wu,wAv\rangle_1 \\
&= \langle wu,wAw^{-1}wv\rangle_1 \\
&= \langle (wAw^{-1})^*wu,wv\rangle_1 \\
&= \langle ww^{-1}(wAw^{-1})^*wu,wv\rangle_1 \\
&= \langle w^{-1}(wAw^{-1})^*wu,v\rangle_w,
\end{aligned}$$

whence

$$A^\dagger = w^{-1}(wAw^{-1})^*w = w^{-1}w^{-1*}A^*w^*w. \tag{4.122}$$

Theorem 4.7.1. Let $p = 2$. Then $\mathscr{A}(PC(\dot{\Gamma}),S_\Gamma,w)$ is a C^*-algebra.

Proof. $\mathscr{A}(PC(\dot{\Gamma}),S_\Gamma,w)$ is a closed subalgebra of the C^*-algebra $\mathscr{L}(L^2(\Gamma,w))$. So we just have to show that the algebra $\mathscr{A}(PC(\dot{\Gamma}),S_\Gamma,w)$ is symmetric. From (4.122) it is obvious that for a piecewise continuous function a, the adjoint operator $(aI)^\dagger$ of aI coincides with $\bar{a}I$, whence $(aI)^\dagger \in \mathscr{A}(PC(\dot{\Gamma}),S_\Gamma,w)$. To prove that $S_\Gamma^\dagger$ belongs to $\mathscr{A}(PC(\dot{\Gamma}),S_\Gamma,w)$, we proceed as in the proof of Corollary 4.4.7.

Let $\tilde{\mathscr{A}}(PC(\dot{\Gamma}),S_\Gamma,w)$ be the smallest closed C^*-subalgebra of $\mathscr{L}(L^2(\Gamma,w))$ which contains the algebra $\mathscr{A}(PC(\dot{\Gamma}),S_\Gamma,w)$, and let $\tilde{\mathscr{A}}^\mathscr{K}(PC(\dot{\Gamma}),S_\Gamma,w)$ denote its image in the Calkin algebra. The algebra $\Phi(C(\dot{\Gamma})I)$ belongs to the center of $\tilde{\mathscr{A}}^\mathscr{K}(PC(\dot{\Gamma}),S_\Gamma,w)$. So we can apply Theorem 2.3.4 with $\mathscr{A} = \tilde{\mathscr{A}}^\mathscr{K}(PC(\dot{\Gamma}),S_\Gamma,w)$, $\mathscr{B} = \Phi(C(\dot{\Gamma})I)$ and $\mathscr{C} = \mathscr{A}(PC(\dot{\Gamma}),S_\Gamma,w)$. Let Φ_z denote the canonical homomorphism from $\tilde{\mathscr{A}}^\mathscr{K}(PC(\dot{\Gamma}),S_\Gamma,w)$ onto the corresponding local algebra at $z \in \dot{\Gamma} = M(\mathscr{B})$.

By Theorem 2.3.4, the inclusion $S_\Gamma^\dagger \in \mathscr{A}(PC(\dot{\Gamma}),S_\Gamma,w)$ will follow once we have shown that, for each $z \in \dot{\Gamma}$, there is an operator $A_z \in \mathscr{A}(PC(\dot{\Gamma}),S_\Gamma,w)$ such that $\Phi_z(S_\Gamma^\dagger - A_z) = 0$. Define the local curve Γ_z and the local weight $w_z(t) = |t|^{\alpha(z)}$ as in Section 4.3 and consider first the operator S_{Γ_z} on $L^2(\Gamma_z,w_z)$.

Write Γ_z as $\cup_{j=1}^{k} e^{i\beta_j} \mathbb{R}^+$ with $0 \leq \beta_1 < \ldots < \beta_k < 2\pi$. By [73, Chap. I, Theorem 7.1], the Banach space adjoint $S_{\Gamma_z}^* \in \mathscr{L}(L^2(\Gamma_z, w_z^{-1}))$ of S_{Γ_z} is given by

$$S_{\Gamma_z}^* = -h_z^{-1} C S_{\Gamma_z} C h_z I \tag{4.123}$$

where C is the operator of complex conjugation and where the function h_z on Γ_z is defined by $h_z(t) := e^{i\beta_j}$ if $t \in e^{i\beta_j} \mathbb{R}^+$. By (4.122), the adjoint of S_{Γ_z} with respect to the weighted inner product $\langle \cdot, \cdot \rangle_w$ is then

$$S_{\Gamma_z}^\dagger = -w_z^{-2} h_z^{-1} C S_{\Gamma_z} C h_z w_z^2 I. \tag{4.124}$$

Let χ_j denote the characteristic function of $e^{i\beta_j} \mathbb{R}^+$ and set $\chi := \{\chi_1, \ldots, \chi_k\}$. From Proposition 4.6.1 we infer that $C S_{\Gamma_z} C \in \mathscr{A}(\chi, S_\Gamma, -\alpha)$. The function h_z is piecewise continuous and, thus, in $\mathscr{A}(\chi, S_\Gamma, -\alpha)$. Since h_z does not vanish on Γ_z, we have $h_z^{-1} C S_{\Gamma_z} C h_z I \in \mathscr{A}(\chi, S_\Gamma, -\alpha)$. Now (4.124), Proposition 4.2.16 and Corollary 4.2.21 imply that $S_{\Gamma_z}^\dagger \in \mathscr{A}(\chi, S_\Gamma, \alpha)$. With this information, the proof can be finished by the same arguments as in the proofs of Corollary 4.4.7 and Proposition 4.6.1. ∎

Corollary 4.7.2. *Let Γ be an admissible curve, and let the weight function w be given by (4.1). Let S_Γ^* denote the Banach space adjoint of $S_\Gamma \in \mathscr{L}(L^2(\Gamma, w))$ with respect to the bilinear form (4.120). Then the operator $S_\Gamma^* - S_\Gamma \in \mathscr{L}(L^2(\Gamma, w^{-1})) = \mathscr{L}(L^2(\Gamma, w)^*)$ is compact if and only if Γ is a simple Lyapunov curve.*

The proof follows from (4.123) in a similar way as the proof of Corollary 4.6.3. Note that this result also holds when $p \neq 2$. The proof is the same.

Corollary 4.7.3. *Let Γ and w be as in the preceding corollary, and let $S_\Gamma^\dagger$ refer to the adjoint of $S_\Gamma \in \mathscr{L}(L^2(\Gamma, w))$ with respect to the inner product (4.121). Then the operator $S_\Gamma^\dagger - S_\Gamma \in \mathscr{L}(L^2(\Gamma, w))$ is compact if and only if $w \equiv 1$ and Γ is a simple Lyapunov curve.*

Proof. One easily checks that the operator standing in the left upper corner of the matrix $\mathrm{smb}(S_\Gamma^\dagger - S_\Gamma, z)$ is equal to $S_{\mathbb{R}^+} - w_z^{-2} S_{\mathbb{R}^+} w_z^2 I = S_{\mathbb{R}^+} - S_{2\alpha(z)}$. From Proposition 4.2.11 and the definition of S_σ in Section 4.2.3 it is immediate that $S_{\mathbb{R}^+} - S_{2\alpha(z)}$ is the zero operator if and only if $\alpha(z) = 0$. The point is that $S_{\mathbb{R}^+} - S_{2\alpha(z)}$ is a Mellin convolution with generating function $s_{1/2 + \alpha(z)} - s_{1/2 - \alpha(z)}$ (see Proposition 4.2.11). Thus, the weight function must be identically 1, and the remainder of the proof follows as in Corollary 4.7.2. ∎

We emphasize that Theorem 4.7.1 is not a triviality. The fact is the more surprising since the algebra $\mathscr{A}(C(\dot{\Gamma}), S_\Gamma, w) \subseteq \mathscr{L}(L^2(\Gamma, w))$ which is generated by the singular integral operator S_Γ and by the operators of multiplication by *continuous* functions on $\dot{\Gamma}$ is *not* a C^*-algebra in general.

Theorem 4.7.4. *Let Γ be an admissible closed curve, and let the weight function w be given by (4.1). Then the algebra $\mathscr{A}(C(\dot{\Gamma}), S_\Gamma, w)$ is a C^*-algebra if and only if $w \equiv 1$ and Γ is a simple Lyapunov curve.*

Proof. Let T be the finite set which contains the weight points $t_1, \ldots, t_n$ and all corners and endpoints of Γ. From (4.124) one concludes easily that $\Phi_z(S_\Gamma - S_\Gamma^\dagger) = 0$ for $z \in \dot{\Gamma} \setminus T$. Now assume that $S_\Gamma^\dagger \in \mathscr{A}(C(\dot{\Gamma}), S_\Gamma, w)$. Then there are continuous functions f, g on $\dot{\Gamma}$ such that $\Phi_z(S_\Gamma^\dagger - fI - gS_\Gamma) = 0$ for all $z \in \dot{\Gamma}$. Thus, $\Phi_z(S_\Gamma - fI - gS_\Gamma) = \Phi_z((1 - g(z))S_\Gamma - f(z)I) = 0$ for all $z \in \dot{\Gamma} \setminus T$. This implies that $f \equiv 0$ and $g \equiv 1$, hence, $S_\Gamma - S_\Gamma^\dagger$ is a compact operator. Then $\Phi(S_\Gamma)$ is self-adjoint, and the norm of $\Phi(S_\Gamma)$ must be equal to its spectral radius, which in turn equals 1. On the other hand, we know from Corollary 4.3.13 that $\|\Phi(S_\Gamma)\| > 1$ if $w \not\equiv 1$ or if Γ has corners. The reverse direction follows from Corollary 4.7.3. ∎

Proposition 4.7.5. *The algebra $\mathscr{A}(C(\dot{\mathbb{R}}^+), S_{\mathbb{R}^+}, t^\alpha)$ is a C^*-algebra for $0 < 1/2 + \alpha < 1$.*

Proof. By equations (4.124), (4.117) and Proposition 4.2.16, one has $S_{\mathbb{R}^+}^\dagger \in \mathscr{E}_{2,\alpha}$. Thus, both $\mathscr{A}(C(\dot{\mathbb{R}}^+), S_{\mathbb{R}^+}, t^\alpha)$ and $\mathscr{E}_{2,\alpha}$ are C^*-algebras. ∎

Theorem 4.7.6. *Let Γ be a suitable curve and let the weight function w be given by $w(t) = |t - t_0|^{\alpha_0} \Pi_{j=1}^n |t - t_j|^{\alpha_j}$. Then the algebra $\mathscr{A}(C(\dot{\Gamma}), S_\Gamma, w) \subseteq \mathscr{L}(L^2(\Gamma, w))$ is a C^*-algebra if and only if Γ is a simple Lyapunov curve and if the weight points t_j have order 1 for $j = 1, \ldots, n$.*

Proof. It follows, as in the proof of Theorem 4.7.4, that a point of order greater than 1 can be neither a corner nor a weight point. The reverse direction follows from Corollary 4.7.3 and Proposition 4.7.5. ∎

4.8 Appendix: Interpolation theorems

We use two interpolation theorems in this text. The first one is due to Stein and Weiss [191, Chap. V, Theorem 1.3] and concerns the boundedness of operators on interpolation spaces. Its proof can also be found in [7, Chapter 4, Theorem 3.6]. Let $1 \leq p_1, p_2 < \infty$ and $n \geq 1$.

Theorem 4.8.1 (Stein–Weiss). *Let A be a bounded operator acting on the spaces $L^{p_1}(\mathbb{R}^n, w_1)$ and $L^{p_2}(\mathbb{R}^n, w_2)$. If*

$$\frac{1}{p} = \frac{1 - \theta}{p_1} + \frac{\theta}{p_2}, \quad 0 \leq \theta \leq 1$$

and $w = w_1^{1-\theta} w_2^\theta$ then A is bounded on $L^p(\mathbb{R}^n, w)$ and

$$\|A\|_{\mathscr{L}(L^p(\mathbb{R}^n, w))} \leq \|A\|_{\mathscr{L}(L^{p_1}(\mathbb{R}^n, w_1))}^{1-\theta} \|A\|_{\mathscr{L}(L^{p_2}(\mathbb{R}^n, w_2))}^\theta.$$

Concerning the compactness of operators on interpolation spaces, one has the following interpolation theorem by Krasnoselskii (see [103, Section 3.4]) and Cwikel [31].

Theorem 4.8.2 (Krasnoselskii). *Let A be a bounded operator acting on the space* $L^{p_1}(\mathbb{R}^n, w_1)$ *and a bounded and compact operator acting on* $L^{p_2}(\mathbb{R}^n, w_2)$. *If*

$$\frac{1}{p} = \frac{1-\theta}{p_1} + \frac{\theta}{p_2} \quad \text{for some } \theta \in \,]0, 1[$$

and $w = w_1^{1-\theta} w_2^{\theta}$, *then A is compact on* $L^p(\mathbb{R}^n, w)$.

4.9 Notes and comments

The theory of one-dimensional singular integral operators has a long and rich history. We only mention the now classical monographs [114, 120, 125, 197] and the two volumes of [73, 74], where the foundations of the theory of singular integral operators were laid and the theory of singular operators with piecewise coefficients on L^p-spaces over Lyapunov curves and power weight was developed. Our presentation also owes a lot to [28, 29, 30].

C^*-algebras generated by singular integral operators with discontinuous coefficients on composed curves were studied in [140, 141], with special emphasis paid to the construction of a matrix-valued symbol calculus and to the determination of the irreducible representations of the algebra. See also the monograph [139]. One-dimensional singular integral operators can also be viewed as pseudo-differential operators of order zero. For a comprehensive account on this topic see [27].

A spectacular breakthrough came with the advent of the monograph by Böttcher and Yu. Karlovich [14], which is devoted to singular integral operators on general Carleson curves and general Muckenhoupt weights and to their Fredholm properties. This monograph was considered as the definitive work in this field by many people, but recently there has been an increasing interest in singular integral operators on more general spaces like Orlicz spaces and L^p-spaces with variable p, see for instance [96, 97, 98, 100, 157, 172].

Our exposition is heavily based on the use of homogenization techniques and Mellin operators. It is hard to say to whom this idea belongs, but Dynin and Eskin must certainly be mentioned here. A first systematic use of this method to study singular integral operators with piecewise continuous coefficients on L^p-spaces over Lyapunov curves appeared in the lovely booklet [189]. The case of admissible curves and L^p-spaces with power weight was treated by two of the authors in [167, 168].

In these papers, one also finds a treatment of singular operators with Carleman shift changing the orientation and of singular operators with complex conjugation. The investigation of algebras generated by singular integral operators with piecewise continuous coefficients and a Carleman shift changing the orientation has its origin in [69, 71]. The paper [30] can be considered as a continuation of these papers. Note that the results presented in these papers are less explicit in comparison with the results of Section 4.4. For a comprehensive account on operators with Carleman shift see, besides the already mentioned [114], the more recent texts [95, 104]. For equations involving singular operators with complex conjugation we also refer to [36].

A modified homogenization/Mellin approach does also work very well for singular integral operators, the coefficients of which are combinations of piecewise continuous and slowly oscillating functions. Also, the curves and the weights can be allowed to have singularities of a slowly oscillating type. The point is that the idea of homogenization can (and must) be replaced by more subtle limit operator techniques. For more in this direction see [15, 16, 153, 154, 155] and [156, Section 4.6].

Besides the homogenization/Mellin approach to study the (local) Fredholmness of singular integral operators, there is an alternate approach via two (or more) projections theorems. The idea of marrying local principles with Halmos' two projections theorem goes back to Douglas [40]. It has proved to be extremely successful for several classes of operators on Hilbert spaces. See [11, 13, 19, 21, 22, 48, 76, 148, 149, 167, 182] for applications to algebras of singular integral operators, Wiener-Hopf operators, Toeplitz plus Hankel operators, and more general Fourier integral operators with piecewise continuous or piecewise quasicontinuous coefficients.

Both approaches have their merits and demerits. In this text, we have mainly used the Mellin approach, simply because it provides some additional insights which would be hard to get with projections theorems. For example, it implies without effort that the Calkin images of the algebras considered in Sections 4.3–4.6 have a trivial radical, which in turn implies that these algebras are algebras with a polynomial identity. On the other hand, the real power of the projections theorems seem to lie in a field which is beyond the scope of this text: They apply to the study of algebras of singular integral operators on L^p-spaces over general Carleson curves and with general Muckenhoupt weights (see [14, Chapter 9]), a situation where homogenization techniques must definitely fail.

The idea to use the two projections theorem for the study of Toeplitz plus Hankel operators with piecewise continuous generating functions is taken from [182]. More about the study of related algebras can be found in the comments on Section 5.6 at the end of Chapter 5, and more about Toeplitz and Hankel operators in the monographs [21, 128, 138].

The computation of norms of SIOs is a long and still developing story. In 1968, Gohberg and Krupnik obtained lower estimates for the essential norms of $S_\mathbb{R}$, $P_\mathbb{R}$, and $Q_\mathbb{R}$. They also calculated the norm of $S_\mathbb{R}$ for $p = 2^n$ and $p = 2^n/(2^n - 1)$ with $n \in \mathbb{Z}^+$. Then, in 1973, Pichorides obtained the upper estimate for the norm of $S_\mathbb{R}$ and solved, thus, the problem of norm calculation for $S_\mathbb{R}$ on $L^p(\mathbb{R})$. This fact and its weighted analogs can be also found in [74, Chapter 13] and [108, Chapter 2]. The calculation of the norm of projections $P_\mathbb{R}$ and $Q_\mathbb{R}$ remained open for a long time. It was solved only in 2000 by Hollenbeck and Verbitsky [86]; see also [126]. Further,

Krupnik and Spigel [111] proved that, for any piecewise Lyapunov curve, there exists a power weight w such that the essential norm of S_Γ in the space $L^2(\Gamma, w)$ does not depend on angles of the curve. Essential norms of S_Γ in $L^2(\Gamma, w)$ over more complex curves and weights were the subject of various works (see [45, 46, 51, 58, 113], for instance).

As noted on Remark 4.1.1 the curves considered in Chapter 4 do not have cusps. Regarding singular integral operators with complex conjugation on curves with cusps, we refer to [47] where it is shown that both the angles at the corners of the curve and the order of the cusps strongly affect the Fredholm properties of the operator. The case of cusps of arbitrary order (on weighted L^2 spaces) was considered in [46].

Also connected with the results of Section 4.6, Böttcher, Karlovich and Rabinovich [17] established Fredholm property criteria and index formulas for operators in the algebra generated by singular integral operators with slowly oscillating coefficients and the operator of complex conjugation, in the case of slowly oscillating composed curves Γ and slowly oscillating Muckenhoupt weights w. In particular, they were able to consider curves with whirl points, in which case massive local spectra may emerge even for constant coefficients and weights. This was done using local principles and is a direct and far reaching generalization of the results of Section 4.6, in the spirit of this chapter.

One-dimensional singular integral operators can also be considered on Hölder-Zygmund spaces H^s with $s > 0$ and on weighted Hölder spaces. The work on these spaces involves not only some unpleasant technical subtleties (in contrast to L^p-spaces, Hölder spaces are neither reflexive nor separable), but also some more serious obstacles. A main point is that one does not get faithful localizing pairs when working with respect to the common essential norm. One way to overcome this difficulty is to replace the essential norm by the Kuratowski measure of non-compactness. The result of this substitution is an equivalent norm on the Calkin algebra with respect to which the coset containing the operator aI of multiplication by a function $a \in H^s$ again has norm $\|a\|_\infty$. This observation simplifies the theory of singular integral operators on Hölder spaces significantly. In particular, it allows the application of the two projections theorem to the study of the Fredholmness of singular integral operators on weighted Hölder spaces. For approaches to the study of Fredholm properties of singular integral operators by means of the Kuratowski measure of non-compactness, see [25, 142, 177].

The interpolation Theorem 4.8.2 was proved first in 1960 by Krasnoselskii for L^p-spaces over finite measure domains (see [103, Section 3.4]). In 1992, Cwikel [31] proved that if $T : E_0 \to F_0$ is a compact linear operator and if $T : E_1 \to F_1$ is a bounded linear operator, then $T : (E_0, E_1)_{\theta, q} \to (F_0, F_1)_{\theta, q}$ is also a compact linear operator, with $(E_0, E_1)_{\theta, q}$ and $(F_0, F_1)_{\theta, q}$ being the real interpolation spaces.

Chapter 5
Convolution operators

Having focused on algebras of singular integral operators in the last chapter, in this chapter convolution operator algebras will be treated. The idea is to give a general perspective of how the material in the first part of the book has been applied in the context of convolution operator algebras in recent decades, while at the same time including some previously unpublished material.

Throughout this chapter, let $1 < p < \infty$ and let w be a power weight on $\mathbb{R}$, i.e., w is of the form (4.1) with $t_i \in \mathbb{R}$ for $i = 1, \ldots, n$. We always assume that w belongs to the class $\mathcal{A}_p(\mathbb{R})$.

5.1 Multipliers and commutators

We will start this section by developing the subject begun in Section 4.2, focusing on specific classes of multipliers on the real line.

A function $a \in L^\infty(\mathbb{R})$ is called *piecewise linear* if there is a partition $-\infty = t_0 < t_1 < \ldots < t_n = +\infty$ of the real line and complex constants c_k, d_k such that $a(t) = c_0 \chi_{]-\infty, t_1[} + \sum_{k=1}^{n-2}(c_k + d_k t)\chi_{]t_k, t_{k+1}[} + d_0 \chi_{]t_{n-1}, +\infty[}$. As usual, the function χ_I represents the characteristic function of the set I.

Since $w \in \mathcal{A}_p(\mathbb{R})$, Stechkin's equality (4.13) implies that the multiplier algebra $\mathcal{M}_{p,w}$ contains the (non-closed) algebras C_0 of all continuous and piecewise linear functions on $\mathbb{R}$, and PC_0 of all piecewise constant functions on $\mathbb{R}$ having only finitely many discontinuities (jumps). Let $C_{p,w}$ and $PC_{p,w}$ represent the closure of C_0 and PC_0 in $\mathcal{M}_{p,w}$, respectively. When $w \equiv 1$, abbreviate $C_{p,w}$ and $PC_{p,w}$ to C_p and PC_p, and write C and PC for C_2 and PC_2, respectively. Thus, the algebras C and PC coincide with the algebras denoted by $C(\dot{\mathbb{R}})$ and $PC(\dot{\mathbb{R}})$ in previous chapters.

It is unknown for general weights w (not necessarily of power form) whether the multiplier algebra $\mathcal{M}_{p,w}$ is continuously embedded into $L^\infty(\mathbb{R}) = \mathcal{M}_2$. So it is by no means evident that functions in $PC_{p,w}$ are piecewise continuous again.

S. Roch et al., *Non-commutative Gelfand Theories*, Universitext, DOI 10.1007/978-0-85729-183-7_5, © Springer-Verlag London Limited 2011

Proposition 5.1.1. *Let* $1 < p < \infty$ *and* $w \in \mathcal{A}_p(\mathbb{R})$. *Then the algebra* $PC_{p,w}$ *is continuously embedded into* $L^\infty(\mathbb{R})$ *and, thus, into* PC. *Moreover, for* $a \in PC_{p,w}$,

$$\|a\|_{L^\infty(\mathbb{R})} \leq 3 \|S_\mathbb{R}\|_{p,w} \|a\|_{\mathcal{M}_{p,w}}.$$

Proof. For $a \in PC_{p,w}$, we find a sequence of piecewise constant functions a_n such that

$$\|a - a_n\|_{\mathcal{M}_{p,w}} = \|W^0(a) - W^0(a_n)\|_{\mathscr{L}(L^p(\mathbb{R},w))} \to 0$$

as $n \to \infty$. Given indices $m, n \in \mathbb{N}$ and a point $x \in \mathbb{R}$, choose piecewise constant characteristic functions $\chi_x^\pm$ with "small" support, both having a jump at x and both of total variation 2, such that

$$\chi_x^\pm (a_n - a_m) = \left(a_n(x^\pm) - a_m(x^\pm) \right) \chi_x^\pm.$$

The Stechkin inequality (4.13) yields

$$\begin{aligned}
|a_n(x^\pm) - a_m(x^\pm)| \| \chi_x^\pm \|_{\mathcal{M}_{p,w}} &= \| \left(a_n(x^\pm) - a_m(x^\pm) \right) \chi_x^\pm \|_{\mathcal{M}_{p,w}} \\
&= \| \chi_x^\pm (a_n - a_m) \|_{\mathcal{M}_{p,w}} \\
&\leq 3 \|S_\mathbb{R}\|_{p,w} \|a_n - a_m\|_{\mathcal{M}_{p,w}}. \tag{5.1}
\end{aligned}$$

Since $\chi_x^\pm$ is real-valued, the conjugate operator to $W^0(\chi_x^\pm) \in \mathscr{L}\left(L^p(\mathbb{R}, w)\right)$ is $W^0(\chi_x^\pm) \in \mathscr{L}\left(L^q(\mathbb{R}, w^{-1})\right)$, with $1/p + 1/q = 1$. Thus, by the Stein-Weiss interpolation theorem 4.8.1,

$$\begin{aligned}
1 = \| \chi_x^\pm \|_{L^\infty(\mathbb{R})} &= \|W^0(\chi_x^\pm)\|_{\mathscr{L}(L^2(\mathbb{R}))} \\
&\leq \|W^0(\chi_x^\pm)\|_{\mathscr{L}(L^p(\mathbb{R},w))}^{1/2} \|W^0(\chi_x^\pm)\|_{\mathscr{L}(L^q(\mathbb{R},w^{-1}))}^{1/2} \\
&= \|W^0(\chi_x^\pm)\|_{\mathscr{L}(L^p(\mathbb{R},w))} = \| \chi_x^\pm \|_{\mathcal{M}_{p,w}},
\end{aligned}$$

whence via (5.1)

$$|a_n(x^\pm) - a_m(x^\pm)| \leq 3 \|S_\mathbb{R}\|_{p,w} \|a_n - a_m\|_{\mathcal{M}_{p,w}}. \tag{5.2}$$

So we arrive at the inequality

$$\|a_n - a_m\|_{L^\infty(\mathbb{R})} \leq 3 \|S_\mathbb{R}\|_{p,w} \|a_n - a_m\|_{\mathcal{M}_{p,w}}, \tag{5.3}$$

from which we conclude that $a_n \to a$ in $L^\infty(\mathbb{R})$. Hence, by (5.3),

$$\begin{aligned}
\|a\|_{L^\infty(\mathbb{R})} &\leq \|a_n\|_{L^\infty(\mathbb{R})} + \|a - a_n\|_{L^\infty(\mathbb{R})} \\
&\leq 3 \|S_\mathbb{R}\|_{p,w} \left(\|a_n\|_{\mathcal{M}_{p,w}} + \|a - a_n\|_{\mathcal{M}_{p,w}} \right).
\end{aligned}$$

Letting n go to infinity we get the assertion.   ∎

Once the continuous embedding of $PC_{p,w}$ into $L^\infty(\mathbb{R})$ is established, the following propositions can be proved as in the case $w \equiv 1$ (see [43, Section 2]).

Proposition 5.1.2.

(i) *The Banach algebra $C_{p,w}$ is continuously embedded into C.*

(ii) *The maximal ideal space of $C_{p,w}$ is homeomorphic to $\dot{\mathbb{R}}$. In particular, any multiplier $a \in C_{p,w}$ is invertible in $C_{p,w}$ if and only if $a(t) \neq 0$ for all $t \in \dot{\mathbb{R}}$.*

(iii) *The algebras $C_{p,w}$ and $PC_{p,w} \cap C(\dot{\mathbb{R}})$ coincide with the closure in $\mathcal{M}_{p,w}$ of the set of all continuous functions on $\dot{\mathbb{R}}$ with finite total variation.*

Remark 5.1.3. Note that the inclusion $C_{p,w} \subseteq \mathcal{M}_{p,w} \cap C(\dot{\mathbb{R}})$ is proper for $p \neq 2$, see [53]. □

Proposition 5.1.4.

(i) *The maximal ideal space of $PC_{p,w}$ is homeomorphic to $\dot{\mathbb{R}} \times \{0,1\}$. In particular, a multiplier $a \in PC_{p,w}$ is invertible in $PC_{p,w}$ if and only if $a(t^{\pm}) \neq 0$ for all $t \in \dot{\mathbb{R}}$.*

(ii) *The algebra $PC_{p,w}$ coincides with the closure in $\mathcal{M}_{p,w}$ of the set of all piecewise continuous functions with finite total variation.*

5.2 Wiener-Hopf and Hankel operators

Denote, as before, the characteristic functions of the positive and negative half axis by χ_+ and χ_-, respectively, and let J stand for the operator $(Ju)(t) = u(-t)$ which is bounded and has norm 1 on $L^p(\mathbb{R}, w)$ if the weight function is symmetric, i.e., if $w(t) = w(-t)$ for all $t \in \mathbb{R}$.

Let $a \in \mathcal{M}_{p,w}$. The restriction of the operator $\chi_+ W^0(a) \chi_+ I$ onto the weighted Lebesgue space $L^p(\mathbb{R}^+, \chi_+ w)$ is called a *Wiener-Hopf operator* and will be denoted by $W(a)$. If, moreover, the weight on $\mathbb{R}$ is symmetric, then the restriction of the operator $\chi_+ W^0(a) \chi_- J$ onto $L^p(\mathbb{R}^+, \chi_+ w)$ is a *Hankel operator* and will be denoted by $H(a)$.

For symmetric weights, it is easy to see that for $a \in \mathcal{M}_{p,w}$ the function $\tilde{a} := Ja$, $\tilde{a}(t) = a(-t)$, is also a multiplier on $L^p(\mathbb{R}, w)$, and that the restriction of the operator $J\chi_- W^0(a) \chi_+ I$ onto $L^p(\mathbb{R}^+, \chi_+ w)$ coincides with the Hankel operator $H(\tilde{a})$. For $a, b \in \mathcal{M}_{p,w}$ one then has the fundamental identity

$$W(ab) = W(a)W(b) + H(a)H(\tilde{b}) \tag{5.4}$$

which, in a similar way to Proposition 4.5.1, follows easily from

$$\begin{aligned}
W(ab) &= \chi_+ W^0(ab)\chi_+ I = \chi_+ W^0(a)W^0(b)\chi_+ I \\
&= \chi_+ W^0(a)(\chi_+ + \chi_- JJ\chi_-)W^0(b)\chi_+ I \\
&= \chi_+ W^0(a)\chi_+ \cdot \chi_+ W^0(b)\chi_+ I + \chi_+ W^0(a)\chi_- J \cdot J\chi_- W^0(b)\chi_+ I.
\end{aligned}$$

The following theorems collect some basic properties of Wiener-Hopf operators with continuous generating functions. These results are the analogs of Theorems 4.1.5 and 4.1.8 for singular integral operators. Detailed proofs can be found in [21, 9.9. and 9.10].

Theorem 5.2.1. *The smallest closed subalgebra of $\mathscr{L}(L^p(\mathbb{R}^+))$ which contains all Wiener-Hopf operators $W(f)$ with $f \in C_p$ contains the ideal of the compact operators on $L^p(\mathbb{R}^+)$.*

Theorem 5.2.2 (Krein, Gohberg). *Let $f \in C_p$. Then the Wiener-Hopf operator $W(f)$ is Fredholm on $L^p(\mathbb{R}^+)$ if and only if $f(x) \neq 0$ for all $x \in \dot{\mathbb{R}}$. If this condition is satisfied, then the Fredholm index of $W(f)$ is the negative winding number of the curve $f(\dot{\mathbb{R}})$ with respect to the origin. If the index of $W(f)$ is zero, then $W(f)$ is invertible.*

We digress for a moment and return to the context of Chapter 3. The point is that we can now give an example of a sufficiently simple algebra which is generated by three idempotents, but which does not possess a finite-dimensional invertibility symbol.

Example 5.2.3. Let $\mathscr{A}$ denote the smallest closed unital subalgebra of $\mathscr{L}(L^2(\mathbb{R}))$ which contains the operator $P_{\mathbb{R}} = (I + S_{\mathbb{R}})/2$ and the operators $\chi_{\mathbb{R}^+}I$ and $\chi_{[0,1]}I$ of multiplication by the characteristic functions of the intervals $\mathbb{R}^+$ and $[0, 1]$, respectively. Thus, $\mathscr{A}$ is generated by three idempotents (actually, three orthogonal projections). Further, let $\mathscr{B}$ refer to the smallest closed subalgebra of $\mathscr{A}$ which contains all operators

$$\chi_{[0,1]}(\chi_{\mathbb{R}^+}S_{\mathbb{R}}\chi_{\mathbb{R}^+}I)^k \chi_{[0,1]}I \quad \text{with } k \in \mathbb{N}.$$

We identify $\mathscr{B}$ with a unital subalgebra of $\mathscr{L}(L^2([0, 1]))$ in the natural way. From Proposition 4.2.17 we conclude that the algebra $\mathscr{B}$ contains all operators $\chi_{[0,1]}M^0(h)\chi_{[0,1]}I$ with $h \in C(\dot{\mathbb{R}})$ with $h(\pm\infty) = 0$. The definition (4.24) of a Mellin convolution further entails that $\mathscr{B}$ contains all operators $E_2^{-1}W(a)E_2$ with $a \in C(\dot{\mathbb{R}})$. But then $\mathscr{B}$ must contain all compact operators on $L^2([0, 1])$ by Theorem 5.2.1. Since the ideal of all compact operators contains a copy of $\mathbb{C}^{l \times l}$ for all l, it is immediate that $\mathscr{B}$ (hence, $\mathscr{A}$) cannot possess a matrix symbol of any finite order.

Thus, even if the three idempotents are projections, and even if two of them commute, a matrix symbol does not need to exist. $\qquad\square$

5.3 Commutators of convolution operators

Now we turn our attention to commutators of convolution and related operators. The commutator $AB - BA$ of two operators A and B will be denoted by $[A, B]$.

Let $\bar{L}^\infty(\mathbb{R})$ denote the set of all functions $a \in L^\infty(\mathbb{R})$ for which the essential limits at infinity exist, i.e., for which there are complex numbers $a(-\infty)$ and $a(+\infty)$ such that

$$\lim_{t \to -\infty} \operatorname*{esssup}_{s \le t} |a(s) - a(-\infty)| = 0,$$

$$\lim_{t \to +\infty} \operatorname*{esssup}_{s \ge t} |a(s) - a(+\infty)| = 0$$

and write $\dot{L}^\infty(\mathbb{R})$ for the set of all functions $a \in \bar{L}^\infty(\mathbb{R})$ such that $a(-\infty) = a(+\infty)$.

Next we define the analogous classes $\bar{\mathcal{M}}_{p,w}$ and $\dot{\mathcal{M}}_{p,w}$ of multipliers. Let Q_t denote the characteristic function of the interval $\mathbb{R} \setminus [-t, t]$. Then we let $\bar{\mathcal{M}}_{p,w}$ refer to the set of all multipliers $a \in \mathcal{M}_{p,w}$ for which there are numbers $a(-\infty)$ and $a(+\infty)$ such that

$$\lim_{t \to \infty} \|Q_t(a - a(-\infty)\chi_- - a(+\infty)\chi_+)\|_{\mathcal{M}_{p,w}} = 0. \tag{5.5}$$

Notice that this definition makes sense since, by the Stechkin inequality (4.13), the characteristic functions Q_t, χ_+ and χ_- of $\mathbb{R} \setminus [-t, t]$, $\mathbb{R}^+$ and $\mathbb{R}^-$, respectively, belong to $\mathcal{M}_{p,w}$. Also notice that the numbers $a(-\infty)$ and $a(+\infty)$ are uniquely determined by a and that $\bar{L}^\infty(\mathbb{R}) = \bar{M}_2$. Further, let $\dot{\mathcal{M}}_{p,w}$ denote the class of all multipliers $a \in \bar{\mathcal{M}}_{p,w}$ such that $a(-\infty) = a(+\infty)$. Via Proposition 5.1.1 one easily gets that

$$PC_{p,w} \subseteq \bar{\mathcal{M}}_{p,w} \text{ and } C_{p,w} \subseteq \dot{\mathcal{M}}_{p,w}.$$

Recall finally that $\mathscr{K}(X)$ stands for the ideal of all compact operators on the Banach space X.

Proposition 5.3.1.

(i) *If $a \in \bar{L}^\infty(\mathbb{R})$, $b \in \dot{\mathcal{M}}_{p,w}$, and $a(\pm\infty) = b(\pm\infty) = 0$, then $aW^0(b)$ and $W^0(b)aI$ are in $\mathscr{K}(L^p(\mathbb{R}, w))$.*

(ii) *If one of the conditions*

1. *$a \in C(\dot{\mathbb{R}})$ and $b \in \bar{\mathcal{M}}_{p,w}$, or*
2. *$a \in \bar{L}^\infty(\mathbb{R})$ and $b \in C_{p,w}$, or*
3. *$a \in C(\bar{\mathbb{R}})$ and $b \in C(\bar{\mathbb{R}}) \cap PC_{p,w}$*

is fulfilled, then $[aI, W^0(b)] \in \mathscr{K}(L^p(\mathbb{R}, w))$.

Proof. (i) Since, by assumption, $\|Q_t a\|_\infty \to 0$ and $\|Q_t b\|_{\mathcal{M}_{p,w}} \to 0$, we can assume without loss of generality that a and b have compact support. Choose functions $u, v \in C_0^\infty(\mathbb{R})$, the space of infinitely differentiable functions with compact support, such that $u|_{\operatorname{supp} a} = 1$ and $v|_{\operatorname{supp} b} = 1$. Then

$$aW^0(b) = (au)W^0(vb) = auW^0(v)W^0(b),$$

and the assertion follows once we have shown that $uW^0(v)$ is compact. Put $k = F^{-1}v$. Then, for $f \in L^p(\mathbb{R}, w)$,

$$\left(uW^0(v)f\right)(t) = \int_{-\infty}^{+\infty} u(t)k(t-s)f(s)\,ds.$$

Since k is an infinitely differentiable function for which the function $t \mapsto t^m k(t)$ is bounded for any $m \in \mathbb{N}$, we have (with $\frac{1}{p} + \frac{1}{q} = 1$)

$$\int_{-\infty}^{+\infty} \left(\int_{-\infty}^{+\infty} |u(t)k(t-s)w^{-1}(s)|^q \, ds \right)^{p/q} dt < \infty,$$

whence the compactness of $uW^0(v)$ and, hence, of $aW^0(b)$ follows. Similarly, the compactness of $W^0(b)aI$ can be established.

(ii) 1. Write $b = b(-\infty)\chi_- + b(+\infty)\chi_+ + b'$ with $b' \in \dot{\mathcal{M}}_{p,w}$ and $b'(\pm\infty) = 0$. Then $[aI, W^0(b)] = K_1 + K_2 + K_3$ with

$$K_1 = (a - a(+\infty))W^0(b'), \qquad K_2 = -W^0(b')(a - a(+\infty))I,$$

and

$$K_3 = [aI, W^0(b(-\infty)\chi_- + b(+\infty)\chi_+)]$$
$$= \frac{b(+\infty) - b(-\infty)}{2}[aI, W^0(\text{sgn})].$$

The compactness of K_1 and K_2 is a consequence of part (i), and the compactness of K_3 follows from Theorem 4.1.4, since $W^0(\text{sgn}) = S_{\mathbb{R}}$.

2. Write $a = a(-\infty)\chi_- + a(+\infty)\chi_+ + a'$ with $a' \in \dot{L}^\infty(\mathbb{R})$ and $a'(\pm\infty) = 0$, and write b as $b' + b(\pm\infty)$. Then $[aI, W^0(b)] = K_1 + K_2 + K_3$ with

$$K_1 = a'W^0(b'), \qquad K_2 = -W^0(b')a'I,$$

and

$$K_3 = [(a(-\infty)\chi_- + a(+\infty)\chi_+)I, W^0(b)]$$
$$= (a(+\infty) - a(-\infty))(\chi_+ W^0(b)\chi_- I - \chi_- W^0(b)\chi_+ I).$$

By (i), the operators K_1 and K_2 are compact. To get the compactness of K_3 note that, by Proposition 5.1.2 (iii), we can assume, without loss of generality, that b has finite total variation. Hence, the operator $K_4 := \chi_+ W^0(b)\chi_- - \chi_- W^0(b)\chi_+$ is bounded on each space $L^p(\mathbb{R}, w)$ with $1 < p < \infty$ and $w \in A_p(\mathbb{R})$. By Krasnoselskii's interpolation theorem 4.8.2, it is sufficient to verify the compactness of K_4 in $L^2(\mathbb{R})$. Since the Fourier transform is a unitary operator on $L^2(\mathbb{R})$, the operator K_4 is compact if and only if the operator

$$FK_4F^{-1} = (F\chi_+ F^{-1})b(F\chi_- F^{-1}) - (F\chi_- F^{-1})b(F\chi_+ F^{-1})$$
$$= \frac{(F\text{sgn}F^{-1})bI - b(F\text{sgn}F^{-1})}{2}$$
$$= \frac{S_{\mathbb{R}}bI - bS_{\mathbb{R}}}{2}$$

is compact. The compactness of this operator has been established in Theorem 4.1.4.

3. As in 2. above we can assume without loss of generality that b is of finite total variation. By Krasnoselskii's interpolation theorem, we just have to verify the compactness of $[aI, W^0(b)]$ in $L^2(\mathbb{R})$. Put $b_\pm := (b(+\infty) \pm b(-\infty))/2$. Then

$$b(z) = b_+ + b_- \coth\left((z + \mathbf{i}/2)\pi\right) + b_0(z)$$

with $b_0 \in C(\mathbb{R})$ and $b_0(\pm\infty) = 0$. By (ii) 2., the commutator $[aI, W^0(b_0)]$ is compact, and it remains to show the compactness of $[aI, W^0(\coth(\cdot + \mathbf{i}/2)\pi)]$. Let E_2 be the operator defined in (4.21) for the weight $w \equiv 1$. Since $M^0(c) = E_2^{-1} W^0(c) E_2$ for all $c \in \mathcal{M}_p$, one has

$$E_2^{-1}\left[aI, W^0\left(\coth((\cdot + \mathbf{i}/2)\pi)\right)\right] E_2 = \left[a'I, M^0\left(\coth((\cdot + \mathbf{i}/2)\pi)\right)\right] = [a'I, S_{\mathbb{R}^+}]$$

by Proposition 4.2.11, where we wrote $a'(t) := a(\frac{1}{2\pi} \ln t)$ for $t \in \mathbb{R}^+$. The compactness of $[a'I, S_{\mathbb{R}^+}]$ on $L^2(\mathbb{R}^+)$ follows from Theorem 4.1.4 by applying the operator $\chi_+ I$ to both sides of the commutator $[cI, S_\mathbb{R}]$ where $c \in C(\mathbb{R})$ is such that $c\chi_+ = a'$. ∎

Let $\bar{L}^\infty(\mathbb{R}^+)$ refer to the Banach space of all measurable functions which possess essential limits at 0 and at ∞, write $\dot{L}^\infty(\mathbb{R}^+)$ for the set of these functions a from $\bar{L}^\infty(\mathbb{R}^+)$ with $a(0) = a(\infty)$, and put $\bar{C}(\mathbb{R}^+) := C(\mathbb{R}^+) \cap \bar{L}^\infty(\mathbb{R}^+)$ and $\dot{C}(\mathbb{R}^+) := C(\mathbb{R}^+) \cap \dot{L}^\infty(\mathbb{R}^+)$.

Proposition 5.3.2.
  (i) If $a \in \bar{L}^\infty(\mathbb{R}^+)$, $b \in \mathcal{M}_p$ and $a(0) = a(\infty) = b(\pm\infty) = 0$, then $aM^0(b)$ and $M^0(b)aI$ are in $\mathcal{K}(L^p(\mathbb{R}^+, w))$.
  (ii) If one of the conditions

    1. $a \in \dot{C}(\mathbb{R}^+)$ and $b \in \bar{\mathcal{M}}_p$,
    2. $a \in \bar{L}^\infty(\mathbb{R}^+)$ and $b \in C_p$, or
    3. $a \in C(\bar{\mathbb{R}})$ and $b \in C(\bar{\mathbb{R}}) \cap PC_p$

  is fulfilled, then $[aI, M^0(b)] \in \mathcal{K}(L^p(\mathbb{R}^+, w))$.

Proof. An operator A on $L^p(\mathbb{R}^+, w)$ is compact if and only if $wAw^{-1} \in \mathcal{L}(L^p(\mathbb{R}^+))$ is compact. Since $waw^{-1} = a$ for all $a \in L^\infty(\mathbb{R}^+)$ and since $wM^0(b)w^{-1}I = M^0(b)$ is bounded on $L^p(\mathbb{R}^+)$, it suffices to prove the compactness of $aM^0(b)$, $M^0(b)aI$ and $[aI, M^0(b)]$ on $L^p(\mathbb{R}^+)$ without weight. For this case, the assertions are immediate consequences of Proposition 5.3.1 via the isomorphism E_p. ∎

Proposition 5.3.3. Let $a, b \in PC_{p,w}$. Then:

  (i) if a and b have no common discontinuities,

$$W(ab) - W(a)W(b) = H(a)H(\tilde{b}) \in \mathcal{K}(L^p(\mathbb{R}^+, w));$$

  (ii) $[W(a), W(b)] \in \mathcal{K}(L^p(\mathbb{R}^+, w))$.

Proof. By definition, the functions a and b are $\mathfrak{M}_{p,w}$-limits of piecewise constant functions. So we can assume without loss of generality that a and b are piecewise constant. Since, moreover, every piecewise constant function is a finite sum of functions with one discontinuity, we can assume that a and b have at most one discontinuity. Finally, it is a clear consequence of Krasnoselskii's interpolation theorem, that it is sufficient to prove the result for $p = 2$ and $w \equiv 1$.

(i) Working on $L^2(\mathbb{R}^+)$, it is sufficient for us to consider, instead of $K_1 := W(ab) - W(a)W(b)$, the unitarily equivalent operator

$$K_2 := FK_1F^{-1} = F\chi_+F^{-1}abF\chi_+F^{-1} - F\chi_+F^{-1}aF\chi_+F^{-1}bF\chi_+F^{-1}$$
$$= Q_{\mathbb{R}}abQ_{\mathbb{R}} - Q_{\mathbb{R}}aQ_{\mathbb{R}}bQ_{\mathbb{R}} = Q_{\mathbb{R}}aP_{\mathbb{R}}bQ_{\mathbb{R}}$$

with $P_{\mathbb{R}} = (I + S_{\mathbb{R}})/2$ and $Q_{\mathbb{R}} = I - P_{\mathbb{R}}$. Note that if one of the functions a and b is in $C(\dot{\mathbb{R}})$, then the result follows immediately from Theorem 4.1.4. So let both a and b be discontinuous. Let $t_a, t_b \in \dot{\mathbb{R}}$ denote the (only) points of discontinuity of a and b, respectively, and assume for definiteness that $t_a < t_b$. If χ stands for the characteristic function of the interval $[t_a, t_b]$, then a and b can be written as $a = a_1\chi + a_2$, $b = b_1\chi + b_2$, respectively, with continuous functions a_1, a_2, b_1 and b_2 such that

$$a_1(t_b) = b_1(t_a) = 0. \tag{5.6}$$

Then

$$K_2 = Q_{\mathbb{R}}aP_{\mathbb{R}}bQ_{\mathbb{R}}$$
$$= Q_{\mathbb{R}}a_1\chi P_{\mathbb{R}}b_1\chi Q_{\mathbb{R}} + Q_{\mathbb{R}}a_1\chi P_{\mathbb{R}}b_2Q_{\mathbb{R}} + Q_{\mathbb{R}}a_2P_{\mathbb{R}}b_1\chi Q_{\mathbb{R}} + Q_{\mathbb{R}}a_2P_{\mathbb{R}}b_2Q_{\mathbb{R}},$$

and it remains to verify that the operator $Q_{\mathbb{R}}a_1\chi P_{\mathbb{R}}b_1\chi Q_{\mathbb{R}}$ is compact. But

$$Q_{\mathbb{R}}a_1\chi P_{\mathbb{R}}b_1\chi Q_{\mathbb{R}} = Q_{\mathbb{R}}\chi a_1 P_{\mathbb{R}}b_1\chi Q_{\mathbb{R}}$$
$$= Q_{\mathbb{R}}\chi P_{\mathbb{R}}a_1b_1\chi Q_{\mathbb{R}} + K_3$$
$$= Q_{\mathbb{R}}\chi a_1b_1\chi P_{\mathbb{R}}Q_{\mathbb{R}} + K_4 = K_4$$

with certain compact operators K_3 and K_4, because $a_1b_1\chi$ is continuous due to (5.6).

(ii) If a and b have no common points of discontinuity, then the assertion is an immediate consequence of the preceding one. So let a and b have common discontinuities. As above, we may assume that both a and b have exactly one point $s \in \mathbb{R}$ of discontinuity. Then there exist a constant β and a continuous function f such that $a = \beta b + f$. Thus,

$$[W(a), W(b)] = [W(\beta b + f), W(b)] = [W(f), W(b)],$$

and the assertion follows from part (i) of this proposition.   ∎

Proposition 5.3.4. *Let the weight function w_α be given by $w_\alpha(t) = |t|^\alpha$. Then:*

(i) *if $a \in C_{p,w_\alpha}$, $b \in \dot{\mathfrak{M}}_p$ and $a(0) = a(\pm\infty) = b(\pm\infty) = 0$, we have $W(a)M^0(b) \in \mathcal{K}(L^p(\mathbb{R}^+, w_\alpha))$ and $M^0(b)W(a) \in \mathcal{K}(L^p(\mathbb{R}^+, w_\alpha))$;*

(ii) *each of the following conditions is sufficient for the compactness of the commutator $[W(a), M^0(b)]$ on $L^p(\mathbb{R}^+, w_\alpha)$:*

1. *$a \in PC_{p,w_\alpha}$ and $b \in C(\overline{\mathbb{R}}) \cap PC_p$;*
2. *$a \in C_{p,w_\alpha}$ with $a(\pm\infty) = a(0)$ and $b \in \dot{\mathfrak{M}}_p$.*

Proof. (i) Since $\|Q_t b\|_{\mathfrak{M}_{p,w_\alpha}} \to 0$ as $t \to \infty$, we can assume that b has compact support. Let f_b be a continuous function with total variation 2 and such that $f_b(t) = 1$ for $t \in$ supp b. Then

$$M^0(b) = M^0(f_b b) = M^0(f_b)M^0(b),$$

so that it remains to show the compactness of $W(a)M^0(f_b)$. Due to Proposition 4.2.10, we can approximate f_b by functions of the form

$$f(t) = \sum_{k=0}^{n} \beta_k \coth^k \left((t + \mathbf{i}(1/p + \alpha))\pi\right), \tag{5.7}$$

for which

$$M^0(f) = \sum_{k=0}^{n} \beta_k S_{\mathbb{R}^+}^k = \sum_{k=0}^{n} \beta_k (W(\mathrm{sgn}))^k.$$

Since $a \cdot \mathrm{sgn}$ is a continuous function, we deduce from Proposition 5.3.3 (i) that

$$W(a)M^0(f) = \sum_{k=0}^{n} \beta_k W(a)(W(\mathrm{sgn}))^k$$

$$= \sum_{k=0}^{n} \beta_k W(a(\mathrm{sgn})^k) + K_1$$

$$= W\left(a \sum_{k=0}^{n} \beta_k (\mathrm{sgn})^k\right) + K_2$$

with compact operators K_1 and K_2. The assumption $b(\pm\infty) = 0$ and (5.7) imply that $\sum_{k=0}^{n} \beta_k(\pm 1)^k = 0$. Thus,

$$\sum_{k=0}^{n} \beta_k (\mathrm{sgn})^k \equiv 0,$$

which gives our claim. The inclusion $M^0(b)W(a) \in \mathcal{K}(L^p(\mathbb{R}^+, w_\alpha))$ can be proved similarly.

(ii) By Proposition 4.2.10, every Mellin convolution $M^0(b)$ with $b \in C(\overline{\mathbb{R}}) \cap PC_p$ belongs to the algebra $\mathcal{E}_{p,\alpha}$ generated by the operators I and $S_{\mathbb{R}^+}$. Thus, $M^0(b)$ is contained in the algebra generated by all Wiener-Hopf operators $W(a)$ with $a \in PC_p$. The latter algebra is commutative modulo the compact operators by Proposition 5.3.3 (ii), which implies the first assertion of (ii). For a proof of the second assertion, write $b_\pm = (b(+\infty) \pm b(-\infty))/2$ again. Then

$$b(z) = b_+ + b_- \coth\left(\left(z + \mathbf{i}(1/p + \alpha)\right)\pi\right) + b_0(z)$$

with $b_0 \in \dot{\mathcal{M}}_p$ and $b_0(\pm\infty) = 0$. By part (i) of this assertion, the operators

$$W(a) \quad \text{and} \quad M^0\left(b_+ + b_- \coth\left((z + \mathbf{i}(1/p + \alpha))\pi\right)\right)$$

commute modulo a compact operator, and we have only to deal with the commutator $[W(a), M^0(b_0)]$. Put $a' = a - a(\infty)$. Then $a' \in \dot{C}_{p,w}$ with $a'(\pm\infty) = a'(0) = 0$ and $[W(a), M^0(b_0)] = [W(a'), M^0(b_0)]$. Part (i) now gives that $[W(a'), M^0(b_0)]$ is compact, and the proof is finished. $\blacksquare$

Proposition 5.3.5. *Let* $w_\alpha(t) = t^\alpha$, $a \in \bar{L}^\infty(\mathbb{R}^+)$, $b \in \dot{C}_{p,w}$ *and* $c \in \bar{\mathcal{M}}_p$. *Then each of the conditions:*

  (i) $a(0) = b(0) = c(\pm\infty) = 0$;
 (ii) $a(0) = c(-\infty) = 0$ *and* $b(t) = 0$ *for all* $t > 0$;
(iii) $a(0) = c(+\infty) = 0$ *and* $b(t) = 0$ *for all* $t < 0$

is sufficient for the compactness of $aW(b)M^0(c)$ *on* $L^p(\mathbb{R}^+, w_\alpha)$.

Proof. First we show that it is sufficient to prove the assertion in the case when $a \in \bar{C}(\mathbb{R}^+)$ and $c \in C(\bar{\mathbb{R}}) \cap PC_p$. Let the function $a' \in \bar{C}(\mathbb{R}^+)$ have the same limits at 0 and ∞ as the function a, and let $c' \in C(\bar{\mathbb{R}}) \cap PC_p$ have the same limits at $\pm\infty$ as the function c. Then

$$
\begin{aligned}
aW(b)M^0(c) &= a'W(b)M^0(c) + (a - a')W(b)M^0(c) \\
&= a'W(b)M^0(c') + a'W(b)M^0(c - c') \\
&\quad + (a - a')W(b)M^0(c') + (a - a')W(b)M^0(c - c').
\end{aligned}
$$

The functions $a - a'$ and $c - c'$ can be approximated (in the supremum and the multiplier norm, respectively) by functions $a_0 \in \bar{L}^\infty(\mathbb{R}^+)$ and $c_0 \in \bar{\mathcal{M}}_p$ with compact support in $[0, +\infty[$ and $]-\infty, +\infty[$, respectively. Thus, $aW(b)M^0(c)$ can be approximated (in the operator norm) as closely as desired by operators of the form

$$a'W(b)M^0(c') + a'W(b)M^0(c_0) + a_0W(b)M^0(c') + a_0W(b)M^0(c_0). \tag{5.8}$$

Choose continuous functions f_0 and g_0 with total variation 2 such that $f_0 \equiv 1$ on supp a_0 and $g_0 \equiv 1$ on supp c_0. Then the operator (5.8) can be written as

$$
\begin{aligned}
a'W(b)M^0(c') &+ a'W(b)M^0(g_0)M^0(c_0) \\
&+ a_0 f_0 W(b)M^0(c') + a_0 f_0 W(b)M^0(g_0)M^0(c_0).
\end{aligned}
$$

It is easy to check that if the triple (a,b,c) satisfies the conditions of the proposition, then so does each of the triples

$$(a',b,g_0), \ (a',b,c'), \ (f_0,b,g_0) \ \text{and} \ (f_0,b,c').$$

Since each of the functions in these triples is continuous, we are indeed left with the proof of the assertion in the continuous setting.

We start the proof in the continuous setting by proving the assertion for a special choice of the functions a, b, c. Let

$$a_1(t) := e^{-1/t^2}, \qquad b_1(t) := \frac{t^4}{(1+t^2)^2}, \qquad c_1(t) := e^{-t^2/4}.$$

Then $c_1 = Mk$ with

$$k(x) = \frac{2}{\pi} x^{-1/p-\alpha} e^{-\ln^2 x}.$$

Indeed, since the function $z \mapsto e^{-z^2}$ is analytic in any strip $-m < \Im(z) < m$ and vanishes as $z \to \infty$ in that strip, one has, by the Cauchy integral theorem,

$$\begin{aligned} (Mk)(t) &= \frac{2}{\pi} \int_0^{+\infty} s^{-it} e^{-\ln^2 s} s^{-1} \, ds = \frac{2}{\pi} \int_{-\infty}^{+\infty} e^{-iyt-y^2} \, dy \\ &= \frac{2}{\pi} e^{-t^2/4} \int_{-\infty}^{+\infty} e^{-(y+\frac{i}{2}t)^2} \, dy = \frac{2}{\pi} e^{-t^2/4} \int_{\mathbb{R}+\frac{i}{2}i} e^{-z^2} \, dz \\ &= \frac{2}{\pi} e^{-t^2/4} \int_{-\infty}^{+\infty} e^{-z^2} \, dz = e^{-t^2/4} = c_1(t). \end{aligned}$$

Let

$$g(t) := \left(F^{-1}(1-b_1)\right)(t) = \int_{-\infty}^{+\infty} e^{2\pi i \lambda t} \frac{1+2\lambda^2}{(1+\lambda^2)^2} \, d\lambda.$$

For $u \in L^p(\mathbb{R}^+, w_\alpha)$, we then have

$$(W(b_1)u)(t) = u(t) - \int_0^\infty g(t-s)u(s) \, ds$$

and

$$\left(M^0(c_1)u\right)(t) = \int_0^\infty k\left(\frac{t}{s}\right) u(s)s^{-1} \, ds,$$

and the kernel $\mathfrak{K}$ of the integral operator $T := a_1 W(b_1) M^0(c_1) a_1 I$ is given by

$$\begin{aligned} \mathfrak{K}(x,y) &= a_1(x)a_1(y) \left(y^{-1} k\left(\frac{x}{y}\right) - \int_0^{+\infty} k\left(\frac{t}{y}\right) g(x-t) y^{-1} \, dt \right) \\ &= a_1(x)a_1(y) \left(y^{-1} h\left(\frac{x}{y}\right) - \int_{-\infty}^{+\infty} h\left(\frac{t}{y}\right) g(x-t) y^{-1} \, dt \right) \end{aligned}$$

with

$$h(z) = \begin{cases} k(z) & \text{if } z > 0, \\ 0 & \text{if } z < 0. \end{cases}$$

It is easy to see that h belongs to the Schwartz space $\mathscr{S}(\mathbb{R})$ of the rapidly decreasing infinitely differentiable functions on $\mathbb{R}$. Set $d := Fh \in \mathscr{S}(\mathbb{R})$. A simple calculation shows that then

$$\frac{1}{y} h\left(\frac{x}{y}\right) = (F^{-1}d_y)(x)$$

where $d_y(z) := d(yz)$ for $y > 0$ and $z \in \mathbb{R}$. Taking into account that $g = F^{-1}(1 - b_1)$, we can write $\mathfrak{K}(x,y)$ as

$$a_1(x)a_1(y)\left(\int_{-\infty}^{+\infty} e^{2\pi i z x} d(yz)\, dz - \int_{-\infty}^{+\infty} (F^{-1}d_y)(t) \cdot (F^{-1}(1 - b_1))(x - t)\, dt \right)$$

or, equivalently,

$$a_1(x)a_1(y)\left((F^{-1}d_y)(x) - (F^{-1}d_y) * (F^{-1}(1 - b_1))(x) \right).$$

By the convolution theorem, we thus get

$$\begin{aligned} \mathfrak{K}(x,y) &= a_1(x)a_1(y)\left((F^{-1}d_y)(x) - F^{-1}(d_y(1 - b_1))(x) \right) \\ &= a_1(x)a_1(y)F^{-1}(d_y b_1)(x) \\ &= a_1(x)a_1(y) \int_{-\infty}^{+\infty} e^{2\pi i z x} d(yz) b_1(z)\, dz. \end{aligned}$$

Integrating twice by parts we find

$$\mathfrak{K}(x,y) = -\frac{a_1(x)a_1(y)}{(2\pi i x)^2} \int_{-\infty}^{+\infty} e^{2\pi i z x} \frac{\partial^2}{\partial z^2}\left(d(yz)\frac{z^4}{(1 + z^2)^2} \right) dz.$$

Thus, there are constants C_1, C_2 and functions $d_1, d_2, d_3 \in \mathscr{S}(\mathbb{R})$ such that

$$\begin{aligned} |\mathfrak{K}(x,y)| &\leq C_1 \frac{|a_1(x)||a_1(y)|}{x^2} \sum_{m=0}^{2} y^m \int_{-\infty}^{+\infty} |d_m(yz)||z|^{m+2}\, dz \\ &= C_1 \frac{|a_1(x)|}{x^2} \frac{|a_1(y)|}{y^2} \sum_{m=0}^{2} y^m \int_{-\infty}^{+\infty} |d_m(yz)|(yz)^{m+2}\, dz \\ &\leq C_2 \frac{|a_1(x)|}{x^2} \frac{|a_1(y)|}{y^2}. \end{aligned}$$

Now insert $a_1(x) = e^{-1/x^2}$ and $q = p/(p-1)$ to obtain

$$\int_0^{+\infty} \left(\int_0^{+\infty} |\mathfrak{K}(x,y)|x^{-\alpha}|^q dx \right)^{p/q} dy < \infty.$$

This estimate implies that the operator $a_1W(b_1)M^0(c_1)$ is compact on $L^p(\mathbb{R}^+, w_\alpha)$, which settles the assertion for the specific functions a_1, b_1 and c_1.

Now we return to the general case when $a \in \bar{C}(\mathbb{R}^+)$, $b \in \dot{C}_{p,w}$ and $c \in C(\bar{\mathbb{R}}) \cap PC_p$.

(i) Let $a(0) = b(0) = c(\pm\infty) = 0$. Standard approximation arguments show that it is sufficient to consider functions with $a(t) = 0$ for $0 < t < \delta$ with some $\delta > 0$, $b(t) = 0$ for $|t| < \varepsilon$ with some $\varepsilon > 0$, and $c(t) = 0$ for $|t| > N$ with some $N > 0$. Due to Propositions 5.3.1–5.3.4,

$$aW(b)M^0(c) = \frac{a}{a_1^2}W(b/b_1)M^0(c/c_1)\left(a_1W(b_1)M^0(c_1)a_1I\right) + K_1$$

with a compact operator K_1. Since the operator $a_1W(b_1)M^0(c_1)$ is compact, the assertion follows.

(ii) Write

$$c(t) = c(+\infty)\frac{1 + \coth\left((t + \mathbf{i}(1/p + \alpha))\pi\right)}{2} + c'(t).$$

Then

$$aW(b)M^0(c) = aW(b)M^0(c') + c(+\infty)aW(b)M^0\left(\frac{1 + \coth\left((t + \mathbf{i}(1/p + \alpha))\pi\right)}{2}\right)$$

$$= aW(b)M^0(c') + c(+\infty)aW(b\chi_+) + K_2$$

with a compact operator K_2 (here we took into account Proposition 4.2.11). Since $b\chi_+ \equiv 0$ by assumption, and since $c'(\pm\infty) = 0$, this reduces our claim to the case previously considered in part (i). The proof of part (iii) proceeds analogously. ∎

The next proposition concerns commutators with Hankel operators. Since the flip operator is involved, we assume the weight to be symmetric.

Proposition 5.3.6. *Let w be a symmetric weight on $\mathbb{R}$, i.e., $w(x) = w(-x)$. Then:*

(i) *if $b \in C_{p,w}$ then $H(b) \in \mathcal{K}(L^p(\mathbb{R}^+, w))$;*
(ii) *if $a \in \bar{L}^\infty(\mathbb{R}^+)$ and $b \in \dot{M}_{p,w}$ with $a(+\infty) = 0$ and $b(\pm\infty) = 0$, then $aH(b)$ and $H(b)aI$ are in $\mathcal{K}(L^p(\mathbb{R}^+, w))$;*
(iii) *if $a \in \bar{C}(\mathbb{R}^+)$ and $b \in \dot{M}_{p,w}$, then $[aI, H(b)] \in \mathcal{K}(L^p(\mathbb{R}^+, w))$;*
(iv) *if $a \in C_{p,w}$ and $b \in M_{p,w}$ with a even, then $[W(a), H(b)] \in \mathcal{K}(L^p(\mathbb{R}^+, w))$;*
(v) *if $a \in C_{p,w}$ and $b \in M_{p,w}$, then $[W(a), W(b)] \in \mathcal{K}(L^p(\mathbb{R}^+, w))$.*

Proof. (i) As in the proof of Proposition 5.3.3, we can restrict ourselves to the case when $p = 2$ and $w \equiv 1$. Then $H(b)$ is unitarily equivalent to the operator

$$FH(b)F^{-1} = F\chi_+F^{-1}bFJ\chi_+F^{-1} = F\chi_+F^{-1}bFJ\chi_-F^{-1}J = Q_{\mathbb{R}}bP_{\mathbb{R}}J,$$

which is compact.

(ii) Extend a symmetrically onto the whole axis. Then $a \in \dot{L}^\infty(\mathbb{R})$ with $a(\pm\infty) = 0$, and from Proposition 5.3.1 (i) we conclude

$$a\chi_+ W^0(b)\chi_- J = \chi_+ aW^0(b)\chi_- J \in \mathcal{K}\left(L^p(\mathbb{R},w)\right)$$

and

$$\chi_+ W^0(b)\chi_- JaI = \chi_+ W^0(b)a\chi_- J \in \mathcal{K}\left(L^p(\mathbb{R},w)\right).$$

(iii) This follows from Proposition 5.3.1 (ii) via the same arguments as in (ii).

(iv) The identity

$$H(ab) = W(a)H(b) + H(a)W(\tilde{b}) \tag{5.9}$$

can be shown as (5.4). Consequently, if $\tilde{a} = a$, then

$$W(a)H(b) = H(ab) - H(a)W(\tilde{b}) = H(ba) - H(a)W(\tilde{b})$$
$$= W(b)H(a) + H(b)W(a) - H(a)W(\tilde{b})$$

which implies the assertion since $H(a)$ is compact by (i).

(v) Finally, by (5.4),

$$W(a)W(b) - W(b)W(a) = H(b)H(\tilde{a}) - H(a)H(\tilde{b}).$$

Since a and $\tilde{a}$ are continuous, the assertion follows from (i).   ∎

5.4 Homogenization of convolution operators

Here we continue the technical preparation for the local study of algebras of convolution operators. In particular we show that the homogenization technique from Section 4.2.5 applies to large classes of convolutions. Recall the definitions of the operators U_t, V_s and Z_τ in (4.19), (4.20) and (4.37), respectively.

For $s \in \mathbb{R}$ and $-1/p < \alpha < 1 - 1/p$, let $w_{\alpha,s}$ be the weight on $\mathbb{R}$ defined by $w_{\alpha,s}(t) := |t - s|^\alpha$ and write w_α for $w_{\alpha,0}$. Let $A \in \mathcal{L}(L^p(\mathbb{R},w_{\alpha,s}))$. If the strong limit

$$\underset{\tau \to +\infty}{\text{s-lim}}\ Z_\tau V_{-s} A V_s Z_\tau^{-1} \tag{5.10}$$

exists, we denote it by $\mathsf{H}_{s,\infty}(A)$. Analogously, if $A \in \mathcal{L}(L^p(\mathbb{R},w_\alpha))$ and if the strong limit

$$\underset{\tau \to +\infty}{\text{s-lim}}\ Z_\tau^{-1} U_t A U_{-t} Z_\tau \tag{5.11}$$

exists for some $t \in \mathbb{R}$, we denote it by $\mathsf{H}_{\infty,t}(A)$. It is easy to see that the set of all operators for which the strong limits $\mathsf{H}_{s,\infty}(A)$ (resp. $\mathsf{H}_{\infty,t}(A)$) exist forms a Banach algebra, that

$$\|\mathsf{H}_{s,\infty}(A)\|_{\mathcal{L}\left(L^p(\mathbb{R},w_{\alpha,0})\right)} \leq \|A\|_{\mathcal{L}\left(L^p(\mathbb{R},w_{\alpha,s})\right)} \tag{5.12}$$

respectively

$$\|\mathsf{H}_{\infty,t}(A)\|_{\mathscr{L}(L^p(\mathbb{R},w_{\alpha,0}))} \leq \|A\|_{\mathscr{L}(L^p(\mathbb{R},w_{\alpha,0}))} \tag{5.13}$$

for all operators in these algebras and that, hence, the operators $\mathsf{H}_{s,\infty}$ and $\mathsf{H}_{\infty,t}$ are bounded homomorphisms.

For $x \in \mathbb{R}$, let $b(x^{\pm})$ denote the right/left one-sided limit of the piecewise continuous function b at x.

Proposition 5.4.1. *Let* $a \in \bar{L}^{\infty}(\mathbb{R})$, $b \in PC_{p,w_{\alpha}}$ *and* $c \in \tilde{\mathsf{M}}_p$. *Then, for* $t \in \mathbb{R}$:

(i) $\mathsf{H}_{\infty,t}(aI) = a(-\infty)\chi_- I + a(+\infty)\chi_+ I$;

(ii) $\mathsf{H}_{\infty,t}(W^0(b)) = b(t^-)Q_{\mathbb{R}} + b(t^+)P_{\mathbb{R}}$;

(iii) $\mathsf{H}_{\infty,t}(\chi_+ M^0(c)\chi_+ I + \chi_- I) = \begin{cases} c(+\infty)\chi_+ I + \chi_- I & \text{if } t > 0, \\ \chi_+ M^0(c)\chi_+ I + \chi_- I & \text{if } t = 0, \\ c(-\infty)\chi_+ I + \chi_- I & \text{if } t < 0. \end{cases}$

Proof. Assertion (i) is immediate from Proposition 4.2.22 (ii) since $U_t a U_{-t} = aI$.

(ii) It is sufficient to prove the assertion for $t = 0$. Write b as $b(0^-)\chi_- + b(0^+)\chi_+ + b_0$ where the function $b_0 \in PC_{p,w_{\alpha}}$ is continuous at 0 and $b_0(0) = 0$. Since

$$W^0\left(b(0^-)\chi_- + b(0^+)\chi_+\right) = b(0^-)Q_{\mathbb{R}} + b(0^+)P_{\mathbb{R}}$$

and the operators $P_{\mathbb{R}}$ and $Q_{\mathbb{R}}$ commute with Z_{τ}, it remains to show that

$$Z_{\tau}^{-1}W^0(b_0)Z_{\tau} \to 0 \quad \text{strongly as } \tau \to \infty.$$

By the definition of the class $PC_{p,w_{\alpha}}$ and by Proposition 5.1.1, we can approximate the function b_0 in the multiplier norm as closely as desired by a piecewise constant function b_{00} which is zero in an open neighborhood U of 0. It is thus sufficient to show that

$$Z_{\tau}^{-1}W^0(b_{00})Z_{\tau} \to 0 \quad \text{strongly as } \tau \to \infty.$$

Since the operators on the left-hand side are uniformly bounded with respect to τ, it is further sufficient to show that

$$Z_{\tau}^{-1}W^0(b_{00})Z_{\tau}u \to 0$$

for all functions u in a certain dense subset of $L^p(\mathbb{R}, w_{\alpha})$. For, consider the set of all functions in the Schwartz space $\mathscr{S}(\mathbb{R})$, the Fourier transform of which has compact support. This space is indeed dense in $L^p(\mathbb{R}, w_{\alpha})$ since the space $\mathscr{D}(\mathbb{R})$ of the compactly supported infinitely differentiable functions is dense in $\mathscr{S}(\mathbb{R})$ ([171, Theorem 7.10]), since the Fourier transform F is a continuous bijection on $\mathscr{S}(\mathbb{R})$, and since $\mathscr{S}(\mathbb{R})$ is dense in $L^p(\mathbb{R}, w_{\alpha})$. In this special setting, the latter fact can easily be proved by hand. Note in this connection that already $\mathscr{D}(\mathbb{R})$ is dense in $L^p(\mathbb{R}, w)$ for *every* Muckenhoupt weight w; see [80, Exercise 9.4.1]. If u is a function with these properties, then

$$Z_{\tau}^{-1}W^0(b_{00})Z_{\tau}u = F^{-1}Z_{\tau}b_{00}Z_{\tau}^{-1}Fu. \tag{5.14}$$

If τ is sufficiently large, then the support of Fu is contained in U; hence, the function on the right-hand side of (5.14) is the zero function.

(iii) For $t = 0$, the assertion follows immediately from Lemma 4.2.13 and the fact that $Z_t\chi_- I = \chi_- Z_t$. Let $t \neq 0$. Then, clearly, $U_{-t}Z_\tau = Z_\tau U_{-t\tau}$ and $Z_\tau^{-1}U_t = U_{t\tau}Z_\tau^{-1}$, whence

$$Z_\tau^{-1}U_t(\chi_+ M^0(c)\chi_+ I + \chi_- I)U_{-t}Z_\tau = U_{t\tau}Z_\tau^{-1}(\chi_+ M^0(c)\chi_+ I + \chi_- I)Z_\tau U_{-t\tau}$$
$$= U_{t\tau}(\chi_+ M^0(c)\chi_+ I + \chi_- I)U_{-t\tau}$$

due to Lemma 4.2.13. Thus,

$$H_{\infty,t}(\chi_+ M^0(c)\chi_+ I + \chi_- I)$$
$$= \begin{cases} \text{s-lim}_{\tau \to +\infty} U_\tau(\chi_+ M^0(c)\chi_+ I + \chi_- I)U_{-\tau} & \text{if } t > 0, \\ \text{s-lim}_{\tau \to -\infty} U_\tau(\chi_+ M^0(c)\chi_+ I + \chi_- I)U_{-\tau} & \text{if } t < 0. \end{cases} \tag{5.15}$$

To deal with the strong limits (5.15), suppose first that c is a polynomial in the function $\coth\left((\cdot + \mathbf{i}(1/p + \alpha))\pi\right)$, i.e.

$$c(t) = \sum_{k=0}^{n} c_k \coth^k\left((t + \mathbf{i}(1/p + \alpha))\pi\right),$$

with certain constants c_k. Then

$$\chi_+ M^0(c)\chi_+ I + \chi_- I = \chi_+ \left(\sum_{k=0}^{n} c_k(W(\text{sgn}))^k\right)\chi_+ I + \chi_- I. \tag{5.16}$$

So it remains to consider the strong limits

$$\text{s-lim}_{\tau \to \pm\infty} U_\tau(\chi_+ W^0(\text{sgn})\chi_+ I + \chi_- I)U_{-\tau}.$$

Since $U_\tau W^0(\text{sgn})U_{-\tau} = W^0(V_{-\tau}\text{sgn}V_\tau)$ by Lemma 4.2.4, we have just to check the strong convergence of $V_{-\tau}\text{sgn}V_\tau$ as $\tau \to \pm\infty$. One has

$$V_{-\tau}aV_\tau \to \begin{cases} a(+\infty)I & \text{as } \tau \to +\infty, \\ a(-\infty)I & \text{as } \tau \to -\infty \end{cases} \tag{5.17}$$

for every function $a \in \tilde{\mathcal{M}}_{p,w_\alpha}$. Thus,

$$U_\tau W^0(\text{sgn})U_{-\tau} \to \pm I \quad \text{as } \tau \to \pm\infty$$

whence, via (5.16),

$$\text{s-lim}_{\tau \to \pm\infty} U_\tau(\chi_+ M^0(c)\chi_+ I + \chi_- I)U_{-\tau} = \chi_+ \sum_{k=0}^{n} c_k(\pm 1)^k \chi_+ I + \chi_- I = c(\pm\infty)\chi_+ I + \chi_- I$$

for each polynomial c. Since the set of all polynomials is dense in $\bar{C}_p$, we obtain assertion (iii) for functions $c \in \bar{C}_p$. To treat the general case, let $c \in \tilde{M}_p$ and write c as

$$c(t) = c_+ + c_- \coth\big((t + \mathbf{i}(1/p + \alpha))\pi\big) + c'(t)$$

with

$$c_\pm := (c(+\infty) \pm c(-\infty))/2$$

and with a function $c' \in \tilde{M}_p$ with $c'(\pm\infty) = 0$. After what has just been proved, we are left to verify that

$$H_{\infty,t}(\chi_+ M^0(c')\chi_+ I + \chi_- I) = \chi_- I \quad \text{for } t \neq 0.$$

Without loss of generality, we can assume that the support of c' is compact. Choose a function $u \in C_0^\infty(\mathbb{R})$ with total variation 2 which is identically 1 on the support of c'. Then

$$U_\tau \chi_+ M^0(c')\chi_+ U_{-\tau} = U_\tau \chi_+ M^0(c'u)\chi_+ U_{-\tau}$$
$$= \big(U_\tau \chi_+ M^0(c')\chi_+ U_{-\tau}\big)\big(U_\tau \chi_+ M^0(u)\chi_+ U_{-\tau}\big).$$

The operators in the first parentheses are uniformly bounded with respect to τ, whereas the operators in the second ones tend strongly to zero as $\tau \to \pm\infty$ by what we have already shown. ∎

The assertions of the following lemma are either taken directly from the preceding proof, or they follow by repeating some arguments of that proof.

Lemma 5.4.2.
  (i) *If $a \in \bar{L}^\infty(\mathbb{R})$, then $V_{-\tau} a V_\tau \to a(\pm\infty)I$ as $\tau \to \pm\infty$.*
  (ii) *If $b \in \tilde{M}_{p,w_\alpha}$, then $U_\tau W^0(b)U_{-\tau} \to b(\pm\infty)I$ as $\tau \to \pm\infty$.*
  (iii) *If $c \in \tilde{M}_p$, then $U_\tau M^0(c)U_{-\tau} \to c(\pm\infty)I$ as $\tau \to \pm\infty$.*

The following is the analog of Proposition 5.4.1 for the second family of strong limits.

Proposition 5.4.3. *Let $s \in \mathbb{R}$, $a \in PC$, $b \in \tilde{M}_{p,w_{\alpha,s}}$ and $c \in \tilde{M}_p$. Then:*

  (i) $H_{s,\infty}(aI) = a(s^-)\chi_- I + a(s^+)\chi_+ I$;
  (ii) $H_{s,\infty}(W^0(b)) = b(-\infty)Q_\mathbb{R} + b(+\infty)P_\mathbb{R}$;
  (iii) $H_{s,\infty}(\chi_+ M^0(c)\chi_+ I + \chi_- I) = \begin{cases} c(-\infty)Q_\mathbb{R} + c(+\infty)P_\mathbb{R} & \text{if } s > 0, \\ \chi_+ M^0(c)\chi_+ I + \chi_- I & \text{if } s = 0, \\ I & \text{if } s < 0. \end{cases}$

Proof. Assertion (i) is immediate from Proposition 4.2.22 (ii) since $(V_{-s} a V_s)(t) = a(t+s)$. For assertion (ii), one uses Lemma 4.2.4 and Proposition 4.2.22 (ii) as in the proof of Proposition 5.4.1(ii). Note that $V_{-s} W^0(b)V_s = W^0(b)$. For $s = 0$, assertion (iii) is a consequence of Lemma 4.2.13 and the commutativity of $\chi_- I$ and Z_τ, and for $s < 0$ it follows from the already proved part (i). For $s > 0$, write

$$c(t) = c_+ + c_- \coth\big((t + \mathbf{i}(1/p + \alpha))\pi\big) + c'(t)$$

with $c_\pm := (c(+\infty) \pm c(-\infty))/2$. First note that

$$\mathsf{H}_{s,\infty}\left(\chi_+ M^0\left(c_+ + c_- \coth\big((\cdot + \mathbf{i}(1/p + \alpha))\pi\big) + c'(t)\right)\chi_+ I + \chi_- I\right)$$

$$= c_+ I + c_- S_{\mathbb{R}} = c(-\infty)\frac{I - S_{\mathbb{R}}}{2} + c(+\infty)\frac{I + S_{\mathbb{R}}}{2}.$$

Indeed, since

$$M^0\left(c_+ + c_- \coth\big((\cdot + \mathbf{i}(1/p + \alpha))\pi\big)\right) = c_+ \chi_+ I + c_- S_{\mathbb{R}^+} = c_+ \chi_+ I + c_- W(\mathrm{sgn}),$$

this follows easily from what we proved in parts (i) and (ii). Next, similarly to the proof of part (iii) of Proposition 5.4.1, one verifies the assertion for c being a polynomial in $\coth\big((\cdot + \mathbf{i}(1/p + \alpha))\pi\big)$. Finally, one shows that $\mathsf{H}_{s,\infty}(\chi_+ M^0(c')\chi_+ I + \chi_- I) = 0$, again by employing the approximation arguments from the proof of Proposition 5.4.1. ∎

Proposition 5.4.4.
   (i) *If K is a compact operator on $L^p(\mathbb{R}, w_\alpha)$, then $\mathsf{H}_{\infty,s}(K) = 0$ for all $s \in \mathbb{R}$.*
   (ii) *If K is compact on $L^p(\mathbb{R}, w_{\alpha,s})$, then $\mathsf{H}_{s,\infty}(K) = 0$ for all $s \in \mathbb{R}$.*

The proof runs as that of Proposition 4.2.22 (iii).

5.5 Algebras of multiplication, Wiener-Hopf and Mellin operators

Given subsets $X \subseteq L^\infty(\mathbb{R}^+)$, $Y \subseteq \mathcal{M}_{p,w_\alpha}$ and $Z \subseteq \mathcal{M}_p$, we let $\mathscr{A}(X,Y,Z)$ denote the smallest closed subalgebra of the algebra of all bounded linear operators on $L^p(\mathbb{R}^+, w_\alpha)$ which contains all multiplication operators aI with $a \in X$, all Wiener-Hopf operators $W(b)$ with $b \in Y$, and all Mellin convolutions $M^0(c)$ with $c \in Z$. By $\mathscr{A}^{\mathcal{K}}(X,Y,Z)$ we denote the image of $\mathscr{A}(X,Y,Z)$ in the Calkin algebra over $L^p(\mathbb{R}^+, w_\alpha)$, and we write Φ for the corresponding canonical homomorphism.

The invertibility of elements of the algebra $\mathscr{A}^{\mathcal{K}}(X,Y,Z)$ will again be studied by using Allan's local principle. Thus we must single out central subalgebras of this algebra which are suitable for localization. For special choices of X, Y and Z, this is done in the next proposition, which follows immediately from Propositions 5.3.1–5.3.6.

Proposition 5.5.1. *In each of the cases below, $\mathscr{B}$ is a central subalgebra of $\mathscr{A}$:*

   (i) $\mathscr{A} = \mathscr{A}^{\mathcal{K}}\left(L^\infty(\mathbb{R}^+), \tilde{\mathcal{M}}_{p,w_\alpha}, \tilde{\mathcal{M}}_p\right)$ *and* $\mathscr{B} = \mathscr{A}^{\mathcal{K}}\left(\dot{C}(\mathbb{R}^+), C^0_{p,w_\alpha}, C(\overline{\mathbb{R}}) \cap PC_p\right);$
   (ii) $\mathscr{A} = \mathscr{A}^{\mathcal{K}}\left(PC(\mathbb{R}^+), PC_{p,w_\alpha}, PC_p\right)$ *and* $\mathscr{B} = \mathscr{A}^{\mathcal{K}}\left(\dot{C}(\mathbb{R}^+), C^0_{p,w_\alpha}, C_p\right);$

(iii) $\mathscr{A} = \mathscr{A}^{\mathscr{K}}\left(PC(\mathbb{R}^+), PC_{p,w_\alpha}, PC_p\right)$ and $\mathscr{B} = \mathscr{A}^{\mathscr{K}}\left(\dot{C}(\mathbb{R}^+), C^0_{p,w_\alpha}, \emptyset\right)$;

(iv) $\mathscr{A} = \mathscr{A}^{\mathscr{K}}\left(\bar{C}(\mathbb{R}^+), PC_{p,w_\alpha}, C(\overline{\mathbb{R}}) \cap PC_p\right)$ and $\mathscr{B} = \mathscr{A}$;

(v) $\mathscr{A} = \mathscr{A}^{\mathscr{K}}\left(PC(\mathbb{R}^+), PC_{p,w_\alpha}, \emptyset\right)$ and $\mathscr{B} = \mathscr{A}^{\mathscr{K}}\left(\bar{C}(\mathbb{R}^+), C_{p,w_\alpha}, \emptyset\right)$.

It is evident that the setting of case (i) is too general for a successful analysis. It will be our goal in this section to examine cases (ii) and (iii). One peculiarity of the present context is that, in general, $(\mathscr{A}, \mathscr{B})$ is not a faithful localizing pair unless $p = 2$ (since one has to localize over algebras of multipliers). Thus, one cannot expect the same elegant and complete results as for the algebra $\mathscr{A}^{\mathscr{K}}(\Gamma, w)$ considered in the previous chapter. The objectives of this section are quite modest when compared with Chapter 4: We will only derive necessary and sufficient conditions for the invertibility of cosets in $\mathscr{A}^{\mathscr{K}}$, and we will show that this algebra is inverse-closed in the Calkin algebra, i.e., that invertibility in $\mathscr{A}^{\mathscr{K}}$ is equivalent to the Fredholm property. On the other hand, we will at least be able to establish *isometrically isomorphic* representations of the *local algebras* that arise. If $p = 2$, the localizing pairs become faithful, and one gets an isometrically isomorphic representation of the (global) algebra $\mathscr{A}^{\mathscr{K}}$.

We derive the maximal ideal spaces for some algebras $\mathscr{B}$ which appear in the above proposition. Let us start with two very simple situations. Corollary 1.4.9 and Proposition 1.4.11 (which we need here for weighted L^p-spaces) imply that the maximal ideal space of the commutative Banach algebra $\mathscr{A}^{\mathscr{K}}\left(\dot{C}(\mathbb{R}^+), \emptyset, \emptyset\right)$ is homeomorphic to the one-point compactification $\dot{\mathbb{R}}^+$ of $\mathbb{R}^+$ by the point $\infty = 0$ (and, thus, homeomorphic to a circle). The maximal ideal which corresponds to $s \in \dot{\mathbb{R}}^+$ is equal to $\{\Phi(aI) : a \in \dot{C}(\mathbb{R}^+), a(s) = 0\}$.

Taking into account Proposition 5.1.2 (i), it is also not hard to see that the maximal ideal space of the commutative Banach algebra $\mathscr{A}^{\mathscr{K}}\left(\emptyset, C^0_{p,w_\alpha}, \emptyset\right)$ is homeomorphic to the compactification of $\mathbb{R}$ which arises by identifying the three points $-\infty$, 0, and $+\infty$. We denote this compactification by $\dot{\mathbb{R}}_0$. One can think of $\dot{\mathbb{R}}_0$ as the

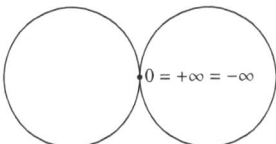

Fig. 5.1 The maximal ideal space of the algebra $\mathscr{A}^{\mathscr{K}}\left(\emptyset, C^0_{p,w_\alpha}, \emptyset\right)$.

the union of two circles which have exactly one point, ∞ say, in common (see Figure 5.1). The maximal ideal of $\mathscr{A}^{\mathscr{K}}\left(\emptyset, C^0_{p,w_\alpha}, \emptyset\right)$ which corresponds to the point $s \in \dot{\mathbb{R}}_0$ is then

$$\{\Phi(W(a)) : a \in C^0_{p,w_\alpha}, a(s) = 0\} \qquad \text{if } s \neq \infty,$$

$$\{\Phi(W(a)) : a \in C^0_{p,w_\alpha}, a(0) = a(\pm\infty) = 0\} \quad \text{if } s = \infty.$$

For the next result, we have to combine the maximal ideal spaces $\dot{\mathbb{R}}^+$ and $\dot{\mathbb{R}}_0$ of these algebras.

Proposition 5.5.2. *The maximal ideal space of the commutative Banach algebra $\mathscr{A}^{\mathscr{K}}\left(\dot{C}(\mathbb{R}^+), C^0_{p,w_\alpha}, \emptyset\right)$ is homeomorphic to that subset of the "double torus" $\dot{\mathbb{R}}^+ \times \dot{\mathbb{R}}_0$ which consists of the circle $\dot{\mathbb{R}}^+ \times \{\infty\}$ and the "double circle" $\{\infty\} \times \dot{\mathbb{R}}_0$. In particular, the value of the Gelfand transform of the coset $\Phi(aW(b))$ with $a \in \dot{C}(\mathbb{R}^+)$ and $b \in C^0_{p,w_\alpha}$ at the point $(s,t) \in (\dot{\mathbb{R}}^+ \times \{\infty\}) \cup (\{\infty\} \times \dot{\mathbb{R}}_0)$ is $a(s)b(t)$.*

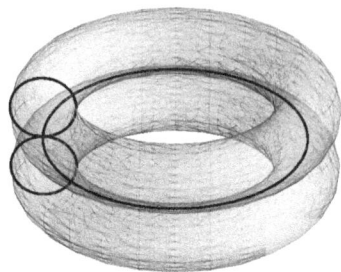

Fig. 5.2 The maximal ideal space of the algebra $\mathscr{A}^{\mathscr{K}}\left(\dot{C}(\mathbb{R}^+), C^0_{p,w_\alpha}, \emptyset\right)$. The intersection point is $\infty \times \infty$ (or 0×0); the points on the single circle are of the form $s \times \infty$, and the ones on the double circle of the form $\infty \times t$.

Proof. Let $\mathscr{J}$ be a maximal ideal of $\mathscr{A}^{\mathscr{K}}\left(\dot{C}(\mathbb{R}^+), C^0_{p,w_\alpha}, \emptyset\right)$. By Proposition 2.2.1, $\mathscr{J} \cap \mathscr{A}^{\mathscr{K}}\left(\dot{C}(\mathbb{R}^+), \emptyset, \emptyset\right)$ and $\mathscr{J} \cap \mathscr{A}^{\mathscr{K}}\left(\emptyset, C^0_{p,w_\alpha}, \emptyset\right)$ are maximal ideals of $\mathscr{A}^{\mathscr{K}}\left(\dot{C}(\mathbb{R}^+), \emptyset, \emptyset\right)$ and $\mathscr{A}^{\mathscr{K}}\left(\emptyset, C^0_{p,w_\alpha}, \emptyset\right)$, respectively. Thus, there are points $s \in \dot{\mathbb{R}}^+$ and $t \in \dot{\mathbb{R}}_0$ such that the value of the Gelfand transform of the coset $\Phi(aW(b))$ at the ideal $\mathscr{J}$ equals $a(s)b(t)$ for each choice of $a \in \dot{C}(\mathbb{R}^+)$ and $b \in C^0_{p,w_\alpha}$. Hence, the maximal ideal space of the algebra $\mathscr{A}^{\mathscr{K}}\left(\dot{C}(\mathbb{R}^+), C^0_{p,w_\alpha}, \emptyset\right)$ can be identified with a subset of the double torus $\dot{\mathbb{R}}^+ \times \dot{\mathbb{R}}_0$.

Now let $s \in \dot{\mathbb{R}}^+ \setminus \{\infty\}$ and $t \in \dot{\mathbb{R}}_0 \setminus \{\infty\}$. Given functions $a \in \dot{C}(\mathbb{R}^+)$ and $b \in C^0_{p,w_\alpha}$, choose functions $a' \in C^\infty_0(\mathbb{R}^+)$ and $b' \in C^\infty_0(\mathbb{R})$ of finite total variation such that $a(s) = a'(s)$, $b(t) = b'(t)$ and $0 \notin \operatorname{supp} b'$. Then,

$$aW(b) = (a - a')W(b - b') + (a - a')W(b') + a'W(b - b') + a'W(b').$$

The first three items of the sum on the right-hand side belong to the ideal $\mathscr{J} = (s,t)$, whereas while the fourth item is compact by Proposition 5.3.1. Thus, the smallest closed ideal of $\mathscr{A}^{\mathscr{K}}\left(\dot{C}(\mathbb{R}^+), C^0_{p,w_\alpha}, \emptyset\right)$ which corresponds to (s,t) with $s \in \dot{\mathbb{R}}^+ \setminus \{\infty\}$ and $t \in \dot{\mathbb{R}}_0 \setminus \{\infty\}$ coincides with the whole algebra. So, the maximal ideals of the algebra under consideration can only correspond to points (s,t) from $(\dot{\mathbb{R}}^+ \times \{\infty\}) \cup (\{\infty\} \times \dot{\mathbb{R}}_0)$. On the other hand, each of these points gives a maximal ideal of $\mathscr{A}^{\mathscr{K}}\left(\dot{C}(\mathbb{R}^+), C^0_{p,w_\alpha}, \emptyset\right)$, which is a consequence of Theorem 2.1.9 (ii). Since the

Shilov boundaries of the algebras $\mathscr{A}^{\mathscr{K}}\left(\dot{C}(\mathbb{R}^+),\emptyset,\emptyset\right)$ and $\mathscr{A}^{\mathscr{K}}\left(\emptyset,C^0_{p,w_\alpha},\emptyset\right)$ coincide with $\dot{\mathbb{R}}^+$ and $\dot{\mathbb{R}}_0$, respectively (recall Exercise 2.1.7), the assertion follows. ∎

For $s \in \dot{\mathbb{R}}^+ \times \{\infty\}$ and $t \in \{\infty\} \times \dot{\mathbb{R}}_0$, let $\mathscr{I}_{s,t}$ denote the smallest closed ideal of the Banach algebra $\mathscr{A}^{\mathscr{K}}\left(PC(\mathbb{R}^+),PC_{p,w_\alpha},PC_p\right)$ which contains the ideal (s,t), and let $\Phi^{\mathscr{K}}_{s,t}$ refer to the canonical homomorphism from $\mathscr{A}\left(PC(\mathbb{R}^+),PC_{p,w_\alpha},PC_p\right)$ onto the local quotient algebra

$$\mathscr{A}^{\mathscr{K}}_{s,t}\left(PC(\mathbb{R}^+),PC_{p,w_\alpha},PC_p\right) := \mathscr{A}^{\mathscr{K}}\left(PC(\mathbb{R}^+),PC_{p,w_\alpha},PC_p\right)/\mathscr{I}_{s,t}.$$

As in Section 4.2.3, $\mathscr{E}_{p,\alpha}$ denotes the smallest closed subalgebra of $\mathscr{L}\left(L^p(\mathbb{R}^+,t^\alpha)\right)$ which contains the identity operator $I = \chi_+$ and the operator $S_{\mathbb{R}^+}$. We provide a description of the local algebras of $\mathscr{A}^{\mathscr{K}}\left(PC(\mathbb{R}^+),PC_{p,w_\alpha},PC_p\right)$ in a couple of separate statements.

Theorem 5.5.3. *Let $s = \infty$ and $t \in \dot{\mathbb{R}}_0 \setminus \{\infty\}$. Then the local algebra*

$$\mathscr{A}^{\mathscr{K}}_{s,t}\left(PC(\mathbb{R}^+),PC_{p,w_\alpha},PC_p\right)$$

is isometrically isomorphic to $\mathscr{E}_{p,\alpha}$. The isomorphism is given by

$$\Phi^{\mathscr{K}}_{s,t}(A) \mapsto \mathsf{H}_{s,t}(A) \tag{5.18}$$

for each operator $A \in \mathscr{A}\left(PC(\mathbb{R}^+),PC_{p,w_\alpha},PC_p\right)$. In particular, for $a \in PC(\mathbb{R}^+)$, $b \in PC_{p,w_\alpha}$ and $c \in PC_p$,

$$\mathsf{H}_{s,t}(aI) = a(+\infty)I,$$

$$\mathsf{H}_{s,t}(W(b)) = b(t^-)\frac{I - S_{\mathbb{R}^+}}{2} + b(t^+)\frac{I + S_{\mathbb{R}^+}}{2},$$

$$\mathsf{H}_{s,t}(M^0(c)) = \begin{cases} c(+\infty)I & \text{if } t > 0, \\ c(-\infty)I & \text{if } t < 0. \end{cases}$$

Proof. Let $A \in \mathscr{A}\left(PC(\mathbb{R}^+),PC_{p,w_\alpha},PC_p\right)$. First we show that the mapping (5.18) is correctly defined in the sense that the operator $\mathsf{H}_{s,t}(A)$ depends only on the local coset $\Phi^{\mathscr{K}}_{s,t}(A)$ of A. Indeed, by Proposition 5.4.4, the ideal $\mathscr{K}\left(L^p(\mathbb{R}^+,w_\alpha)\right)$ is contained in the kernel of the operator $\mathsf{H}_{s,t}$. Hence, $\mathsf{H}_{s,t}(A)$ depends only on the coset $\Phi(A)$ of A. Moreover, if $b \in C_{p,w_\alpha}$ and $b(t) = 0$ then, by Proposition 5.4.1 (ii), $\mathsf{H}_{s,t}\left(\Phi(W(b))\right) = 0$. Consequently, the operator $\mathsf{H}_{s,t}(A)$ depends only on the coset $\Phi^{\mathscr{K}}_{s,t}(A)$ of A in the local algebra.

It follows from the definition of $\mathsf{H}_{s,t}$ that (5.18) is a bounded algebra homomorphism with a norm not greater than 1. The images of the operators aI, $W(b)$ and $M^0(c)$ under this homomorphism were studied in Proposition 5.4.1. From the concrete form of these images, one concludes that $\mathsf{H}_{s,t}$ is in fact a mapping onto $\mathscr{E}_{p,\alpha}$.

It remains to show that the homomorphism (5.18) is an isometry and, hence, an isomorphism. To that end we prove that, for each $A \in \mathscr{A}\left(PC(\mathbb{R}^+),PC_{p,w_\alpha},PC_p\right)$,

$$\Phi_{s,t}^{\mathcal{H}}(A) = \Phi_{s,t}^{\mathcal{H}}(U_{-t}\mathsf{H}_{s,t}(A)U_t). \tag{5.19}$$

Once this equality is verified, the assertion will follow from

$$\|\Phi_{s,t}^{\mathcal{H}}(A)\| = \|\Phi_{s,t}^{\mathcal{H}}(U_{-t}\mathsf{H}_{s,t}(A)U_t)\| \le \|U_{-t}\mathsf{H}_{s,t}(A)U_t\| \le \|\mathsf{H}_{s,t}(A)\| \le \|\Phi_{s,t}^{\mathcal{H}}(A)\|$$

by (5.12). So we are left to verify the identity (5.19). Since $\Phi_{s,t}^{\mathcal{H}}$ and $\mathsf{H}_{s,t}$ are continuous homomorphisms, it suffices to check (5.19) with A replaced by the operators $aI, W(b)$ and $M^0(c)$. Let $A = aI$ with $a \in PC(\mathbb{R}^+)$. Then (5.19) reduces to

$$\Phi_{s,t}^{\mathcal{H}}(aI) = \Phi_{s,t}^{\mathcal{H}}(a(\infty)I). \tag{5.20}$$

Choose $f \in \dot{C}(\mathbb{R}^+)$ with $f(\infty) = f(0) = 1$ and such that the support of f is contained in $[0,1] \cup [N,\infty]$ with N large enough, and write f as $f_0 + f_\infty$ with

$$f_0(t) = \begin{cases} f(t) & \text{if } t \in [0,1], \\ 0 & \text{if } t \in\,]1,\infty] \end{cases} \quad \text{and} \quad f_\infty(t) = \begin{cases} f(t) & \text{if } t \in [N,\infty], \\ 0 & \text{if } t \in [0,N[. \end{cases}$$

Further choose $g \in C_{p,w_\alpha}^0$ with $g(t) = 1$ and $g(\infty) = g(0) = 0$. Then, obviously, $\Phi_{s,t}^{\mathcal{H}}(fW(g)) = \Phi_{s,t}^{\mathcal{H}}(I)$. From this equality and from

$$fW(g) = f_0 W(g) + f_\infty W(g) = f_\infty W(g) + \text{compact},$$

by Proposition 5.3.1 (i) we obtain

$$\begin{aligned}
\|\Phi_{s,t}^{\mathcal{H}}((a-a(\infty))I)\| &= \|\Phi_{s,t}^{\mathcal{H}}((a-a(\infty))fW(g))\| \\
&= \|\Phi_{s,t}^{\mathcal{H}}((a-a(\infty))f_\infty W(g))\| \\
&\le \|(a-a(\infty))f_\infty\|_\infty \|W(g)\|.
\end{aligned}$$

The right-hand side of this estimate can be made as small as desired if N is chosen large enough. Now let $A = W(b)$ with $b \in PC_{p,w_\alpha}$. Let χ_t refer to the characteristic function of the interval $[t,+\infty]$, and choose the function g as above, but with the additional property that g has total variation 2. Using Proposition 5.4.1 (ii), we then conclude that

$$\begin{aligned}
\left\|\Phi_{s,t}^{\mathcal{H}}\left(W(b) - U_{-t}\mathsf{H}_{s,t}(W(b))U_t\right)\right\| \\
= \left\|\Phi_{s,t}^{\mathcal{H}}\left(W(b) - U_{-t}(W(b(t^-)\chi_- + b(t^+)\chi_+))U_t W(g)\right)\right\| \\
= \left\|\Phi_{s,t}^{\mathcal{H}}\left(W(b - (b(t^-)(1-\chi_t) + b(t^+)\chi_t))W(g)\right)\right\| \\
\le \left\|((b - (b(t^-)(1-\chi_t) + b(t^+)\chi_t))g\right\|_{\mathcal{M}_{p,w_\alpha}}.
\end{aligned}$$

The right-hand side of this estimate becomes as small as desired if the support of g is chosen small enough. Finally, let $A = M^0(c)$ with $c \in PC_p$. For definiteness, let $t > 0$. Then (5.19) reduces to

$$\Phi_{s,t}^{\mathscr{K}}\left(M^0(c)\right) = \Phi_{s,t}^{\mathscr{K}}\left(c(+\infty)I\right).$$

To verify this equality, choose f, f_0, f_∞ and g as above and suppose that supp $g \subseteq \mathbb{R}^+$. Then

$$\Phi_{s,t}^{\mathscr{K}}\left(M^0(c - c(+\infty))\right) = \Phi_{s,t}^{\mathscr{K}}\left(fW(g)M^0(c - c(+\infty))\right)$$
$$= \Phi_{s,t}^{\mathscr{K}}\left(f_\infty W(g)M^0(c - c(+\infty))\right) = 0$$

since $f_\infty W(g)M^0(c - c(+\infty))$ is a compact operator by Proposition 5.3.5 (iii). ∎

Let alg$\{I, \chi_+ I, S_\mathbb{R}\}$ denote the smallest closed subalgebra of $\mathscr{L}(L^p(\mathbb{R}))$ which contains the operators $I, \chi_+ I$ and $S_\mathbb{R}$. The following theorem identifies a second family of the local algebras.

Theorem 5.5.4. *Let $s \in \dot{\mathbb{R}}^+ \setminus \{\infty\}$ and $t = \infty$. Then the local algebra*

$$\mathscr{A}_{s,t}^{\mathscr{K}}\left(PC(\mathbb{R}^+), PC_{p,w_\alpha}, PC_p\right)$$

is isometrically isomorphic to the subalgebra alg$\{I, \chi_+ I, S_\mathbb{R}\}$ *of $\mathscr{L}(L^p(\mathbb{R}))$. The isomorphism is given by*

$$\Phi_{s,t}^{\mathscr{K}}(A) \mapsto H_{s,t}(A) \tag{5.21}$$

for each operator $A \in \mathscr{A}\left(PC(\mathbb{R}^+), PC_{p,w_\alpha}, PC_p\right)$. In particular, for $a \in PC(\mathbb{R}^+)$, $b \in PC_{p,w_\alpha}$ and $c \in PC_p$,

$$H_{s,t}(aI) = a(s^-)\chi_- I + a(s^+)\chi_+ I,$$
$$H_{s,t}(W(b)) = b(-\infty)\frac{I - S_\mathbb{R}}{2} + b(+\infty)\frac{I + S_\mathbb{R}}{2},$$
$$H_{s,t}(M^0(c)) = c(-\infty)\frac{I - S_\mathbb{R}}{2} + c(+\infty)\frac{I + S_\mathbb{R}}{2}.$$

Proof. Taking into account that the weight function w_α is locally non-trivial only at the points 0 and ∞, one can show by repeating the arguments of the proof of Proposition 4.3.2 that the local algebras $\mathscr{A}_{s,t}^{\mathscr{K}}\left(PC(\mathbb{R}^+), PC_{p,w_\alpha}, PC_p\right)$ generated by operators acting on $L_p(\mathbb{R}^+, w_\alpha)$ and $\mathscr{A}_{s,t}^{\mathscr{K}}\left(PC(\mathbb{R}^+), PC_p, PC_p\right)$ generated by operators on $L_p(\mathbb{R}^+)$ are isometrically isomorphic. So we shall only deal with the latter algebra.

The correctness of the definition (5.21) as well as the fact that it defines a bounded algebra homomorphism with norm not greater than 1 can be checked as in the preceding proof. The values of this homomorphism at the operators aI, $W(b)$ and $M^0(c)$ are a consequence of Proposition 5.4.3, from which we also conclude that (5.21) maps the local algebra onto alg$\{I, \chi_+ I, S_\mathbb{R}\}$. That this homomorphism is an isometry and, hence, an isomorphism, will follow once we have shown that

$$\Phi_{s,t}^{\mathscr{K}}(A) = \Phi_{s,t}^{\mathscr{K}}\left(\chi_+ V_s H_{s,t}(A) V_{-s} \chi_+ I|_{L^p(\mathbb{R}^+)}\right) \tag{5.22}$$

for all operators A in $\mathscr{A}\left(PC(\mathbb{R}^+), PC_{p,w_\alpha}, PC_p\right)$. To verify this identity, it is again sufficient to check it for the generating operators aI, $W(b)$ and $M^0(c)$ in place of A.

Let $A = aI$ with $a \in PC(\mathbb{R}^+)$. Then (5.22) states that

$$\Phi_{s,t}^{\mathscr{K}}(aI) = \Phi_{s,t}^{\mathscr{K}}\left(a(s^-)(1-\chi_s)\chi_+ I + a(s^+)\chi_s I\right). \tag{5.23}$$

To prove this equality, choose a function $f \in \dot{C}(\mathbb{R}^+)$ with $f(s) = 1$ and compact support and a function $g \in C_p^0$ with $g(\infty) = 1$ with support in $[-\infty, -N] \cup [-1,1] \cup [N, +\infty]$ where N is chosen sufficiently large. Then

$$\left\| \Phi_{s,t}^{\mathscr{K}}\left(aI - a(s^-)(1-\chi_s)\chi_+ I + a(s^+)\chi_s I\right) \right\|$$
$$= \left\| \Phi_{s,t}^{\mathscr{K}}\left((aI - a(s^-)(1-\chi_s)\chi_+ + a(s^+)\chi_s)fW(g)\right) \right\|$$
$$\leq \left\| \left((aI - a(s^-)(1-\chi_s)\chi_+ + a(s^+)\chi_s)f\right) \right\|_\infty \|W(g)\|,$$

and the right-hand side of this estimate becomes as small as desired if supp f is chosen small enough.

Now let $A = W(b)$ with $b \in PC_{p,w_\alpha}$. Choose f and g as above and write g as $g_0 + g_\infty$ with a function g_0 vanishing outside the interval $[-1,1]$. Then, according to Proposition 5.3.1 (i), $\Phi_{s,t}^{\mathscr{K}}(I) = \Phi_{s,t}^{\mathscr{K}}(fW(g)) = \Phi_{s,t}^{\mathscr{K}}(fW(g_\infty))$, whence

$$\left\| \Phi_{s,t}^{\mathscr{K}}\left(W(b - b(-\infty)\chi_- - b(+\infty)\chi_+)\right) \right\|$$
$$= \left\| \Phi_{s,t}^{\mathscr{K}}\left(W\left((b - b(-\infty)\chi_- - b(+\infty)\chi_+)g_\infty\right)fI\right) \right\|$$
$$\leq \left\| (b - b(-\infty)\chi_- - b(+\infty)\chi_+)g_\infty \right\|_{\mathcal{M}_p} \|f\|_\infty.$$

Again, the norm on the right-hand side becomes arbitrarily small if N is chosen large enough. Hence,

$$\Phi_{s,t}^{\mathscr{K}}(W(b)) = \Phi_{s,t}^{\mathscr{K}}\left(b(-\infty)W(\chi_-) + b(+\infty)W(\chi_+)\right) \tag{5.24}$$

which verifies (5.22) for $A = W(b)$. Finally, let $A = M^0(c)$ with $c \in PC_p$. Now one has to show that

$$\Phi_{s,t}^{\mathscr{K}}(M^0(c)) = \Phi_{s,t}^{\mathscr{K}}\left(c(-\infty)\frac{I - S_{\mathbb{R}^+}}{2} + c(+\infty)\frac{I + S_{\mathbb{R}^+}}{2}\right). \tag{5.25}$$

For, write c as

$$c(t) = c(-\infty)\frac{1 - \coth\left((t + \mathbf{i}/p)\pi\right)}{2} + c(+\infty)\frac{1 + \coth\left((t + \mathbf{i}/p)\pi\right)}{2} + c'(t).$$

Then $M^0(c) = c(-\infty)\frac{I - S_{\mathbb{R}^+}}{2} + c(+\infty)\frac{I + S_{\mathbb{R}^+}}{2} + M^0(c')$, and it remains to show that $\Phi_{s,t}^{\mathscr{K}}(M^0(c'))$ is the zero coset. For, choose f and $g = g_0 + g_\infty$ as above and take into account that $\Phi_{s,t}^{\mathcal{K}}(fW(g_\infty)) = \Phi_{s,t}^{\mathcal{K}}(I)$ and that the operator $fW(g_\infty)M^0(c')$ is compact by Proposition 5.3.5. ∎

One can show by the same arguments as above that Theorems 5.5.3 and 5.5.4 remain valid if the algebra $\mathscr{A}^{\mathcal{K}}\left(PC(\mathbb{R}^+), PC_{p,w_\alpha}, PC_p\right)$ is replaced by the larger algebra $\mathscr{A}^{\mathcal{K}}\left(PC(\mathbb{R}^+), PC_{p,w_\alpha}, \bar{\mathcal{M}}_p\right)$ and if the same subalgebra $\mathscr{A}^{\mathcal{K}}\left(\dot{C}(\mathbb{R}^+), C^0_{p,w_\alpha}, \emptyset\right)$ is used for localizing both algebras.

Let us now turn to the local algebra at (∞, ∞), which has a more involved structure than the local algebras already studied. For this reason we start with analyzing the smaller algebra $\mathscr{A}^{\mathcal{K}}_{\infty,\infty}\left(PC(\mathbb{R}^+), PC_{p,w_\alpha}, C_p\right)$ before dealing with the full local algebra $\mathscr{A}^{\mathcal{K}}_{\infty,\infty}\left(PC(\mathbb{R}^+), PC_{p,w_\alpha}, PC_p\right)$.

We shall need a few more strong limit operators. For $A \in \mathscr{L}\left(L^p(\mathbb{R}, w)\right)$, let

$$\mathsf{H}^{+\pm}(A) := \underset{t\to\pm\infty}{\text{s-lim}} \; \underset{s\to+\infty}{\text{s-lim}} \; U_t V_{-s} A V_s U_{-t}, \tag{5.26}$$

provided that this strong limit exists.

Proposition 5.5.5. *For $A \in \mathscr{A}\left(PC(\dot{\mathbb{R}}), PC_{p,w}, C_p\right)$, the strong limits (5.26) exist, and the mappings*

$$\mathsf{H}^{+\pm} : \mathscr{A}\left(PC(\dot{\mathbb{R}}), PC_{p,w}, C_p\right) \to \mathbb{C}I$$

are algebra homomorphisms. In particular, for $a \in PC(\dot{\mathbb{R}})$, $b \in PC_{p,w}$, $c \in C_p$ and $K \in \mathcal{K}\left(L^p(\mathbb{R}, w)\right)$,

$$\mathsf{H}^{+\pm}(aI) = a(+\infty)I, \qquad \mathsf{H}^{+\pm}(W^0(b)) = b(\pm\infty)I,$$

$$\mathsf{H}^{+\pm}(M^0(c)) = c(\pm\infty)I \quad and \quad \mathsf{H}^{+\pm}(K) = 0.$$

Proof. The first assertion comes from Lemma 5.4.2. The multiplicativity of $\mathsf{H}^{+\pm}$ is due to the uniform boundedness of $U_{-t}V_{-s}AV_sU_t$. The existence of the first three of the strong limits was established in Lemma 5.4.2. The last assertion follows from Lemma 1.4.6 since the V_s tend weakly to zero and the V_{-s} are uniformly bounded. ∎

We have to introduce some new notation in order to give a description of the local algebras at (∞, ∞). Let f be a function in $\dot{C}(\mathbb{R}^+)$ with $f(\infty) = f(0) = 1$ the support of which is contained in $[0,1] \cup [N,\infty]$ with some sufficiently large N, and write f as $f_0 + f_\infty$ with

$$f_0(t) = \begin{cases} f(t) & \text{if } t \in [0,1], \\ 0 & \text{if } t \in]1,\infty] \end{cases} \quad \text{and} \quad f_\infty(t) = \begin{cases} f(t) & \text{if } t \in [N,\infty], \\ 0 & \text{if } t \in [0,N[. \end{cases}$$

Further, let $g \in C^0_{p,w_\alpha}$ with $g(\infty) = g(0) = 1$ and supp $g \subseteq [-\infty, -N] \cup [-1,1] \cup [N,+\infty]$, and write $g = g_0 + g_\infty$ with

$$g_0(t) = \begin{cases} g(t) & \text{if } t \in [-1,1], \\ 0 & \text{if } t \in \mathbb{R} \setminus [-1,1] \end{cases} \quad \text{and} \quad g_\infty(t) = \begin{cases} 0 & \text{if } t \in [-N,N], \\ g(t) & \text{if } t \in \mathbb{R} \setminus [-N,N]. \end{cases}$$

Set $g_\infty^\pm := \chi_\pm g_\infty$. Since the operator $f_0 W(g_0)$ is compact by Proposition 5.3.1, one gets

$$\Phi_{\infty,\infty}^{\mathcal{K}}(I) = \Phi_{\infty,\infty}^{\mathcal{K}}(fW(g))$$
$$= \Phi_{\infty,\infty}^{\mathcal{K}}(f_0 W(g_\infty)) + \Phi_{\infty,\infty}^{\mathcal{K}}(f_\infty W(g_0)) + \Phi_{\infty,\infty}^{\mathcal{K}}(f_\infty W(g_\infty)).$$

Denote the first, second and third item in the sum of the right-hand side by $P_{0,\infty}$, $P_{\infty,0}$ and $P_{\infty,\infty}$, respectively, and define for $(x,y) \in \{(0,\infty),(\infty,0),(\infty,\infty)\}$,

$$\mathscr{A}_{\infty,\infty}^{x,y} := P_{x,y} \mathscr{A}_{\infty,\infty}^{\mathcal{K}}\big(PC(\mathbb{R}^+), PC_{p,w_\alpha}, C_p\big) P_{x,y}.$$

Theorem 5.5.6. *Let $s = \infty$ and $t = \infty$. Then:*

(i) *the sets $\mathscr{A}_{\infty,\infty}^{0,\infty}$, $\mathscr{A}_{\infty,\infty}^{\infty,0}$ and $\mathscr{A}_{\infty,\infty}^{\infty,\infty}$ are Banach algebras, and*

$$\mathscr{A}_{\infty,\infty}^{\mathcal{K}}\big(PC(\mathbb{R}^+), PC_{p,w_\alpha}, C_p\big) = \mathscr{A}_{\infty,\infty}^{0,\infty} \dotplus \mathscr{A}_{\infty,\infty}^{\infty,0} \dotplus \mathscr{A}_{\infty,\infty}^{\infty,\infty}$$

where the sums are direct;

(ii) *the algebra $\mathscr{A}_{\infty,\infty}^{0,\infty}$ is isometrically isomorphic to $\mathscr{E}_{p,\alpha}$, and the isomorphism is given by*

$$P_{0,\infty} \Phi_{\infty,\infty}^{\mathcal{K}}(A) P_{0,\infty} \mapsto H_{0,\infty}(A)$$

for each $A \in \mathscr{A}_{\infty,\infty}^{\mathcal{K}}\big(PC(\mathbb{R}^+), PC_{p,w_\alpha}, C_p\big)$;

(iii) *the algebra $\mathscr{A}_{\infty,\infty}^{\infty,0}$ is isometrically isomorphic to $\mathscr{E}_{p,\alpha}$, and the isomorphism is given by*

$$P_{\infty,0} \Phi_{\infty,\infty}^{\mathcal{K}}(A) P_{\infty,0} \mapsto H_{\infty,0}(A)$$

for each $A \in \mathscr{A}\big(PC(\mathbb{R}^+), PC_{p,w_\alpha}, C_p\big)$;

(iv) *the algebra $\mathscr{A}_{\infty,\infty}^{\infty,\infty}$ is commutative and finitely generated. Its generators are the cosets $\Phi_{\infty,\infty}^{\mathcal{K}}(f_\infty W(g_\infty^\pm))$. For $A \in \mathscr{A}_{\infty,\infty}^{\mathcal{K}}\big(PC(\mathbb{R}^+), PC_{p,w_\alpha}, C_p\big)$, the coset $P_{\infty,\infty} \Phi_{\infty,\infty}^{\mathcal{K}}(A) P_{\infty,\infty}$ is invertible if and only if the operators $H^{+\pm}(A)$, which are constant multiples of the identity, are invertible.*

Proof. (i) It is easy to see that $P_{\infty,0}$, $P_{0,\infty}$ and $P_{\infty,\infty}$ are idempotents which satisfy

$$P_{\infty,0} + P_{0,\infty} + P_{\infty,\infty} = \Phi_{\infty,\infty}^{\mathcal{K}}(I),$$

and

$$P_{x_1,y_1} P_{x_2,y_2} = 0 \quad \text{if} \quad (x_1,y_1) \neq (x_2,y_2).$$

Hence, assertion (i) will follow once we have shown that $P_{\infty,0}$, $P_{0,\infty}$ and $P_{\infty,\infty}$ belong to the center of the local algebra $\mathscr{A}_{\infty,\infty}^{\mathcal{K}}\big(PC(\mathbb{R}^+), PC_{p,w_\alpha}, C_p\big)$. By Proposition 4.2.10, every Mellin convolution $M^0(c)$ with $c \in C_p$ can be approximated by a polynomial in $S_{\mathbb{R}^+}$. Since $S_{\mathbb{R}^+} = W(\text{sgn})$, it thus suffices to check whether $P_{\infty,0}$, $P_{0,\infty}$ and $P_{\infty,\infty}$

commute with $\Phi_{\infty,\infty}^{\mathscr{K}}(aI)$ and $\Phi_{\infty,\infty}^{\mathscr{K}}(W(b))$ for all functions $a \in PC(\mathbb{R}^+)$ and $b \in PC_{p,w_\alpha}$.

First consider the commutator $[P_{0,\infty}, \Phi_{\infty,\infty}^{\mathscr{K}}(aI)]$. A little thought shows that there is a function $a_\infty \in \bar{C}(\mathbb{R}^+)$ such that $\Phi_{\infty,\infty}^{\mathscr{K}}(aI) = \Phi_{\infty,\infty}^{\mathscr{K}}(a_\infty I)$. So the assertion follows immediately from the compactness of $[W(g_\infty), a_\infty]$, which we infer from Proposition 5.3.1 (ii). Now consider $[P_{0,\infty}, \Phi_{\infty,\infty}^{\mathscr{K}}(W(b))]$ where $b \in PC_{p,w_\alpha}$. Since g_∞ is continuous on $\mathbb{R}$, we conclude via Proposition 5.3.3 (i) that

$$W(g_\infty)W(b) = W(g_\infty b) + K_1 = W(b)W(g_\infty) + K_2$$

with compact operators K_1 and K_2. As above, one finds a function $b_\infty \in PC_{p,w_\alpha} \cap C(\bar{\mathbb{R}})$ such that $\Phi_{\infty,\infty}^{\mathscr{K}}(W(b)) = \Phi_{\infty,\infty}^{\mathscr{K}}(W(b_\infty))$. Since the commutator $[f_0 W(b_\infty)]$ is compact by Proposition 5.3.1 (ii), it follows that $P_{0,\infty}$ commutes with all elements of the local algebra.

To get that the commutator $[P_{\infty,0}, \Phi_{\infty,\infty}^{\mathscr{K}}(aI)]$ vanishes, one can argue as above. So we are left to verify that $[P_{\infty,0}, \Phi_{\infty,\infty}^{\mathscr{K}}(W(b))] = 0$ for all $b \in PC_{p,w_\alpha}$. Using Proposition 5.3.3 again, we obtain that $[W(g_0), W(b)]$ is compact, and from Proposition 5.3.1 (ii) we infer that $[f_\infty, W(b)]$ is compact. Thus, $P_{\infty,0}$ is also in the center of the local algebra. Since $P_{\infty,\infty} = I - P_{\infty,0} - P_{0,\infty}$, the coset $P_{\infty,\infty}$ belongs to the center, too.

(ii) Propositions 5.4.1 and 5.4.3 imply that the operator $H_{0,\infty}(A)$ depends only on the coset $P_{0,\infty}\Phi_{\infty,\infty}^{\mathscr{K}}(A)P_{0,\infty}$. The specific form of $H_{0,\infty}(A)$ is also a consequence of Propositions 5.4.1 and 5.4.3. The identity,

$$P_{0,\infty}\Phi_{\infty,\infty}^{\mathscr{K}}(A)P_{0,\infty} = P_{0,\infty}\Phi_{\infty,\infty}^{\mathscr{K}}(H_{0,\infty}(A))P_{0,\infty}$$

can be checked by repeating arguments from the proofs of Theorems 5.5.3 and 5.5.4. This proves assertion (ii), and assertion (iii) of the theorem follows in a similar way.

(iv) Let $a \in PC(\mathbb{R}^+)$ and $b \in PC_{p,w_\alpha}$. Then

$$\begin{aligned}
&P_{\infty,\infty}\Phi_{\infty,\infty}^{\mathscr{K}}\big(aW(b)\big)P_{\infty,\infty} \\
&= P_{\infty,\infty}\Phi_{\infty,\infty}^{\mathscr{K}}\big(a(\infty)f_\infty W(b(-\infty)g_\infty^- + b(+\infty)g_\infty^+)\big)P_{\infty,\infty} \\
&= a(\infty)b(-\infty)\Phi_{\infty,\infty}^{\mathscr{K}}\big(f_\infty W(g_\infty^-)\big) + a(\infty)b(+\infty)\Phi_{\infty,\infty}^{\mathscr{K}}\big(f_\infty W(g_\infty^+)\big).
\end{aligned} \tag{5.27}$$

Taking into account that $\Phi_{\infty,\infty}^{\mathscr{K}}\big(f_\infty W(g_\infty^-)\big) + \Phi_{\infty,\infty}^{\mathscr{K}}\big(f_\infty W(g_\infty^+)\big) = P_{\infty,\infty}$ is the identity element in $\mathscr{A}_{\infty,\infty}^{\infty,\infty}$ and that

$$\begin{aligned}
\Phi_{\infty,\infty}^{\mathscr{K}}\big(f_\infty W(g_\infty^-)\big)\Phi_{\infty,\infty}^{\mathscr{K}}\big(f_\infty W(g_\infty^+)\big) &= \Phi_{\infty,\infty}^{\mathscr{K}}\big(f_\infty W(g_\infty^-)f_\infty W(g_\infty^+)f_\infty I\big) \\
&= \Phi_{\infty,\infty}^{\mathscr{K}}\big(f_\infty W^0(g_\infty^-)f_\infty W^0(g_\infty^+)f_\infty I\big) \\
&= \Phi_{\infty,\infty}^{\mathscr{K}}\big(f_\infty W^0(g_\infty^-)W^0(g_\infty^+)f_\infty I\big) = 0
\end{aligned}$$

by Proposition 5.3.1 (ii), we conclude that every element B of $\mathscr{A}_{\infty,\infty}^{\infty,\infty}$ can be written in the form

$$B = \alpha_-(B)\Phi_{\infty,\infty}^{\mathscr{K}}\big(f_\infty W(g_\infty^-)\big) + \alpha_+(B)\Phi_{\infty,\infty}^{\mathscr{K}}\big(f_\infty W(g_\infty^+)\big)$$

with uniquely determined complex numbers $\alpha_\pm(B)$. Since the existence of the strong limits follows from Proposition 5.5.5, it remains to show that

$$\mathsf{H}_{+\pm}(A) = \alpha_\pm\big(P_{\infty,\infty}\Phi_{\infty,\infty}^{\mathscr{K}}(A)P_{\infty,\infty}\big)I. \tag{5.28}$$

The mappings $A \mapsto \mathsf{H}_{+\pm}(A)$ and $A \mapsto \alpha_\pm\big(P_{\infty,\infty}\Phi_{\infty,\infty}^{\mathscr{K}}(A)P_{\infty,\infty}\big)$ are continuous homomorphisms. It is thus sufficient to verify (5.28) with A replaced by aI and $W(b)$. For these operators, the assertion follows immediately from Proposition 5.5.5 and equality (5.27). ∎

Now we turn our attention to the larger algebra $\mathscr{A}_{\infty,\infty}^{\mathscr{K}}\big(PC(\mathbb{R}^+),PC_{p,w_\alpha},PC_p\big)$. Again one can show that the idempotent $P_{\infty,\infty}$ belongs to the center of this algebra, but the idempotents $P_{0,\infty}$ and $P_{\infty,0}$ no longer possess this property. Therefore, this larger local algebra does not admit as simple a decomposition as that one observed in Theorem 5.5.6. We shall study the local algebra $\mathscr{A}_{\infty,\infty}^{\mathscr{K}}\big(PC(\mathbb{R}^+),PC_{p,w_\alpha},PC_p\big)$ via a second localization. To that end notice that, for $c \in C(\overline{\mathbb{R}}) \cap PC_p$, the coset $\Phi_{\infty,\infty}^{\mathscr{K}}(M^0(c))$ belongs to the center of this algebra by Propositions 5.3.2 (ii) and 5.3.4 (ii) (take into account that, for each $a \in PC(\mathbb{R}^+)$, there is an $a_\infty \in \bar{C}(\mathbb{R}^+)$ such that $\Phi_{\infty,\infty}^{\mathscr{K}}(aI - a_\infty I) = 0$). Thus, using Allan's local principle, we can localize $\mathscr{A}_{\infty,\infty}^{\mathscr{K}}\big(PC(\mathbb{R}^+),PC_{p,w_\alpha},PC_p\big)$ with respect to the maximal ideal space of the Banach algebra

$$\big\{\Phi_{\infty,\infty}^{\mathscr{K}}(M^0(c)) : c \in C(\overline{\mathbb{R}}) \cap PC_p\big\},$$

which can be identified with the two-point compactification $\overline{\overline{\mathbb{R}}}$ of the real axis in an obvious way. For $x \in \overline{\mathbb{R}}$, let $\mathscr{A}_{\infty,\infty,x}^{\mathscr{K}}$ denote the corresponding bilocal algebra, and write $\Phi_{\infty,\infty,x}^{\mathscr{K}}$ for the canonical homomorphism from $\mathscr{A}$ onto $\mathscr{A}_{\infty,\infty,x}^{\mathscr{K}}$. Further, let the functions f_0, f_∞, g_0 and g_∞ be defined as before Theorem 5.5.6. For $x \in \{\pm\infty\}$ and $(y,z) \in \{(0,\infty),(\infty,0),(\infty,\infty)\}$, set

$$P_x^{y,z} := \Phi_{\infty,\infty,x}^{\mathscr{K}}\big(f_y W(g_z)\big)$$

and abbreviate

$$\mathscr{A}_{\infty,\infty,x}^{y,z} := P_x^{y,z}\mathscr{A}_{\infty,\infty,x}^{\mathscr{K}}P_x^{y,z}.$$

Finally, let $\mathscr{B}_{p,\alpha}$ denote the smallest closed subalgebra of $\mathscr{L}\big(L^p(\mathbb{R},w_\alpha)\big)$ which contains $S_{\mathbb{R}}$ and $\chi_+ I$. The following theorem identifies the local algebras $\mathscr{A}_{\infty,\infty,x}^{\mathscr{K}}$.

Theorem 5.5.7.

(i) *Let $x \in \mathbb{R}$. For each $A \in \mathscr{A}\big(PC(\mathbb{R}^+),PC_{p,w_\alpha},PC_p\big)$, there is an operator $A_\infty \in \mathscr{A}\big(PC(\mathbb{R}^+),\emptyset,PC_p\big)$ such that $\Phi_{\infty,\infty,x}^{\mathscr{K}}(A - A_\infty) = 0$. The local algebra $\mathscr{A}_{\infty,\infty,x}^{y,z}$ is isometrically isomorphic to the algebra $\mathscr{B}_{p,\alpha}$, and the isomorphism is given by*

$$\mathsf{H}_{\infty,\infty,x} : \Phi_{\infty,\infty,x}^{\mathscr{K}}(A) \mapsto \mathsf{H}_{\infty,x}(E_{p,w_\alpha}A_\infty E_{p,w_\alpha}^{-1}).$$

In particular,

$$\mathsf{H}_{\infty,\infty,x}\big(\Phi^{\mathscr{K}}_{\infty,\infty,x}(aI)\big) = a(+\infty)\chi_- I + a(0^+)\chi_+ I,$$

$$\mathsf{H}_{\infty,\infty,x}\big(\Phi^{\mathscr{K}}_{\infty,\infty,x}(W(b))\big) = \left(b(-\infty)\frac{1-d(x)}{2} + b(+\infty)\frac{1+d(x)}{2}\right)\chi_- I$$
$$+ \left(b(0^-)\frac{1-d(x)}{2} + b(0^+)\frac{1+d(x)}{2}\right)\chi_+ I$$

with $d(x) := \coth\big((x+\mathbf{i}(1/p+\alpha))\pi\big)$, *and*

$$\mathsf{H}_{\infty,\infty,x}\big(\Phi^{\mathscr{K}}_{\infty,\infty,x}(M^0(c))\big) = c(x^-)Q_{\mathbb{R}} + c(x^+)P_{\mathbb{R}}.$$

(ii) *Let* $x \in \{\pm\infty\}$. *Then* $\mathscr{A}^{0,\infty}_{\infty,\infty,x}$, $\mathscr{A}^{\infty,0}_{\infty,\infty,x}$ *and* $\mathscr{A}^{\infty,\infty}_{\infty,\infty,x}$ *are Banach algebras, and the algebra* $\mathscr{A}^{\mathscr{K}}_{\infty,\infty,x}$ *decomposes into the direct sum*

$$\mathscr{A}^{\mathscr{K}}_{\infty,\infty,x} = \mathscr{A}^{0,\infty}_{\infty,\infty,x} \dotplus \mathscr{A}^{\infty,0}_{\infty,\infty,x} \dotplus \mathscr{A}^{\infty,\infty}_{\infty,\infty,x}.$$

Moreover, for $(y,z) \in \{(0,\infty),(\infty,0),(\infty,\infty)\}$, *there is an isomorphism* $\mathsf{H}^{y,z}_x$ *from* $\mathscr{A}^{y,z}_{\infty,\infty,x}$ *onto* $\mathbb{C}$. *In particular,*

$$\mathsf{H}^{y,z}_x\big(P^{y,z}_{\pm\infty}\Phi^{\mathscr{K}}_{\infty,\infty,\pm\infty}(aW(b)M^0(c))P^{y,z}_{\pm\infty}\big) = a(y)b(z^{\pm})c(x)$$

where $\infty^{\pm} := \pm\infty$.

Proof. (i) Choose $c_x \in C(\dot{\mathbb{R}}) \cap PC_p$ so that supp c_x is compact and $c_x(x) = 1$. Then the operator $f_\infty W(g_\infty)M^0(c_x)$ is compact by Proposition 5.3.5 (i). Hence, for every function $b \in PC_{p,w_\alpha}$, which is continuous at the point 0 and satisfies $b(0) = 0$, we obtain

$$\Phi^{\mathscr{K}}_{\infty,\infty,x}(W(b)) = \Phi^{\mathscr{K}}_{\infty,\infty,x}\big(W(b)\big(f_0 W(g_\infty) + f_\infty W(g_0) + f_\infty W(g_\infty)\big)M^0(c_x)\big)$$
$$= \Phi^{\mathscr{K}}_{\infty,\infty,x}\big(W(bg_\infty)f_0 M^0(c_x) + W(bg_0)f_\infty M^0(c_x)\big)$$
$$= \Phi^{\mathscr{K}}_{\infty,\infty,x}\big(W(bg_\infty)f_0 M^0(c_x)\big) \qquad (5.29)$$
$$= \Phi^{\mathscr{K}}_{\infty,\infty,x}\big(W(b(-\infty)\chi_- + b(+\infty)\chi_+)f_0 M^0(c_x)\big)$$
$$= \Phi^{\mathscr{K}}_{\infty,\infty,x}\big((b(-\infty)W(\chi_-) + b(+\infty)W(\chi_+))f_0 M^0(c_x)\big).$$

If now b is an arbitrary function in PC_{p,w_α}, then we write

$$W(b) = b(0^-)W(\chi_-) + b(0^+)W(\chi_+) + W(b') \qquad (5.30)$$

with b' being continuous at zero and $b'(0) = 0$. Then the first part of assertion (i) follows, since the $W(\chi_\pm)$ are also Mellin operators.

Since the images of aI and $M^0(c)$ under the mapping $A \mapsto E_{p,w_\alpha}AE^{-1}_{p,w_\alpha}$ are the operators $\breve{a}I$ with $\breve{a}(t) = a(e^{2\pi t})$ and $W^0(c)$, respectively, the strong limits

$H_{\infty,x}(E_{p,w_\alpha}A_\infty E_{p,w_\alpha}^{-1})$ exist for each operator $A_\infty \in \mathscr{A}(PC(\mathbb{R}^+),0,PC_p)$ by Proposition 5.4.1.

Let $\mathscr{J}_{\infty,\infty,x}$ stand for the closed ideal generated by all cosets $\Phi_{\infty,\infty}^{\mathscr{K}}(M^0(c))$ with $c \in C(\bar{\mathbb{R}}) \cap PC_p$ and $c(x) = 0$. In order to get the correctness of the definition of $H_{\infty,\infty,x}$, we must show that if $A \in \mathscr{A}(PC(\mathbb{R}^+),PC_{p,w_\alpha},PC_p)$ and $\Phi_{\infty,\infty,x}^{\mathscr{K}}(A) = 0$, then the strong limit $H(A) := H_{\infty,x}(E_{p,w_\alpha}A_\infty E_{p,w_\alpha}^{-1})$ exists and is equal to 0.

To see this, note first that the ideal $\mathscr{K}$ of the compact operators belongs to the kernel of H. Thus, H depends only on the coset $\Phi(A)$. Further, since $H(aI) = 0$ for each continuous function a with $a(0) = a(\infty) = 0$ by Proposition 5.4.1 (i), the local ideal $\mathscr{J}_{\infty,\infty}$ lies in the kernel of H. Notice that for this conclusion we do *not* need to know whether the strong limit $H(\Phi(W(b))$ exists: indeed, each operator A with $\Phi(A) \in \mathscr{J}_{\infty,\infty}$ can be approximated by finite sums

$$\sum_j A_j a_j I + K$$

where $a_j(0) = a_j(\infty) = 0$ and K is compact. If A is of this form then

$$
\begin{aligned}
H(A) &= \operatorname*{s-lim}_{\tau \to +\infty} Z_\tau^{-1}U_{-x}E_{p,w_\alpha}AE_{p,w_\alpha}^{-1}U_x Z_\tau \\
&= \operatorname*{s-lim}_{\tau \to +\infty} \sum_j (Z_\tau^{-1}U_{-x}E_{p,w_\alpha}A_j E_{p,w_\alpha}^{-1}U_x Z_\tau)(Z_\tau^{-1}U_{-x}E_{p,w_\alpha}a_j E_{p,w_\alpha}^{-1}U_x Z_\tau),
\end{aligned}
$$

from which the conclusion follows since $Z_\tau^{-1}U_{-x}E_{p,w_\alpha}a_j E_{p,w_\alpha}^{-1}U_x Z_\tau \to 0$ and since the norms of $Z_\tau^{-1}U_{-x}E_{p,w_\alpha}A_j E_{p,w_\alpha}^{-1}U_x Z_\tau$ are uniformly bounded with respect to τ. Hence, $H(A)$ depends only on $\Phi_{\infty,\infty}^{\mathscr{K}}(A)$. The same arguments show that the local ideal $\mathscr{J}_{\infty,\infty,x}$ is contained in the kernel of the mapping $\Phi_{\infty,\infty}^{\mathscr{K}}(A) \mapsto H(A)$ (take into account that $H(M^0(c)) = 0$ whenever $c \in C(\bar{\mathbb{R}}) \cap PC_p$ and $c(x) = 0$). This observation establishes the correctness of the definition of $H_{\infty,\infty,x}$.

We further have to show that the invertibility of

$$H_{\infty,\infty,x}(\Phi_{\infty,\infty,x}^{\mathscr{K}}(A)) = H_{\infty,x}(E_{p,w_\alpha}A_\infty E_{p,w_\alpha}^{-1})$$

implies the invertibility of $\Phi_{\infty,\infty,x}^{\mathscr{K}}(A)$. But this is an easy consequence of the identity

$$\Phi_{\infty,\infty,x}^{\mathscr{K}}(A) = \Phi_{\infty,\infty,x}^{\mathscr{K}}(E_{p,w_\alpha}^{-1}U_x H_{\infty,x}(E_{p,w_\alpha}A_\infty E_{p,w_\alpha}^{-1})U_x E_{p,w_\alpha})$$

which can be verified in a similar way as the corresponding identity in the proof of Theorem 5.5.3. Finally, the special values of $H_{\infty,\infty,x}$ at the generators of the algebra follow from the equalities (4.21), (5.29) and (5.30) and from Proposition 5.4.1.

(ii) Since $\Phi_{\infty,\infty,x}^{\mathscr{K}}(M^0(c)) = c(x)\Phi_{\infty,\infty,x}^{\mathscr{K}}(I)$, the proof of the first part of this assertion runs as that of Theorem 5.5.6. The second part of the assertion will follow immediately from the identity

$$P_{\pm\infty}^{y,z}\Phi_{\infty,\infty,\pm\infty}^{\mathscr{K}}(aW(b)M^0(c)) = a(y)b(z^\pm)c(\pm\infty)P_{\pm\infty}^{y,z}$$

which we shall verify only for the basic case when $a \equiv 1$ and $c \equiv 1$. For definiteness, let $(y,z) = (\infty,0)$. Then

$$
\begin{aligned}
P_{\pm\infty}^{\infty,0} \Phi_{\infty,\infty,\pm\infty}^{\mathcal{H}}\big(W(b)\big) &= \Phi_{\infty,\infty,\pm\infty}^{\mathcal{H}}\big(W(bg_0)f_\infty I\big) \\
&= \Phi_{\infty,\infty,\pm\infty}^{\mathcal{H}}\Big(W\big(b(0^-)\chi_- + b(0^+)\chi_+\big)\Big)P_{\pm\infty}^{\infty,0} \\
&= \Phi_{\infty,\infty,\pm\infty}^{\mathcal{H}}\Big(b(0^-)W(\chi_-) + b(0^+)W(\chi_+)\Big)P_{\pm\infty}^{\infty,0} \\
&= \Phi_{\infty,\infty,\pm\infty}^{\mathcal{H}}\Big(b(0^-)M^0\Big(\frac{1-d}{2}\Big) + b(0^+)M^0\Big(\frac{1+d}{2}\Big)\Big)P_{\pm\infty}^{\infty,0} \\
&= \Big(b(0^-)\frac{1-d(\pm\infty)}{2} + b(0^+)\frac{1+d(\pm\infty)}{2}\Big)P_{\pm\infty}^{\infty,0} \\
&= b(0^\pm)P_{\pm\infty}^{\infty,0}.
\end{aligned}
$$

Similarly one gets that, for every operator $A \in \mathscr{A}\big(PC(\mathbb{R}^+),PC_{p,w_\alpha},PC_p\big)$, there is a function $c \in C(\overline{\mathbb{R}}) \cap PC_p$ such that

$$
P_{\pm\infty}^{y,z} \Phi_{\infty,\infty,\pm\infty}^{\mathcal{H}}(A) = \Phi_{\infty,\infty,\pm\infty}^{\mathcal{H}}(M^0(c))P_{\pm\infty}^{y,z} = c(\pm\infty)P_{\pm\infty}^{y,z}.
$$

This observation finishes the proof. ∎

We summarize the results obtained in this section in the following theorem.

Theorem 5.5.8. *Let $A \in \mathscr{A}\big(PC(\mathbb{R}^+),PC_{p,w_\alpha},PC_p\big)$. The coset $A + \mathscr{K}\big(L^p(\mathbb{R}^+,w_\alpha)\big)$ is invertible in $\mathscr{A}^{\mathscr{K}}\big(PC(\mathbb{R}^+),PC_{p,w_\alpha},PC_p\big)$ if and only if the operators*

$$
\begin{aligned}
&\mathsf{H}_{\infty,t}(A) \in \mathscr{E}_{p,\alpha} &&\text{for } r \in \dot{\mathbb{R}}_0 \setminus \{\infty\}, \\
&\mathsf{H}_{s,\infty}(A) \in \mathscr{B}_p &&\text{for } s \in \dot{\mathbb{R}}^+ \setminus \{\infty\}, \\
&\mathsf{H}_{\infty,\infty,x}\big(\Phi_{\infty,\infty,x}^{\mathcal{H}}(A)\big) \in \mathscr{B}_{p,\alpha} &&\text{for } x \in \mathbb{R}
\end{aligned}
$$

are invertible in the respective algebras and if the complex numbers

$$
\mathsf{H}_x^{y,z}\big(P_{\pm\infty}^{y,z}\Phi_{\infty,\infty,\pm\infty}^{\mathcal{H}}(A)\big) \qquad \text{for } (y,z) \in \{(0,\infty),(\infty,0),(\infty,\infty)\}
$$

are not zero.

The following theorem establishes the relation of this result to the Fredholm property of operators in $\mathscr{A}\big(PC(\mathbb{R}^+),PC_{p,w_\alpha},PC_p\big)$.

Theorem 5.5.9. *The algebra $\mathscr{A}^{\mathscr{K}}\big(PC(\mathbb{R}^+),PC_{p,w_\alpha},PC_p\big)$ is inverse-closed in the Calkin algebra $\mathscr{L}(L^p(\mathbb{R}^+,w_\alpha))/\mathscr{K}(L^p(\mathbb{R}^+,w_\alpha))$.*

Proof. There are several ways to verify the inverse-closedness. One way is to consider the smallest (non-closed) subalgebra $\mathscr{A}_0$ of $\mathscr{A}\big(PC(\mathbb{R}^+),PC_{p,w_\alpha},PC_p\big)$ which contains all operators aI, $W(b)$ and $M^0(c)$ with piecewise constant functions a, b and c. Applying Theorem 5.5.8 to an operator $A \in \mathscr{A}_0$, we find that the spectrum of the coset $A + \mathscr{K}(L^p(\mathbb{R}^+, w_\alpha))$ in $\mathscr{A}^{\mathscr{K}}(PC(\mathbb{R}^+), PC_{p,w_\alpha}, PC_p)$ is a thin subset of the complex plane. Since $\mathscr{A}_0$ is dense in $\mathscr{A}(PC(\mathbb{R}^+), PC_{p,w_\alpha}, PC_p)$, the assertion follows from Corollary 1.2.32.

For another proof, one shows that, for every Fredholm operator A in the algebra $\mathscr{A}(PC(\mathbb{R}^+), PC_{p,w_\alpha}, PC_p)$, the H-limits quoted in Theorem 5.5.8 are invertible (as operators on the respective Banach spaces). Then one employs the inverse-closedness of the algebras $\mathscr{E}_{p,\alpha}$ and $\mathscr{B}_{p,\alpha}$ in the algebra $\mathscr{L}(L^p(\mathbb{R}^+, w_\alpha))$ and applies Theorem 5.5.8.   ∎

Corollary 5.5.10. *Let* $A \in \mathscr{A}(PC(\mathbb{R}^+), PC_{p,w_\alpha}, PC_p)$. *Then* A *is a Fredholm operator on* $L^p(\mathbb{R}^+, w_\alpha)$ *if and only if the operators*

$$\mathsf{H}_{\infty,t}(A) \in \mathscr{E}_{p,\alpha} \qquad\qquad \text{for } r \in \dot{\mathbb{R}}_0 \setminus \{\infty\}$$

$$\mathsf{H}_{s,\infty}(A) \in \mathscr{B}_p \qquad\qquad \text{for } s \in \dot{\mathbb{R}}^+ \setminus \{\infty\}$$

$$\mathsf{H}_{\infty,\infty,x}\big(\Phi^{\mathscr{K}}_{\infty,\infty,x}(A)\big) \in \mathscr{B}_{p,\alpha} \qquad \text{for } x \in \mathbb{R}$$

are invertible (as operators on the respective Banach spaces) and if the complex numbers

$$\mathsf{H}^{y,z}_x\big(\mathsf{P}^{y,z}_{\pm\infty}\Phi^{\mathscr{K}}_{\infty,\infty,\pm\infty}(A)\big) \qquad\qquad \text{for } (y,z) \in \{(0,\infty),(\infty,0),(\infty,\infty)\}$$

are not zero.

Combining this result with the results of Section 4.2 one easily gets a matrix-valued symbol for the Fredholmness of operators in $\mathscr{A}(PC(\mathbb{R}^+), PC_{p,w_\alpha}, PC_p)$.

Remark 5.5.11. In this section we constructed representations of the local algebras by employing a basic property of the operators which constitute the local algebras: their local homogeneity. This property enabled us to identify the local algebras via homogenizing strong limits. It would also have been possible to identify the local algebras by means of the concepts developed in Section 2.6 and Chapter 3: PI-algebras and, in particular, algebras generated by idempotents. We will illustrate the use of those concepts in Section 5.7 to identify some of the local algebras that appear there.   □

5.6 Algebras of multiplication and Wiener-Hopf operators

Let the weight function w be given by (4.8). In this section we address the smallest closed subalgebra of $\mathscr{L}(L^p(\mathbb{R}, w))$ which contains all operators aI of multiplication by a function $a \in PC(\dot{\mathbb{R}})$ and all Fourier convolutions $W^0(b)$ where $b \in PC_{p,w}$. We denote this algebra by $\mathscr{A}(PC(\dot{\mathbb{R}}), PC_{p,w})$, and we write $\mathscr{A}^{\mathscr{K}}(PC(\dot{\mathbb{R}}), PC_{p,w})$ for the image of this algebra in the Calkin algebra and Φ for the canonical homomorphism from $\mathscr{A}\left(PC(\dot{\mathbb{R}}),PC_{p,w}\right)$ onto $\mathscr{A}^{\mathscr{K}}\left(PC(\dot{\mathbb{R}}),PC_{p,w}\right)$.

If $f \in C(\dot{\mathbb{R}})$ and $g \in C_{p,w}$ then the coset $\Phi\left(fW^0(g)\right)$ belongs to the center of $\mathscr{A}^{\mathscr{K}}\left(PC(\dot{\mathbb{R}}),PC_{p,w}\right)$ by Proposition 5.3.1. So we can localize this algebra with respect to the maximal ideal space of $\mathscr{A}^{\mathscr{K}}\left(C(\dot{\mathbb{R}}),C_{p,w}\right)$, which is homeomorphic to the subset $(\dot{\mathbb{R}} \times \{\infty\}) \cup (\{\infty\} \times \dot{\mathbb{R}})$ of the torus $\dot{\mathbb{R}} \times \dot{\mathbb{R}}$. The proof of the latter fact is similar to the proof of Proposition 5.5.2.

Given $(s,t) \in (\dot{\mathbb{R}} \times \{\infty\}) \cup (\{\infty\} \times \dot{\mathbb{R}})$, let $\mathscr{I}_{s,t}$ denote the smallest closed ideal of the Banach algebra $\mathscr{A}^{\mathscr{K}}\left(PC(\dot{\mathbb{R}}),PC_{p,w}\right)$ which contains the point (s,t), and let $\Phi_{s,t}^{\mathscr{K}}$ refer to the canonical homomorphism from $\mathscr{A}^{\mathscr{K}}\left(PC(\dot{\mathbb{R}}),PC_{p,w}\right)$ onto the local quotient algebra

$$\mathscr{A}_{s,t}^{\mathscr{K}} := \mathscr{A}^{\mathscr{K}}\left(PC(\dot{\mathbb{R}}),PC_{p,w}\right)/\mathscr{I}_{s,t}.$$

Further, for each weight w of the form (4.8) and for each $x \in \mathbb{R}$, define the local weight $w_{\alpha(x)}$ at x by $w_{\alpha(x)}(t) := |t|^{\alpha(x)}$ with

$$\alpha(x) := \begin{cases} 0 & \text{if } x \notin \{t_1,\ldots,t_n,\infty\}, \\ \alpha_j & \text{if } x = t_j \text{ for some } (j = 1,\ldots,n), \\ \sum_{j=0}^{n} \alpha_j & \text{if } x = \infty. \end{cases} \tag{5.31}$$

To describe the local algebras $\mathscr{A}_{s,t}^{\mathscr{K}}$, we have to introduce some new strong limit operators. For $A \in \mathscr{L}\left(L^p(\mathbb{R},w)\right)$, let

$$\mathsf{H}^{\pm\pm}(A) := \operatorname*{s\text{-}lim}_{t \to \pm\infty} \operatorname*{s\text{-}lim}_{s \to \pm\infty} U_t V_{-s} A V_s U_{-t} \tag{5.32}$$

provided that the strong limits exist. Here, by convention, the first superscript in $\mathsf{H}^{\pm\pm}$ refers to the strong limit with respect to $s \to \pm\infty$ and the second one to $t \to \pm\infty$.

Proposition 5.6.1. *The strong limits* (5.32) *exist for* $A \in \mathscr{A}\left(PC(\dot{\mathbb{R}}),PC_{p,w}\right)$, *and the mappings* $\mathsf{H}^{\pm\pm}$ *are algebra homomorphisms from* $\mathscr{A}\left(PC(\dot{\mathbb{R}}),PC_{p,w}\right)$ *onto the algebra* $\mathbb{C}I$. *In particular, for* $a \in PC(\dot{\mathbb{R}})$ *and* $b \in PC_{p,w}$,

$$\mathsf{H}^{+\pm}(aI) = a(+\infty)I, \qquad \mathsf{H}^{-\pm}(aI) \quad = a(-\infty)I, \tag{5.33}$$

$$\mathsf{H}^{\pm+}(W^0(b)) = b(+\infty)I, \qquad \mathsf{H}^{\pm-}(W^0(b)) = b(-\infty)I, \tag{5.34}$$

and

$$\mathsf{H}^{\pm\pm}(K) = 0 \qquad \text{for} \quad K \in \mathscr{K}\left(L^p(\mathbb{R},w)\right). \tag{5.35}$$

Proof. The first assertion comes from Lemma 5.4.2. The multiplicativity of $\mathsf{H}^{\pm\pm}$ is due to the uniform boundedness of $U_t V_{-s} A V_s U_{-t}$. Finally, if K is compact then, by Lemma 1.4.6, KV_s goes strongly to zero, as V_s tends weakly to zero. The result then follows from the uniform boundedness of V_{-s}. ∎

Theorem 5.6.2. *Let $A \in \mathscr{A}\left(PC(\dot{\mathbb{R}}), PC_{p,w}\right)$.*

(i) *The coset $A + \mathscr{K}\left(L^p(\mathbb{R}, w)\right)$ is invertible in $\mathscr{A}^{\mathscr{K}}\left(PC(\dot{\mathbb{R}}), PC_{p,w}\right)$ if and only if the coset $\Phi_{s,t}^{\mathscr{K}}(A)$ is invertible in $\mathscr{A}_{s,t}^{\mathscr{K}}$ for each $(s,t) \in (\dot{\mathbb{R}} \times \{\infty\}) \cup (\{\infty\} \times \mathbb{R})$.*

(ii) *For $s \in \mathbb{R}$, the local algebra $\mathscr{A}_{s,\infty}^{\mathscr{K}}$ is isometrically isomorphic to the subalgebra $\mathrm{alg}\{I, \chi_+ I, S_{\mathbb{R}}\}$ of $\mathscr{L}\left(L^p(\mathbb{R}, w_{\alpha(s)})\right)$, and the isomorphism is given by*

$$\Phi_{s,\infty}^{\mathscr{K}}(A) \mapsto \mathsf{H}_{s,\infty}(A) \qquad (5.36)$$

for each operator $A \in \mathscr{A}\left(PC(\dot{\mathbb{R}}), PC_{p,w}\right)$.

(iii) *For $t \in \mathbb{R}$, the local algebra $\mathscr{A}_{\infty,t}^{\mathscr{K}}$ is isometrically isomorphic to the subalgebra $\mathrm{alg}\{I, \chi_+ I, S_{\mathbb{R}}\}$ of $\mathscr{L}\left(L^p(\mathbb{R}, w_{\alpha(\infty)})\right)$, and the isomorphism is given by*

$$\Phi_{\infty,t}^{\mathscr{K}}(A) \mapsto \mathsf{H}_{\infty,t}(A) \qquad (5.37)$$

for each operator $A \in \mathscr{A}\left(PC(\dot{\mathbb{R}}), PC_{p,w}\right)$.

(iv) *The local algebra $\mathscr{A}_{\infty,\infty}^{\mathscr{K}}$ is generated by the four idempotent elements*

$$\Phi_{\infty,\infty}^{\mathscr{K}}(W(\chi_\pm)\chi_\pm I),$$

and the coset $\Phi_{\infty,\infty}^{\mathscr{K}}(A)$ is invertible if and only if the four operators

$$\mathsf{H}^{\pm\pm}(A),$$

which are complex multiples of the identity operator, are invertible.

Proof. Assertion (i) is just a reformulation of Allan's local principle. For the proof of assertion (ii), one employs the same arguments as in the first and second step of the proof of Proposition 4.3.2 to obtain that the algebras $\mathscr{A}_{s,\infty}^{\mathscr{K}}$ corresponding to the spaces $L^p(\mathbb{R}, w)$ and $L^p(\mathbb{R}, w_s)$ with $w_s(x) = |x - s|^{\alpha(s)}$ are isometrically isomorphic. The remainder of the proof of assertion (ii) can be done as in Theorem 5.5.4.

The proof of part (iii) runs parallel to that of Theorem 5.5.3. One only has to take into account that the local algebras related to $L^p(\mathbb{R}, w)$ and $L^p(\mathbb{R}, w_\infty)$ with $w_\infty(x) = |x|^{\alpha(\infty)}$ are isometrically isomorphic. To prove assertion (iv), note that there are functions $f \in C(\overline{\mathbb{R}})$ and $g \in C(\overline{\mathbb{R}}) \cap PC_{p,w}$ such that

$$\Phi_{\infty,\infty}^{\mathscr{K}}(fI - \chi_+ I) = 0 \quad \text{and} \quad \Phi_{\infty,\infty}^{\mathscr{K}}(W^0(g - \chi_+)) = 0.$$

From Proposition 5.3.1(ii) we infer that the commutator $[fI, W^0(g)]$ is compact. Thus, the cosets $\Phi_{\infty,\infty}^{\mathscr{K}}(\chi_+ I)$ and $\Phi_{\infty,\infty}^{\mathscr{K}}(W^0(\chi_+))$ commute. Since

$$\Phi_{\infty,\infty}^{\mathscr{K}}(aI) = \Phi_{\infty,\infty}^{\mathscr{K}}(a(-\infty)\chi_- I + a(+\infty)\chi_+ I)$$

for every $a \in PC(\dot{\mathbb{R}})$ and

$$\Phi_{\infty,\infty}^{\mathscr{K}}(W^0(b)) = \Phi_{\infty,\infty}^{\mathscr{K}}(b(-\infty)W^0(\chi_-) + b(+\infty)W^0(\chi_+)),$$

for every $b \in PC_{p,w}$, we find that, for each $A \in \mathscr{A}\left(PC(\dot{\mathbb{R}}), PC_{p,w}\right)$, the coset $\Phi_{\infty,\infty}^{\mathscr{K}}(A)$ can be represented in the form

$$\Phi_{\infty,\infty}^{\mathscr{K}}\left(h_{--}W(\chi_-)\chi_- I + h_{-+}W(\chi_+)\chi_- I + h_{+-}W(\chi_-)\chi_+ I + h_{++}W(\chi_+)\chi_+ I\right)$$

with uniquely determined complex numbers $h_{\pm\pm} = h_{\pm\pm}(A)$. Thereby,

$$h_{+\pm}(aI) = a(+\infty)I, \qquad h_{-\pm}(aI) \quad = a(-\infty)I, \tag{5.38}$$

$$h_{\pm+}(W^0(b)) = b(+\infty)I, \qquad h_{\pm-}(W^0(b)) = b(-\infty)I, \tag{5.39}$$

and the coset $\Phi_{\infty,\infty}^{\mathscr{K}}(A)$ is invertible if and only if the numbers $h_{\pm\pm}(A)$ are not zero. Since $\mathsf{H}^{\pm\pm}(A) = h_{\pm\pm}(A)I$ by Proposition 5.6.1, the result follows. $\blacksquare$

The following corollary can be proved by repeating the arguments from the proof of Theorem 5.5.9.

Corollary 5.6.3. *The algebra* $\mathscr{A}^{\mathscr{K}}\left(PC(\dot{\mathbb{R}}), PC_{p,w}\right)$ *is inverse-closed in the Calkin algebra* $\mathscr{L}(L^p(\mathbb{R},w))/\mathscr{K}(L^p(\mathbb{R},w))$, *and an operator* $A \in \mathscr{A}\left(PC(\dot{\mathbb{R}}), PC_{p,w}\right)$ *is Fredholm if and only if the operators* $\mathsf{H}_{s,\infty}(A)$, $\mathsf{H}_{\infty,t}(A)$ *and* $\mathsf{H}^{\pm\pm}(A)$ *are invertible for all* $s,t \in \mathbb{R}$.

To illustrate the previous results, we consider a particular class of operators, the so-called paired convolution operators. These are operators of the form

$$A = a_1 W^0(b_1) + a_2 W^0(b_2) \tag{5.40}$$

with $a_1, a_2 \in PC(\dot{\mathbb{R}})$ and $b_1, b_2 \in PC_{p,w}$. The following result is an immediate consequence of Corollary 5.6.3.

Theorem 5.6.4. *The operator* A *in* (5.40) *is Fredholm on* $L^p(\mathbb{R},w)$ *if and only if the following three conditions are fulfilled:*

(i) *the operator* $c_+ P_{\mathbb{R}} + c_- Q_{\mathbb{R}}$ *with*

$$\begin{aligned} c_\pm(s) := & \left(a_1(s^-)b_1(\pm\infty) + a_2(s^-)b_2(\pm\infty)\right)\chi_- I \\ & + \left(a_1(s^+)b_1(\pm\infty) + a_2(s^+)b_2(\pm\infty)\right)\chi_+ I \end{aligned}$$

is invertible on $L^p(\mathbb{R}, w_{\alpha(s)})$ *for each* $s \in \mathbb{R}$;

(ii) *the operator* $d_+ P_{\mathbb{R}} + d_- Q_{\mathbb{R}}$ *with*

$$\begin{aligned} d_\pm(t) := & \left(a_1(-\infty)b_1(t^\pm) + a_2(-\infty)b_2(t^\pm)\right)\chi_- I \\ & + \left(a_1(+\infty)b_1(t^\pm) + a_2(+\infty)b_2(t^\pm)\right)\chi_+ I \end{aligned}$$

is invertible on $L^p(\mathbb{R}, w_{\alpha(\infty)})$ *for each* $t \in \mathbb{R}$;

(iii) *none of the following numbers is zero:*

$$a_1(+\infty)b_1(\pm\infty) + a_2(+\infty)b_2(\pm\infty), \quad a_1(-\infty)b_1(\pm\infty) + a_2(-\infty)b_2(\pm\infty).$$

Of particular interest are paired operators of the form

$$A = a_1 W^0(\chi_+) + a_2 W^0(\chi_-) = a_1 P_{\mathbb{R}} + a_2 Q_{\mathbb{R}} \tag{5.41}$$

with $a_1, a_2 \in PC(\dot{\mathbb{R}})$, which can also be written as the singular integral operator

$$\frac{a_1 + a_2}{2} I + \frac{a_1 - a_2}{2} S_{\mathbb{R}}.$$

For these operators, Corollary 5.6.3 implies the following.

Corollary 5.6.5. *Let* $a_1, a_2 \in PC(\dot{\mathbb{R}})$. *The singular integral operator* $a_1 P_{\mathbb{R}} + a_2 Q_{\mathbb{R}}$ *is Fredholm on* $L^p(\mathbb{R}, w)$ *if and only if*

(i) *the operator* $(a_2(s^-)\chi_- + a_2(s^+)\chi_+)Q_{\mathbb{R}} + (a_1(s^-)\chi_- + a_1(s^+)\chi_+)P_{\mathbb{R}}$ *is invertible on* $L^p(\mathbb{R}, w_{\alpha(s)})$ *for each* $s \in \mathbb{R}$ *and*
(ii) *the operator* $(a_2(-\infty)\chi_- + a_2(+\infty)\chi_+)Q_{\mathbb{R}} + (a_1(-\infty)\chi_- + a_1(+\infty)\chi_+)P_{\mathbb{R}}$ *is invertible on* $L^p(\mathbb{R}, w_{\alpha(\infty)})$.

The corresponding result for operators on the semi-axis reads as follows.

Corollary 5.6.6. *Let* $a_1, a_2 \in PC(\mathbb{R}^+)$. *The singular integral operator* $a_1 P_{\mathbb{R}^+} + a_2 Q_{\mathbb{R}^+}$ *is Fredholm on* $L^p(\mathbb{R}^+, w)$ *if and only if*

(i) *the operator* $a_1(0^+)P_{\mathbb{R}^+} + a_2(0^+)Q_{\mathbb{R}^+}$ *is invertible on* $L^p(\mathbb{R}, w_{\alpha(0)})$,
(ii) *the operator* $(a_2(s^-)\chi_- + a_2(s^+)\chi_+)Q_{\mathbb{R}} + (a_1(s^-)\chi_- + a_1(s^+)\chi_+)P_{\mathbb{R}}$ *is invertible on* $L^p(\mathbb{R}, w_{\alpha(s)})$ *for each* $s \in \mathbb{R}^+ \setminus \{0\}$, *and*
(iii) *the operator* $a_1(+\infty)P_{\mathbb{R}^+} + a_2(+\infty)Q_{\mathbb{R}^+}$ *is invertible on* $L^p(\mathbb{R}^+, w_{\alpha(\infty)})$.

Proof. This follows by applying Corollary 5.6.3 to the operator

$$\left(a_1 W^0(\chi_+) + a_1 W^0(\chi_-)\right)\chi_+ I + \chi_- I \in \mathscr{L}(L^p(\mathbb{R}, w))$$

with a_1 and a_2 extended to the whole line by zero. This operator is equivalent to the singular integral operator $a_1 P_{\mathbb{R}^+} + a_2 Q_{\mathbb{R}^+}$ in the sense that these operators are Fredholm, or not, simultaneously. (Of course, one could also apply Corollary 5.5.10 directly.) ∎

Note that Propositions 4.2.11 and 4.2.19 combined with the above results give a matrix-valued symbol for the Fredholmness of the operators considered. Note further that one can derive similar results for operators of the form $a_1 M(b_1) + a_2 M(b_2)$ with $a_1, a_2 \in PC([0,1])$ and $b_1, b_2 \in PC_{p,w_\alpha}$ considered on the space $L^p([0,1], w_\alpha)$. The easiest way to do this is to reduce them to the operators considered above via the mapping $A \mapsto E_{p,w}^{-1} A E_{p,w}$.

5.7 Algebras of multiplication, convolution and flip operators

Let $\tilde{w}$ be a weight function on $\mathbb{R}^+$ of the form (4.8), and let w denote its symmetric extension to $\mathbb{R}$, i.e.

$$w(t) := \begin{cases} \tilde{w}(t) & \text{if } t \geq 0, \\ \tilde{w}(-t) & \text{if } t < 0. \end{cases}$$

The symmetry of the weight implies that the flip operator J given by $(Ju)(t) := u(-t)$ is bounded on $L^p(\mathbb{R}, w)$. It thus makes sense to consider the smallest closed subalgebra of $\mathcal{L}(L^p(\mathbb{R}, w))$ which contains all operators aI of multiplication by a function $a \in PC(\dot{\mathbb{R}})$, all Fourier convolutions $W^0(b)$ where $b \in PC_{p,w}$, and the flip J. We denote this algebra by $\mathcal{A}(PC(\dot{\mathbb{R}}), PC_{p,w}, J)$. Note that this algebra contains the Hankel operators $H(b) := \chi_+ W^0(b) J \chi_+ I$ with $b \in PC_{p,w}$. Further, we let $\mathcal{A}^{\mathcal{K}}(PC(\dot{\mathbb{R}}), PC_{p,w}, J)$ refer to the image of $\mathcal{A}(PC(\dot{\mathbb{R}}), PC_{p,w}, J)$ in the Calkin algebra and write Φ for the corresponding canonical homomorphism.

Let $\tilde{C}(\dot{\mathbb{R}})$ and $\tilde{C}_{p,w}$ denote the subalgebras of $C(\dot{\mathbb{R}})$ and $C_{p,w}$, respectively, which are constituted by the even functions, i.e., by the functions f with $Jf = f$.

Proposition 5.7.1. *If $f \in \tilde{C}(\dot{\mathbb{R}})$ and $g \in \tilde{C}_{p,w}$, then the coset $\Phi(fW^0(g))$ belongs to the center of $\mathcal{A}^{\mathcal{K}}(PC(\dot{\mathbb{R}}), PC_{p,w}, J)$.*

Proof. It easy to see that $fW^0(g)J = JfW^0(g)$. From

$$fW^0(g) = (f - f(\infty))W^0(g - g(\infty)) + (f - f(\infty))W^0(g(\infty))$$
$$+ f(\infty)W^0(g - g(\infty)) + f(\infty)W^0(g(\infty))$$
$$= (f - f(\infty))W^0(g - g(\infty)) + (f - f(\infty))g(\infty)I$$
$$+ f(\infty)W^0(g - g(\infty)) + f(\infty)g(\infty)I$$

and from Proposition 5.3.1 it becomes clear that $fW^0(g)$ also commutes with the other generators of the algebra modulo compact operators. ∎

Let $\overline{\mathbb{R}}^+$ denote the compactification of $\mathbb{R}^+$ by the point $\{\infty\}$, i.e., $\overline{\mathbb{R}}^+$ is homeomorphic to $[0, 1]$. The maximal ideal space of the algebra generated by all cosets $\Phi(fW^0(g))$ with $f \in \tilde{C}(\dot{\mathbb{R}})$ and $g \in \tilde{C}_{p,w}$ is homeomorphic to the subset $(\overline{\mathbb{R}}^+ \times \{\infty\}) \cup (\{\infty\} \times \overline{\mathbb{R}}^+)$ of the square $\overline{\mathbb{R}}^+ \times \overline{\mathbb{R}}^+$, which can be checked as in the proof of Proposition 5.5.2. The maximal ideal corresponding to $(s,t) \in (\overline{\mathbb{R}}^+ \times \{\infty\}) \cup (\{\infty\} \times \overline{\mathbb{R}}^+)$ is just the class of all cosets $\Phi(fW^0(g))$ where $f \in \tilde{C}(\dot{\mathbb{R}})$ with $f(s) = 0$ and $g \in \tilde{C}_{p,w}$ with $g(t) = 0$.

We proceed by localization over this maximal ideal space. Let $\mathscr{I}_{s,t}$ stand for the smallest closed ideal of the Banach algebra $\mathcal{A}^{\mathcal{K}}(PC(\dot{\mathbb{R}}), PC_{p,w}, J)$ which contains the maximal ideal $(s,t) \in (\overline{\mathbb{R}}^+ \times \{\infty\}) \cup (\{\infty\} \times \overline{\mathbb{R}}^+)$, and write $\mathcal{A}_{s,t}^{\mathcal{K}}$ for the quotient algebra

$$\mathcal{A}^{\mathcal{K}}(PC(\dot{\mathbb{R}}), PC_{p,w}, J) / \mathscr{I}_{s,t}$$

and $\Phi_{s,t}^{\mathscr{K}}$ for the canonical homomorphism from $\mathscr{A}\left(PC(\dot{\mathbb{R}}), PC_{p,w}, J\right)$ onto $\mathscr{A}_{s,t}^{\mathscr{K}}$.
For $x \in \overline{\mathbb{R}}^+$, let $\alpha(x)$ be the local exponent defined by (5.31).

Let $A \in \mathscr{A}\left(PC(\dot{\mathbb{R}}), PC_{p,w}, J\right)$. Allan's local principle implies that the coset $A + \mathscr{K}(L^p(\mathbb{R},w))$ is invertible in $\mathscr{A}^{\mathscr{K}}\left(PC(\dot{\mathbb{R}}), PC_{p,w}, J\right)$ if and only if the local cosets $\Phi_{s,t}^{\mathscr{K}}(A)$ are invertible for all $(s,t) \in (\overline{\mathbb{R}}^+ \times \{\infty\}) \cup (\{\infty\} \times \overline{\mathbb{R}}^+)$. We are thus left to analyze the local algebras $\mathscr{A}_{s,t}^{\mathscr{K}}$. Some care is in order since, in contrast to the previous sections, the strong limits $H_{s,\infty}(J)$ and $H_{\infty,t}(J)$ exist only at $s = 0$ and $t = 0$, respectively.

We are going to start with the local algebras $\mathscr{A}_{0,\infty}^{\mathscr{K}}$ and $\mathscr{A}_{\infty,0}^{\mathscr{K}}$.

Proposition 5.7.2. *The local algebra $\mathscr{A}_{0,\infty}^{\mathscr{K}}$ is isometrically isomorphic to the closed subalgebra* $\mathrm{alg}\{I, \chi_+ I, P_{\mathbb{R}}, J\}$ *of* $\mathscr{L}(L^p(\mathbb{R}^+, w_{\alpha(0)}))$, *with the isomorphism given by* $\Phi_{0,\infty}^{\mathscr{K}}(A) \mapsto H_{0,\infty}(A)$. *In particular, for $a \in PC(\dot{\mathbb{R}})$ and $b \in PC_{p,w}$,*

$$
\begin{aligned}
\Phi_{0,\infty}^{\mathscr{K}}(aI) &\mapsto a(0^-)\chi_- I + a(0^+)\chi_+ I, \\
\Phi_{0,\infty}^{\mathscr{K}}(W(b)) &\mapsto b(-\infty)Q_{\mathbb{R}} + b(+\infty)P_{\mathbb{R}}, \\
\Phi_{0,\infty}^{\mathscr{K}}(J) &\mapsto J.
\end{aligned}
$$

Proof. First note that the algebras $\mathscr{A}_{0,\infty}^{\mathscr{K}}$ related to $L^p(\mathbb{R}^+, w)$ and to $L^p(\mathbb{R}^+, |t|^{\alpha(0)})$, respectively, are isometrically isomorphic, as can be seen by the same arguments as in Proposition 4.3.2, steps 1 and 2. From this fact one deduces the independence of $H_{0,\infty}(A)$ of operators belonging to the local ideal $\mathscr{I}_{0,\infty}$, whence the correctness of the definition of the homomorphism follows. The concrete form of the values of the homomorphism at the generators comes from Proposition 5.4.3 and the fact that the operator J is homogeneous. Thus we conclude that $\Phi_{0,\infty}^{\mathscr{K}}(A) \mapsto H_{0,\infty}(A)$ is a mapping onto $\mathrm{alg}\{I, \chi_+ I, P_{\mathbb{R}}, J\}$. Finally, since $\Phi_{0,\infty}^{\mathscr{K}}(A) = \Phi_{0,\infty}^{\mathscr{K}}(H_{0,\infty}(A))$, this mapping is an isometry. ∎

In a similar way, one gets the following description of the local algebra at $(\infty, 0)$.

Proposition 5.7.3. *The local algebra $\mathscr{A}_{\infty,0}^{\mathscr{K}}$ is isometrically isomorphic to the closed subalgebra* $\mathrm{alg}\{I, \chi_+ I, P_{\mathbb{R}}, J\}$ *of* $\mathscr{L}(L^p(\mathbb{R}^+, w_{\alpha(\infty)}))$, *and the isomorphism is given by* $\Phi_{\infty,0}^{\mathscr{K}}(A) \mapsto H_{\infty,0}(A)$. *In particular, for $a \in PC(\dot{\mathbb{R}})$ and $b \in PC_{p,w}$,*

$$
\begin{aligned}
\Phi_{\infty,0}^{\mathscr{K}}(aI) &\mapsto a(-\infty)\chi_- I + a(+\infty)\chi_+ I, \\
\Phi_{\infty,0}^{\mathscr{K}}(W(b)) &\mapsto b(0^-)Q_{\mathbb{R}} + b(0^+)P_{\mathbb{R}}, \\
\Phi_{\infty,0}^{\mathscr{K}}(J) &\mapsto J.
\end{aligned}
$$

Now we turn to the local algebras $\mathscr{A}_{s,\infty}^{\mathscr{K}}$ and $\mathscr{A}_{\infty,t}^{\mathscr{K}}$ where $s, t > 0$. As already mentioned, the mappings $H_{s,\infty}$ and $H_{\infty,t}$ are not well defined on $\mathscr{A}\left(PC(\dot{\mathbb{R}}), PC_{p,w}, J\right)$ for $s, t \neq 0$. So we will have to use a modified approach which is based on the fact that every operator A in $\mathscr{A}\left(PC(\dot{\mathbb{R}}), PC_{p,w}, J\right)$ can be approximated as closely as desired by operators of the form $A_1 + JA_2$ where A_1 and A_2 belong to the algebra $\mathscr{A}\left(PC(\dot{\mathbb{R}}), PC_{p,w}\right)$ without flip. To be precise, $\mathscr{A}\left(PC(\dot{\mathbb{R}}), PC_{p,w}\right)$ stands for the smallest closed subalgebra of $\mathscr{L}(L^p(\mathbb{R}, w))$ which contains all multiplication operators aI with $a \in PC(\dot{\mathbb{R}})$ and all convolution operators $W^0(b)$ with $b \in PC_{p,w}$. Note that the decomposition of an operator A in the form $A_1 + JA_2$ is not unique in general.

We start with verifying that the homomorphisms $\mathsf{H}_{s,\infty}$ and $\mathsf{H}_{\infty,t}$ *are* well defined on the elements of the ideals $\mathscr{I}_{s,t}$. Note that $\mathsf{H}_{s,\infty}(K) = 0$ and $\mathsf{H}_{\infty,t}(K) = 0$ for every compact operator K.

Proposition 5.7.4. *If $A + \mathscr{K} \in \mathscr{I}_{s,t}$, then $\mathsf{H}_{s,\infty}(A + \mathscr{K}) = 0$ and $\mathsf{H}_{\infty,t}(A + \mathscr{K}) = 0$.*

Proof. It is evident from the definition of the ideal $\mathscr{I}_{s,t}$ that each of its elements can be approximated as closely as desired by operators of the form $A_1 + JA_2$ with $A_1, A_2 \in \mathscr{A}\left(PC(\dot{\mathbb{R}}), PC_{p,w}\right)$ belonging to the smallest closed ideal of that algebra which contains all cosets $\Phi(fW^0(g))$ with $f \in \tilde{C}(\dot{\mathbb{R}})$ with $f(s) = 0$ and $g \in \tilde{C}_{p,w}$ with $g(t) = 0$. So we can assume, without loss of generality, that A is of this form. Then

$$
\begin{aligned}
\mathsf{H}_{s,\infty}(A + \mathscr{K}) &= \underset{\tau \to +\infty}{\text{s-lim}}\, Z_\tau V_{-s} A V_s Z_\tau^{-1} \\
&= \underset{\tau \to +\infty}{\text{s-lim}}\, Z_\tau V_{-s}(A_1 + JA_2) V_s Z_\tau^{-1} \\
&= \underset{\tau \to +\infty}{\text{s-lim}}\, Z_\tau V_{-s} A_1 V_s Z_\tau^{-1} + \underset{\tau \to +\infty}{\text{s-lim}}\, (Z_\tau V_{-s} JV_s Z_\tau^{-1})(Z_\tau V_{-s} A_2 V_s Z_\tau^{-1}).
\end{aligned}
$$

For $i = 1, 2$, one has $\text{s-lim}\, Z_\tau V_{-s} A_i V_s Z_\tau^{-1} = 0$, and the operators $Z_\tau V_{-s} JV_s Z_\tau^{-1}$ are uniformly bounded with respect to τ. Thus, $\mathsf{H}_{s,\infty}(A + \mathscr{K}) = 0$. The proof of the second assertion is similar. ■

Let f_s be a continuous function with support in $\mathbb{R}^+$ and such that $f_s(s) = 1$. Set $p := \Phi^{\mathscr{K}}_{s,\infty}(f_s I)$, $j := \Phi^{\mathscr{K}}_{s,\infty}(J)$, and $e := \Phi^{\mathscr{J}}_{s,\infty}(I)$. Then $p^2 = p$, p commutes with all generators of the algebra except with j, and $jpj = e - p$. Thus, by Corollary 1.1.20, every element of $\mathscr{A}^{\mathscr{K}}_{s,\infty}$ can be (uniquely) written as $a = a_1 + a_2 j$, where the a_i belong to the corresponding local algebra without flip, and we can employ this corollary to eliminate the flip by doubling the dimension. Let L denote the mapping defined before Proposition 1.1.19 and consider the mapping

$$
\dot{\mathsf{H}}_{s,\infty} := \mathsf{H}_{s,\infty} L : \mathscr{A}^{\mathscr{K}}_{s,\infty} \to [\text{alg}\{I, \chi_+ I, P_{\mathbb{R}}\}]^{2 \times 2}, \tag{5.42}
$$

where $\mathsf{H}_{s,\infty}$ now refers to the canonical (diagonal) extension for matrix operators of the strong limit defined in (5.10). The mapping $\dot{\mathsf{H}}_{s,\infty}$ is well defined due to Proposition 5.7.4, and it acts as an homomorphism between the algebras mentioned. In what follows, the notation $-s^\pm$ is understood as $(-s)^\pm$.

Proposition 5.7.5. *Let $s > 0$. The local algebra $\mathscr{A}^{\mathscr{K}}_{s,\infty}$ is isomorphic to the matrix algebra $[\text{alg}\{I, \chi_+ I, P_{\mathbb{R}}\}]^{2 \times 2}$ with entries acting on $L^p(\mathbb{R}, w_{\alpha(s)})$. The isomorphism* *is given by $\Phi^{\mathscr{K}}_{s,\infty}(A) \mapsto \overset{\bullet}{\mathsf{H}}_{s,\infty}(A)$. In particular, for $a \in PC(\dot{\mathbb{R}})$ and $b \in PC_{p,w}$,*

$$\Phi^{\mathscr{K}}_{s,\infty}(aI) \quad \mapsto \quad \begin{bmatrix} a(s^-)\chi_- I + a(s^+)\chi_+ I & 0 \\ 0 & a(-s^+)\chi_- I + a(-s^-)\chi_+ I \end{bmatrix},$$

$$\Phi^{\mathscr{K}}_{s,\infty}(W^0(b)) \quad \mapsto \quad \begin{bmatrix} b(-\infty)Q_{\mathbb{R}} + b(+\infty)P_{\mathbb{R}} & 0 \\ 0 & b(+\infty)Q_{\mathbb{R}} + b(-\infty)P_{\mathbb{R}} \end{bmatrix},$$

$$\Phi^{\mathscr{K}}_{s,\infty}(J) \quad \mapsto \quad \begin{bmatrix} 0 & I \\ I & 0 \end{bmatrix}.$$

Proof. The mapping $\overset{\bullet}{\mathsf{H}}_{s,\infty}$ is well defined on $\mathscr{A}^{\mathscr{K}}_{s,\infty}$ by Proposition 5.7.4. The values of the homomorphism can be derived from Corollary 1.1.20 and Proposition 5.4.3. To see that the homomorphism $\overset{\bullet}{\mathsf{H}}_{s,\infty}$ is injective, define

$$\overset{\bullet}{\mathsf{H}}{}'_{s,\infty}: [\mathrm{alg}\{I,\ \chi_+ I,\ P_{\mathbb{R}}\}]^{2\times 2} \to \mathscr{A}^{\mathscr{K}}_{s,\infty},$$
$$\left(\begin{bmatrix} A_{11} & A_{12} \\ A_{21} & A_{22} \end{bmatrix} \right) \mapsto L^{-1} \left(\begin{bmatrix} p\Phi^{\mathscr{K}}_{s,\infty}(V_s A_{11} V_{-s}) & p\Phi^{\mathscr{K}}_{s,\infty}(V_s A_{12} V_{-s}) \\ p\Phi^{\mathscr{K}}_{s,\infty}(V_s A_{21} V_{-s}) & p\Phi^{\mathscr{K}}_{s,\infty}(V_s A_{22} V_{-s}) \end{bmatrix} \right). \tag{5.43}$$

The injectivity will follow once we have shown that

$$\overset{\bullet}{\mathsf{H}}{}'_{s,\infty} \left(\overset{\bullet}{\mathsf{H}}_{s,\infty}(\Phi^{\mathscr{K}}_{s,\infty}(A)) \right) = \Phi^{\mathscr{K}}_{s,\infty}(A) \quad \text{for all} \quad \Phi^{\mathscr{K}}_{s,\infty}(A) \in \mathscr{A}^{\mathscr{K}}_{s,\infty}.$$

It is sufficient to check this equality for the generating cosets of $\mathscr{A}^{\mathscr{K}}_{s,\infty}$, i.e., for $\Phi^{\mathscr{K}}_{s,\infty}(I)$, $\Phi^{\mathscr{K}}_{s,\infty}(J)$, $\Phi^{\mathscr{K}}_{s,\infty}(P_{\mathbb{R}})$, and $\Phi^{\mathscr{K}}_{s,\infty}(\chi_s I)$, where χ_s stands for the characteristic function of $]-\infty, s]$. This check is straightforward.

Finally, to verify the surjectivity of the homomorphism $\overset{\bullet}{\mathsf{H}}_{s,\infty}$, we again rely on $\overset{\bullet}{\mathsf{H}}{}'_{s,\infty}$. Indeed, this mapping is well defined on all of $[\mathrm{alg}\{I,\ \chi_+ I,\ P_{\mathbb{R}}\}]^{2\times 2}$, and one has $\overset{\bullet}{\mathsf{H}}_{s,\infty} \left(\overset{\bullet}{\mathsf{H}}{}'_{s,\infty}(A) \right) = A$ for all $A \in [\mathrm{alg}\{I,\ \chi_+ I,\ P_{\mathbb{R}}\}]^{2\times 2}$. ∎

Now let $t > 0$. For the local algebras $\mathscr{A}^{\mathscr{K}}_{\infty,t}$, we again apply Corollary 1.1.20 to eliminate the flip by doubling the dimension. Let f_t be a continuous function with support in $\mathbb{R}^+$ such that $f_t(t) = 1$, and put $p := \Phi^{\mathscr{K}}_{\infty,t}(W^0(f_t))$, $j := \Phi^{\mathscr{K}}_{\infty,t}(J)$, and $e := \Phi^{\mathscr{K}}_{\infty,t}(I)$. Then p is an idempotent which commutes with all generators of the algebra except with j, for which one has $jpj = e - p$. Every element of $\mathscr{A}^{\mathscr{K}}_{\infty,t}$ can be (uniquely) written as $a = a_1 + a_2 j$, where the a_i belong to the corresponding local algebra without flip. Define the homomorphism

$$\overset{\bullet}{\mathsf{H}}_{\infty,t} := \mathsf{H}_{\infty,t} L : \mathscr{A}^{\mathscr{K}}_{\infty,t} \to [\mathrm{alg}\{I,\ \chi_+ I,\ P_{\mathbb{R}}\}]^{2\times 2} \tag{5.44}$$

where $\mathsf{H}_{\infty,t}$ now refers to the canonical (diagonal) extension for matrix operators of the strong limit defined in (5.11), and where L is again the mapping defined before Proposition 1.1.19.

The proof of the next result is the same as that of Proposition 5.7.5.

Proposition 5.7.6. *Let $t > 0$. The local algebra $\mathscr{A}^{\mathscr{K}}_{\infty,t}$ is isomorphic to the matrix algebra $[\mathrm{alg}\{I, \chi_+ I, P_{\mathbb{R}}\}]^{2\times 2}$ with entries acting on $L^p(\mathbb{R}, w_{\alpha(\infty)})$. The isomorphism is given by $\Phi^{\mathscr{K}}_{\infty,t}(A) \mapsto \dot{\mathsf{H}}_{\infty,t}(A)$. In particular, for $a \in PC(\dot{\mathbb{R}})$ and $b \in PC_{p,w}$,*

$$\Phi^{\mathscr{K}}_{\infty,t}(aI) \quad \mapsto \quad \begin{bmatrix} a(-\infty)\chi_- I + a(+\infty)\chi_+ I & 0 \\ 0 & a(+\infty)\chi_- I + a(-\infty)\chi_+ I \end{bmatrix},$$

$$\Phi^{\mathscr{K}}_{\infty,t}(W^0(b)) \quad \mapsto \quad \begin{bmatrix} b(t^-)Q_{\mathbb{R}} + b(t^+)P_{\mathbb{R}} & 0 \\ 0 & b(-t^+)Q_{\mathbb{R}} + b(-t^-)P_{\mathbb{R}} \end{bmatrix},$$

$$\Phi^{\mathscr{K}}_{\infty,t}(J) \quad \mapsto \quad \begin{bmatrix} 0 & I \\ I & 0 \end{bmatrix}.$$

Our final concern is the local algebra $\mathscr{A}^{\mathscr{K}}_{\infty,\infty}$. We are going to show that $\mathscr{A}^{\mathscr{K}}_{\infty,\infty}$ is a unital algebra generated by two commuting projections and a flip.

Proposition 5.7.7. *The local algebra $\mathscr{A}^{\mathscr{K}}_{\infty,\infty}$ is generated by the elements $e := \Phi^{\mathscr{K}}_{\infty,\infty}(I)$, $p := \Phi^{\mathscr{K}}_{\infty,\infty}(\chi_+ I)$, $r := \Phi^{\mathscr{K}}_{\infty,\infty}(W^0(\chi_+))$ and $j := \Phi^{\mathscr{K}}_{\infty,\infty}(J)$.*

Proof. For $a \in PC(\dot{\mathbb{R}})$, write $\Phi^{\mathscr{K}}_{\infty,\infty}(aI)$ as

$$\Phi^{\mathscr{K}}_{\infty,\infty}(a(-\infty)\chi_- I + a(+\infty)\chi_+ I) - \Phi^{\mathscr{K}}_{\infty,\infty}((a - a(-\infty)\chi_- - a(+\infty)\chi_+)I).$$

Since the function $a - a(-\infty)\chi_- - a(+\infty)\chi_+$ is continuous at infinity and has the value 0 there, we obtain

$$\Phi^{\mathscr{K}}_{\infty,\infty}(aI) = \Phi^{\mathscr{K}}_{\infty,\infty}((a(-\infty)\chi_- + a(+\infty)\chi_+)I) = 0.$$

For $b \in PC_{p,w}$, one gets similarly

$$\Phi^{\mathscr{K}}_{\infty,\infty}(W^0(b)) = \Phi^{\mathscr{K}}_{\infty,\infty}(b(-\infty)W^0(\chi_-) + b(+\infty)W^0(\chi_+)).$$

For the other generators, the result is obvious. ∎

The generators of the algebra $\mathscr{A}^{\mathscr{K}}_{\infty,\infty}$ satisfy the relations

$$rp = pr, \qquad jrj = e - r \quad \text{and} \quad jpj = e - p.$$

Only the first of these relations is not completely evident. It can be verified by repeating arguments from the proof of Theorem 5.6.2 (iv). Thus, the algebra $\mathscr{A}^{\mathscr{K}}_{\infty,\infty}$ is generated by two commuting projections and a flip.

To get a matrix-valued symbol for the invertibility in the algebra $\mathscr{A}^{\mathscr{K}}_{\infty,\infty}$, one can apply Proposition 1.1.19 to eliminate the flip by doubling the dimension, or, one refers formally to Theorem 3.3.13,. For the latter, note that the elements b and c defined in (3.53) and (3.54) are given by $b = pr + (e - p)(e - r)$ and $c = (pr - rp)j = 0$ in the present context, and that the spectrum of b is $\{0, 1\}$ since b is a non-trivial idempotent. Thus, Theorem 3.3.13 applies with $y = 0$ and $x = \pm 1$. In each case, we arrive at the following.

Proposition 5.7.8. *The local algebra $\mathscr{A}^{\mathscr{K}}_{\infty,\infty}$ is generated by the commuting projections $p = \Phi^{\mathscr{K}}_{\infty,\infty}(\chi_+ I)$ and $r = \Phi^{\mathscr{K}}_{\infty,\infty}(W^0(\chi_+))$ and by the flip $j = \Phi^{\mathscr{K}}_{\infty,\infty}(J)$. There is a symbol mapping which assigns with e, p, j and r a matrix-valued function on $\{0,1\}$ by*

$$(\mathrm{smb}\,e)(x) = \begin{bmatrix} 1 & 0 \\ 0 & 1 \end{bmatrix}, \quad (\mathrm{smb}\,p)(x) = \begin{bmatrix} 1 & 0 \\ 0 & 0 \end{bmatrix},$$

$$(\mathrm{smb}\,j)(x) = \begin{bmatrix} 0 & 1 \\ 1 & 0 \end{bmatrix}, \quad (\mathrm{smb}\,r)(x) = \begin{bmatrix} x & 0 \\ 0 & 1-x \end{bmatrix}.$$

Combining the previous results with Allan's local principle, we arrive at the following.

Theorem 5.7.9. *Let $A \in \mathscr{A}\left(PC(\dot{\mathbb{R}}), PC_{p,w}, J\right)$. The coset $A + \mathscr{K}\left(L^p(\mathbb{R},w)\right)$ is invertible in the quotient algebra $\mathscr{A}^{\mathscr{K}}\left(PC(\dot{\mathbb{R}}), PC_{p,w}, J\right)$ if and only if:*

(i) *the operator $\mathsf{H}_{0,\infty}(A)$ is invertible in the subalgebra $\mathrm{alg}\{I, \chi_+ I, P_{\mathbb{R}}, J\}$ of $\mathscr{L}(L^p(\mathbb{R}, w_{\alpha(0)}))$;*

(ii) *the operator $\mathsf{H}_{\infty,0}(A)$ is invertible in the subalgebra $\mathrm{alg}\{I, \chi_+ I, P_{\mathbb{R}}, J\}$ of $\mathscr{L}(L^p(\mathbb{R}, w_{\alpha(\infty)}))$;*

(iii) *the operator $\overset{\bullet}{\mathsf{H}}_{s,\infty}(A)$ is invertible in the subalgebra $[\mathrm{alg}\{I, \chi_+ I, P_{\mathbb{R}}\}]^{2 \times 2}$ of $[\mathscr{L}(L^p(\mathbb{R}, w_{\alpha(s)}))]^{2 \times 2}$ for every $s > 0$;*

(iv) *the operator $\overset{\bullet}{\mathsf{H}}_{\infty,t}(A)$ is invertible in the subalgebra $[\mathrm{alg}\{I, \chi_+ I, P_{\mathbb{R}}\}]^{2 \times 2}$ of $[\mathscr{L}(L^p(\mathbb{R}, w_{\alpha(\infty)}))]^{2 \times 2}$ for every $t > 0$;*

(v) *the matrix $\mathrm{smb}\,\Phi^{\mathscr{K}}_{\infty,\infty}(A)$ is invertible in $\mathbb{C}^{2 \times 2}$.*

The following corollary can be proved by repeating the arguments from the proof of Theorem 5.5.9.

Corollary 5.7.10. *The algebra $\mathscr{A}^{\mathscr{K}}\left(PC(\dot{\mathbb{R}}), PC_{p,w}, J\right)$ is inverse-closed in the Calkin algebra $\mathscr{L}(L^p(\mathbb{R},w))/\mathscr{K}(L^p(\mathbb{R},w))$. An operator $A \in \mathscr{A}\left(PC(\dot{\mathbb{R}}), PC_{p,w}, J\right)$ is Fredholm if and only if:*

(i) *the operator $\mathsf{H}_{0,\infty}(A)$ is invertible on $L^p(\mathbb{R}, w_{\alpha(0)})$;*

(ii) *the operator $\mathsf{H}_{\infty,0}(A)$ is invertible on $L^p(\mathbb{R}, w_{\alpha(\infty)})$;*

(iii) *the operator $\overset{\bullet}{\mathsf{H}}_{s,\infty}(A)$ is invertible on $L^p_2(\mathbb{R}, w_{\alpha(s)})$ for every $s > 0$;*

(iv) *the operator $\overset{\bullet}{\mathsf{H}}_{\infty,t}(A)$ is invertible on $L^p_2(\mathbb{R}, w_{\alpha(\infty)})$ for every $t > 0$;*

(v) *the matrix $\mathrm{smb}\,\Phi^{\mathscr{K}}_{\infty,\infty}(A)$ is invertible in $\mathbb{C}^{2 \times 2}$.*

In the remainder of this section we are going to apply this corollary to a class of operators of particular interest: the Wiener-Hopf plus Hankel operators. These are the operators $W(b) + H(c)$ on $L^p(\mathbb{R}^+, w)$ with $b, c \in PC_{p,w}$. Equivalently, one can think of a Wiener-Hopf plus Hankel operator $W(b) + H(c)$ as the operator

$$\chi_+ W^0(b)\chi_+ I + \chi_+ W^0(c)J\chi_+ I + \chi_- I, \tag{5.45}$$

acting on $L^p(\mathbb{R}, w)$.

Theorem 5.7.11. *Let* $b, c \in PC_{p,w}$. *The operator* (5.45) *is Fredholm on* $L^p(\mathbb{R}, w)$ *if and only if* $b(\pm\infty) \neq 0$ *and if the functions*

(i) $y \mapsto (b(+\infty) + b(-\infty)) \sinh\big((y + i\upsilon)\pi\big) + (b(+\infty) - b(-\infty)) \cosh\big((y + i\upsilon)\pi\big)$
   $+ c(+\infty) - c(-\infty)$ *with* $\upsilon := 1/p + \alpha(0)$,

(ii) $y \mapsto (b(0^+) + b(0^-)) \sinh\big((y + i\upsilon)\pi\big) + (b(0^+) - b(0^-)) \cosh\big((y + i\upsilon)\pi\big)$
   $+ c(0^+) - c(0^-)$ *with* $\upsilon := 1/p + \alpha(\infty)$, *and*

(iii) $y \mapsto b_t^+ b_{-t}^+ + (b_t^- b_{-t}^+ + b_t^+ b_{-t}^-) \coth\big((y + i\upsilon)\pi\big) + b_t^- b_{-t}^- \Big(\coth\big((y + i\upsilon)\pi\big)\Big)^2$
   $- c_t c_{-t} \Big(\sinh\big((y + i\upsilon)\pi\big)\Big)^{-2}$ *with* $\upsilon := 1/p + \alpha(\infty)$, $b_t^\pm := b(t^+) \pm b(t^-)$,
   $b_{-t}^\pm := b(-t^+) \pm b(-t^-)$ *and* $c_{\pm t} := c(\pm t^+) - c(\pm t^-)$ *for* $t > 0$

do not vanish on $\overline{\mathbb{R}}$.

Proof. By Corollary 5.7.10, the operator (5.45) is Fredholm if and only if a collection of related operators, labeled by the points of the set $(\overline{\mathbb{R}}^+ \times \{\infty\}) \cup (\{\infty\} \times \overline{\mathbb{R}}^+)$, is invertible. We are going to examine the invertibility of the related operators for each point in this set.

For the point $(0, \infty)$, the related operator is

$$\chi_+ \big(b(-\infty)Q_\mathbb{R} + b(+\infty)P_\mathbb{R} + (c(-\infty)Q_\mathbb{R} + c(+\infty)P_\mathbb{R})J\big) \chi_+ I + \chi_- I.$$

This operator is invertible on $L^p(\mathbb{R}, w_{\alpha(0)})$ if and only if the operator

$$\frac{b(+\infty) + b(-\infty)}{2} I + \frac{b(+\infty) - b(-\infty)}{2} S_{\mathbb{R}^+} + \frac{c(+\infty) - c(-\infty)}{2} H_\pi$$

is invertible on $L^p(\mathbb{R}^+, w_{\alpha(0)})$. The latter condition is equivalent to condition (i) which can easily be seen by inserting the Mellin symbols of the operators $S_{\mathbb{R}^+}$ and H_π quoted in Section 4.2.2.

For the point $(\infty, 0)$, the related operator

$$\chi_+ \big(b(0^-)Q_\mathbb{R} + b(0^+)P_\mathbb{R} + (c(0^-)Q_\mathbb{R} + c(0^+)P_\mathbb{R})J\big) \chi_+ I + \chi_- I$$

is invertible on $L^p(\mathbb{R}, w_{\alpha(\infty)})$ if and only if the operator

$$\frac{b(0^+) + b(0^-)}{2} I + \frac{b(0^+) - b(0^-)}{2} S_{\mathbb{R}^+} + \frac{c(0^+) - c(0^-)}{2} H_\pi$$

is invertible on $L^p(\mathbb{R}^+, w_{\alpha(\infty)})$. Condition (ii) states the conditions for the invertibility of the Mellin symbol of this operator.

For (s, ∞) with $s > 0$, we have

$$b(-\infty)Q_\mathbb{R} + b(+\infty)P_\mathbb{R}$$

as the related operator. This operator is invertible on $L^p(\mathbb{R}, w_{\alpha(s)})$ if and only if $b(\pm\infty) \neq 0$.

For (∞, t) with $t > 0$, the invertibility of the related operator

$$\begin{bmatrix} \chi_+(b(t^-)Q_{\mathbb{R}}+b(t^+)P_{\mathbb{R}})\chi_+I+\chi_-I & \chi_+(c(t^-)Q_{\mathbb{R}}+c(t^+)P_{\mathbb{R}})\chi_-I \\ \chi_-(c(-t^+)Q_{\mathbb{R}}+c(-t^-)P_{\mathbb{R}})\chi_+I & \chi_-(b(-t^+)Q_{\mathbb{R}}+b(-t^-)P_{\mathbb{R}})\chi_-I+\chi_+I \end{bmatrix}$$

on $L_2^p(\mathbb{R}, w_{\alpha(\infty)})$ is equivalent to the invertibility of the operator

$$\begin{bmatrix} b(t^-)Q_{\mathbb{R}^+}+b(t^+)P_{\mathbb{R}^+} & \frac{c(t^+)-c(t^-)}{2}H_\pi \\ \frac{c(-t^+)-c(-t^-)}{2}H_\pi & b(-t^+)P_{\mathbb{R}^+}+b(-t^-)Q_{\mathbb{R}^+} \end{bmatrix}.$$

The Mellin symbol of this operator is

$$\begin{bmatrix} b_t^+ + b_t^-\coth\left((y+i\upsilon)\pi\right) & \frac{c_t}{2}\left(\sinh\left((y+i\upsilon)\pi\right)\right)^{-1} \\ \frac{c_{-t}}{2}\left(\sinh\left((y+i\upsilon)\pi\right)\right)^{-1} & b_{-t}^+ + b_{-t}^-\coth\left((y+i\upsilon)\pi\right) \end{bmatrix}, \tag{5.46}$$

and condition (iii) states exactly the conditions for the invertibility of this matrix function. Finally, the matrix related to the point (∞,∞) is invertible if and only if $b(\pm\infty) \neq 0$.   ∎

What the above results tell us about the essential spectrum of the Wiener-Hopf plus Hankel operator is in some sense expected. The local spectrum at each point where both b and c are continuous corresponds to the value of the function b at that point. If only b is discontinuous at some point, then the local spectrum corresponds to the left and right one-sided limits, joined by a circular arc, the shape of which depends on the space and weight (see Figure 4.4). If only c is discontinuous at some point t, but not at $-t$, there is no effect on the essential spectrum. If both b and c share a point of discontinuity at 0 or ∞, the effect on the essential spectrum is the "sum" of the circular arc with the "water drop" arc (see Figure 4.5 (a)). The more complex effects occur when both b and c share points of discontinuity on $\pm t$, $t \in \mathbb{R}^+$. In this case, the essential spectrum is given by the spectrum of the matrix function (5.46).

Example 5.7.12. Let the weight w be such that $1/p + \alpha(\infty) \in \{1/2, 1/2 + 0.01, 2/3\}$. Define $b \in PC_p$ by

$$b(t) = \begin{cases} \frac{t+10}{10} + \frac{t+10}{2}i & \text{for } -20 < t < 0, \\ \frac{t-10}{10} + \frac{t-10}{2}i & \text{for } 0 < t < 20, \\ -3i & \text{for all other } t \end{cases}$$

and consider a function c which is continuous at all points except the integer points in $[-19, 19] \setminus \{-10, 0, 10\}$, where it satisfies $c(t^+) - c(t^-) = 1$, $c(-t^+) - c(-t^-) = 1$ if $1 \leq t \leq 9$ and $c(t^+) - c(t^-) = 1$, $c(-t^+) - c(-t^-) = -1$ in the case $11 \leq t \leq 19$. Then the essential spectrum of the operator $W(b) + H(c)$ on the space $\mathscr{L}(L^p(\mathbb{R}, w))$ is given by Figure 5.3. The three arcs joining the points of discontinuity of the function b are clear, as is the variation of the size of the "water drop" arcs derived from the distance between $b(-t)$ and $b(t)$. When the jumps of c have the same sign, they will actually interfere with one another and form other geometric figures. If we change p or the weight, such that $\upsilon = 1/p + \alpha(\infty)$ approaches $1/2$, all curves from discontinuities turn into line segments.   □

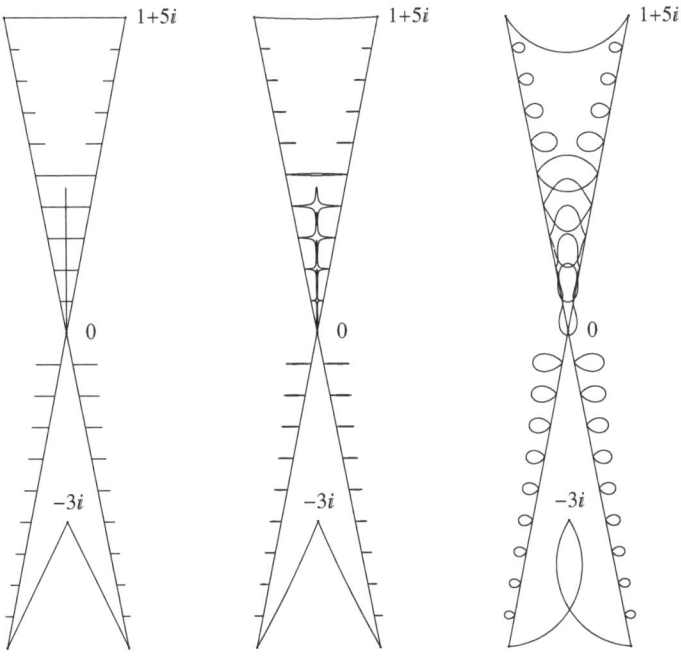

Fig. 5.3 The essential spectrum of $W(b) + H(c)$ for $\upsilon = 1/2$, $1/2 + 0.01$ and $2/3$.

Remark 5.7.13. In contrast to Section 4.5, the proof of Theorem 5.7.11 yields immediately a 2-symbol for the Wiener-Hopf plus Hankel operators with piecewise generating functions due to the finer localization used to obtain Theorem 5.7.9.   □

5.8 Multidimensional convolution type operators

Now we turn our attention to the Fredholm property of multidimensional convolution type operators on $\mathbb{R}^N$. We will see that the techniques developed so far –

localization and homogenization – work well also in the multidimensional context, but that a new difficulty appears if $N > 1$: already the generators of the algebra will have massive spectra. Hence, Corollary 1.2.32 does not apply to prove the inverse-closedness of the operator algebra under consideration (and, actually, we still do not know if this algebra is inverse-closed). In this section, we will point out one way to overcome this difficulty.

Let N be a positive integer. We denote the Euclidean norm on $\mathbb{R}^N$ by $|\cdot|$ and write $\langle \cdot, \cdot \rangle$ for the related scalar product on $\mathbb{R}^N$. Thus, $|x|^2 = \langle x, x \rangle$ for $x \in \mathbb{R}^N$. The unit sphere in $\mathbb{R}^N$ will be denoted by $\mathbb{S}^{N-1}$, and the open unit ball by $\mathbb{B}^N$.

It is easy to see that the mapping

$$\xi : \mathbb{B}^N \to \mathbb{R}^N, \quad x \mapsto \frac{x}{1 - |x|}, \tag{5.47}$$

is a homeomorphism with inverse

$$\xi^{-1} : \mathbb{R}^N \to \mathbb{B}^N, \quad x \mapsto \frac{x}{1 + |x|}. \tag{5.48}$$

In particular, a function f on $\mathbb{R}^N$ is continuous if and only if the function $f \circ \xi$ is continuous on $\mathbb{B}^N$. We denote by $C(\overline{\mathbb{R}^N})$ the set of all continuous complex-valued functions f on $\mathbb{R}^N$ for which the (continuous) function $f \circ \xi$ on $\mathbb{B}^N$ possesses a continuous extension $f^{\sim}$ onto the closed ball $\overline{\mathbb{B}^N}$. Provided with pointwise operations and the supremum norm, $C(\overline{\mathbb{R}^N})$ forms a commutative C^*-algebra, and this algebra is isomorphic to $C(\overline{\mathbb{B}^N})$. Thus, the maximal ideal spaces of these algebras are homeomorphic. The maximal ideal space of $C(\overline{\mathbb{B}^N})$ is the closed unit ball $\overline{\mathbb{B}^N}$, which is a union of the open ball $\mathbb{B}^N$ and the unit sphere $\mathbb{S}^{N-1}$. Analogously, one can think of the maximal ideal space of $C(\overline{\mathbb{R}^N})$ as the union of $\mathbb{R}^N$ and of an "infinitely distant" sphere. More precisely, every multiplicative linear functional on $C(\overline{\mathbb{R}^N})$ is either of the form

$$f \mapsto f(x) \quad \text{with} \quad x \in \mathbb{R}^N$$

or of the form

$$f \mapsto (f \circ \xi)^{\sim}(\theta) \quad \text{with} \quad \theta \in \mathbb{S}^{N-1}.$$

We denote the latter functional by θ_∞ and write $f(\theta_\infty)$ in place of $\theta_\infty(f)$. Clearly, $f(\theta_\infty) = \lim_{t \to \infty} f(t\theta)$, and a sequence $h \in \mathbb{R}^N$ converges to θ_∞ if $\xi^{-1}(h_n)$ converges to θ. A basis of neighborhoods of θ_∞ is provided by the sets of the form

$$U_{R,\varepsilon}(\theta_\infty) := \quad \{ |x| \psi \in \mathbb{R}^N : |x| > R, \ \psi \in \mathbb{S}^{N-1} \text{ and } |\psi - \theta| < \varepsilon \}$$

$$\bigcup \{ \psi_\infty : \psi \in \mathbb{S}^{N-1} \text{ and } |\psi - \theta| < \varepsilon \}. \tag{5.49}$$

We denote the maximal ideal space of $C(\overline{\mathbb{R}^N})$ by $\overline{\mathbb{R}^N}$.

Every function $a \in L^1(\mathbb{R}^N)$ defines an operator W_a^0 of convolution by a by

$$W_a^0 : L^p(\mathbb{R}^N) \to L^p(\mathbb{R}^N), \quad g \mapsto \int_{\mathbb{R}^N} a(t - s) g(s) \, ds. \tag{5.50}$$

The operator of convolution by $a \in L^1(\mathbb{R}^N)$ is bounded on $L^p(\mathbb{R}^N)$, and

$$\|W_a^0\|_{\mathscr{L}(L^p(\mathbb{R}^N))} \leq \|a\|_{L^1(\mathbb{R}^N)}.$$

The goal of this section is to study the Fredholm property of operators which belong to the smallest closed subalgebra $\mathscr{A}_p$ of $\mathscr{L}(L^p(\mathbb{R}^N))$ which contains

- all convolution operators W_a^0 with $a \in L^1(\mathbb{R}^N)$,
- all operators of multiplication by a function in $C(\overline{\mathbb{R}^N})$,
- all operators $\chi_{\mathbb{H}(\theta)}I$ of multiplication by the characteristic function of a half-space

$$\mathbb{H}(\theta) := \{x \in \mathbb{R}^N : \langle x, \theta \rangle \geq 0\} \quad \text{with} \quad \theta \in \mathbb{S}^{N-1}.$$

Proposition 5.8.1. *The algebra $\mathscr{A}_p$ contains the ideal $\mathscr{K}(L^p(\mathbb{R}^N))$ of the compact operators.*

Proof. Let $\mathscr{A}_p'$ denote the smallest closed subalgebra of $\mathscr{L}(L^p(\mathbb{R}^N))$ which contains all operators W_a^0 with $a \in L^1(\mathbb{R}^N)$ and all operators of multiplication by a function in $C_0^\infty(\mathbb{R}^N)$. We will show $\mathscr{K}(L^p(\mathbb{R}^N))$ is already contained in $\mathscr{A}_p'$, which implies the assertion .

It is sufficient to show that $\mathscr{A}_p'$ contains all operators of rank one. Every operator of rank one on $L^p(\mathbb{R}^N)$ has the form

$$(Ku)(t) = a(t) \int_{\mathbb{R}^N} b(s)u(s)ds, \quad t \in \mathbb{R}^N, \tag{5.51}$$

where $a \in L^p(\mathbb{R}^N)$ and $b \in L^q(\mathbb{R}^N)$ with $1/p + 1/q = 1$. Since $C_0^\infty(\mathbb{R}^N)$ is dense in $L^p(\mathbb{R}^N)$ and in $L^q(\mathbb{R}^N)$ (with respect to the corresponding norms), it is further sufficient to show that every operator (5.51) with $a, b \in C_0^\infty(\mathbb{R}^N)$ belongs to $\mathscr{A}_p'$.

Let $a, b \in C_0^\infty(\mathbb{R}^N)$, and choose a function $k \in L^1(\mathbb{R}^N)$ which is 1 on the compact set $\{t - s : t \in \operatorname{supp} f, s \in \operatorname{supp} g\}$. Then the operator (5.51) can be written as

$$(Ku)(t) = a(t) \int_{\mathbb{R}^N} k(t - s)b(s)u(s)ds, \quad t \in \mathbb{R}^N.$$

Evidently, this operator belongs to $\mathscr{A}_p'$. ∎

As already mentioned, we do not know if the algebra $\mathscr{A}_p^{\mathscr{K}} := \mathscr{A}_p/\mathscr{K}(L^p(\mathbb{R}^N))$ is inverse-closed in the Calkin algebra. Therefore we have to apply the local principle in a larger algebra which we are going to introduce in a couple of steps.

Let $\Lambda_p \subset \mathscr{L}(L^p(\mathbb{R}^N))$ denote the Banach algebra of all operators of local type with respect to the algebra $C(\overline{\mathbb{R}^N})$, that is, an operator $A \in \mathscr{L}(L^p(\mathbb{R}^N))$ belongs to Λ_p if and only if

$$fA - AfI \in \mathscr{K}(L^p(\mathbb{R}^N)) \quad \text{for every} \quad f \in C(\overline{\mathbb{R}^N}).$$

In order to show that $\mathscr{A}_p \subset \Lambda_p$, we need the following lemma.

Lemma 5.8.2. *Let F_1 and F_2 be disjoint closed subsets of $\overline{\mathbb{R}^N}$. Then there exists a $\delta > 0$ such that $|x - y| > (R+1)\delta$ for all $R > 0$, all $x \in F_1 \cap \mathbb{R}^N$ with $|x| > R$ and all $y \in F_2 \cap \mathbb{R}^N$ with $|y| > R$.*

Proof. Let $\widetilde{\xi}$ stand for the homeomorphism from $\overline{\mathbb{B}^N}$ onto $\overline{\mathbb{R}^N}$ which coincides with ξ on $\mathbb{B}^N$, and set $\tilde{F}_1 := \widetilde{\xi}^{-1}(F_1)$ and $\tilde{F}_2 := \widetilde{\xi}^{-1}(F_2)$. Then $\tilde{F}_1$ and $\tilde{F}_2$ are disjoint compact subsets of $\overline{\mathbb{B}^N}$; hence, $\delta := \text{dist}(\tilde{F}_1, \tilde{F}_2) > 0$. Thus, for $x \in F_1$ and $y \in F_2$ one has $|\xi^{-1}(x) - \xi^{-1}(y)| > \delta$ or, equivalently,

$$\left| \frac{x}{1 + |x|} - \frac{y}{1 + |y|} \right| > \delta. \tag{5.52}$$

Let $y \neq 0$ and consider the function $f(t) := |x - ty|$ on $\mathbb{R}$. This function attains its minimum at the point $t^* := \langle x, y \rangle / \langle y, y \rangle$ and is therefore monotonically increasing on the interval $[t^*, \infty[$. Since $t^* \leq |x| |y| / |y|^2 = |x|/|y|$ and

$$\frac{|x|}{|y|} \leq \frac{1 + |x|}{1 + |y|} \leq 1 \quad \text{if} \quad |x| \leq |y|,$$

we conclude that $f(\frac{1+|x|}{1+|y|}) \leq f(1)$. Thus, by (5.52),

$$|x - y| \geq \left| x - \frac{1+|x|}{1+|y|} y \right| = (1 + |x|) \left| \frac{x}{1+|x|} - \frac{y}{1+|y|} \right| \geq (1 + |x|)\delta \tag{5.53}$$

for $|x| \leq |y|$. Analogously, if $|y| \leq |x|$, then $|x - y| \geq (1 + |y|)\delta$. So one gets

$$|x - y| \geq \min\{1 + |x|, 1 + |y|\}\, \delta,$$

which implies the assertion. ∎

Proposition 5.8.3. *If $f \in C(\overline{\mathbb{R}^N})$ and $a \in L^1(\mathbb{R}^N)$, then $W_a^0 fI - fW_a^0$ is a compact operator. Thus, $\mathscr{A}_p \subset \Lambda_p$.*

Proof. By Krasnoselskii's interpolation theorem, it is sufficient to verify the compactness of $W_a^0 fI - fW_a^0$ on $L^2(\mathbb{R})$. By Theorem 2.5.6, this operator is compact if and only if the operator $\chi_{F_1} W_a^0 \chi_{F_2} I$ is compact for each choice of closed disjoint subsets F_1 and F_2 of $\overline{\mathbb{R}^N}$. For g in $L^2(\mathbb{R}^N)$, one has

$$(\chi_{F_1} W_a^0 \chi_{F_2} g)(s) = \int_{\mathbb{R}^N} \chi_{F_1}(s) a(s - t) \chi_{F_2}(t) g(t)\, dt.$$

Since functions in $L^1(\mathbb{R}^N)$ can be approximated by continuous functions with compact support, we can assume that a is a continuous function with support contained in the centered ball of radius M. Set $\widetilde{a}(s,t) := \chi_{F_1}(s) a(s - t) \chi_{F_2}(t)$. By the previous lemma, there exists $\delta > 0$ such that $|s - t| > R\delta$ for every $R > 0$ and for arbitrary points $s \in F_1 \cap \mathbb{R}^N$ and $t \in F_2 \cap \mathbb{R}^N$ with $|s| > R$ and $|t| > R$. Choose R such that

$R > M$ and $R\delta > M$. Then $\widetilde{a}(s,t) = 0$ if $s \in F_1$ and $|s| > 2R$, or if $t \in F_2$ and $|t| > 2R$. Indeed, if $|s| > 2R$ and $|t| \le R$, then $|s-t| \ge |s| - |t| > R > M$, and if $|s| > 2R$ and $|t| > R$, then $|s-t| > R\delta > M$ due to the choice of R. Similarly, $|t| > 2R$ implies that $|s-t| > M$. Hence, $\widetilde{a}$ is a compactly supported bounded function, whence $\widetilde{a} \in L^2(\mathbb{R}^N \times \mathbb{R}^N)$. Therefore, $\chi_{F_1} W_a^0 \chi_{F_2} I$ is a Hilbert-Schmidt operator, and thus compact. ∎

Our next goal is to introduce certain strong limit operators which will be used later to identify local algebras. For $k \in \mathbb{R}^N$, we define the shift operator

$$V_k : L^p(\mathbb{R}^N) \to L^p(\mathbb{R}^N), \quad (V_k u)(s) = u(s-k),$$

and for $t > 0$, the dilation operator

$$Z_t : L^p(\mathbb{R}^N) \to L^p(\mathbb{R}^N), \quad (Z_t u)(s) := t^{-N/p} u(s/t).$$

Both operators act as bijective isometries, with inverses given by $V_k^{-1} = V_{-k}$ and $Z_t^{-1} = Z_{t^{-1}}$.

Proposition 5.8.4. *Let $x \in \mathbb{R}^N$. Then the strong limit*

$$H_x(A) := \operatorname*{s-lim}_{t \to \infty} Z_t V_{-x} A V_x Z_t^{-1} \tag{5.54}$$

exists for every operator $A \in \mathscr{A}_p$, the mapping H_x defines a homomorphism on $\mathscr{A}_p$, and

(i) $H_x(W_a^0) = 0$ *for* $a \in L^1(\mathbb{R}^N)$;
(ii) $H_x(fI) = f(x)I$ *for* $f \in C(\mathbb{R}^N)$;
(iii) $H_x(\chi_{\mathbb{H}(\theta)} I)$ *is 0, I or $\chi_{\mathbb{H}(\theta)} I$, depending on whether x is outside, in the interior or on the boundary of $\mathbb{H}(\theta)$, respectively, where $\theta \in \mathbb{S}^{N-1}$;*
(iv) $H_x(K) = 0$ *for K compact.*

The proof follows as that of Proposition 5.4.3 (but is actually much simpler since all the functions are continuous). The details are left to the reader.

For $\theta \in \mathbb{S}^{N-1}$, consider the sequence $h^\theta : \mathbb{N} \to \mathbb{R}^N$ defined by $h^\theta(n) := n\theta$.

Proposition 5.8.5. *Let $\theta \in \mathbb{S}^{N-1}$. Then the strong limit*

$$H_\theta^\circ(A) := \operatorname*{s-lim}_{n \to \infty} V_{-h^\theta(n)} A V_{h^\theta(n)} \tag{5.55}$$

exists for every operator $A \in \mathscr{A}_p$, the mapping H_θ° defines a homomorphism on $\mathscr{A}_p$, and

(i) $H_\theta^\circ(W_a^0) = W_a^0$ *for* $a \in L^1(\mathbb{R}^N)$;
(ii) $H_\theta^\circ(fI) = f(\theta_\infty)I$ *for* $f \in C(\overline{\mathbb{R}^N})$;
(iii) $H_\theta^\circ(\chi_{\mathbb{H}(\psi)} I)$ *is 0, I or $\chi_{\mathbb{H}(\psi)} I$, depending on whether θ is outside, inside or on the boundary of $\mathbb{H}(\psi)$, respectively, where $\psi \in \mathbb{S}^{N-1}$;*
(iv) $H_\theta^\circ(K) = 0$ *for K compact.*

Proof. Assertion (i) is evident since convolution operators are shift invariant. For assertion (ii), let g be a function in $L^p(\mathbb{R}^N)$ the support of which is in the ball $B_M(0)$ with radius M for some $M > 0$. By definition, we have

$$(V_{-h^\theta(n)} f V_{h^\theta(n)} g)(x) = f(x + h^\theta(n)) g(x).$$

Since f is continuous at θ_∞ there is, for every $\varepsilon > 0$, a neighborhood $U_{R,\delta}(\theta_\infty)$ of θ_∞ as in (5.49) such that $|f(x) - f(\theta_\infty)| < \varepsilon$ for every $x \in U_{R,\delta}(\theta_\infty)$. The compactness of $B_M(0)$ guarantees that there is an $n_0 \in \mathbb{N}$ such that $B_M(0) + h^\theta(n)$ is contained in $U_{R,\delta}(\theta_\infty)$ whenever $n \geq n_0$. Thus, for $n \geq n_0$,

$$\|(V_{-h^\theta(n)} f V_{h^\theta(n)} - f(\theta_\infty)) g\|_{L^p}$$
$$\leq \sup_{x \in B_M(0)} |f(x + h^\theta(n)) - f(\theta_\infty)| \, \|g\|_{L^p} \leq \varepsilon \|g\|_{L^p}.$$

Since the functions with compact support are dense in $L^p(\mathbb{R}^N)$, we get

$$\|(V_{-h^\theta(n)} f V_{h^\theta(n)} - f(\theta_\infty)) g\|_{L^p} \to 0$$

for every g in $L^p(\mathbb{R}^N)$.

Assertion (iii) is again evident, since $V_{-h^\theta(n)} \chi_{\mathbb{H}(\psi)} V_{h^\theta(n)}$ is the operator of multiplication by the characteristic function of the shifted half space $-h^\theta(n) + \mathbb{H}(\psi)$. Assertion (iv) follows from the compactness of T and from the weak convergence of the operators $V_{h(n)}$ to zero as $|h(n)| \to \infty$.   ∎

Let Λ_p^{hom} stand for the set of all operators $A \in \mathscr{L}(L^p(\mathbb{R}^N))$ which are subject to the following conditions:

- A is of local type, i.e., $A \in \Lambda_p$;
- the strong limits $H_x(A)$ on $L^p(\mathbb{R}^N)$ and $H_x(A^*)$ on $(L^p(\mathbb{R}^N))^*$ defined by (5.54) exist for every $x \in \mathbb{R}^N$;
- the strong limits $H_\theta^\circ(A)$ on $L^p(\mathbb{R}^N)$ and $H_\theta^\circ(A^*)$ on $(L^p(\mathbb{R}^N))^*$ defined by (5.55) exist for every $\theta \in \mathbb{S}^{N-1}$.

Proposition 5.8.6.
(i) Λ_p^{hom} is a closed subalgebra of $\mathscr{L}(L^p(\mathbb{R}^N))$ which contains $\mathscr{A}_p$;
(ii) the algebra Λ_p^{hom} is inverse-closed in $\mathscr{L}(L^p(\mathbb{R}^N))$; and
(iii) the quotient algebra $\Lambda_p^{hom}/\mathscr{K}(L^p(\mathbb{R}^N))$ is inverse-closed in the Calkin algebra $\mathscr{L}(L^p(\mathbb{R}^N))/\mathscr{K}(L^p(\mathbb{R}^N))$.

Proof. The proof of the first part of assertion (i) is straightforward, and the second part is a consequence of Propositions 5.8.3, 5.8.4 and 5.8.5. Assertion (ii) follows from assertion (iii) via Lemma 1.2.33 (note that the ideal of the compact operators is included in $\mathscr{A}_p \subseteq \Lambda_p^{hom}$ by Proposition 5.8.1).

So we are left with verifying assertion (iii). Let A be an operator in Λ_p^{hom} which has the Fredholm property, i.e., the coset $A + \mathscr{K}(L^p(\mathbb{R}^N))$ is invertible in the Calkin algebra $\mathscr{L}(L^p(\mathbb{R}^N))/\mathscr{K}(L^p(\mathbb{R}^N))$. Let $R \in \mathscr{L}(L^p(\mathbb{R}^N))$ be an operator such that $RA - I =: K$ and $AR - I =: L$ are compact. We have to show that $R \in \Lambda_p^{hom}$. Let $f \in C(\overline{\mathbb{R}^N})$. Then the operator

$$fR - RfI = (RA - K)fR - Rf(AR - L) = R(AfI - fA)R - KfR + RfL$$

is compact, whence $R \in \Lambda_p$. It remains to show that all required strong limits of R exist. We will verify this for the strong limit H_0; the proof for the other limits proceeds analogously.

First we show that $\mathsf{H}_0(A)$ is an invertible operator. Since A is Fredholm, there is a positive number c and a compact operator T such that

$$\|Au\| + \|Tu\| \geq c\|u\| \quad \text{for all} \quad u \in L^p(\mathbb{R}^N)$$

(see Exercise 1.4.7). Since the operators Z_t are isometries, this estimate implies

$$\|Z_t A Z_t^{-1} u\| + \|Z_t T Z_t^{-1} u\| \geq c\|u\|$$

for all $u \in L^p(\mathbb{R}^N)$. Passing to the strong limit as $t \to \infty$ we finally obtain

$$\|\mathsf{H}_0(A)u\| \geq c\|u\| \quad \text{for all} \quad u \in L^p(\mathbb{R}^N).$$

Thus, $\mathsf{H}_0(A)$ is bounded below. Applying the same argument to the adjoint operator A^* (which is Fredholm, too) we find that $\mathsf{H}_0(A^*) = \mathsf{H}_0(A)^*$ is also bounded below. Hence, $\mathsf{H}_0(A)$ is invertible.

Now we show that the strong limit $\mathsf{H}_0(R)$ exists and that $\mathsf{H}_0(R) = \mathsf{H}_0(A)^{-1}$. Indeed, let $RA - I =: K$ as before. Then, for each $u \in L^p(\mathbb{R}^N)$,

$$
\begin{aligned}
&\|(Z_t R Z_t^{-1} - \mathsf{H}_0(A)^{-1})u\| \\
&= \|(Z_t R Z_t^{-1} - Z_t(RA - K)Z_t^{-1}\mathsf{H}_0(A)^{-1})u\| \\
&= \|(Z_t R Z_t^{-1} - (Z_t R Z_t^{-1} Z_t A Z_t^{-1} - Z_t K Z_t^{-1})\mathsf{H}_0(A)^{-1})u\| \\
&\leq \|Z_t R Z_t^{-1}\|\,\|u - Z_t A Z_t^{-1}\mathsf{H}_0(A)^{-1}u\| + \|Z_t K Z_t^{-1}\mathsf{H}_0(A)^{-1}u\| \\
&\leq \|R\|\,\|\mathsf{H}_0(A)v - Z_t A Z_t^{-1}v\| + \|Z_t K Z_t^{-1}v\|
\end{aligned}
$$

with $v := \mathsf{H}_0(A)^{-1}u$. Since the right-hand side of this estimate tends to zero as $t \to \infty$, the assertion follows. ∎

Thus, an operator $A \in \Lambda_p^{hom}$ is Fredholm if and only if its coset modulo compact operators is invertible in $\Lambda_p^{hom}/\mathscr{K}(L^p(\mathbb{R}^N))$. For operators $A \in \mathscr{A}_p$, we will study the invertibility of the coset $A + \mathscr{K}(L^p(\mathbb{R}^N))$ in this quotient algebra by localizing the algebra $\Lambda_p^{hom}/\mathscr{K}(L^p(\mathbb{R}^N))$ by Allan's local principle over its central subalgebra which consists of all cosets $fI + \mathscr{K}(L^p(\mathbb{R}^N))$ with $f \in C(\overline{\mathbb{R}^N})$.

Proposition 5.8.7. *The algebra* $\mathscr{C} := \{fI + \mathscr{K}(L^p(\mathbb{R}^N)) : f \in C(\overline{\mathbb{R}^N})\}$ *is isometrically isomorphic to the algebra* $C(\overline{\mathbb{R}^N})$ *in a natural way.*

Proof. One has only to prove that

$$\|f\| = \|fI + \mathcal{K}(L^p(\mathbb{R}^N))\| := \inf_{K \in \mathcal{K}(L^p(\mathbb{R}^N))} \|fI + K\|$$

for every $f \in C(\overline{\mathbb{R}^N})$. Proposition 5.8.4 (ii), (iv) ensure that $\|fI + K\| \geq |f(x)|$ for every $f \in C(\overline{\mathbb{R}^N})$, $K \in \mathcal{K}(L^p(\mathbb{R}^N))$ and $x \in \mathbb{R}^N$. Hence, $\|fI + \mathcal{K}(L^p(\mathbb{R}^N))\| \geq |f(x)|$ for all $x \in \mathbb{R}^N$. Since $\mathbb{R}^N$ is dense in $\overline{\mathbb{R}^N}$ with respect to the Gelfand topology, we get

$$\|f\| \geq \|fI + \mathcal{K}(L^p(\mathbb{R}^N))\| \geq \|f\|,$$

which is the assertion.   ∎

In particular, the maximal ideal space of the algebra $\mathscr{C}$ is homeomorphic to $\overline{\mathbb{R}^N}$. The maximal ideal which corresponds to $x \in \overline{\mathbb{R}^N}$ is the set of all cosets $fI + \mathcal{K}(L^p(\mathbb{R}^N))$ with $f(x) = 0$. We denote this maximal ideal by x and let $\mathscr{J}_x$ stand for the smallest closed ideal of $\Lambda_p^{hom}/\mathcal{K}(L^p(\mathbb{R}^N))$ which contains x. Further we write Φ_x for the canonical homomorphism

$$\Lambda_p^{hom} \to (\Lambda_p^{hom}/\mathcal{K}(L^p(\mathbb{R}^N)))/\mathscr{J}_x, \quad A \mapsto (A + \mathcal{K}(L^p(\mathbb{R}^N))) + \mathscr{J}_x.$$

Note that the compact operators lie in the kernel of each homomorphism H_x and H_θ° with $x \in \mathbb{R}^N$ and $\theta \in \mathbb{S}^{N-1}$ by (iv) in Propositions 5.8.4 and 5.8.5. Thus, the mappings

$$A + \mathcal{K}(L^p(\mathbb{R}^N)) \mapsto \mathsf{H}_x(A) \quad \text{and} \quad A + \mathcal{K}(L^p(\mathbb{R}^N)) \mapsto \mathsf{H}_\theta^\circ(A)$$

are correctly defined for each operator $A \in \Lambda_p^{hom}$. We denote them again by H_x and H_θ°, respectively. Further, by (ii) in Propositions 5.8.4 and 5.8.5, the local ideal $\mathscr{J}_x$ lies in the kernel of $\mathsf{H}_x : \Lambda_p^{hom}/\mathcal{K}(L^p(\mathbb{R}^N)) \to \mathscr{L}(L^p(\mathbb{R}^N))$ for every $x \in \mathbb{R}^N$, and the local ideal $\mathscr{J}_{\theta_\infty}$ lies in the kernel of $\mathsf{H}_\theta^\circ : \Lambda_p^{hom}/\mathcal{K}(L^p(\mathbb{R}^N)) \to \mathscr{L}(L^p(\mathbb{R}^N))$ for every $\Theta \in \mathbb{S}^{N-1}$. Hence, the mappings

$$(A + \mathcal{K}(L^p(\mathbb{R}^N))) + \mathscr{J}_x \mapsto \mathsf{H}_x(A) \quad \text{and} \quad (A + \mathcal{K}(L^p(\mathbb{R}^N))) + \mathscr{J}_{\theta_\infty} \mapsto \mathsf{H}_\theta^\circ(A)$$

are correctly defined for each $A \in \Lambda_p^{hom}$, and we denote them again by H_x and H_θ°, respectively. The following propositions identify the algebras $\Phi_x(\mathscr{A}_p)$ for $x \in \overline{\mathbb{R}^N}$.

Proposition 5.8.8. *Let $x = 0 \in \mathbb{R}^N$. Then:*

(i) *the local algebra $\Phi_0(\mathscr{A}_p)$ is isometrically isomorphic to the smallest closed subalgebra $PC(\mathbb{S}^{N-1})$ of $\mathscr{L}(L^p(\mathbb{R}^N))$ which contains all operators $\chi_{\mathbb{H}(\psi)}$ with $\psi \in \mathbb{S}^{N-1}$;*

(ii) *for every operator $A \in \mathscr{A}_p$, the coset $\Phi_0(A)$ is invertible in the local algebra $(\Lambda_p^{hom}/\mathcal{K}(L^p(\mathbb{R}^N)))/\mathscr{J}_0$ if and only if the operator $\mathsf{H}_0(A)$ is invertible (in $\mathscr{L}(L^p(\mathbb{R}^N)))$;*

(iii) *the algebra $\Phi_0(\mathscr{A}_p)$ is inverse-closed in $(\Lambda_p^{hom}/\mathcal{K}(L^p(\mathbb{R}^N)))/\mathscr{J}_0$.*

The notation $PC(\mathbb{S}^{N-1})$ has been chosen since $PC(\mathbb{S}^1)$ can be identified (by restriction) with the algebra of all piecewise continuous functions on the (one-dimensional) unit sphere.

Proof. (i) It follows from Proposition 5.8.4 that H_0 is a homomorphism from $\Phi_0(\mathscr{A}_p)$ into $PC(\mathbb{S}^{N-1})$. This homomorphism is onto since $PC(\mathbb{S}^{N-1})$ is a subalgebra of $\mathscr{A}_p$ and $H_0(A) = A$ for every operator $A \in PC(\mathbb{S}^{N-1})$. Further, since the Z_t are isometries, it is clear that the mapping

$$H_0 : \Phi_0(\mathscr{A}_p) \rightarrow PC(\mathbb{S}^{N-1})$$

is a contraction. In order to show that this mapping is an isometric isomorphism, we claim that

$$\Phi_0(A) = \Phi_0(H_0(A)) \quad \text{for every} \quad A \in \mathscr{A}_p. \tag{5.56}$$

Since the mappings Φ_0 and H_0 are continuous homomorphisms, it is sufficient to check (5.56) for the generating operators of the algebra $\mathscr{A}_p$.

For the operators $A = fI$ with $f \in C(\overline{\mathbb{R}^N})$ one has $H_0(fI) = f(0)I$ by Proposition 5.8.4; so one has to check that $\Phi_0(fI) = \Phi_0(f(0)I)$, which is immediate from the definition of the local ideals. For $A = \chi_{\mathbb{H}(\psi)}I$ with $\psi \in \mathbb{S}^{N-1}$, the claim (5.56) is evident.

So we are left with the case when $A = W_a^0$ with $a \in L^1(\mathbb{R}^N)$. Then we have to show that $\Phi_0(W_a^0) = 0$. Let f be a continuous function on $\mathbb{R}^N$ with compact support and with $f(0) = 1$. The operator $W_a^0 fI$ which acts on $L^p(\mathbb{R}^N)$ by

$$[W_a^0 f(g)](x) = \int_{\mathbb{R}^N} a(x-t)f(t)g(t)dt, \quad x \in \mathbb{R}^N,$$

is compact. Indeed, we can suppose, without loss of generality, that a is a continuous function with compact support, because the functions with these properties are dense in $L^1(\mathbb{R}^N)$. Further, by Krasnoselskii's interpolation theorem, we can also suppose that $p = 2$. Since then the kernel of the integral operator $W_a^0 fI$ is a continuous and compactly supported function, we conclude that $W_a^0 fI$ is a Hilbert-Schmidt operator, and therefore compact. Thus, $\Phi_x(W_a^0 fI) = 0$. Since $\Phi_x(fI)$ is the identity element of the local algebra, we have $\Phi_x(W_a^0) = 0$, which proves the claim.

(ii) Let A be an operator in $\mathscr{A}_p$ for which the coset $\Phi_0(A)$ is invertible in $(\Lambda_p^{hom}/\mathscr{K}(L^p(\mathbb{R}^N)))/\mathscr{J}_0$. Since H_0 acts as a homomorphism on that algebra, we conclude that $H_0(A)$ is an invertible operator.

Conversely, let the operator $H_0(A)$ be invertible (in $\mathscr{L}(L^p(\mathbb{R}^N))$). From part (i) we know that $H_0(A)$ belongs to the algebra Λ_p^{hom}, and from Proposition 5.8.6 (ii) we infer that the algebra Λ_p^{hom} is inverse-closed in $\mathscr{L}(L^p(\mathbb{R}^N))$. Hence, the inverse operator $H_0(A)^{-1}$ belongs to Λ_p^{hom}. Applying the local mapping Φ_0 to the equality

$$H_0(A)^{-1}H_0(A) = H_0(A)H_0(A)^{-1} = I$$

and recalling (5.56) we conclude that $\Phi_0(A)$ is invertible.

(iii) This is the same proof as before if one takes into account that the algebra

$PC(\mathbb{S}^{N-1})$ is inverse-closed in $\mathscr{L}(L^p(\mathbb{R}^N))$. The latter fact can easily be proved via Corollary 1.2.32; it follows also from Theorem 2.2.8. ∎

Proposition 5.8.9. *Let $x \in \mathbb{R}^N \setminus \{0\}$. Then:*

(i) *the local algebra $\Phi_x(\mathscr{A}_p)$ is isometrically isomorphic to the smallest closed subalgebra $\mathscr{B}_x$ of $\mathscr{L}(L^p(\mathbb{R}^N))$ which contains all operators $\chi_{\mathbb{H}(\psi)}I$ for which x lies on the boundary of $\mathbb{H}(\psi)$;*

(ii) *for every operator $A \in \mathscr{A}_p$, the coset $\Phi_x(A)$ is invertible in the local algebra $(\Lambda_p^{hom}/\mathscr{K}(L^p(\mathbb{R}^N)))/\mathscr{J}_x$ if and only if the operator $\mathsf{H}_x(A)$ is invertible (in $\mathscr{L}(L^p(\mathbb{R}^N))$);*

(iii) *the algebra $\Phi_x(\mathscr{A}_p)$ is inverse-closed in $(\Lambda_p^{hom}/\mathscr{K}(L^p(\mathbb{R}^N)))/\mathscr{J}_x$.*

Clearly, if $N = 2$, there are only two values of ψ such that x lies on the boundary of $\mathbb{H}(\psi)$. If $\psi(x)$ is one of these values, then $-\psi(x)$ is the other one, and the algebra $\mathscr{B}_x$ consists of all linear combinations of $\chi_{\mathbb{H}(\psi(x))}I$ and $\chi_{\mathbb{H}(-\psi(x))}I$. Thus, $\mathscr{B}_x \cong \mathbb{C}^2$ in this case. If $N > 2$, then x lies on the boundary of each half space $\mathbb{H}(\psi)$ with ψ being orthogonal to x. The set of these ψ can be identified with $\mathbb{S}^{N-2}$.

Proof. The proof proceeds as that of the preceding proposition. In place of (5.56), one now has to verify that

$$\Phi_x(A) = \Phi_x(\mathsf{H}_x(A)) \quad \text{for every} \quad A \in \mathscr{A}_p. \tag{5.57}$$

We only note that $\chi_{\mathbb{H}(\psi)}$ is continuous and equal to one in a neighborhood of x if x is in the interior of $\mathbb{H}(\psi)$. Thus, $\Phi_x(\chi_{\mathbb{H}(\psi)}I)$ is the local identity element in this case. Similarly, if x is in the exterior of $\mathbb{H}(\psi)$, then $\Phi_x(\chi_{\mathbb{H}(\psi)}I)$ is the local zero element. In the case x lies on the boundary of $\mathbb{H}(\psi)$ then $\Phi_x(\chi_{\mathbb{H}(\psi)}I)$ is a proper idempotent (i.e., the spectrum of this local coset is $\{0, 1\}$), by Proposition 5.8.4 (iii). ∎

Proposition 5.8.10. *Let $\theta \in \mathbb{S}^{N-1}$. Then:*

(i) *the local algebra $\Phi_{\theta_\infty}(\mathscr{A}_p)$ is isometrically isomorphic to the smallest closed subalgebra $\mathscr{B}_{\theta_\infty}$ of $\mathscr{L}(L^p(\mathbb{R}^N))$ which contains all convolutions W_a^0 with $a \in L^1(\mathbb{R}^N)$ and all operators $\chi_{\mathbb{H}(\psi)}I$ for which θ lies on the boundary of $\mathbb{H}(\psi)$;*

(ii) *for every operator $A \in \mathscr{A}_p$, the coset $\Phi_{\theta_\infty}(A)$ is invertible in the local algebra $(\Lambda_p^{hom}/\mathscr{K}(L^p(\mathbb{R}^N)))/\mathscr{J}_{\theta_\infty}$ if and only if the operator $\mathsf{H}_\theta^\circ(A)$ is invertible in $\mathscr{L}(L^p(\mathbb{R}^N))$;*

(iii) *for every operator $A \in \mathscr{A}_p$, the coset $\Phi_{\theta_\infty}(A)$ is invertible in the local algebra $\Phi_{\theta_\infty}(\mathscr{A}_p)$ if and only if the operator $\mathsf{H}_\theta^\circ(A)$ is invertible in $\mathscr{B}_{\theta_\infty}$.*

Proof. The proof follows the same lines as that of the preceding propositions, where one has now to check that

$$\Phi_{\theta_\infty}(A) = \Phi_{\theta_\infty}(\mathsf{H}_\theta^\circ(A)) \quad \text{for every} \quad A \in \mathscr{A}_p. \tag{5.58}$$

Note that in the case at hand, we do not know if the algebras $\mathscr{B}_{\theta_\infty}$ are inverse-closed in $\mathscr{L}(L^p(\mathbb{R}^N))$. That is why we give two invertibility criteria: one for invertibility in $(\Lambda_p^{hom}/\mathcal{K}(L^p(\mathbb{R}^N)))/\mathcal{J}_{\theta_\infty}$, and one for invertibility in $\Phi_{\theta_\infty}(\mathcal{A}_p)$. Assertion (iii) can be proved as assertion (ii) of Proposition 5.8.8.   ∎

Now one can formulate and prove the main result of this section.

Theorem 5.8.11. *An operator $A \in \mathcal{A}_p$ is Fredholm if and only if all operators $\mathsf{H}_x(A)$ with $x \in \mathbb{R}^N$ and all operators $\mathsf{H}_\theta^\circ(A)$ with $\theta \in \mathbb{S}^{N-1}$ are invertible (as operators on $L^p(\mathbb{R}^N)$).*

Proof. The proof follows immediately from Allan's local principle and from the criteria for invertibility in the corresponding local algebras which are stated in assertions (ii) of Propositions 5.8.8, 5.8.9 and 5.8.10.   ∎

For completeness, let us mention that the coset $A + \mathcal{K}(L^p(\mathbb{R}^N))$ of an operator $A \in \mathcal{A}_p$ is invertible in the quotient algebra $\mathcal{A}_p/\mathcal{K}(L^p(\mathbb{R}^N))$ if and only if the operator $\mathsf{H}_x(A)$ is invertible in $\mathcal{L}(L^p(\mathbb{R}^N))$ for every $x \in \mathbb{R}^N$ and if the operator $\mathsf{H}_\theta^\circ(A)$ is invertible in $\mathcal{B}_{\theta_\infty}$ for every $\theta \in \mathbb{S}^{N-1}$. This follows again from Allan's local principle, but now applied in $\mathcal{A}_p/\mathcal{K}(L^p(\mathbb{R}^N))$, and from assertions (iii) of Propositions 5.8.8, 5.8.9 and 5.8.10.

To illustrate the previous results we let $N = 2$ and consider restrictions of convolution operators to half-planes and cones. By a *cone* in $\mathbb{R}^2$ with vertex at the origin we mean a set of the form $\mathbb{K}(\psi_1, \psi_2) := \mathbb{H}(\psi_1) \cap \mathbb{H}(\psi_2)$ with $\psi_1, \psi_2 \in \mathbb{S}^1$. To avoid trivialities, we assume that neither $\psi_1 = \psi_2$ nor $\psi_1 = -\psi_2$. Thus, $\mathbb{K}(\psi_1, \psi_2)$ is neither a half-plane nor a line.

Let χ_M refer to the characteristic function of a measurable subset M of $\mathbb{R}^2$. The following is an immediate consequence of the Fredholm criterion in Theorem 5.8.11 and of Propositions 5.8.4 and 5.8.5.

Corollary 5.8.12. *Let $a \in L^1(\mathbb{R}^2)$ and $f \in C(\overline{\mathbb{R}^2})$, and let $\psi, \psi_1, \psi_2 \in \mathbb{S}^1$ be subject to the above agreement.*

(i) *The operator $\chi_{\mathbb{H}(\psi)}(W_a^0 + fI)\chi_{\mathbb{H}(\psi)}I + (1 - \chi_{\mathbb{H}(\psi)})I$ is Fredholm on $L^p(\mathbb{R}^2)$ if and only if*

- $f(x) \neq 0$ *for all $x \in \mathbb{H}(\psi)$,*
- $W_a^0 + f(\theta_\infty)I$ *is invertible for every $\theta \in \mathbb{S}^1$ in the interior of $\mathbb{H}(\psi)$,*
- $\chi_{\mathbb{H}(\psi)}(W_a^0 + f(\theta_\infty)I)\chi_{\mathbb{H}(\psi)}I + (1 - \chi_{\mathbb{H}(\psi)})I$ *is invertible for every $\theta \in \mathbb{S}^1$ on the boundary of $\mathbb{H}(\psi)$.*

(ii) *The operator $\chi_{\mathbb{K}(\psi_1,\psi_2)}(W_a^0 + fI)\chi_{\mathbb{K}(\psi_1,\psi_2)}I + (1 - \chi_{\mathbb{K}(\psi_1,\psi_2)})I$ is Fredholm on $L^p(\mathbb{R}^2)$ if and only if*

- $f(x) \neq 0$ *for all $x \in \mathbb{K}(\psi_1, \psi_2)$,*
- $W_a^0 + f(\theta_\infty)I$ *is invertible for every $\theta \in \mathbb{S}^1$ in the interior of $\mathbb{K}(\psi_1, \psi_2)$,*
- $\chi_{\mathbb{H}(\psi_i)}(W_a^0 + f(\theta_\infty)I)\chi_{\mathbb{H}(\psi_i)}I + (1 - \chi_{\mathbb{H}(\psi_i)})I$ *is invertible for every $\theta \in \mathbb{S}^1$ on the boundary of $\mathbb{K}(\psi_1, \psi_2)$.*

Note that the invertibility of the half-plane operators in (i) and (ii) can be effectively checked by means of a result by Goldenstein and Gohberg [77] which states that the following conditions are equivalent for $a \in L^1(\mathbb{R}^2)$, $\lambda \in \mathbb{C}$ and $\psi \in \mathbb{S}^1$:

(i) the operator $\chi_{\mathbb{H}(\psi)}(W_a^0 + \lambda I)\chi_{\mathbb{H}(\psi)}I + (1 - \chi_{\mathbb{H}(\psi)})I$ is invertible on $L^p(\mathbb{R}^2)$,
(ii) the operator $W_a^0 + \lambda I$ is invertible on $L^p(\mathbb{R}^2)$,
(iii) $\lambda \neq 0$, and the function $Fa + \lambda$, with F standing for the Fourier transform on $\mathbb{R}^2$, does not vanish on $\mathbb{R}^2$.

With this additional information, Corollary 5.8.12 can be reformulated as follows.

Corollary 5.8.13. *Let $a \in L^1(\mathbb{R}^2)$ and $f \in C(\overline{\mathbb{R}^2})$, and let $\psi, \psi_1, \psi_2 \in \mathbb{S}^1$ be subject to the above agreement.*

(i) *The operator $\chi_{\mathbb{H}(\psi)}(W_a^0 + fI)\chi_{\mathbb{H}(\psi)}I + (1 - \chi_{\mathbb{H}(\psi)})I$ is Fredholm on $L^p(\mathbb{R}^2)$ if and only if*

  • $f(x) \neq 0$ for all $x \in \mathbb{H}(\psi)$,
  • $W_a^0 + f(\theta_\infty)I$ is invertible for every $\theta \in \mathbb{S}^1 \cap \mathbb{H}(\psi)$.

(ii) *The operator $\chi_{\mathbb{K}(\psi_1, \psi_2)}(W_a^0 + fI)\chi_{\mathbb{K}(\psi_1, \psi_2)}I + (1 - \chi_{\mathbb{K}(\psi_1, \psi_2)})I$ is Fredholm on $L^p(\mathbb{R}^2)$ if and only if*

  • $f(x) \neq 0$ for all $x \in \mathbb{K}(\psi_1, \psi_2)$,
  • $W_a^0 + f(\theta_\infty)I$ is invertible for every $\theta \in \mathbb{S}^1 \cap \mathbb{K}(\psi_1, \psi_2)$.

Note in this connection also that the following assertions are equivalent for $a \in L^1(\mathbb{R}^2)$, $\lambda \in \mathbb{C}$ and $\psi_1, \psi_2 \in \mathbb{S}^1$ (see [21, Section 9.53]):

(i) $\chi_{\mathbb{K}(\psi_1, \psi_2)}(W_a^0 + \lambda I)\chi_{\mathbb{K}(\psi_1, \psi_2)}I + (1 - \chi_{\mathbb{K}(\psi_1, \psi_2)})I$ is Fredholm on $L^p(\mathbb{R}^2)$,
(ii) $\chi_{\mathbb{H}(\psi_i)}(W_a^0 + \lambda I)\chi_{\mathbb{H}(\psi_i)}I + (1 - \chi_{\mathbb{H}(\psi_i)})I$ is invertible on $L^p(\mathbb{R}^2)$ for $i \in \{1, 2\}$.

Corollary 5.8.14. *Let A belong to the smallest closed subalgebra on $\mathscr{L}(L^p(\mathbb{R}^2))$ which contains all operators W_a^0 with $a \in L^1(\mathbb{R}^2)$ and the operator $\chi_{\mathbb{H}(\theta)}I$ for a fixed $\theta \in \mathbb{S}^1$. Then A is Fredholm if and only if A is invertible.*

Proof. If A is Fredholm then by (the easy half of) Theorem 5.8.11, the limit operator $H_\theta^\circ(A)$ is invertible. From Proposition 5.8.5 we infer that $H_\theta^\circ(A) = A$. Thus, A is invertible. ∎

Let us finally mention that the algebras $\Lambda_p/\mathscr{K}(L^p(\mathbb{R}^N))$ and $C(\overline{\mathbb{R}^N})$ constitute a faithful localization pair by Theorem 2.5.13. Thus, the machinery of norm-preserving localization and local enclosement theorems as well as Simonenko's theory of local operators work in the present setting.

5.9 Notes and comments

Wiener-Hopf integral equations of the type

$$cu(t) + \int_0^\infty k(t-s)u(s)\,ds = v(t) \quad (t > 0),$$

with $k \in L^1(\mathbb{R})$, had been the subject of detailed studies by many people, including Wiener and Hopf [201], Paley and Wiener [133], Smithies [190], Reissner [161], Fock [57], Titchmarsh [193], Rapoport [158] and Noble [129]. The fundamental 1958 paper [105] of Krein, translated into English by the American Mathematical Society in 1962, presented a clear and complete theory of this topic at the time. The case of systems of Wiener-Hopf integral equations with kernels belonging to $L^1(\mathbb{R})$ was studied by Gohberg and Krein in [68]. Gohberg and Feldman's book [66], published originally in Russian in 1971, is devoted to a unified approach to different kinds of convolution equations with continuous generating functions. But Duduchava's book [43] marked the start of a new era in the topic, with the study of convolution operators with piecewise continuous generating functions and of algebras generated by such operators.

The results of Sections 5.1, 5.2 and 5.3 are taken from Duduchava's works [43, 44] with exception of Proposition 5.1.2 which is due to Schneider [176]. The results from Section 5.4 go back to two of the authors [168]. In Sections 5.3 and 5.4 some of the proofs are streamlined with respect to their original versions.

Duduchava [44] studied the algebra $\mathscr{A} := \mathscr{A}(X,Y,Z)$ in the particular case $X = C(\mathbb{R}^+)$, $Y = PC_{p,w_\alpha}$ and $Z = C_p$. These restrictions imply the commutativity of the related quotient algebra $\mathscr{A}^{\mathscr{K}}$. Duduchava further wrote that ...*the same methods make it possible to investigate a more complicated algebra...*, namely the algebra $\mathscr{A}^{\mathscr{K}}(PC(\mathbb{R}^+), PC_{p,w_\alpha}, PC_p)$ in our notation. As far as we know, he never published these results. Moreover, even in the case of continuous generating functions, the approach presented above in this chapter gives a more precise information than Duduchava's approach: it allows us to characterize the local algebras up to isometry as algebras of Mellin convolution operators.

The algebra $\mathscr{A}(PC(\mathbb{R}^+), PC_{p,w}, \emptyset)$ studied in Section 5.5 was the subject of Duduchava's investigations in the particular case of unweighted spaces or for weights $w(t) := |t|^\alpha$. For Khvedelidze weights, Schneider proved a criterion for the Fredholm property of the Wiener-Hopf operator $W(a)$ with $a \in PC_{p,w}$.

The results of Section 5.6 are again taken from [168]. These results can be extended to algebras generated by Wiener-Hopf and Hankel operators with piecewise continuous generating functions. But in Section 5.7 we present more general results based on considering the flip as an independent generator for the algebra. That possibility was used initially in [165] by the authors to analyze algebras resulting from approximating methods, the theme of Chapter 6. Algebras generated by Wiener-Hopf and Hankel operators have an analog in algebras generated by Toeplitz and Hankel operators with piecewise continuous generating function (defined on the unit circle $\mathbb{T}$). That problem was studied by Power in [147, 149] and by one of the authors [182] by different means, but only for the C^*-case. Interestingly, the approach of [182] is based upon the two projections theorem. Moreover, Theorem 4.4.5 leads to the description of the Fredholm properties of operators belonging to the Banach algebras generated by Toeplitz and Hankel operators with piecewise continuous generating functions and acting on Hardy spaces with weight. In particular, the essential spectrum of Hankel operators with piecewise continuous generating functions acting on a variety of Banach spaces has been known since 1990 [168]. Some of these results were reproved in [94].

Section 5.8 is devoted to the reproduction of some results obtained by Simonenko [187] who derived them by using the local principle named after him. Our exposition is based on a combination of Allan's local principle and limit operator techniques and is in the spirit of the previous sections.

The methods described in Chapter 5 also apply to other classes of operators such as multidimensional singular integral operators, singular integral operators with fixed singularities, singular integro-differential operators and certain classes of boundary integral operators (e.g., single and double layer potential operators). Concerning the investigation of multidimensional operators on $L^p(\mathbb{R}^n)$ by local principles see also the nice recent book by Simonenko [188]. It contains a complete study of shift-invariant operators as well as a description of a few important subclasses of such operators. Simonenko also considered Banach algebras generated by operators which are locally equivalent to operators in one of the mentioned subclasses.

Chapter 6
Algebras of operator sequences

Now we change our topic and move from operator theory to numerical analysis. In this chapter, X is a Banach space (which will be separable and of infinite dimension in all actual settings that we consider), I is the identity operator on X, $\mathscr{L}(X)$ is the Banach algebra of all bounded linear operators on X, and $\mathscr{K}(X)$ is the ideal of the compact operators in $\mathscr{L}(X)$.

Let $A \in \mathscr{L}(X)$. It is a basic problem of numerical mathematics (if not *the* basic problem of numerical mathematics) to provide approximations to the solution of the operator equation

$$Au = v \quad \text{for given } v \in X. \tag{6.1}$$

The standard procedure (which occurs in thousands of variations in the literature) is to choose a sequence of projections P_n (often assumed to be of finite rank) which converge strongly to the identity operator, and a sequence of operators $A_n : \operatorname{Im} P_n \to \operatorname{Im} P_n$ such that the $A_n P_n$ converge strongly to A, and to replace equation (6.1) by the sequence of linear systems

$$A_n u_n = P_n v \quad \text{for } n = 1, 2, \ldots, \tag{6.2}$$

the solutions u_n of which are sought in $\operatorname{Im} P_n$.

6.1 Approximation methods and sequences of operators

Definition 6.1.1. We say that the approximation method given by the sequence of the equations (6.2) *applies* or *converges* to A if

(i) there is an $n_0 \in \mathbb{N}$ such that, for any $n > n_0$ and any $v \in X$, there exists a unique solution u_n of the equation $A_n u_n = P_n v$ and if
(ii) for any $v \in X$, the sequence $(u_n)_{n \geq n_0}$ of these solutions converges in the norm of X to a solution of the equation $Au = v$.

S. Roch et al., *Non-commutative Gelfand Theories*, Universitext, 317
DOI 10.1007/978-0-85729-183-7_6, © Springer-Verlag London Limited 2011

Note that condition (ii) implies that A is surjective. Note also that we suppose the sequence $(A_n P_n)$ to be strongly convergent. The point is that the norm convergence of this sequence together with the finite rank property of the $A_n P_n$ would imply that its limit operator is compact. Since we want to consider equations $Au = v$ with non-compact A, strong convergence is the right choice here.

On the other hand, it is exactly this choice of convergence which involves serious difficulties. Indeed, if a sequence (B_n) tends to an invertible operator B in the norm, then the operators B_n are invertible for large n and the sequence of their inverses tends to B^{-1} also in the norm (which follows from Theorem 1.2.2). With respect to strong convergence, it is in general no longer true that the invertibility of B guarantees the invertibility of the B_n for large n and the strong convergence of their inverses to B^{-1}.

The identification of necessary and sufficient conditions which ensure the applicability of an approximation method is thus one of the main problems in numerical analysis. While the strong convergence of the sequence $(A_n P_n)$ to A is often easy to show, the core of the applicability problem is to verify the stability of the method in the following sense.

Definition 6.1.2. The sequence (A_n) is said to be *stable* if there is an $n_0 \in \mathbb{N}$ such that, for any $n > n_0$, the operators $A_n : \operatorname{Im} P_n \to \operatorname{Im} P_n$ are invertible and the norms of their inverses are uniformly bounded.

Note that the invertibility of the A_n is equivalent to condition (i) of the previous definition, whereas the uniform boundedness of the A_n^{-1} together with the strong convergence of $A_n P_n$ to A and the strong convergence of the adjoint sequence guarantee condition (ii). We will therefore pay our main attention to the stability problem in what follows.

There are some instances where the stability problem is easy to solve. For a simple example, let H be a Hilbert space, A a positive definite operator, and $(P_n)_{n \geq 1}$ be a sequence of self-adjoint projections which converges strongly to the identity operator. By the positive definiteness, there is a positive constant C such that

$$\langle Ax, x \rangle \geq C \|x\|^2 \quad \text{for all } x \in H.$$

Then it is immediate that

$$\langle P_n A P_n x_n, x_n \rangle = \langle A P_n x_n, P_n x_n \rangle \geq C \|x_n\|^2 \quad \text{for all } x_n \in \operatorname{Im} P_n.$$

Thus, the operators $A_n : \operatorname{Im} P_n \to \operatorname{Im} P_n$ are invertible for each $n \geq 1$, and the norms of their inverses are not greater that $1/C$. In other words, the sequence $(P_n A P_n)$ is stable under these conditions, and one even has $n_0 = 1$. An argument of similar simplicity shows the more general fact that if A is the sum of a strongly elliptic operator and a compact operator, then the invertibility of A already implies the stability of the sequence $(P_n A P_n)$. But, for most operators A (including operators considered in this chapter), the stability of $(P_n A P_n)$ or of any other approximation sequence (A_n) does not follow from the invertibility of A simply by applying some general principles.

To study the stability of such sequences, one thus needs techniques which are able to exploit the special asymptotic properties of these sequences. We shall discuss one of these techniques in this chapter: the formulation of the stability problem as an invertibility problem in a suitably constructed Banach algebra, and the use of local principles (or other Banach and C^*-algebraic techniques) to study this invertibility problem. The observation that stability is equivalent to invertibility goes back to A. Kozak [102], who used Simonenko's local principle to study the related invertibility problem. We will use the local principle by Allan and Douglas instead. Note that the above mentioned "asymptotic properties" of the sequence (A_n) are expressed in this approach by saying that certain strong limits exist for that sequence, or that this sequences commutes with others modulo sequences in a certain ideal.

It is thus important to emphasize that there are many settings, also of practical importance, where the stability follows from some general principles. But if they do not, then the algebraic approach discussed below can sometimes offer one way to study stability.

There is another point which we would like to emphasize: for many approximation methods, the Banach algebraic (local) approach is not just the only known approach to study stability; it has also some other striking advantages. Indeed, if the answer to the stability question is formulated in the correct (= algebraic) way, then it provides a lot of additional information. For example, without any additional effort it becomes evident then, that several spectral quantities of the A_n (e.g., the spectra of the A_n if these operators are self-adjoint, the ε-pseudospectra, the numerical ranges, the sets of the singular values) converge with respect to the Hausdorff metric, and one can even describe the corresponding limit sets. We can not discuss all these applications here and refer to [82] instead.

There is a general procedure for treating approximation problems with the algebraic approach, which we will use throughout this chapter. Suppose we are interested in deriving a stability criterion for all sequences in a certain algebra $\mathscr{A}$. Then we usually proceed according to the following steps.

(i) **Algebraization:** Find a unital Banach algebra $\mathscr{E}$ which contains $\mathscr{A}$ and a closed ideal $\mathscr{G} \subset \mathscr{E}$ such that the stability problem becomes equivalent to an invertibility problem in the quotient algebra $\mathscr{E}/\mathscr{G}$.

(ii) **Essentialization:** Find a unital subalgebra $\mathscr{F}$ of $\mathscr{E}$ which contains $\mathscr{A}$ and $\mathscr{G}$, and a closed ideal $\mathscr{J}$ of $\mathscr{F}/\mathscr{G}$ such that the algebra $\mathscr{A}^{\mathscr{J}} := (\mathscr{A}/\mathscr{G} + \mathscr{J})/\mathscr{J}$ has a large center.

To be able to use the latter property, one has to guarantee that no essential information is lost regarding the invertibility of a coset of a sequence $\mathbf{A} \in \mathscr{F}$ in the algebra $\mathscr{E}/\mathscr{G}$ and the invertibility of the same sequence modulo the ideal $\mathscr{J}$. This is done employing inverse-closedness results and a lifting theorem.

(iii) **Localization:** Use a local principle to translate the invertibility problem in the algebra $\mathscr{A}^{\mathscr{J}}$ to a family of simpler invertibility problems in local algebras.

(iv) **Identification:** Find necessary and sufficient conditions for the invertibility of elements in the local algebras.

This can hopefully be done by using results from Chapter 3, or by employing the homogenization technique as in the previous chapter. Occasionally, in order to treat more involved local algebras, a repeated localization can be useful.

We start with a closer look at the first two steps.

6.2 Algebraization

Sometimes it proves convenient to label the approximating operators by quantities other than the positive integers. So we will allow here for an arbitrary unbounded set $\mathbb{I}$ of non-negative real numbers in place of $\mathbb{N}$. Recall from Section 1.4.2 the basic convergence properties of (generalized) sequences $(A_\tau)_{\tau \in \mathbb{I}}$ of operators in $\mathscr{L}(X)$ labeled by the points in $\mathbb{I}$. Of course, notions like the *stability* of a (generalized) sequence and the *applicability* of the corresponding (generalized) approximation method are defined as in the case $\mathbb{I} = \mathbb{N}$.

We will also occasionally assume that the approximation operators A_n act on the whole space X rather than on the range of certain projections P_n. This does not imply any restriction, since we can identify an operator A_n acting on $\mathrm{Im}\, P_n$ with the operator $A_n P_n + (I - P_n)$ acting on X. Evidently, the sequences (A_n) and $(A_n P_n + (I - P_n))$ are stable (or not) simultaneously.

Let $\mathscr{E}$ be the set of all bounded sequences $(A_\tau)_{\tau \in \mathbb{I}}$ of operators $A_\tau \in \mathscr{L}(X)$. Provided with the operations

$$(A_\tau) + (B_\tau) := (A_\tau + B_\tau), \quad \alpha(A_\tau) := (\alpha A_\tau), \quad (A_\tau)(B_\tau) := (A_\tau B_\tau)$$

and with the norm $\|(A_\tau)\| := \sup_\tau \|A_\tau\|_{\mathscr{L}(X)}$, the set $\mathscr{E}$ becomes a unital Banach algebra. Note that the constant sequences (A) are included in $\mathscr{E}$ for every operator $A \in \mathscr{L}(X)$. If X is a Hilbert space, then we equip $\mathscr{E}$ with the involution $(A_\tau)^* := (A_\tau^*)$ which makes $\mathscr{E}$ a C^*-algebra.

Let $\mathscr{G}$ be the set of all sequences $(A_\tau) \in \mathscr{E}$ with $\lim_{\tau \to \infty} \|A_\tau\| = 0$. This set forms a closed ideal in $\mathscr{E}$.

Definition 6.2.1. We say that a sequence $(A_\tau) \in \mathscr{E}$ is *stable* if there is a constant $\tau_0 \in \mathbb{I}$ such that A_τ is invertible for all $\tau > \tau_0$ and $\sup_{\tau > \tau_0} \|A_\tau^{-1}\| < \infty$.

The next result establishes the relationship between applicability, stability and invertibility. It is the basis for the use of Banach algebra techniques in numerical analysis.

Theorem 6.2.2. *Let $(A_\tau) \in \mathscr{E}$ and $A \in \mathscr{L}(X)$, and suppose that $A_\tau \to A$ strongly as $\tau \to \infty$. Then the following assertions are equivalent:*

(i) *the approximation method (6.2) applies to A;*
(ii) *(A_τ) is stable and A is invertible;*
(iii) *A is invertible, and the coset $(A_\tau) + \mathscr{G}$ is invertible in the quotient algebra $\mathscr{E}/\mathscr{G}$.*

Proof. (i) $\Rightarrow$ (ii): Assertion (i) implies that there exists a constant τ_0 such that A_τ is invertible for $\tau > \tau_0$ and the sequence $(A_\tau^{-1})_{\tau \geq \tau_0}$ converges strongly. By the Banach-Steinhaus theorem, there is a $\tau_1 > \tau_0$ such that $\sup_{\tau > \tau_1} \|A_\tau^{-1}\| < \infty$, whence the stability of (A_τ). To prove the invertibility of A, note that

$$\|u - A_\tau^{-1} A u\| \leq \|A_\tau^{-1}\| \|A_\tau u - A u\|$$

for $\tau \geq \tau_1$. The first factor on the right-hand side is uniformly bounded, and the second one tends to zero as τ increases. Thus, if $u \in \operatorname{Ker} A$, then necessarily $\|u\| = 0$, which shows that $\operatorname{Ker} A = \{0\}$. Since $\operatorname{Im} A = X$ by the definition of an applicable method, A is invertible.

(ii) $\Rightarrow$ (iii): Let $\tau_0 \in \mathbb{I}$ be such that A_τ is invertible for $\tau \geq \tau_0$ and that the norms of the inverses are uniformly bounded. Set $B_\tau := A_\tau^{-1}$ if $\tau > \tau_0$ and $B_\tau := I$ if $\tau \leq \tau_0$. Then, clearly,

$$(B_\tau)(A_\tau) = (A_\tau)(B_\tau) \in (I) + \mathcal{G},$$

whence the invertibility of the coset $(A_\tau) + \mathcal{G}$.

(iii) $\Rightarrow$ (i): Let $(B_\tau) + \mathcal{G}$ be the inverse of $(A_\tau) + \mathcal{G}$. Thus,

$$B_\tau A_\tau = I + C_\tau \quad \text{and} \quad A_\tau B_\tau = I + D_\tau$$

with sequences (C_τ) and (D_τ) in $\mathcal{G}$. Choose τ_0 such that $\|C_\tau\| < 1/2$ and $\|D_\tau\| < 1/2$ for $\tau > \tau_0$. Then $I + C_\tau$ and $I + D_\tau$ are invertible for $\tau > \tau_0$, and the norms of their inverses are uniformly bounded by 2. Thus, the operators A_τ are invertible for $\tau > \tau_0$, and the norms of their inverses are uniformly bounded. In particular, the equations $A_\tau u_\tau = v$ possess unique solutions

$$u_\tau = A_\tau^{-1} v = (I + C_\tau)^{-1} B_\tau v,$$

and it remains to prove that $\|u - u_\tau\| \to 0$ where u is the solution of $Au = v$. This follows easily from the estimate

$$\|u - u_\tau\| = \|u - A_\tau^{-1} A u\| \leq \|A_\tau^{-1}\| \|A_\tau u - A u\|;$$

with the last term going to zero because of the strong convergence of (A_τ) to A and the uniform boundedness of the operators A_τ^{-1}. $\blacksquare$

6.3 Essentialization and lifting theorems

The application of local principles requires a certain algebraic property which can serve as a substitute for commutativity: the algebra must have a sufficiently large center, or it must fulfill a polynomial identity, for instance. If the algebra under consideration does not possess this property (which, unfortunately, happens quite often when studying the stability of approximation methods), a prior *lifting* process can prove useful. The goal of this section is to discuss several lifting theorems. Our starting point is a purely algebraic result.

Lemma 6.3.1 (*N ideals lemma*). *Let $\mathscr{A}$ be an algebra with identity e, and let $\mathscr{J}_1, \ldots, \mathscr{J}_N$ be ideals of $\mathscr{A}$ such that $\mathscr{J}_1 \cdot \ldots \cdot \mathscr{J}_N$ is in the radical of $\mathscr{A}$. Then an element $a \in \mathscr{A}$ is invertible if and only if its cosets $a + \mathscr{J}_1, \ldots, a + \mathscr{J}_N$ are invertible in the corresponding quotient algebras.*

Proof. The invertibility of a implies the invertibility of every coset $a + \mathscr{J}_i$. Conversely, let all cosets $a + \mathscr{J}_i$ be invertible. Then, for every i, there are elements $c_i \in \mathscr{A}$ and $j_i \in \mathscr{J}_i$ such that $c_i a = e - j_i$. Hence,

$$j := j_1 \ldots j_N = (e - c_1 a) \ldots (e - c_N a) = e - ca$$

with a certain element $c \in \mathscr{A}$. Since j is in the radical of $\mathscr{A}$ by hypothesis, the element $e - j = ca$ is invertible. Thus, a is invertible from the left, and the invertibility of a from the right-hand side follows analogously. ∎

Let $\mathscr{A}$ and $\mathscr{B}$ be unital algebras, $\mathscr{J}$ an ideal of $\mathscr{A}$, and $W : \mathscr{A} \to \mathscr{B}$ a unital homomorphism. We say that W *lifts the ideal* $\mathscr{J}$ if the intersection $\operatorname{Ker} W \cap \mathscr{J}$ lies in the radical of $\mathscr{A}$. If $\mathscr{A}$ is semi-simple, then this is equivalent to the fact that the restriction of W to $\mathscr{J}$ is injective. For example, every homomorphism lifts the radical, and the identical homomorphism lifts every ideal. For another example, let

$$\operatorname{Ann} \mathscr{J} := \{a \in \mathscr{A} : a\mathscr{J} \cup \mathscr{J} a \subseteq \operatorname{Rad} \mathscr{A}\}$$

denote the annulator of the ideal $\mathscr{J}$ of $\mathscr{A}$. Then $\operatorname{Ann} \mathscr{J}$ is an ideal of $\mathscr{A}$, and the canonical homomorphism from $\mathscr{A}$ onto $\mathscr{A}/\operatorname{Ann} \mathscr{J}$ lifts the ideal $\mathscr{J}$.

Now let $\{W_t\}_{t \in T}$ be a (finite or infinite) family of homomorphisms W_t which lift certain ideals $\mathscr{J}_t$. The lifting theorem states that the homomorphisms W_t and the ideals $\mathscr{J}_t$ can be glued into a homomorphism W and an ideal $\mathscr{J}$, respectively, such that W lifts $\mathscr{J}$. We first derive a general algebraic version of the lifting theorem, and then embark upon the special settings where $\mathscr{A}$ is a Banach or a C^*-algebra.

Theorem 6.3.2 (*General lifting theorem*). *Let $\mathscr{A}$ be a unital algebra and, for every element t of a certain set T, let $\mathscr{J}_t$ be an ideal of $\mathscr{A}$ which is lifted by a unital homomorphism $W_t : \mathscr{A} \to \mathscr{B}_t$. Further, let $\mathscr{J}$ stand for the smallest ideal of $\mathscr{A}$ which contains all ideals $\mathscr{J}_t$. Then an element $a \in \mathscr{A}$ is invertible if and only if the coset $a + \mathscr{J}$ is invertible in $\mathscr{A}/\mathscr{J}$ and if all elements $W_t(a)$ are invertible in $W_t(\mathscr{A})$.*

Proof. If a is invertible, then $a + \mathscr{J}$ and all $W_t(a)$ are invertible. Conversely, if $a + \mathscr{J}$ is invertible, then there are elements $a' \in \mathscr{A}$ and $j \in \mathscr{J}$ such that $a'a = e + j$. Further, by the definition of $\mathscr{J}$, there are finitely many elements $j_{t_i} \in \mathscr{J}_{t_i}$ such that $j = j_{t_1} + \ldots + j_{t_m}$. Thus, a is invertible modulo the ideal $\widehat{\mathscr{J}} := \mathscr{J}_{t_1} + \ldots + \mathscr{J}_{t_n} \subseteq \mathscr{J}$.

Further, since $b + \operatorname{Ker} W_t \mapsto W_t(b)$ is an isomorphism from $\mathscr{A}/\operatorname{Ker} W_t$ onto $W_t(\mathscr{A})$, we get the invertibility of the cosets $a + \operatorname{Ker} W_{t_i}$ in $\mathscr{A}/\operatorname{Ker} W_{t_i}$ for $i =$

$1, \ldots, n$. Since $\mathscr{J}_{t_i} \cap \mathrm{Ker}\, W_{t_i} \subseteq \mathrm{Rad}\, \mathscr{A}$ by hypothesis, one has

$$\begin{aligned}
\widehat{\mathscr{J}} \cdot \mathrm{Ker}\, W_{t_1} \cdot \ldots \cdot \mathrm{Ker}\, W_{t_n} &= (\mathscr{J}_{t_1} + \ldots + \mathscr{J}_{t_n}) \cdot \mathrm{Ker}\, W_{t_1} \cdot \ldots \cdot \mathrm{Ker}\, W_{t_n} \\
&\subseteq \mathscr{J}_{t_1} \cdot \mathrm{Ker}\, W_{t_1} + \ldots + \mathscr{J}_{t_n} \cdot \mathrm{Ker}\, W_{t_n} \subseteq \mathrm{Rad}\, \mathscr{A}.
\end{aligned}$$

The $N = n + 1$ ideals lemma, applied to the ideals $\widehat{\mathscr{J}}$, $\mathrm{Ker}\, W_{t_1}, \ldots, \mathrm{Ker}\, W_{t_n}$, yields the assertion. ∎

The family $\{T_t\}_{t \in T}$ induces a homomorphism W from $\mathscr{A}$ into the product $\prod_{t \in T} \mathscr{B}_t$ via

$$W : a \mapsto (t \mapsto W_t(a)). \tag{6.3}$$

Corollary 6.3.3. *Let the notation be as in the general lifting theorem with the W_t being surjective, and let W be the homomorphism (6.3). Then W lifts the ideal $\mathscr{J}$.*

Proof. Let $k \in \mathrm{Ker}\, W \cap \mathscr{J}$, and let a be an invertible element of $\mathscr{A}$. Then the following assertions are equivalent:

   (i) a is invertible;
  (ii) $W(a)$ and $a + \mathscr{J}$ are invertible;
 (iii) $a + \mathrm{Ker}\, W$ and $a + \mathscr{J}$ are invertible;
 (iv) $a + k + \mathrm{Ker}\, W$ and $a + k + \mathscr{J}$ are invertible;
  (v) $W(a + k)$ and $a + k + \mathscr{J}$ are invertible;
 (vi) $a + k$ is invertible.

The equivalences (i) $\Leftrightarrow$ (ii) and (v) $\Leftrightarrow$ (vi) are consequences of the general lifting theorem, the equivalences (ii) $\Leftrightarrow$ (iii) and (iv) $\Leftrightarrow$ (v) follow via the isomorphy theorem $\mathscr{A} / \mathrm{Ker}\, W \cong W(\mathscr{A})$, and (iii) $\Leftrightarrow$ (iv) is obvious. Thus, k belongs to the radical of $\mathscr{A}$. Since W is unital, the assertion follows. ∎

Now we turn over to the Banach algebra setting. Here one wishes to work with closed ideals and continuous homomorphisms.

Theorem 6.3.4 (Lifting theorem, Banach algebra version). *Let $\mathscr{A}$ be a unital Banach algebra and, for every element t of a certain set T, let $\mathscr{J}_t$ be a closed ideal of $\mathscr{A}$ which is lifted by a unital and continuous homomorphism W_t from $\mathscr{A}$ into a unital Banach algebra $\mathscr{B}_t$. Further, let $\mathscr{J}$ stand for the smallest closed ideal of $\mathscr{A}$ which contains all ideals $\mathscr{J}_t$. Then an element $a \in \mathscr{A}$ is invertible if and only if the coset $a + \mathscr{J}$ is invertible in $\mathscr{A} / \mathscr{J}$ and if all elements $W_t(a)$ are invertible in $W_t(\mathscr{A})$.*

The proof follows immediately from Theorem 6.3.2 and the following simple observation.

Lemma 6.3.5. *Let $\mathscr{A}$ be a unital Banach algebra and $\mathscr{J}_0$ an ideal in $\mathscr{A}$. Then an element $a \in \mathscr{A}$ is invertible modulo $\mathscr{J}_0$ if and only if it is invertible modulo the closure $\mathscr{J}$ of $\mathscr{J}_0$.*

Proof. If $a \in \mathscr{A}$ is invertible modulo $\mathscr{J}$, then there exists a $b \in \mathscr{A}$ and a $j \in \mathscr{J}$ such that $ba = e + j$ where e denotes the identity element of $\mathscr{A}$. Choose $j_0 \in \mathscr{J}_0$ such that $\|j - j_0\| < 1$. Then $e + j - j_0$ is invertible (Neumann series), and

$$(e + j - j_0)^{-1}ba = e + (e + j - j_0)^{-1}j_0 \in e + \mathscr{J}_0.$$

Thus, a is invertible modulo $\mathscr{J}_0$ from the left-hand side. The right-sided invertibility follows analogously. ∎

In practice, it will often prove hard to check the invertibility of the elements $W_t(a)$ in the subalgebra $W_t(\mathscr{A})$ of $\mathscr{B}_t$. If $W_t(\mathscr{A})$ is inverse-closed in $\mathscr{B}_t$, it is possible to guaranty that if $W_t(a)$ is invertible in $\mathscr{B}_t$, then it is invertible in $W_t(\mathscr{A})$. That is the case, if $\mathscr{A}$ is a C^*-algebra.

Theorem 6.3.6 (Lifting theorem, C^*-algebra version). *Let $\mathscr{A}$ be a unital C^*-algebra and, for every element t of a certain set T, let $\mathscr{J}_t$ be a closed ideal of $\mathscr{A}$ which is lifted by a unital $*$-homomorphism W_t from $\mathscr{A}$ into a unital $*$-algebra $\mathscr{B}_t$. Further, let $\mathscr{J}$ stand for the smallest closed ideal of $\mathscr{A}$ which contains all ideals $\mathscr{J}_t$. Then an element $a \in \mathscr{A}$ is invertible if and only if the coset $a + \mathscr{J}$ is invertible in $\mathscr{A}/\mathscr{J}$ and if all elements $W_t(a)$ are invertible in $\mathscr{B}_t$.*

Let us come back to the Banach version of the lifting theorem. Also in this case one would clearly prefer a version of this theorem where only the invertibility of $W_t(a)$ in $\mathscr{B}_t$ is needed. We will see now that one can formulate the lifting theorem in the desired way under the additional (but moderate) assumption that $W_t(\mathscr{J}_t)$ is an ideal of $\mathscr{B}_t$. We refer to these versions of the lifting theorems as the *IC-versions*, with IC referring to Inverse-Closed.

Theorem 6.3.7 (General lifting theorem, IC-version). *Let $\mathscr{A}$ be a unital algebra and, for every element t of a certain set T, let $\mathscr{J}_t$ be an ideal of $\mathscr{A}$ which is lifted by a unital homomorphism $W_t : \mathscr{A} \to \mathscr{B}_t$. Suppose furthermore that $W_t(\mathscr{J}_t)$ is an ideal of $\mathscr{B}_t$. Let $\mathscr{J}$ stand for the smallest ideal of $\mathscr{A}$ which contains all ideals $\mathscr{J}_t$. Then an element $a \in \mathscr{A}$ is invertible if and only if the coset $a + \mathscr{J}$ is invertible in $\mathscr{A}/\mathscr{J}$ and if all elements $W_t(a)$ are invertible in $\mathscr{B}_t$.*

Proof. The proof starts as that of Theorem 6.3.2. If a is invertible, then $a + \mathscr{J}$ and all $W_t(a)$ are invertible. Conversely, let $a + \mathscr{J}$ be invertible in $\mathscr{A}/\mathscr{J}$ and let $W_t(a)$ be invertible in $\mathscr{B}_t$ for every $t \in T$. Then there are elements $b \in \mathscr{A}$ and $j \in \mathscr{J}$ such that $ba = e + j$, and there are finitely many elements $j_{t_i} \in \mathscr{J}_{t_i}$ such that $j = j_{t_1} + \ldots + j_{t_m}$. Thus,

$$ba = e + j_{t_1} + \ldots + j_{t_m}. \tag{6.4}$$

Let $W_{t_i}(a)^{-1}$ stand for the inverse of $W_{t_i}(a)$ in $\mathscr{B}_{t_i}$. By assumption, $W_{t_i}(j_{t_i})W_{t_i}(a)^{-1}$ belongs to the ideal $W_{t_i}(\mathscr{J}_{t_i})$ of $\mathscr{B}_{t_i}$. Choose elements $k_{t_i} \in \mathscr{J}_{t_i}$ such that $W_{t_i}(k_{t_i}) = W_{t_i}(j_{t_i})W_{t_i}(a)^{-1}$, and define $b' := b - k_{t_1} - \ldots - k_{t_m}$. Then

$$b'a = e + j_{t_1} - k_{t_1}a + \ldots + j_{t_m} - k_{t_m}a. \tag{6.5}$$

Due to the choice of the k_{t_i}, one has $W_{t_i}(j_{t_i} - k_{t_i}a) = 0$ and $j_{t_i} - k_{t_i}a \in \mathscr{J}_{t_i}$ for every i. Thus, by the lifting property of W_{t_i}, the elements $j_{t_i} - k_{t_i}a$ belong to the radical of $\mathscr{A}$. So we conclude from (6.5) that $b'a - e$ is in the radical of $\mathscr{A}$. Similarly, $ab' - e$ is in the radical. Hence, a is invertible.   ∎

Note that this proof did not rely on the N ideals lemma. One could prove Theorem 6.3.7 also by means of that lemma if another, again moderate, additional assumption is fulfilled: The homomorphisms W_t have to be *separating* in the sense that $W_t(\mathscr{J}_s) = \{0\}$ whenever $s \neq t$. Indeed, in this case we obtain from (6.5), by applying W_{t_i} to both sides of this equality that

$$W_{t_i}(b')W_{t_i}(a) = W_{t_i}(e) + W_{t_i}(j_{t_i} - k_{t_i}a) = W_{t_i}(e).$$

Hence, $W_{t_i}(a)$ is invertible in $W_{t_i}(\mathscr{A})$, whence the invertibility of $a + \mathrm{Ker}\,W_{t_i}$ in $\mathscr{A}/\mathrm{Ker}\,W_{t_i}$ follows. The remainder of the proof runs as that of Theorem 6.3.2.

Theorem 6.3.8 (Lifting theorem, Banach algebra IC-version). *Let $\mathscr{A}$ be a unital Banach algebra and, for every element t of a certain set T, let $\mathscr{J}_t$ be a closed ideal of $\mathscr{A}$ which is lifted by a unital and continuous homomorphism W_t from $\mathscr{A}$ into a unital Banach algebra $\mathscr{B}_t$. Assume that $W_t(\mathscr{J}_t)$ is a closed ideal of $\mathscr{B}_t$. Further, let $\mathscr{J}$ stand for the smallest closed ideal of $\mathscr{A}$ which contains all ideals $\mathscr{J}_t$. Then an element $a \in \mathscr{A}$ is invertible if and only if the coset $a + \mathscr{J}$ is invertible in $\mathscr{A}/\mathscr{J}$ and if all elements $W_t(a)$ are invertible in $\mathscr{B}_t$.*

In applications, we will often use this theorem with $\mathscr{B}_t$ being the Banach algebra of all bounded linear operators on some Banach space X_t and with $W_t(\mathscr{J}_t)$ being the ideal of the compact operators on X_t.

6.4 Finite sections of Wiener-Hopf operators

As a first illustration of the concepts discussed above, we consider the finite sections method for Wiener-Hopf operators on $L^p(\mathbb{R}^+)$ with continuous generating functions. In this section, let $1 < p < \infty$.

Let I denote the identity operator on $L^p(\mathbb{R}^+)$. For every positive real number τ, define operators

$$(P_\tau u)(t) := \begin{cases} u(t) & \text{if } t < \tau, \\ 0 & \text{if } t > \tau, \end{cases} \qquad Q_\tau := I - P_\tau \tag{6.6}$$

on $L^p(\mathbb{R}^+)$. The operators P_τ are projections with $\|P_\tau\| = 1$. We consider finite sections with respect to these projections.

For $f \in C_p$ and $v \in L^p(\mathbb{R}^+)$, consider the approximation method

$$P_\tau W(f) P_\tau u_\tau = P_\tau v, \tag{6.7}$$

to find an approximate solution $u_\tau \in \operatorname{Im} P_\tau$ to the Wiener-Hopf equation

$$W(f)u = v. \tag{6.8}$$

According to (6.7), we introduce the Banach algebra $\mathscr{E}$ of all sequences $(A_\tau)_{\tau>0}$ of bounded linear operators $A_\tau : \operatorname{Im} P_\tau \to \operatorname{Im} P_\tau$ for which $\sup_\tau \|A_\tau\|_{\mathscr{L}(L^p(\mathbb{R}^+))} < \infty$. Making the natural definition for the ideal $\mathscr{G}$, Theorem 6.2.2 holds.

6.4.1 Essentialization

For $\tau > 0$, define the operators acting on $L^p(\mathbb{R}^+)$,

$$(R_\tau u)(t) = \begin{cases} u(\tau - t) & \text{if } t < \tau, \\ 0 & \text{if } t > \tau. \end{cases} \tag{6.9}$$

It is easy to check that $R_\tau^2 = P_\tau$ and $\|R_\tau\| = 1$.

Let $\mathscr{F} \subset \mathscr{E}$ be the set of all sequences $\mathbf{A} := (A_\tau)$ for which there exist operators $W_0(\mathbf{A})$ and $W_1(\mathbf{A})$ such that

$$A_\tau P_\tau \to W_0(\mathbf{A}), \qquad\qquad A_\tau^* P_\tau \to W_0(\mathbf{A})^*,$$
$$R_\tau A_\tau R_\tau \to W_1(\mathbf{A}), \qquad\qquad (R_\tau A_\tau R_\tau)^* \to W_1(\mathbf{A})^*,$$

where "$\to$" again refers to strong convergence as $\tau \to \infty$.

Proposition 6.4.1. *The set $\mathscr{F}$ forms a closed subalgebra of $\mathscr{E}$ that contains $\mathscr{G}$, and the mappings W_0 and W_1 are continuous homomorphisms on $\mathscr{F}$.*

Proof. Let $\mathbf{A} := (A_\tau)$ and $\mathbf{B} := (B_\tau)$ be sequences in $\mathscr{F}$. Then $W_0(\mathbf{AB}) = W_0(\mathbf{A})W_0(\mathbf{B})$ by Lemma 1.4.4. Also,

$$R_\tau A_\tau B_\tau R_\tau = R_\tau A_\tau P_\tau B_\tau R_\tau = R_\tau A_\tau R_\tau R_\tau B_\tau R_\tau,$$

whence $W_1(\mathbf{AB}) = W_1(\mathbf{A})W_1(\mathbf{B})$. Thus, $\mathscr{F}$ is a subalgebra of $\mathscr{E}$. It is further trivial to check that $\mathscr{F}$ contains the ideal $\mathscr{G}$.

To prove the closedness of $\mathscr{F}$ in $\mathscr{E}$, let $(\mathbf{A}_k)_{k \in \mathbb{N}}$ with $\mathbf{A}_k := (A_\tau^{(k)})_{\tau>0}$ be a sequence in $\mathscr{F}$ which converges to a sequence $\mathbf{A} := (A_\tau)_{\tau>0}$ in $\mathscr{E}$. Since $(\mathbf{A}_k)_{k \in \mathbb{N}}$ is a Cauchy sequence and $\|W_0(\mathbf{B})\| \le \|\mathbf{B}\|$ for every sequence $\mathbf{B} \in \mathscr{F}$, we conclude that $(W_0(\mathbf{A}_k))_{k \in \mathbb{N}}$ is a Cauchy sequence in $\mathscr{L}(L^p(\mathbb{R}^+))$. Let A^0 denote the limit of that sequence. We show that A_0 is the strong limit of the sequence $\mathbf{A}$. For let $u \in L^p(\mathbb{R}^+)$. For every $\varepsilon > 0$, there exist a $\tau_0 > 0$ and a $k_0 \in \mathbb{N}$ such that, for $\tau > \tau_0$,

$$\begin{aligned} \|(A^0 - A_\tau)P_\tau u\| &\le \|(A^0 - A_\tau^{(k_0)} P_\tau)u\| + \|A_\tau^{(k_0)} P_\tau - A_\tau P_\tau\|\|u\| \\ &\le \|(A^0 - A_\tau^{(k_0)} P_\tau)u\| + \|\mathbf{A} - \mathbf{A}_{k_0}\|\|u\| < \varepsilon, \end{aligned}$$

which establishes the existence of the strong limit $W_0(\mathbf{A})$. In a similar way, the existence of the strong limit $W_1(\mathbf{A})$ follows. Thus, $\mathbf{A} \in \mathscr{F}$, whence the closedness of $\mathscr{F}$.   ∎

Proposition 6.4.2. *The algebra* $\mathscr{F}/\mathscr{G}$ *is inverse-closed in the algebra* $\mathscr{E}/\mathscr{G}$.

Proof. Let $\mathbf{A} := (A_\tau) \in \mathscr{F}$, and suppose that $\mathbf{A} + \mathscr{G} \in \mathscr{F}/\mathscr{G}$ is invertible in $\mathscr{E}/\mathscr{G}$. Then there exists a sequence $\mathbf{B} := (B_\tau) \in \mathscr{E}$ and a sequence $(G_\tau) \in \mathscr{G}$ such that $B_\tau A_\tau = P_\tau + G_\tau$ for every $\tau > 0$. Let $u \in L^p(\mathbb{R}^+)$. Then

$$\|P_\tau u\| = \|(B_\tau A_\tau - G_\tau)u\| \le c\|A_\tau u\| + \|G_\tau u\|$$

with a constant $c := \|\mathbf{B}\| > 0$. Taking the limit as $\tau \to \infty$ we obtain

$$\|u\| \le c\|W_0(\mathbf{A})u\|,$$

which implies that the kernel of $W_0(\mathbf{A})$ is $\{0\}$ and the range of $W_0(\mathbf{A})$ is closed. Applying the same argument to the adjoint sequence we find that the kernel of $W_0(\mathbf{A}^*) = W_0(\mathbf{A})^*$ is $\{0\}$, too. Hence, $W_0(\mathbf{A})$ is invertible. Further, for $u \in L^p(\mathbb{R}^+)$, we have

$$\begin{aligned}
&\|B_\tau P_\tau u - W_0(\mathbf{A})^{-1}u\| \\
&= \|B_\tau P_\tau u - (B_\tau A_\tau P_\tau - G_\tau P_\tau + Q_\tau)W_0(\mathbf{A})^{-1}u\| \\
&\le \|B_\tau\|\|P_\tau u - A_\tau P_\tau W_0(\mathbf{A})^{-1}u\| + \|(-G_\tau P_\tau + Q_\tau)W_0(\mathbf{A})^{-1}u\| \\
&= \|B_\tau\|\|P_\tau W_0(\mathbf{A})v - A_\tau P_\tau v\| + \|(-G_\tau P_\tau + Q_\tau)W_0(\mathbf{A})^{-1}u\|
\end{aligned}$$

with $v = W_0(\mathbf{A})^{-1}u$. Since the right-hand side of this estimate tends to zero as $\tau \to \infty$, the inverse sequence $\mathbf{B}$ is strongly convergent, too. Similarly, one shows the strong convergence of the adjoint sequence $\mathbf{B}^*$. From

$$R_\tau B_\tau R_\tau R_\tau A_\tau R_\tau = R_\tau B_\tau A_\tau R_\tau = R_\tau P_\tau R_\tau + R_\tau G_\tau R_\tau = P_\tau + G'_\tau$$

with $(G'_\tau) \in \mathscr{G}$ one concludes that the sequence $(R_\tau A_\tau R_\tau)$ is invertible in $\mathscr{E}/\mathscr{G}$ also. As above, one concludes that $W_1(\mathbf{A})$ is invertible and that the sequences $(R_\tau B_\tau R_\tau)$ and $(R_\tau B_\tau R_\tau)^*$ are strongly convergent. Thus, the sequence $\mathbf{B}$ belongs to $\mathscr{F}$, whence the inverse-closedness of $\mathscr{F}/\mathscr{G}$.   ∎

Let $\mathscr{K} \subset \mathscr{L}(L^p(\mathbb{R}^+))$ denote the ideal of the compact operators and consider the subsets of $\mathscr{E}$

$$\begin{aligned}
\mathscr{J}_0 &:= \{(P_\tau K P_\tau) + (G_\tau) : K \in \mathscr{K}, (G_\tau) \in \mathscr{G}\}, & (6.10) \\
\mathscr{J}_1 &:= \{(R_\tau K R_\tau) + (G_\tau) : K \in \mathscr{K}, (G_\tau) \in \mathscr{G}\}. & (6.11)
\end{aligned}$$

Proposition 6.4.3. $\mathscr{J}_0$ *and* $\mathscr{J}_1$ *are closed ideals of* $\mathscr{F}$.

Proof. We prove the assertion for $\mathscr{J}_0$. The proof for $\mathscr{J}_1$ is similar. Since $P_\tau \to I$ strongly and $R_\tau \rightharpoonup 0$ weakly, one has $P_\tau K P_\tau \rightrightarrows K$ in the norm and $R_\tau K R_\tau \to 0$ strongly. Hence, $\mathscr{J}_0$ is contained in $\mathscr{F}$, and one easily checks that $\mathscr{J}_0$ is a linear subspace of $\mathscr{F}$. To prove the left ideal property, one has to check that all sequences $\mathbf{AK}$ with $\mathbf{A} := (A_\tau) \in \mathscr{F}$ and $\mathbf{K} := (P_\tau K P_\tau)$ are in $\mathscr{J}_0$. Write

$$\mathbf{AK} = \big(P_\tau W_0(\mathbf{A}) K P_\tau\big) + \big(P_\tau (A_\tau P_\tau - W_0(\mathbf{A})) K P_\tau\big).$$

Clearly, the first term on the right-hand side is in $\mathscr{J}_0$, and the second one is even in $\mathscr{G}$ since $A_\tau \to W_0(\mathbf{A})$ strongly as $\tau \to \infty$ and from Lemma 1.4.7. The right ideal property follows similarly (here we need the strong convergence of the adjoint sequences). We are left with the closedness of $\mathscr{J}_0$. Recall that $(P_\tau K P_\tau + G_\tau)$ converges to K in the norm. Therefore,

$$\|K\| = \lim_{\tau \to \infty} \|P_\tau K P_\tau + G_\tau\| \le \|(P_\tau K P_\tau + G_\tau)\|_{\mathscr{F}}. \qquad (6.12)$$

Consider a sequence $(\mathbf{J}_k)_{k \in \mathbb{N}}$ with $\mathbf{J}_k := (P_\tau K^{(k)} P_\tau + G_\tau^{(k)})_{\tau \in \mathbb{R}^+}$ in $\mathscr{J}_0$ which converges in $\mathscr{F}$. Then $(\mathbf{J}_k)$ is a Cauchy sequence, and (6.12) implies that $(K^{(k)})$ is a Cauchy sequence of compact operators. Hence, there is a compact operator K such that $\|K - K^{(k)}\| \to 0$. Consequently, $\|(P_\tau K P_\tau) - (P_\tau K^{(k)} P_\tau)\|_{\mathscr{F}} \to 0$. But then $\big((G_\tau^{(k)})\big)_{k \in \mathbb{N}}$ is also a Cauchy sequence, and the closedness of $\mathscr{G}$ implies that there exists a sequence $(G_\tau) \in \mathscr{G}$ such that $\lim_{k \to \infty} \|(G_\tau) - (G_\tau^{(k)})\| = 0$. We conclude that the sequence $\mathbf{J} := (P_\tau K P_\tau + G_\tau)$ is the limit of $(\mathbf{J}_k)$. Evidently, $\mathbf{J} \in \mathscr{J}_0$. ∎

Let $i \in \{0,1\}$. Since $\mathscr{G}$ is in the kernel of W_i, one can define the corresponding quotient homomorphism

$$\mathscr{F}/\mathscr{G} \to \mathscr{L}(L^p(\mathbb{R}^+)), \quad \mathbf{A} + \mathscr{G} \mapsto W_i(\mathbf{A})$$

which we denote by W_i again. It turns out that $\mathscr{J}_i/\mathscr{G}$ is a closed ideal of $\mathscr{F}/\mathscr{G}$ which is lifted by the quotient homomorphism W_i since $\operatorname{Ker} W_i \cap \mathscr{J}_i = \mathscr{G}$. In accordance with the lifting theorem, define $\mathscr{J}$ as the smallest closed ideal in $\mathscr{F}/\mathscr{G}$ which contains $\mathscr{J}_0/\mathscr{G}$ and $\mathscr{J}_1/\mathscr{G}$. By repeating arguments from the proof of the preceding proposition, one easily finds that

$$\mathscr{J} = \{(P_\tau K P_\tau) + (R_\tau L R_\tau) + \mathscr{G} : K, L \in \mathscr{K}\}.$$

The equality

$$R_\tau W(f) R_\tau = P_\tau W(\tilde{f}) P_\tau \quad \text{with} \quad \tilde{f}(t) := f(-t)$$

implies that the sequence $(P_\tau W(f) P_\tau)$ associated with the finite sections method (6.7) belongs to $\mathscr{F}$ with $W_0(P_\tau W(f) P_\tau) = W(f)$ and $W_1(P_\tau W(f) P_\tau) = W(\tilde{f})$. The lifting theorem implies the following preliminary result.

Theorem 6.4.4. *Let $f \in C_p$. The finite sections method (6.7) applies to the Wiener-Hopf operator $W(f)$ if and only if the operator $W(f)$ is invertible and the coset $(P_\tau W(f) P_\tau) + \mathscr{J}$ is invertible in $(\mathscr{F}/\mathscr{G})/\mathscr{J}$.*

Proof. By Theorem 6.3.8, the coset $(P_\tau W(f) P_\tau) + \mathscr{G}$ is invertible in $\mathscr{F}/\mathscr{G}$ if and only if the operators $W_0(P_\tau W(f) P_\tau) = W(f)$ and $W_1(P_\tau W(f) P_\tau) = W(\tilde{f})$ are invertible on $L^p(\mathbb{R}^+)$ and if the coset $(P_\tau W(f) P_\tau) + \mathscr{J}$ is invertible in $(\mathscr{F}/\mathscr{G})/\mathscr{J}$. But $W(f)$ is invertible on $L^p(\mathbb{R}^+)$ if and only if $W(\tilde{f})$ is invertible there, which is an immediate consequence of Theorem 5.2.2 (note that the winding number of f is minus the winding number of $\tilde{f}$). $\blacksquare$

6.4.2 Structure and stability

Let $\mathscr{A}$ be the smallest closed subalgebra of $\mathscr{F}/\mathscr{G}$ which contains the cosets $(P_\tau W(f) P_\tau) + \mathscr{G}$ with $f \in C_p$ and the ideal $\mathscr{J}$. Let $\mathscr{A}^{\mathscr{J}}$ stand for the quotient algebra $\mathscr{A}/\mathscr{J}$. We are going to prove that the algebra $\mathscr{A}^{\mathscr{J}}$ is commutative and, thus, subject to the simplest local principle, namely classical Gelfand theory. For $\tau > 0$, let $\tilde{V}_\tau$ and $\tilde{V}_{-\tau}$ be the operators on $L^p(\mathbb{R}^+)$ defined by

$$(\tilde{V}_\tau u)(t) = \begin{cases} 0 & \text{if } t < \tau, \\ u(t - \tau) & \text{if } t \geq \tau, \end{cases} \qquad (\tilde{V}_{-\tau} u)(t) = u(t + \tau). \qquad (6.13)$$

Clearly, $\tilde{V}_{-\tau} \tilde{V}_\tau = I$ and $\tilde{V}_\tau \tilde{V}_{-\tau} = Q_\tau$. It is also not hard to check by straightforward computation that $R_\tau W(f) \tilde{V}_\tau = P_\tau H(\tilde{f})$ and $\tilde{V}_{-\tau} W(f) R_\tau = H(f) P_\tau$, where $H(f)$ refers to the Hankel operator introduced in Section 5.2. We will use these equalities in the proof of the next proposition.

Proposition 6.4.5. *The algebra $\mathscr{A}^{\mathscr{J}}$ is commutative, every element of $\mathscr{A}^{\mathscr{J}}$ has the form $(P_\tau W(f) P_\tau) + \mathscr{J}$ with $f \in C_p$, and the mapping $(P_\tau W(f) P_\tau) + \mathscr{J} \mapsto f$ is an isometric isomorphism from $\mathscr{A}^{\mathscr{J}}$ onto C_p.*

Proof. Let $f, g \in C_p$. Our first goal is the equality

$$(P_\tau W(fg) P_\tau) + \mathscr{J} = (P_\tau W(f) P_\tau)(P_\tau W(g) P_\tau) + \mathscr{J}. \qquad (6.14)$$

Write

$$(P_\tau W(fg) P_\tau) - (P_\tau W(f) P_\tau W(g) P_\tau)$$
$$= (P_\tau (W(fg) - W(f) W(g)) P_\tau) + (P_\tau W(f) Q_\tau W(g) P_\tau).$$

By Proposition 5.3.3, the operator $W(fg) - W(f)W(g)$ is compact. Thus, the sequence $(P_\tau (W(fg) - W(f)W(g)) P_\tau)$ is in $\mathscr{J}$. Further,

$$P_\tau W(f) Q_\tau W(g) P_\tau = R_\tau R_\tau W(f) \tilde{V}_\tau \tilde{V}_{-\tau} W(g) R_\tau R_\tau = R_\tau H(\tilde{f}) H(g) R_\tau.$$

Since $H(\tilde{f})H(g)$ is compact by Proposition 5.3.6, the sequence $(P_\tau W(f)Q_\tau W(g)P_\tau)$ is in $\mathscr{J}$, too. This settles equality (6.14). From this equality, one immediately gets the first two assertions.

Equality (6.14) further implies that the mapping

$$C_p \to \mathscr{A}/\mathscr{J}, \quad f \mapsto (P_\tau W(f)P_\tau) + \mathscr{J} \tag{6.15}$$

is an algebra homomorphism, the norm of which is not greater than one since

$$\|(P_\tau W(f)P_\tau) + \mathscr{J}\|_{\mathscr{F}/\mathscr{J}} \le \|(P_\tau W(f)P_\tau)\|_{\mathscr{F}} \le \|W(f)\| \le \|f\|_{C_p}. \tag{6.16}$$

To show that the mapping (6.15) is an isometry (hence, an isomorphism), we identify the Wiener-Hopf operator $W(f)$ with the operator $\chi_+ W^0(f)\chi_+ I$ acting on $L^p(\mathbb{R})$. With the shift operators on $L^p(\mathbb{R})$ defined by $(V_\tau u)(t) := u(t - \tau)$, one has

$$V_{-\tau}(W(f) + K)V_\tau \to W^0(f) \quad \text{strongly as } \tau \to \infty$$

for arbitrary generating functions $f \in C_p$ and compact operators K. Hence, for arbitrary compact operators K, L and sequences $(G_\tau) \in \mathscr{G}$,

$$\begin{aligned}
\|f\|_{C_p} &= \|W^0(f)\| \le \|W(f) + K\| \\
&= \|W_0(P_\tau W(f)P_\tau + P_\tau K P_\tau + R_\tau L R_\tau + G_\tau)\| \\
&\le \|(P_\tau W(f)P_\tau + P_\tau K P_\tau + R_\tau L R_\tau + G_\tau)\|_{\mathscr{F}}.
\end{aligned}$$

Taking the infimum with respect to K, L and (G_τ) we obtain

$$\|f\|_{C_p} \le \|(P_\tau W(f)P_\tau) + \mathscr{J}\|_{\mathscr{F}/\mathscr{J}}$$

which, together with (6.16) states the desired isometry.   ∎

Since the invertibility of the Wiener-Hopf operator $W(f)$ implies the invertibility of f, the following result is an immediate consequence of Theorem 6.4.4 and Proposition 6.4.5.

Theorem 6.4.6 (Gohberg-Feldman). *Let $f \in C_p$. The finite sections method (6.7) applies to the Wiener-Hopf operator $W(f)$ if and only if this operator is invertible.*

Remark 6.4.7. Thanks to Proposition 6.4.5, a formal localization step was not needed to prove the preceding theorem. Of course, one gets the same result also by describing the commutative Banach algebra $\mathscr{A}^{\mathscr{J}}$ via Gelfand theory, in a similar way to the one in the proof of Theorem 6.4.8, below.

It should be further emphasized that Gelfand theory implies conditions for the invertibility in $\mathscr{A}^{\mathscr{J}}$, whereas Theorem 6.4.4 refers to invertibility in $(\mathscr{F}/\mathscr{G})/\mathscr{J}$. That no inverse-closedness problem occurs in the case at hand is a consequence of the fact that if $(P_\tau W(f)P_\tau) + \mathscr{J}$ is not invertible in $\mathscr{A}^{\mathscr{J}}$ then $W(f)$ is not invertible.

□

Theorem 6.4.8. *Let* $\mathbf{A} := (A_\tau) \in \mathscr{F}$ *be such that* $\mathbf{A} + \mathscr{G} \in \mathscr{A}$. *Then* $\mathbf{A}$ *is stable if and only if the operators* $\mathsf{W}_0(\mathbf{A})$ *and* $\mathsf{W}_1(\mathbf{A})$ *are invertible.*

Proof. By Proposition 6.4.5, there is a (uniquely defined) $f \in C_p$ such that

$$\mathbf{A} = (P_\tau W(f)P_\tau + P_\tau K P_\tau + R_\tau L R_\tau + G_\tau) \tag{6.17}$$

with compact operators K and L, and $(G_\tau) =: \mathbf{G} \in \mathscr{G}$. Clearly, $\mathsf{W}_0(\mathbf{A}) = W(f) + K$, $\mathsf{W}_1(\mathbf{A}) = W(\tilde{f}) + L$. Because $\mathbf{G}$ is in the kernel of both homomorphisms W_0 and W_1, it can be assumed that W_0 and W_1 are also well defined on $\mathscr{A}$ as quotient homomorphisms. If $\mathbf{A}$ is stable then $\mathbf{A} + \mathscr{G}$ is invertible in $\mathscr{A}$ by Theorem 6.3.8, whence it follows that $\mathsf{W}_0(\mathbf{A})$ and $\mathsf{W}_1(\mathbf{A})$ are invertible. Conversely, let $\mathsf{W}_0(\mathbf{A})$ and $\mathsf{W}_1(\mathbf{A})$ be invertible. Then the function $f \in C_p$ does not vanish on $\dot{\mathbb{R}}$ by Theorem 5.2.2. Since the maximal ideal space of C_p can be identified with $\dot{\mathbb{R}}$ in a natural way, this means that then also $(P_\tau W(f)P_\tau) + \mathscr{J}$ is invertible, and Theorem 6.3.8 finally gives that $\mathbf{A} + \mathscr{G}$ is invertible in $\mathscr{A}$. Hence, $\mathbf{A}$ is stable. ∎

6.5 Spline Galerkin methods for Wiener-Hopf operators

The finite sections method for Wiener-Hopf operators discussed in the previous section yields infinite-dimensional approximation equations. In this section, we will examine an example of a fully discretized approximation method that results in finite-dimensional approximation operators, through the use of splines. To keep it simple, we will consider piecewise constant splines on uniform meshes. But the techniques presented in this book allow for far more general settings. We refer to the historical remarks at the end of the chapter for more information. We would also like to turn the reader's attention to the fact that the analysis of the stability conditions for the finite sections method and the spline Galerkin method run completely parallel. We will thus not provide all details here.

For $n \in \mathbb{N}$ consider the mesh sequence (Δ^n) with

$$\Delta^n := \left\{ x_j^{(n)} = j/n, \quad 0 \le j \le n\tau_n \right\}$$

where the τ_n are given, positive numbers which go to ∞ as n increases and such that $n^2 \tau_n$ is integer. Define also

$$J_j^n := \left] \frac{j}{n}, \frac{j+1}{n} \right[,$$

and let χ_{jn} refer to the characteristic function of the interval J_j^n. We consider the subspace S^n of $L^p(\mathbb{R}^+)$ of all functions which are piecewise constant with respect to this mesh,

$$S^n := \left\{ u \in L^p(\mathbb{R}^+) : u_{|J_j^n} \text{ is constant}, \quad u_{|\mathbb{R}^+ \setminus \cup J_j^n} = 0 \right\}.$$

The Galerkin projection of $L^p(\mathbb{R}^+)$ onto S^n is the operator

$$(P^n u)(x) = \sum_j \left(n \int_{J_j^n} u(y)\, dy \right) \chi_{jn}(x).$$

Note that P^n is actually the orthogonal projection of $L^p(\mathbb{R}^+)$ onto S^n in the case $p = 2$ and that the P^n are uniformly bounded with respect to n in $L^p(\mathbb{R}^+)$. We also have that $P^n \to I$ as $n \to \infty$. The complementary projection $I - P^n$ will be denoted by Q^n.

Let $f \in C_p$ and $u, v \in L^p(\mathbb{R}^+)$. We consider the approximation method

$$P^n W(f) P^n u_n = P^n v \tag{6.18}$$

to find an approximate solution u_n to the equation

$$W(f)u = v. \tag{6.19}$$

The initial sequence algebra $\mathscr{E}$ is constituted by all sequences $\mathbf{A} = (A_n)_{n \in \mathbb{N}}$ of bounded linear operators $A_n : \operatorname{Im} P^n \to \operatorname{Im} P^n$ such that $\sup_n \|A_n\|_{\mathscr{L}(X)} < \infty$. Further, $\mathscr{G}$ again stands for the ideal of $\mathscr{E}$ which consists of all sequences tending to zero in the norm.

6.5.1 Essentialization

For every positive integer n, define

$$(R_n u)(t) = \begin{cases} P^n u(\tau_n - t) & \text{if } t < \tau_n, \\ 0 & \text{if } t > \tau_n. \end{cases} \tag{6.20}$$

Clearly, the R_n bounded operators on every space $L^p(\mathbb{R}^+)$, their norm is 1, and $R_n^2 = P^n$. Let $\mathscr{F}$ be the set of all sequences $\mathbf{A} = (A_n) \in \mathscr{E}$ for which there exist operators $W_0(\mathbf{A})$ and $W_1(\mathbf{A})$ such that

$$\begin{array}{ll} A_n \to W_0(\mathbf{A}), & A_n^* \to W_0(\mathbf{A})^*, \\ R_n A_n R_n \to W_1(\mathbf{A}), & (R_n A_n R_n)^* \to W_1(\mathbf{A})^*. \end{array}$$

It is easy to see that $\mathscr{F}$ forms a closed and inverse-closed unital subalgebra of $\mathscr{E}$. The proof of the following proposition is similar to that of Proposition 6.4.2.

Proposition 6.5.1. *The algebra $\mathscr{F}/\mathscr{G}$ is inverse-closed in the algebra $\mathscr{E}/\mathscr{G}$.*

We proceed in analogy to Section 6.4. As there, one can show that the sets

$$\mathscr{J}_0 := \{(P^n K P^n) + (G_n) : K \in \mathscr{K}, (G_n) \in \mathscr{G}\}, \tag{6.21}$$
$$\mathscr{J}_1 := \{(R_n K R_n) + (G_n) : K \in \mathscr{K}, (G_n) \in \mathscr{G}\} \tag{6.22}$$

are closed ideals of $\mathscr{F}$, and $\mathrm{Ker}\,W_i \cap \mathscr{J}_i = \mathscr{G}$ for $i = 0,1$. Thus, for $i = 0,1$, $\mathscr{J}_i/\mathscr{G}$ is a closed ideal of $\mathscr{E}/\mathscr{G}$, and the quotient homomorphism

$$\mathscr{E}/\mathscr{G} \to \mathscr{L}(L^p(\mathbb{R}^+)), \quad \mathbf{A} + \mathscr{G} \mapsto W_i(\mathbf{A})$$

is well defined. We denote it by W_i again. Then W_i lifts the ideal $\mathscr{J}_i/\mathscr{G}$ in the sense of the lifting theorem. Define $\mathscr{J}$ as the smallest closed ideal in $\mathscr{F}/\mathscr{G}$ which contains $\mathscr{J}_0/\mathscr{G}$ and $\mathscr{J}_1/\mathscr{G}$. Thus, $\mathscr{J}$ consists of all cosets

$$(P^n K P^n) + (R_n L R_n) + \mathscr{G} \quad \text{with} \quad K, L \in \mathscr{K}.$$

Let $\mathbf{A} := (P^n W(f) P^n)$ with $f \in C_p$. It is easy to see that the sequence $\mathbf{A}$ belongs to $\mathscr{F}/\mathscr{G}$ and that

$$W_0(\mathbf{A}) = W(f) \quad \text{and} \quad W_1(\mathbf{A}) = W(\tilde{f})$$

with $\tilde{f}(t) := f(-t)$. Via Theorem 6.3.4, we arrive at the following.

Theorem 6.5.2. *The approximation method (6.18) applies to the Wiener-Hopf operator $W(f)$ with $f \in C_p$ if and only if the operator $W(f)$ is invertible and the coset $(P^n W(f) P^n) + \mathscr{J}$ is invertible in $(\mathscr{F}/\mathscr{G})/\mathscr{J}$.*

6.5.2 Structure and stability

We will now have a closer look at the invertibility of the sequence $(P^n W(a) P^n)$ modulo $\mathscr{J}$. Let $\mathscr{A}$ denote the smallest closed subalgebra of $\mathscr{F}/\mathscr{G}$ which contains all cosets $(P^n W(f) P^n) + \mathscr{G}$ with $f \in C_p$ and the ideal $\mathscr{J}$. Further, let $\mathscr{A}^{\mathscr{J}}$ stand for the quotient algebra $\mathscr{A}/\mathscr{J}$.

Lemma 6.5.3. *Let $f \in C_p$ with $f(\infty) = 0$. The norms of the operators $P_{\tau_n} Q^n W(f)$ and $W(f) P_{\tau_n} Q^n$ tend to zero as $n \to \infty$.*

Proof. Approximate $W(f)$ in the operator norm by operators $W(f_m)$ such that $k_m = F^{-1} f_m$ are functions belonging, together with their derivatives, to $L^1(\mathbb{R})$. Repeating the arguments of the proof of [151, Lemma 5.25] we obtain that there exists a constant d such that

$$\|Q^n W(f_m) u\|_{J_j^n} \leq \frac{d}{n} \|DW(f_m) u\|_{J_j^n},$$

the D referring to the operator of differentiation. Consequently,

$$\|P_{\tau_n} Q^n W(f_m) u\|_{L^p} \leq \frac{d}{n} \|DW(f_m) u\|_{L^p} \leq \frac{d'}{n} \|u\|_{L^p}$$

with some constants d, d'. The last inequality is due to the continuity of $W(f_m)$ from $L^p(\mathbb{R}^+)$ to the Sobolev space $L^{p,1}(\mathbb{R}^+)$, the subspace of $L^p(\mathbb{R}^+)$ containing all functions with derivatives in $L^p(\mathbb{R}^+)$. This proves the assertion for the first sequence of operators. The convergence of the second sequence follows by a duality argument, since $(Q^n)^* = Q^n$, the latter acting on the dual space $L^q(\mathbb{R}^+)$, with $q = p/(p-1)$. ∎

Proposition 6.5.4. *The algebra $\mathscr{A}^{\mathscr{I}}$ is commutative.*

Proof. The result will follow once we have shown that

$$(P^n W(fg)P^n) + \mathscr{I} = (P^n W(f)P^n)(P^n W(g)P^n) + \mathscr{I}$$

for arbitrary functions $f, g \in C_p$. Write

$$(P^n W(fg)P^n - P^n W(f)P^n W(g)P^n)$$
$$= (P^n(W(fg) - W(f)W(g))P^n) + (P^n W(f)Q^n W(g)P^n). \qquad (6.23)$$

By Proposition 5.3.3, the operator $W(fg) - W(f)W(g)$ is compact. Therefore, the first part of the above sum is in $\mathscr{I}$. Further, because $Q^n = Q^n(P_{\tau_n} + Q_{\tau_n}) = Q^n P_{\tau_n} + Q_{\tau_n}$ and from Lemma 6.5.3, one has

$$P^n W(f)Q^n W(g)P^n = P^n W(f)P_{\tau_n}Q^n W(g)P^n + R_n R_n W(f)V_{\tau_n}V_{-\tau_n}W(g)R_n R_n$$
$$= G_n + R_n H(\tilde{f})H(g)R_n,$$

where $(G_n) \in \mathscr{G}$. Since the operator $H(\tilde{f})H(g)$ is compact by Proposition 5.3.6, the second part of the sum (6.23) is in $\mathscr{I}$, too. ∎

Being a commutative Banach algebra, $\mathscr{A}^{\mathscr{I}}$ is subject to classical Gelfand theory. The maximal ideal space of $\mathscr{A}^{\mathscr{I}}$ is homeomorphic to the maximal ideal space of C_p which, in turn, is homeomorphic to $\dot{\mathbb{R}}$. Under these conditions, the Gelfand transform of the coset $F = (P^n W(f)P^n) + \mathscr{I}$ with $f \in C_p$ is just the function $\widehat{F}(x) := f(x)$ on $\dot{\mathbb{R}}$. In particular, the coset A is invertible in $\mathscr{A}^{\mathscr{I}}$ if and only if the function f is invertible in $C(\dot{\mathbb{R}})$. Since the invertibility of $W(f)$ implies the invertibility of f, it therefore also implies the invertibility of F. In combination with Theorem 6.5.2, this yields the following.

Theorem 6.5.5. *Let $f \in C_p$. Then the approximation method (6.18) applies to $W(f)$ if and only if this operator is invertible.*

More generally, one has the following analog of Theorem 6.4.8.

Theorem 6.5.6. *A sequence $\mathbf{A} \in \mathscr{A}$ is stable if and only if the operators $W_0(\mathbf{A})$ and $W_1(\mathbf{A})$ are invertible.*

6.6 Finite sections of convolution and multiplication operators on $L^p(\mathbb{R})$

Here we come back to the setting of Section 6.4 and consider it in a larger context. We are going to examine the finite sections method for operators which are constituted by operators of convolution and operators of multiplication by piecewise continuous functions. Of course, this includes the finite sections method for Wiener-Hopf operators. In this section, we let $1 < p < \infty$ and work on the unweighted space $L^p(\mathbb{R})$. That we are now on the whole real line (thus, on a group) will allow us to work with the shift operators

$$V_\tau : L^p(\mathbb{R}) \to L^p(\mathbb{R}), \quad (V_\tau u)(x) := u(x - \tau)$$

in place of the reflections R_τ used in Section 6.4, which will simplify some of the arguments. We consider finite sections $P_\tau A P_\tau$ where $A \in \mathscr{L}(L^p(\mathbb{R}))$ and where the projections P_τ, $\tau > 0$, are defined by

$$P_\tau : L^p(\mathbb{R}) \to L^p(\mathbb{R}), \quad (P_\tau u)(t) := \begin{cases} u(t) & \text{if } |t| \leq \tau, \\ 0 & \text{if } |t| > \tau. \end{cases}$$

It will prove to be convenient to consider *extended* finite sections $P_\tau A P_\tau + Q_\tau$ where $Q_\tau := I - P_\tau$ instead of the usual $P_\tau A P_\tau$. The passage from finite sections to extended finite sections does not involve any complications since, of course, both sequences $(P_\tau A P_\tau + Q_\tau)_{\tau>0}$ and $(P_\tau A P_\tau)_{\tau>0}$ are simultaneously stable or not, and since they have the same strong limit. One advantage of using extended finite sections is that the operator A and its extended finite sections act on the same space.

Let $\mathscr{E}$ stand for the Banach algebra of all bounded sequences $(A_\tau)_{\tau>0}$ of operators $A_\tau \in \mathscr{L}(L^p(\mathbb{R}))$, and let $\mathscr{G}$ denote the closed ideal of $\mathscr{E}$ which consists of all sequences tending in the norm to zero. Of course, it is still true that a sequence $(A_\tau) \in \mathscr{E}$ is stable if and only if its coset is invertible in the quotient algebra $\mathscr{E}/\mathscr{G}$. The sequences we are interested in belong to the smallest closed subalgebra $\mathscr{A}(PC(\dot{\mathbb{R}}), PC_p, (P_\tau))$ of $\mathscr{E}$ which contains all constant sequences (aI) of operators of multiplication by a function $a \in PC(\dot{\mathbb{R}})$, all constant sequences $(W^0(b))$ of operators of convolution by a multiplier $b \in PC_p$, the sequence $(P_\tau)_{\tau>0}$, and the ideal $\mathscr{G}$. This algebra can be seen as an extension of the algebra $\mathscr{A}(PC(\dot{\mathbb{R}}), PC_p)$, studied in Section 5.6, with the addition of the non-constant sequence (P_τ).

6.6.1 Essentialization

Now we start the second step to analyze the stability of sequences in the algebra $\mathscr{A}(PC(\dot{\mathbb{R}}), PC_p, (P_\tau))$: We describe a closed and inverse-closed subalgebra of $\mathscr{E}/\mathscr{G}$ in which we have available all the technical tools we need, i.e., we can apply the lifting theorem, and the corresponding quotient algebra has a useful center.

Let $\mathscr{F}$ denote the set of all sequences $\mathbf{A} := (A_\tau) \in \mathscr{E}$ which have the following properties (all limits are considered with respect to strong convergence as $\tau \to \infty$):

- there is an operator $\mathsf{W}_0(\mathbf{A})$ such that $A_\tau \to \mathsf{W}_0(\mathbf{A})$ and $A_\tau^* \to \mathsf{W}_0(\mathbf{A})^*$;
- there are operators $\mathsf{W}_{\pm 1}(\mathbf{A})$ such that

$$V_{-\tau} A_\tau V_\tau \to \mathsf{W}_1(\mathbf{A}) \quad \text{and} \quad (V_{-\tau} A_\tau V_\tau)^* \to \mathsf{W}_1(\mathbf{A})^*$$

and

$$V_\tau A_\tau V_{-\tau} \to \mathsf{W}_{-1}(\mathbf{A}) \quad \text{and} \quad (V_\tau A_\tau V_{-\tau})^* \to \mathsf{W}_{-1}(\mathbf{A})^*;$$

- for each $y \in \mathbb{R}$, there is an operator $\mathsf{H}_{\infty, y}(\mathbf{A})$ such that

$$Z_\tau^{-1} U_y A_\tau U_{-y} Z_\tau \to \mathsf{H}_{\infty, y}(\mathbf{A}) \quad \text{and} \quad (Z_\tau^{-1} U_y A_\tau U_{-y} Z_\tau)^* \to \mathsf{H}_{\infty, y}(\mathbf{A})^*;$$

- for each $x \in \mathbb{R}$, there is an operator $\mathsf{H}_{x, \infty}(\mathbf{A})$ such that

$$Z_\tau V_{-x} A_\tau V_x Z_\tau^{-1} \to \mathsf{H}_{x, \infty}(\mathbf{A}) \quad \text{and} \quad (Z_\tau V_{-x} A_\tau V_x Z_\tau^{-1})^* \to \mathsf{H}_{x, \infty}(\mathbf{A})^*.$$

Proposition 6.6.1.

(i) *The set $\mathscr{F}$ is a closed subalgebra of $\mathscr{E}$. The mappings W_i with $i \in \{-1, 0, 1\}$, $\mathsf{H}_{\infty, y}$ with $y \in \mathbb{R}$, and $\mathsf{H}_{x, \infty}$ with $x \in \mathbb{R}$ act as bounded homomorphisms on $\mathscr{F}$, and the ideal $\mathscr{G}$ of $\mathscr{F}$ lies in the kernel of each these homomorphisms.*

(ii) *The algebra $\mathscr{F}$ is inverse-closed in $\mathscr{E}$, and the algebra $\mathscr{F}/\mathscr{G}$ is inverse-closed in $\mathscr{E}/\mathscr{G}$.*

All facts stated in this proposition are either evident, or they follow by arguments employed in the previous sections. So we omit the details of the proof.

The W- and the H-homomorphisms will play different roles in what follows. Whereas the W-homomorphisms are needed to define an ideal of $\mathscr{F}$ which is subject to the lifting theorem and for which the quotient algebra has a center which is useful for applying Allan's local principle, the family of H-homomorphisms will be used to identify the corresponding local algebras (of a suitable subalgebra of $\mathscr{F}$).

Let us turn to the lifting theorem. Let $\mathscr{K}$ denote the ideal of the compact operators on $L^p(\mathbb{R})$, and set

$$\mathscr{J} := \{(V_\tau K_1 V_{-\tau}) + (K_0) + (V_{-\tau} K_{-1} V_\tau) + (G_\tau) : K_{-1}, K_0, K_1 \in \mathscr{K}, (G_\tau) \in \mathscr{G}\}.$$

Proposition 6.6.2. *$\mathscr{J}$ is a closed ideal of $\mathscr{F}$.*

Proof. We will only prove that $\mathscr{J}$ is indeed a subset of $\mathscr{F}$. Once this is clear, the remainder of the proof runs completely parallel to that of Proposition 6.4.3.

In order to show that $\mathscr{J} \subset \mathscr{F}$, we have to show that all strong limits required in the definition of $\mathscr{F}$ exist for the sequences (K_0), $(V_{-\tau} K_{-1} V_\tau)$ and $(V_\tau K_1 V_{-\tau})$ with compact operators K_i. This is evident for the W-homomorphisms: One clearly has

$$\mathsf{W}_{-1}(V_{-\tau} K_{-1} V_\tau) = K_{-1}, \quad \mathsf{W}_0(K_0) = K_0, \quad \mathsf{W}_1(V_\tau K_1 V_{-\tau}) = K_1, \quad (6.24)$$

whereas all other W-homomorphisms give 0 when applied to these sequences since the sequences $(V_{-\tau})$ and (V_τ) are uniformly bounded and tend weakly to zero as $\tau \to +\infty$.

We claim that the H-homomorphisms applied to a sequence in $\mathscr{J}$ also give zero. This will follow once we have checked that the sequences

$$(U_{-y}Z_\tau),\ (V_\tau U_{-y}Z_\tau),\ (V_{-\tau}U_{-y}Z_\tau) \quad \text{and} \quad (V_x Z_\tau^{-1}),\ (V_\tau V_x Z_\tau^{-1}),\ (V_{-\tau}V_x Z_\tau^{-1})$$

are uniformly bounded and converge weakly to zero as $\tau \to +\infty$ for every choice of $x, y \in \mathbb{R}$. Since the operators U_{-y} and V_x are independent of τ and since V_x commutes with V_τ, it is sufficient to check these assertions for the sequences

$$(Z_\tau),\ (V_\tau U_{-y}Z_\tau),\ (V_{-\tau}U_{-y}Z_\tau) \quad \text{and} \quad (Z_\tau^{-1}),\ (V_\tau Z_\tau^{-1}),\ (V_{-\tau}Z_\tau^{-1}).$$

For the sequences $(Z_\tau^{\pm 1})$, this is done in Lemma 4.2.12. For the other sequences, the uniform boundedness is evident, and for their weak convergence to zero we can argue similarly to the proof of that lemma. Let B_τ denote any of the operators $V_\tau U_{-y}$ and $V_{-\tau}U_{-y}$ with $y \in \mathbb{R}$. Then

$$\left|(B_\tau \chi_{[\tau a, \tau b]})(x)\right| \le 1 \quad \text{and} \quad \left|(B_\tau^* \chi_{[c,d]})(x)\right| \le 1$$

for every possible choice of B_τ, a, b, c, d, τ, s and x. Hence,

$$\left|\langle \chi_{[c,d]}, B_\tau Z_\tau \chi_{[a,b]}\rangle\right| = \frac{1}{\tau^{1/p}}\left|\langle \chi_{[c,d]}, B_\tau \chi_{[\tau a, \tau b]}\rangle\right|$$
$$= \frac{1}{\tau^{1/p}}\left|\int_c^d (B_\tau \chi_{[\tau a, \tau b]})(x)\,dx\right| \le \frac{1}{\tau^{1/p}}(d-c) \to 0$$

and

$$\left|\langle \chi_{[c,d]}, B_\tau Z_\tau^{-1}\chi_{[a,b]}\rangle\right| = \tau^{1/p}\left|\langle B_\tau^*\chi_{[c,d]}, \chi_{[\frac{a}{\tau},\frac{b}{\tau}]}\rangle\right|$$
$$= \tau^{1/p}\left|\int_{\frac{a}{\tau}}^{\frac{b}{\tau}} (B_\tau^*\chi_{[c,d]})(x)\,dx\right| \le \tau^{1/p}\left(\frac{b}{\tau}-\frac{a}{\tau}\right) \to 0,$$

which proves the claimed weak convergence since the set of all linear combinations of functions of the form $\chi_{[c,d]}$ is dense both in $L^p(\mathbb{R})$ and in its dual space. $\blacksquare$

Now the Lifting Theorem 6.3.8 reduces to the following. Note that the stability of a sequence (A_τ) in $\mathscr{F}$ is equivalent to the invertibility of the coset $(A_\tau) + \mathscr{G}$ in the quotient algebra $\mathscr{F}/\mathscr{G}$ due to the inverse-closedness of $\mathscr{F}/\mathscr{G}$ in $\mathscr{E}/\mathscr{G}$ by Proposition 6.6.1.

Theorem 6.6.3. *Let* $\mathbf{A} := (A_\tau) \in \mathscr{F}$. *The sequence* $\mathbf{A}$ *is stable if and only if the operators* $W_{-1}(\mathbf{A})$, $W_0(\mathbf{A})$ *and* $W_1(\mathbf{A})$ *are invertible in* $\mathscr{L}(L^p(\mathbb{R}))$ *and if the coset* $\mathbf{A} + \mathscr{J}$ *is invertible in the quotient algebra* $\mathscr{F}^{\mathscr{J}} := \mathscr{F}/\mathscr{J}$.

The goal of the following lemmas is to show that all strong limits required in the definition of the algebra $\mathscr{F}$ exist for the generating sequences of the algebra $\mathscr{A}\big(PC(\dot{\mathbb{R}}), PC_p, (P_\tau)\big)$, which implies that this algebra is a closed subalgebra of $\mathscr{F}$.

Lemma 6.6.4. *The strong limit* $W_0(A_\tau)$ [1] *exists for the following elements of* $\mathscr{E}$:

(i) $W_0(P_\tau) = I$;
(ii) $W_0(aI) = aI$ *for* $a \in PC(\dot{\mathbb{R}})$;
(iii) $W_0(W^0(b)) = W^0(b)$ *for* $b \in PC_p$.

Lemma 6.6.5. *The strong limit* $W_{-1}(A_\tau)$ *exists for the following elements of* $\mathscr{E}$:

(i) $W_{-1}(P_\tau) = \chi_+ I$;
(ii) $W_{-1}(aI) = a(-\infty)I$ *for* $a \in PC(\dot{\mathbb{R}})$;
(iii) $W_{-1}(W^0(b)) = W^0(b)$ *for* $b \in PC_p$.

Lemma 6.6.6. *The strong limit* $W_1(A_\tau)$ *exists for the following elements of* $\mathscr{E}$:

(i) $W_1(P_\tau) = \chi_- I$;
(ii) $W_1(aI) = a(+\infty)I$ *for* $a \in PC(\dot{\mathbb{R}})$;
(iii) $W_1(W^0(b)) = W^0(b)$ *for* $b \in PC_p$.

Indeed, for the constant sequences (aI), these assertions are shown in Lemma 5.4.2, and the remaining assertions are evident. For the H-homomorphisms one has the following.

Lemma 6.6.7. *Let* $y \in \mathbb{R}$. *The strong limit* $H_{\infty,y}(A_\tau)$ *exists for the following elements of* $\mathscr{E}$:

(i) $H_{\infty,y}(P_\tau) = P_1$;
(ii) $H_{\infty,y}(aI) = a(-\infty)\chi_- I + a(+\infty)\chi_+ I$ *for* $a \in PC(\dot{\mathbb{R}})$;
(iii) $H_{\infty,y}(W^0(b)) = b(y^-)W^0(\chi_-) + b(y^+)W^0(\chi_+)$ *for* $b \in PC_p$.

Lemma 6.6.8. *Let* $x \in \mathbb{R}$. *The strong limit* $H_{x,\infty}(A_\tau)$ *exists for the following elements of* $\mathscr{E}$:

(i) $H_{x,\infty}(P_\tau) = I$;
(ii) $H_{x,\infty}(aI) = a(x^-)\chi_- I + a(x^+)\chi_+ I$ *for* $a \in PC(\dot{\mathbb{R}})$;
(iii) $H_{x,\infty}(W^0(b)) = b(-\infty)W^0(\chi_-) + b(+\infty)W^0(\chi_+)$ *for* $b \in PC_p$.

Indeed, these assertions are easy to check for the sequence (P_τ), and they were shown in Propositions 5.4.1 and 5.4.3 for the other sequences.

Corollary 6.6.9. $\mathscr{A}\big(PC(\dot{\mathbb{R}}), PC_p, (P_\tau)\big)$ *is a closed subalgebra of the algebra* $\mathscr{F}$.

Let us emphasize that for a pure finite sections sequence $(A_\tau) = (P_\tau A P_\tau + Q_\tau)$ with (A) a constant sequence in $\mathscr{F}$, the strong limits are given by

$$W_{-1}(A_\tau) = \chi_+ W_{-1}(A)\chi_+ I + \chi_- I, \quad W_0(A_\tau) = A, \quad W_1(A_\tau) = \chi_- W_1(A)\chi_- I + \chi_+ I$$

---

[1] We write $W_0(A_\tau)$ and not $W_0((A_\tau))$ to make the notation less heavy, but remember that all homomorphisms act on sequences, and not on particular operators.

and

$$\mathsf{H}_{\infty,y}(A_\tau) = P_1 \mathsf{H}_{\infty,y}(A) P_1 + Q_1, \quad \mathsf{H}_{x,\infty}(A_\tau) = \mathsf{H}_{x,\infty}(A)$$

for all $x, y \in \mathbb{R}$.

6.6.2 Localization

The algebra $\mathscr{F}/\mathscr{J}$ has a trivial center, thus, Allan's local principle is not helpful to study invertibility in this algebra. Therefore we are going to look for a subalgebra $\mathscr{F}_0$ of $\mathscr{F}$ for which $\mathscr{F}_0/\mathscr{G}$ is inverse-closed in $\mathscr{F}/\mathscr{G}$, which contains the ideal $\mathscr{J}$, and which has the property that the center of $\mathscr{F}_0/\mathscr{J}$ includes all cosets $(fI) + \mathscr{J}$ and $(W^0(g)) + \mathscr{J}$ with $f \in C(\dot{\mathbb{R}})$ and $g \in C_p$. Note that the inverse-closedness of $\mathscr{F}_0/\mathscr{G}$ in $\mathscr{F}/\mathscr{G}$ and, thus, in $\mathscr{E}/\mathscr{G}$ is needed to guarantee that the invertibility in $\mathscr{F}_0/\mathscr{G}$ is still equivalent to the stability.

Exercise 1.2.14 offers a way to introduce a subalgebra with the desired properties. Following this exercise, we consider the set $\mathscr{F}_0$ of all sequences in $\mathscr{F}$ which commute with all constant sequences (fI) and $(W^0(g))$ where $f \in C(\dot{\mathbb{R}})$ and $g \in C_p$ modulo sequences in the ideal $\mathscr{J}$.

Proposition 6.6.10.
  (i) *The set $\mathscr{F}_0$ is a closed subalgebra of $\mathscr{F}$ and contains the ideal $\mathscr{J}$.*
  (ii) *The mappings W_i with $i \in \{-1, 0, 1\}$, $\mathsf{H}_{\infty,y}$ with $y \in \mathbb{R}$, and $\mathsf{H}_{x,\infty}$ with $x \in \mathbb{R}$ act as bounded homomorphisms on $\mathscr{F}_0$. The ideal $\mathscr{G}$ of $\mathscr{F}$ lies in the kernel of each these homomorphisms, and the ideal $\mathscr{J}$ lies in the kernel of each of the H-homomorphisms.*
  (iii) *The algebra $\mathscr{F}_0$ is inverse-closed in $\mathscr{E}$, and the algebra $\mathscr{F}_0/\mathscr{G}$ is inverse-closed in $\mathscr{E}/\mathscr{G}$.*

By assertion (i), the lifting theorem applies to study invertibility in $\mathscr{F}_0/\mathscr{G}$.

Theorem 6.6.11. *Let $\mathbf{A} = (A_\tau) \in \mathscr{F}_0$. The sequence $\mathbf{A}$ is stable if and only if the operators $\mathsf{W}_{-1}(\mathbf{A})$, $\mathsf{W}_0(\mathbf{A})$ and $\mathsf{W}_1(\mathbf{A})$ are invertible in $\mathscr{L}(L^p(\mathbb{R}))$ and if the coset $\mathbf{A} + \mathscr{J}$ is invertible in the quotient algebra $\mathscr{F}_0^{\mathscr{J}} := \mathscr{F}_0/\mathscr{J}$.*

The algebra $\mathscr{F}_0$ is still large enough to contain all sequences that interest us.

Proposition 6.6.12. $\mathscr{A}\big(PC(\dot{\mathbb{R}}), PC_p, (P_\tau)\big)$ *is a closed subalgebra of $\mathscr{F}_0$.*

Proof. We have to show that the generators (aI) with $a \in PC(\dot{\mathbb{R}})$, $(W^0(b))$ with $b \in PC_p$, and (P_τ) of $\mathscr{A}\big(PC(\dot{\mathbb{R}}), PC_p, (P_\tau)\big)$ commute with the constant sequences (fI) where $f \in C(\dot{\mathbb{R}})$ and $(W^0(g))$ where $g \in C_p$ modulo sequences in $\mathscr{J}$. For the generators which are constant sequences this follows immediately from Proposition 5.3.1. For instance, one has

$$(fI)(W^0(b)) - (W^0(b))(fI) = (fW^0(b) - W^0(b)fI),$$

which is a constant sequence with a compact entry by Proposition 5.3.1 (ii). Hence, this sequence is in $\mathscr{J}$.

It is evident that (P_τ) commutes with (cI), and it remains to verify that the commutator

$$(P_\tau)(W^0(g)) - (W^0(g))(P_\tau)$$

belongs to $\mathscr{J}$ for every multiplier $g \in C_p$. Write

$$
\begin{aligned}
(P_\tau W^0(g) - W^0(g)P_\tau) &= (P_\tau W^0(g)Q_\tau - Q_\tau W^0(g)P_\tau) \\
&= (P_\tau \chi_+ W^0(g)\chi_+ Q_\tau - Q_\tau \chi_+ W^0(g)\chi_+ P_\tau) \\
&\quad + (P_\tau \chi_+ W^0(g)\chi_- Q_\tau - Q_\tau \chi_+ W^0(g)\chi_- P_\tau) \\
&\quad + (P_\tau \chi_- W^0(g)\chi_+ Q_\tau - Q_\tau \chi_- W^0(g)\chi_+ P_\tau) \\
&\quad + (P_\tau \chi_- W^0(g)\chi_- Q_\tau - Q_\tau \chi_- W^0(g)\chi_- P_\tau).
\end{aligned}
$$

The sequences in the second and third line of the right-hand side of this equality belong to the ideal $\mathscr{G}$ since the operators $\chi_\pm W^0(g)\chi_\mp I$ are compact by Proposition 5.3.1 (ii) and since the Q_τ converge strongly to zero. The sequence in last line can be treated as the sequence in the first line. So we are left to verify that

$$(P_\tau \chi_+ W^0(g)\chi_+ Q_\tau - Q_\tau \chi_+ W^0(g)\chi_+ P_\tau) \in \mathscr{J}.$$

Write this sequence as

$$
\begin{aligned}
&\left(\chi_{[0,\tau]} W^0(g)\chi_{[\tau,\infty[} I - \chi_{[\tau,\infty[} W^0(g)\chi_{[0,\tau]} I\right) \\
&= \left(V_\tau \left(V_{-\tau}\left(\chi_{[0,\tau]} W^0(g)\chi_{[\tau,\infty[} I - \chi_{[\tau,\infty[} W^0(g)\chi_{[0,\tau]} I\right) V_\tau\right) V_{-\tau}\right) \\
&= \left(V_\tau \left(\chi_{[-\tau,0]} W^0(g)\chi_{[0,\infty[} I - \chi_{[0,\infty[} W^0(g)\chi_{[-\tau,0]} I\right) V_{-\tau}\right) \\
&= \left(V_\tau \left(\chi_{[-\tau,0]} \chi_- W^0(g)\chi_+ I - \chi_+ W^0(g)\chi_- \chi_{[-\tau,0]} I\right) V_{-\tau}\right).
\end{aligned}
$$

Since the operators $\chi_\pm W^0(g)\chi_\mp I$ are compact and $\chi_{[-\tau,0]} I \to \chi_- I$ strongly as $\tau \to \infty$, we conclude that the sequence in the last line of this equality is of the form $(V_\tau K V_{-\tau}) + (G_\tau)$ with K compact and $(G_\tau) \in \mathscr{G}$. Hence, this sequence is in $\mathscr{J}$. ∎

We proceed with localization. Repeating arguments from the proof of Proposition 6.4.5, one can easily check that the algebra generated by the cosets of constant sequences $(fI) + \mathscr{J}$ and $(W^0(g)) + \mathscr{J}$ with $f \in C(\dot{\mathbb{R}})$ and $g \in C_p$ is isomorphic to the subalgebra of the Calkin algebra which is generated by $fI + \mathscr{K}$ and $W^0(g) + \mathscr{K}$. From Section 5.6 we infer that the maximal ideal space of the latter algebra and, thus, of our present central subalgebra, is homeomorphic to the subset $(\dot{\mathbb{R}} \times \{\infty\}) \cup (\{\infty\} \times \dot{\mathbb{R}})$ of the torus $\dot{\mathbb{R}} \times \dot{\mathbb{R}}$.

Given $(s,t) \in (\dot{\mathbb{R}} \times \{\infty\}) \cup (\{\infty\} \times \dot{\mathbb{R}})$, let $\mathscr{I}_{s,t}$ denote the smallest, closed, two-sided ideal of the quotient algebra $\mathscr{F}_0^{\mathscr{J}}$ which contains the maximal ideal corresponding to the point (s,t), and let $\Phi_{s,t}^{\mathscr{J}}$ refer to the canonical homomorphism from $\mathscr{F}_0$ onto the quotient algebra $\mathscr{F}_{s,t}^{\mathscr{J}} := \mathscr{F}_0^{\mathscr{J}} / \mathscr{I}_{s,t}$. In order not to burden the notation, we write $\Phi_{s,t}^{\mathscr{J}}(A_\tau)$ instead of $\Phi_{s,t}^{\mathscr{J}}((A_\tau) + \mathscr{J})$ for every sequence $(A_\tau) \in \mathscr{F}_0$.

Let $(s,t) \in (\mathbb{R} \times \{\infty\}) \cup (\{\infty\} \times \mathbb{R})$. One cannot expect that the local algebra $\mathscr{F}_{s,t}^{\mathscr{J}}$ can be identified completely. But we will be able to identify the smallest closed subalgebra $\mathscr{A}_{s,t}^{\mathscr{J}}$ of $\mathscr{F}_{s,t}^{\mathscr{J}}$ which contains all cosets $(P_\tau) + \mathscr{I}_{s,t}$, $(aI) + \mathscr{I}_{s,t}$ with $a \in PC(\mathbb{\dot{R}})$ and $(W^0(b)) + \mathscr{I}_{s,t}$ with $b \in PC_p$, and this identification will be sufficient for our purposes. We will identify the algebras $\mathscr{A}_{s,t}^{\mathscr{J}}$ by means of the family of H-homomorphisms. Note that, by Proposition 6.6.10 (ii), the operators $H_{\infty,y}(\mathbf{A})$ and $H_{x,\infty}(\mathbf{A})$ depend only on the coset of the sequence $\mathbf{A}$ modulo $\mathscr{J}$. Thus, the quotient homomorphisms

$$\mathbf{A} + \mathscr{J} \mapsto H_{\infty,y}(\mathbf{A}) \quad \text{and} \quad \mathbf{A} + \mathscr{J} \mapsto H_{x,\infty}(\mathbf{A})$$

are well defined. We denote them again by $H_{\infty,y}$ and $H_{x,\infty}$, respectively.

6.6.3 The local algebras

We start with describing the local algebras $\mathscr{A}_{s,\infty}^{\mathscr{J}}$.

Theorem 6.6.13. Let $s \in \mathbb{R}$. The algebra $\mathscr{A}_{s,\infty}^{\mathscr{J}}$ is isometrically isomorphic to the subalgebra $\mathrm{alg}\{I, \chi_+ I, W^0(\chi_+)\}$ of $\mathscr{L}(L^p(\mathbb{R}))$, and the isomorphism is given by

$$\Phi_{s,\infty}^{\mathscr{J}}(\mathbf{A}) \mapsto H_{s,\infty}(\mathbf{A}). \tag{6.25}$$

Proof. By definition, $\mathscr{I}_{s,\infty}$ is the smallest two-sided ideal of $\mathscr{F}_0^{\mathscr{J}}$ which contains the cosets $(fW^0(g)) + \mathscr{J}$ with $f(s) = 0$ and $g(\infty) = 0$. From Lemma 6.6.8 we infer that $H_{s,\infty}(\mathscr{I}_{s,\infty}) = 0$. Thus, the homomorphism $H_{s,\infty}$ is well defined on the quotient algebra $\mathscr{A}_{s,\infty}^{\mathscr{J}}$. The same lemma also implies that $H_{s,\infty}$ maps $\mathscr{A}_{s,\infty}^{\mathscr{J}}$ to $\mathrm{alg}\{I, \chi_+ I, W^0(\chi_+)\}$.

We claim that the homomorphism $H_{s,\infty} : \mathscr{A}_{s,\infty}^{\mathscr{J}} \to \mathrm{alg}\{I, \chi_+ I, W^0(\chi_+)\}$ is an isometry. This will follow once we have shown that the identity

$$\Phi_{s,\infty}^{\mathscr{J}}(\mathbf{A}) = \Phi_{s,\infty}^{\mathscr{J}}(V_s H_{s,\infty}(\mathbf{A}) V_{-s}) \tag{6.26}$$

holds for all generators of the algebra $\mathscr{A}(PC(\mathbb{\dot{R}}), PC_p, (P_\tau))$ in place of the sequence $\mathbf{A}$. Note that the right-hand side of (6.26) makes sense since the constant sequence $(V_s H_{s,\infty}(\mathbf{A}) V_{-s})$ belongs to the algebra $\mathscr{F}_0$ by Proposition 6.6.12.

For the generators (aI) and $(W^0(b))$ with $a \in PC(\mathbb{\dot{R}})$ and $b \in PC_p$ of the algebra $\mathscr{A}(PC(\mathbb{\dot{R}}), PC_p, (P_\tau))$, the identity (6.26) can be proved by analogy with Theorem 5.5.4. With the generating sequence (P_τ) in place of $\mathbf{A}$, the right-hand side of (6.26) is the identity element. So we have to show that $\Phi_{s,\infty}^{\mathscr{J}}(P_\tau)$ is the identity element of the local algebra.

Choose $y \in \mathbb{R}$ greater than $|s|$, and let f_s be a continuous function supported on the interval $]-y, y[$ such that $f_s(s) = 1$. Since $\Phi^{\mathscr{J}}_{s,\infty}(f_s I)$ is the identity in the local algebra, we have

$$\Phi^{\mathscr{J}}_{s,\infty}(Q_\tau) = \Phi^{\mathscr{J}}_{s,\infty}(f_s I)\Phi^{\mathscr{J}}_{s,\infty}(Q_\tau) = \Phi^{\mathscr{J}}_{s,\infty}(f_s Q_\tau).$$

Since $f_s Q_\tau = 0$ for τ sufficiently large, the sequence $(f_s Q_\tau)_{\tau>0}$ belongs to the ideal $\mathscr{G}$, which implies $\Phi^{\mathscr{J}}_{s,\infty}(Q_\tau) = 0$. Hence, $\Phi^{\mathscr{J}}_{s,\infty}(P_\tau) = \Phi^{\mathscr{J}}_{s,\infty}(I - Q_\tau)$ is the identity element. ∎

The previous theorem implies that, for sequences $\mathbf{A} \in \mathscr{A}\big(PC(\dot{\mathbb{R}}), PC_p, (P_\tau)\big)$, the coset $\Phi^{\mathscr{J}}_{s,\infty}(\mathbf{A})$ is invertible in the local algebra $\mathscr{A}^{\mathscr{J}}_{s,\infty}$ if and only if the operator $H_{s,\infty}(\mathbf{A})$ is invertible in $\text{alg}\{I, \chi_+ I, W^0(\chi_+)\}$. Of course, one would prefer to check the invertibility of the operator $H_{s,\infty}(\mathbf{A})$ in $\mathscr{L}(L^p(\mathbb{R}))$, not in $\text{alg}\{I, \chi_+ I, W^0(\chi_+)\}$. In the present setting, this causes no problem since the algebra $\text{alg}\{I, \chi_+ I, W^0(\chi_+)\}$ is inverse-closed in $\mathscr{L}(L^p(\mathbb{R}))$ by Corollary 4.2.20. The following proposition and its proof show how the desired invertibility condition can be derived without any a priori information on the inverse-closedness of the local algebras.

Proposition 6.6.14. *Let* $\mathbf{A} \in \mathscr{A}\big(PC(\dot{\mathbb{R}}), PC_p, (P_\tau)\big)$. *Then the coset* $\Phi^{\mathscr{J}}_{s,\infty}(\mathbf{A})$ *is invertible in the local algebra* $\mathscr{F}^{\mathscr{J}}_{s,\infty}$ *if and only if the operator* $H_{s,\infty}(\mathbf{A})$ *is invertible in* $\mathscr{L}(L^p(\mathbb{R}))$.

Proof. For $s \in \mathbb{R}$, let $\mathscr{D}_{s,\infty}$ denote the set of all operators $A \in \mathscr{L}(L^p(\mathbb{R}))$ with the property that the constant sequence $(V_s A V_{-s})$ belongs to the algebra $\mathscr{F}_0$. One easily checks that $\mathscr{D}_{s,\infty}$ is a closed subalgebra of $\mathscr{L}(L^p(\mathbb{R}))$. Moreover, $\mathscr{D}_{s,\infty}$ is inverse-closed in $\mathscr{L}(L^p(\mathbb{R}))$, which can be seen as follows.

Let $A \in \mathscr{D}_{s,\infty}$ be invertible in $\mathscr{L}(L^p(\mathbb{R}))$. The constant sequence $(V_s A V_{-s})$ is invertible in the algebra $\mathscr{E}$ of all bounded sequences, and its inverse is the sequence $(V_s A^{-1} V_{-s})$. Since $(V_s A V_{-s}) \in \mathscr{F}_0$ by hypothesis, and since $\mathscr{F}_0$ is inverse-closed in $\mathscr{E}$ by Proposition 6.6.10 (iii), we conclude that $(V_s A^{-1} V_{-s}) \in \mathscr{F}_0$. Hence, $A^{-1} \in \mathscr{D}_{s,\infty}$.

Now let $\mathbf{A} \in \mathscr{A}\big(PC(\dot{\mathbb{R}}), PC_p, (P_\tau)\big)$. If the coset $\Phi^{\mathscr{J}}_{s,\infty}(\mathbf{A})$ is invertible in $\mathscr{F}^{\mathscr{J}}_{s,\infty}$, then $H_{s,\infty}(\mathbf{A})$ is invertible in $\mathscr{L}(L^p(\mathbb{R}))$, since $H_{s,\infty}$ acts as a homomorphism on that local algebra. Conversely, let $H_{s,\infty}(\mathbf{A})$ be invertible in $\mathscr{L}(L^p(\mathbb{R}))$. We know already that $H_{s,\infty}(\mathbf{A})$ belongs to the algebra $\text{alg}\{I, \chi_+ I, W^0(\chi_+)\}$, and one easily checks that this algebra is contained in $\mathscr{D}_{s,\infty}$. By the inverse-closedness of $\mathscr{D}_{s,\infty}$, the operator $H_{s,\infty}(\mathbf{A})$ possesses an inverse in $\mathscr{D}_{s,\infty}$. Let B denote this inverse. From $BH_{s,\infty}(\mathbf{A}) = I$ we get

$$(V_s B V_{-s})(V_s H_{s,\infty}(\mathbf{A})V_{-s}) = (I). \qquad (6.27)$$

Note that the sequences in (6.27) are constant. Since the operators B and $H_{s,\infty}(\mathbf{A})$ belong to $\mathscr{D}_{s,\infty}$, it is also clear that the sequences in (6.27) belong to $\mathscr{F}_0$. Hence, one can apply the local homomorphism $\Phi^{\mathscr{J}}_{s,\infty}$ to both sides of (6.27), which gives

$$\Phi_{s,\infty}^{\mathscr{J}}(V_s B V_{-s})\Phi_{s,\infty}^{\mathscr{J}}(V_s \mathsf{H}_{s,\infty}(\mathbf{A})V_{-s}) = \Phi_{s,\infty}^{\mathscr{J}}(I).$$

From (6.26) we conclude that $\Phi_{s,\infty}^{\mathscr{J}}(\mathbf{A})$ is invertible in $\mathscr{F}_{s,\infty}^{\mathscr{J}}$. ∎

The following is an immediate consequence of Theorem 6.6.13, Proposition 6.6.14 and the well-known inverse-closedness of the algebra $\mathrm{alg}\{I,\chi_+ I,W^0(\chi_+)\}$ in $\mathscr{L}(L^p(\mathbb{R}))$.

Corollary 6.6.15. *The local algebra $\mathscr{A}_{s,\infty}^{\mathscr{J}}$ is inverse-closed in $\mathscr{F}_{s,\infty}^{\mathscr{J}}$.*

Next we are going to examine the local algebras $\mathscr{A}_{\infty,t}^{\mathscr{J}}$.

Theorem 6.6.16. *Let $t \in \mathbb{R}$. The algebra $\mathscr{A}_{\infty,t}^{\mathscr{J}}$ is isometrically isomorphic to the subalgebra $\mathrm{alg}\{I,\chi_+ I,P_1,W^0(\chi_+)\}$ of $\mathscr{L}(L^p(\mathbb{R}))$, and the isomorphism is given by*

$$\Phi_{\infty,t}^{\mathscr{J}}(\mathbf{A}) \mapsto \mathsf{H}_{\infty,t}(\mathbf{A}). \tag{6.28}$$

Proof. It follows from Lemma 6.6.7, that the operator $\mathsf{H}_{\infty,t}(\mathbf{A})$ belongs to the algebra $\mathrm{alg}\{I,\chi_+ I,P_1,W^0(\chi_+)\}$ and that this operator depends on the coset $\Phi_{\infty,t}^{\mathscr{J}}(\mathbf{A})$ of the sequence $\mathbf{A}$ only. Thus, there is a well-defined homomorphism

$$\mathscr{A}_{\infty,y}^{\mathscr{J}} \to \mathrm{alg}\{I,\chi_+ I,P_1,W^0(\chi_+)\}, \quad \Phi_{\infty,t}^{\mathscr{J}}(\mathbf{A}) \mapsto \mathsf{H}_{\infty,t}(\mathbf{A})$$

which we denote by $\mathsf{H}_{\infty,y}$ again. It will follow that this homomorphism is an isometry once we have verified the identity

$$\Phi_{\infty,t}^{\mathscr{J}}(\mathbf{A}) = \Phi_{\infty,t}^{\mathscr{J}}(U_{-t}Z_\tau \mathsf{H}_{\infty,t}(\mathbf{A})Z_\tau^{-1}U_t) \tag{6.29}$$

for all sequences $\mathbf{A}$ in $\mathscr{A}\big(PC(\dot{\mathbb{R}}),PC_p,(P_\tau)\big)$. This is done as in the proof of Theorem 5.5.3 for the constant generating sequences of $\mathscr{A}\big(PC(\dot{\mathbb{R}}),PC_p,(P_\tau)\big)$, and it is evident for the sequence (P_τ). ∎

The previous theorem can be completed as follows.

Proposition 6.6.17. *Let $\mathbf{A} \in \mathscr{A}\big(PC(\dot{\mathbb{R}}),PC_p,(P_\tau)\big)$. Then the coset $\Phi_{\infty,t}^{\mathscr{J}}(\mathbf{A})$ is invertible in the local algebra $\mathscr{F}_{\infty,t}^{\mathscr{J}}$ if and only if the operator $\mathsf{H}_{\infty,t}(\mathbf{A})$ is invertible in $\mathscr{L}(L^p(\mathbb{R}))$.*

Proof. The proof proceeds as that of Proposition 6.6.14. For $t \in \mathbb{R}$, introduce the algebra $\mathscr{D}_{\infty,t}$ of all operators $A \in \mathscr{L}(L^p(\mathbb{R}))$ with the property that the sequence $(U_{-t}Z_\tau A Z_\tau^{-1}U_t)_{\tau>0}$ belongs to the algebra $\mathscr{F}_0$. Again one easily checks that $\mathscr{D}_{\infty,t}$ is an inverse-closed subalgebra of $\mathscr{L}(L^p(\mathbb{R}))$ and that the algebra $\mathrm{alg}\{I,\chi_+ I,P_1,W^0(\chi_+)\}$ is contained in $\mathscr{D}_{\infty,t}$. ∎

Corollary 6.6.18. *The algebra $\mathrm{alg}\{I,\chi_+ I,P_1,W^0(\chi_+)\}$ is inverse-closed in the algebra $\mathscr{L}(L^p(\mathbb{R}))$ if and only if the local algebra $\mathscr{A}_{\infty,t}^{\mathscr{J}}$ is inverse-closed in $\mathscr{F}_{\infty,t}^{\mathscr{J}}$.*

Our final goal is the local algebra $\mathscr{A}_{\infty,\infty}^{\mathscr{I}}$. It is easy to see that this algebra is generated by the identity element and by the three projections $\Phi_{\infty,\infty}^{\mathscr{I}}(\chi_+ I)$, $\Phi_{\infty,\infty}^{\mathscr{I}}(W^0(\chi_+))$, and $\Phi_{\infty,\infty}^{\mathscr{I}}(P_\tau)$. The following proposition shows that this algebra has a non-trivial center.

Proposition 6.6.19. *The projection $\Phi_{\infty,\infty}^{\mathscr{I}}(\chi_+ I)$ belongs to the center of $\mathscr{A}_{\infty,\infty}^{\mathscr{I}}$.*

Proof. One only has to check the relation

$$\Phi_{\infty,\infty}^{\mathscr{I}}(W^0(\chi_+)\chi_+ I) = \Phi_{\infty,\infty}^{\mathscr{I}}(\chi_+ W^0(\chi_+)). \tag{6.30}$$

Choose a continuous and monotonically increasing function $\chi_+' \in C_p$ which takes the values 0 at $-\infty$ and 1 at $+\infty$. Then, clearly,

$$\Phi_{\infty,\infty}^{\mathscr{I}}(\chi_+ I) = \Phi_{\infty,\infty}^{\mathscr{I}}(\chi_+' I) \quad \text{and} \quad \Phi_{\infty,\infty}^{\mathscr{I}}(W^0(\chi_+)) = \Phi_{\infty,\infty}^{\mathscr{I}}(W^0(\chi_+')).$$

Since $W^0(\chi_+')\chi_+' I - \chi_+' W^0(\chi_+')$ is compact by Proposition 5.3.1 (ii) 3., the equality (6.30) follows.   ∎

Proposition 6.6.19 implies that the local algebra $\mathscr{A}_{\infty,\infty}^{\mathscr{I}}$ splits into the direct sum

$$\mathscr{A}_{\infty,\infty}^{\mathscr{I}} = \mathscr{A}_{\infty,\infty}^{+} \dotplus \mathscr{A}_{\infty,\infty}^{-} \tag{6.31}$$

where $\mathscr{A}_{\infty,\infty}^{\pm} := \Phi_{\infty,\infty}^{\mathscr{I}}(\chi_\pm I)\mathscr{A}_{\infty,\infty}^{\mathscr{I}}\Phi_{\infty,\infty}^{\mathscr{I}}(\chi_\pm I)$. The algebras $\mathscr{A}_{\infty,\infty}^{\pm}$ are unital, and the cosets $\Phi_{\infty,\infty}^{\mathscr{I}}(\chi_\pm I)$ can be considered as their identity elements. It is evident that the invertibility of the coset $\Phi_{\infty,\infty}^{\mathscr{I}}(\mathbf{A})$ in $\mathscr{F}_{\infty,\infty}^{\mathscr{I}}$ for $\mathbf{A} = (A_\tau) \in \mathscr{A}$ is equivalent to the invertibility of the two cosets $\Phi_{\infty,\infty}^{\mathscr{I}}(\chi_\pm A_\tau \chi_\pm I)$ in the algebras $\mathscr{A}_{\infty,\infty}^{\pm}$, respectively. (One obtains the same result by localizing the algebra $\mathscr{A}_{\infty,\infty}^{\mathscr{I}}$ over its central subalgebra described in Proposition 6.6.19; see Exercise 2.2.6.)

Consider the algebra $\mathscr{A}_{\infty,\infty}^{+}$. It is another consequence of Proposition 6.6.19 that this algebra is generated by the two idempotent elements $p := \Phi_{\infty,\infty}^{\mathscr{I}}(P_\tau\chi_+ I)$ and $r := \Phi_{\infty,\infty}^{\mathscr{I}}(W^0(\chi_+)\chi_+ I)$ and by the identity element $e := \Phi_{\infty,\infty}^{\mathscr{I}}(\chi_+ I)$. Thus, the local algebra $\mathscr{A}_{\infty,\infty}^{+}$ is subject to the two projections Theorem 3.1.4. To apply this theorem, we have to determine the spectrum of the element

$$X := prp + (e-p)(e-r)(e-p) = \Phi_{\infty,\infty}^{\mathscr{I}}\left(P_\tau W^0(\chi_+)P_\tau\chi_+ I + Q_\tau W^0(\chi_-)Q_\tau\chi_+ I\right)$$

in the local algebra $\mathscr{A}_{\infty,\infty}^{+}$. The following simple lemma will be useful. Let $\mathscr{H}$ denote the smallest closed subalgebra of $\mathscr{E}$ which contains the sequence (P_τ) and all constant sequences of homogeneous operators in $\mathscr{L}(L_p(\mathbb{R}))$.

Lemma 6.6.20. *Let $(B_\tau) \in \mathscr{H}$. Then (B_τ) is invertible in $\mathscr{E}$ if and only if $(B_\tau) + \mathscr{G}$ is invertible in $\mathscr{E}/\mathscr{G}$ if and only if B_1 is invertible in $\mathscr{L}(L_p(\mathbb{R}))$.*

Indeed, $Z_\tau^{-1}B_\tau Z_\tau = B_1$ for every $\tau > 0$.

Let I be the closed interval between $1/p$ and $1 - 1/p$ and define $\mathfrak{L}_p := \cup_{\alpha \in I} \mathfrak{A}_\alpha$ where $\mathfrak{A}_\alpha$ is defined in (3.44) (see Figure 6.1).

Fig. 6.1 The lens $\mathfrak{L}_p$ for $p = 3$ (or $p = 3/2$). If $p = 2$ it would be just the straight line between the points 0 and 1.

Proposition 6.6.21. *The spectrum of the element X in each of the algebras $\mathscr{F}_{\infty,\infty}^{\mathscr{I}}$, $\mathscr{A}_{\infty,\infty}^{\mathscr{I}}$ and $\mathscr{A}_{\infty,\infty}^+$ is $\mathfrak{L}_p$.*

Proof. Our first goal is to prove that the spectrum of X in $\mathscr{F}_{\infty,\infty}^{\mathscr{I}}$ is $\mathfrak{L}_p$. It is easy to see that this fact will follow once we have shown that both the spectrum of prp and the spectrum of $(e - p)(e - r)(e - p)$ coincides with $\mathfrak{L}_p$. Since both spectra can be determined in the same way, we will only demonstrate that the spectrum of prp is $\mathfrak{L}_p$. Let $\sigma_{s,t}(\mathbf{A})$ refer to the spectrum of the coset $\Phi_{s,t}^{\mathscr{I}}(\mathbf{A})$ in $\mathscr{F}_{s,t}^{\mathscr{I}}$. We claim that

$$\sigma_{\infty,\infty}(P_\tau \chi_+ W^0(\chi_+) \chi_+ P_\tau) = \mathfrak{L}_p. \tag{6.32}$$

By Lemma 6.6.20, the spectrum of the coset

$$(P_\tau \chi_+ W^0(\chi_+) \chi_+ P_\tau) + \mathscr{G} \tag{6.33}$$

in $\mathscr{E}/\mathscr{G}$ is equal to the spectrum of the operator

$$P_1 \chi_+ W^0(\chi_+) \chi_+ P_1 = \chi_{[0,1]} W(\chi_+) \chi_{[0,1]} I$$

on $L^p([0,1])$. The spectrum of this operator is the lentiform domain $\mathfrak{L}_p$ as we infer from [120, Chap. 6, Theorem 6.2]. Since $\mathscr{F}_0/\mathscr{G}$ is inverse-closed in $\mathscr{E}/\mathscr{G}$ by Proposition 6.6.10 (iii), the spectrum of (6.33) is also $\mathfrak{L}_p$, which implies that the spectrum of prp in $\mathscr{F}_{\infty,\infty}^{\mathscr{I}}$ is indeed contained in $\mathfrak{L}_p$.

For the reverse inclusion, we argue as follows. Let a be a multiplier which is continuous on $\mathbb{R}$, has the one-sided limits $a(-\infty) = 0$ and $a(+\infty) = 1$, and which has the circular arc $\mathfrak{A}_{1/p}$ as its range. Then, evidently,

$$\sigma_{\infty,\infty}(P_\tau \chi_+ W^0(\chi_+) \chi_+ P_\tau) = \sigma_{\infty,\infty}(P_\tau \chi_+ W^0(a) \chi_+ P_\tau). \tag{6.34}$$

From [21, Theorem 9.46] we infer that the finite sections method applies to the Wiener-Hopf operator $W(a - \lambda)$ if and only if $\lambda \notin \mathfrak{L}_p$. Thus, the spectrum of the coset $(P_\tau \chi_+ W^0(a) \chi_+ P_\tau) + \mathscr{G}$ is $\mathfrak{L}_p$.

Next we show that the spectrum of $(P_\tau \chi_+ W^0(a) \chi_+ P_\tau) + \mathscr{J}$ is also $\mathfrak{L}_p$. Clearly, this spectrum must be contained in $\mathfrak{L}_p$. Now let $\lambda \in \mathbb{C}$ be such that

$$(P_\tau \chi_+ W^0(a) \chi_+ P_\tau - \lambda I) + \mathscr{J}$$

is invertible. Thus, there are a sequence $(B_\tau) \in \mathscr{F}_0$ and a sequence $(J_\tau) \in \mathscr{J}$ such that

$$(P_\tau \chi_+ W^0(a) \chi_+ P_\tau - \lambda I)) B_\tau = I + J_\tau.$$

We multiply this equality from the left by Z_τ^{-1} and from the right by Z_τ to obtain

$$(\chi_{[0,1]} Z_\tau^{-1} W(a) Z_\tau \chi_{[0,1]} - \lambda I)(Z_\tau^{-1} B_\tau Z_\tau) = I + Z_\tau^{-1} J_\tau Z_\tau.$$

Letting τ *go to zero* we get

$$(\chi_{[0,1]} W(\chi_+) \chi_{[0,1]} - \lambda I) B = I$$

with a certain operator B (see Proposition 5.4.3 (ii)). The invertibility of $\chi_{[0,1]} W(\chi_+) \chi_{[0,1]} - \lambda I$ from the other side follows analogously. Thus, λ does not belong to the spectrum of $\chi_{[0,1]} W(\chi_+) \chi_{[0,1]}$ on $L^p(\mathbb{R})$, which is $\mathfrak{L}_p$.

Abbreviate $\mathbf{A} := (P_\tau \chi_+ W^0(a) \chi_+ P_\tau)$. Since the spectrum if $\mathbf{A} + \mathscr{J}$ is $\mathfrak{L}_p$, Allan's local principle implies that

$$\bigcup_{s \in \mathbb{R}} \sigma_{s,\infty}(\mathbf{A}) \bigcup \bigcup_{t \in \mathbb{R}} \sigma_{\infty,t}(\mathbf{A}) \bigcup \sigma_{\infty,\infty}(\mathbf{A}) = \mathfrak{L}_p. \tag{6.35}$$

The local spectra $\sigma_{s,\infty}(\mathbf{A})$ and $\sigma_{\infty,t}(\mathbf{A})$ (with finite s,t) can be determined via Propositions 6.6.14 and 6.6.17. They imply that, in each case, the local spectrum is contained in the circular arc $\mathfrak{A}_{1/p}$, which is contained in the boundary of $\mathfrak{L}_p$. Hence, and by (6.35), the interior of $\mathfrak{L}_p$ must be contained in $\sigma_{\infty,\infty}(\mathbf{A})$. But then all of $\mathfrak{L}_p$ is contained in that local spectrum, which proves the claim (6.32).

Now the proof of the proposition can be completed as follows. As we have just seen, the spectrum of X in $\mathscr{F}_{\infty,\infty}^{\mathscr{J}}$ is $\mathfrak{L}_p$. Since $\mathfrak{L}_p$ is simply connected in $\mathbb{C}$, the spectrum of X in the subalgebra $\mathscr{A}_{\infty,\infty}^{\mathscr{J}}$ of $\mathscr{F}_{\infty,\infty}^{\mathscr{J}}$ also coincides with $\mathfrak{L}_p$. Finally, the finite sum decomposition (6.31) and the fact that X belongs to the summand $\mathscr{A}_{\infty,\infty}^+$ in this sum imply that the spectrum of X in $\mathscr{A}_{\infty,\infty}^+$ also coincides with $\mathfrak{L}_p$. (The formal argument is that if A is an element of an algebra $\mathscr{A}$ with identity I and if P is an idempotent in that algebra such that $(I-P)A = A(I-P) = 0$, then $A - \lambda I = PAP - \lambda P - \lambda(I-P)$ is invertible in $\mathscr{A}$ if and only if $PAP - \lambda P$ is invertible in $P \mathscr{A} P$ and $\lambda \neq 0$.) ∎

A similar description holds for the algebra $\mathscr{A}_{\infty,\infty}^-$.

The following proposition summarizes the results obtained for the case $(s,t) = (\infty,\infty)$. Define functions $\widehat{P}, \widehat{p}, \widehat{r} : \mathfrak{L}_p \to \mathbb{C}^{4 \times 4}$ by

$$\widehat{P}:x\mapsto\begin{bmatrix}1&0&0&0\\0&1&0&0\\0&0&0&0\\0&0&0&0\end{bmatrix},\quad\widehat{p}:x\mapsto\begin{bmatrix}1&0&0&0\\0&0&0&0\\0&0&1&0\\0&0&0&0\end{bmatrix}$$

and

$$\widehat{r}:x\mapsto\begin{bmatrix}x&\sqrt{x(1-x)}&0&0\\\sqrt{x(1-x)}&1-x&0&0\\0&0&x&\sqrt{x(1-x)}\\0&0&\sqrt{x(1-x)}&1-x\end{bmatrix}.$$

Here $\sqrt{x(1-x)}$ stands for any complex number c with $c^2=x(1-x)$. Note that these 4×4 matrices have a 2×2-block diagonal structure, which reflects the decomposing property of the local algebra at (∞,∞).

Proposition 6.6.22.

(i) *The mapping* Ψ *which sends the local cosets* $\Phi^{\mathscr{J}}_{\infty,\infty}(\chi_+I)$, $\Phi^{\mathscr{J}}_{\infty,\infty}(P_\tau)$ *and* $\Phi^{\mathscr{J}}_{\infty,\infty}(W^0(\chi_+))$ *to the functions* $\widehat{P}$, $\widehat{p}$ *and* $\widehat{r}$ *extends to a homomorphism from the algebra* $\mathscr{A}^{\mathscr{J}}_{\infty,\infty}$ *into the algebra of all bounded,* 4×4 *matrix-valued functions on* $\mathfrak{L}_p$.

(ii) *Let* $\mathbf{A}\in\mathscr{A}\left(PC(\dot{\mathbb{R}}),PC_p,(P_\tau)\right)$. *Then the coset* $\Phi^{\mathscr{J}}_{\infty,\infty}(\mathbf{A})$ *is invertible in* $\mathscr{F}^{\mathscr{J}}_{\infty,\infty}$ *if and only if the associated function* $\Psi(\Phi^{\mathscr{J}}_{\infty,\infty})$ *is invertible.*

Note that the intersection of each of the intervals $]-\infty,0[$ and $]1,\infty[$ with the lens $\mathfrak{L}_p$ is empty. Hence, the values of the function $x\mapsto x(1-x)$ on $\mathfrak{L}_p$ do not meet the negative real axis $]-\infty,0[$. One can therefore choose the square roots $\sqrt{x(1-x)}$ in such a way that $\widehat{r}$ becomes a continuous function on $\mathfrak{L}_p$, and Ψ becomes a homomorphism into $C(\mathfrak{L}_p,\mathbb{C}^{4\times4})$.

6.6.4 The main result

Having identified all local algebras, we can now state the main result of this section. Write $\mathsf{H}_{\infty,\infty}(\mathbf{A})$ for the function $\Psi(\Phi^{\mathscr{J}}_{\infty,\infty})$. Recall also the definition of the algebra $\mathscr{A}:=\mathscr{A}(PC(\dot{\mathbb{R}}),PC_p,(P_\tau))$ as the smallest closed subalgebra of $\mathscr{E}$ which contains the constant sequences (aI) with $a\in PC(\dot{\mathbb{R}})$ and $(W^0(b))$ with $b\in PC_p$, the sequence (P_τ) and the ideal $\mathscr{G}$.

Theorem 6.6.23. *A sequence* $\mathbf{A}\in\mathscr{A}$ *is stable if and only if the operators* $\mathsf{W}_{-1}(\mathbf{A})$, $\mathsf{W}_0(\mathbf{A})$ *and* $\mathsf{W}_1(\mathbf{A})$ *and the operators* $\mathsf{H}_{s,\infty}(\mathbf{A})$ *and* $\mathsf{H}_{\infty,t}(\mathbf{A})$ *with* $s,t\in\mathbb{R}$ *are invertible in* $\mathscr{L}(L^p(\mathbb{R}))$ *and if the matrix function* $\mathsf{H}_{\infty,\infty}(\mathbf{A})$ *is invertible.*

Making Theorem 6.6.23 specific to the case when (A_τ) is a sequence of finite sections yields the following.

Theorem 6.6.24. *Let A be an operator in the smallest subalgebra of $\mathscr{L}(L^p(\mathbb{R}))$ which contains the operators aI with $a \in PC(\mathbb{\dot{R}})$ and $W^0(b)$ with $b \in PC_p$. Then the finite sections method*

$$(P_\tau A P_\tau + Q_\tau)u_\tau = f$$

applies to the operator A if and only if the operators

$$\chi_+ W_{-1}(A)\chi_+ I + \chi_- I, \quad A, \quad and \quad \chi_- W_1(A)\chi_- I + \chi_+ I$$

and the operators

$$\mathsf{H}_{s,\infty}(A) \quad and \quad P_1 \mathsf{H}_{\infty,t}(A)P_1 + Q_1 \quad with\ s, t \in \mathbb{R}$$

are invertible in $\mathscr{L}(L^p(\mathbb{R}))$, and if the function $\mathsf{H}_{\infty,\infty}(P_\tau A P_\tau + Q_\tau)$ is invertible.

Formally, we proved Theorem 6.6.23 for the scalar case. For matrix-valued functions $a \in [PC(\mathbb{R})]^{n \times n}$ and $b \in [PC_p]^{n \times n}$, the proof remains essentially the same. This covers, for example, systems of singular integral equations and systems of Wiener-Hopf operators. Obviously, the operators resulting from homomorphisms will then have matrix coefficients, and it can prove difficult to study the invertibility of these operators. Note that a non-scalar version of the two projections theorem was stated in Section 3.5.1.

6.6.5 Some examples

We are going to present two simple examples where Theorem 6.6.24 works. Consider the singular integral operator

$$A := cW^0(\chi_+) + dW^0(\chi_-) \tag{6.36}$$

with coefficients $c, d \in PC(\mathbb{\dot{R}})$. Clearly, this operator can be written in the form

$$\frac{c+d}{2}I + \frac{c-d}{2}S_\mathbb{R}.$$

Theorem 6.6.25. *The finite sections method* (6.2) *applies to the singular integral operator A in* (6.36) *if and only if the operator A is invertible on $L^p(\mathbb{R})$ and the operator*

$$P_1\left(\left(c(-\infty)\chi_- + c(+\infty)\chi_+\right)W^0(\chi_+) + \left(d(-\infty)\chi_- + d(+\infty)\chi_+\right)W^0(\chi_-)\right)P_1$$

is invertible on $L^p([-1, 1])$.

Proof. Let $\mathbf{A} := (P_\tau A P_\tau + (I - P_\tau))_{\tau > 0}$. By Theorem 6.6.24, the sequence $\mathbf{A}$ is stable and, hence, the finite sections method applies to A, if and only if the following operators are invertible:

(i) $W_0(\mathbf{A}) = cW^0(\chi_+) + dW^0(\chi_-)$;

(ii) $W_{-1}(\mathbf{A}) = \chi_+\left(c(-\infty)W^0(\chi_+) + d(-\infty)W^0(\chi_-)\right)\chi_+I + \chi_-I$;

(iii) $W_1(\mathbf{A}) = \chi_-\left(c(+\infty)W^0(\chi_+) + d(+\infty)W^0(\chi_-)\right)\chi_-I + \chi_+I$;

(iv) $H_{x,\infty}(\mathbf{A}) = (c(x^-)\chi_- + c(x^+)\chi_+)W^0(\chi_+) + (d(x^-)\chi_- + d(x^+)\chi_+)W^0(\chi_-)$

for $x \in \mathbb{R}$;

(v) $H_{\infty,0}(\mathbf{A}) = Q_1 +$

$\quad P_1\left((c(-\infty)\chi_- + c(+\infty)\chi_+)W^0(\chi_+) + (d(-\infty)\chi_- + d(+\infty)\chi_+)W^0(\chi_-)\right)P_1$;

(vi) $H_{\infty,t}(\mathbf{A}) = P_1(c(-\infty)\chi_- + c(+\infty)\chi_+)P_1 + Q_1$ for $t > 0$;

$\quad H_{\infty,t}(A_\tau) = P_1(d(-\infty)\chi_- + d(+\infty)\chi_+)P_1 + Q_1$ for $t < 0$;

(vii) $(H_{\infty,\infty}(\mathbf{A}))(z) = \begin{bmatrix} c(+\infty)z + d(+\infty)(1-z) & 0 & 0 & 0 \\ 0 & 1 & 0 & 0 \\ 0 & 0 & c(-\infty)z + d(-\infty)(1-z) & 0 \\ 0 & 0 & 0 & 1 \end{bmatrix}$

for $z \in \mathfrak{L}_p$.

Thus, the conditions stated in the theorem are necessary: the operators quoted there are $W_0(\mathbf{A})$ and $H_{\infty,0}(\mathbf{A})$, respectively. To prove the sufficiency, we have to show that the invertibility of $W_0(\mathbf{A})$ and $H_{\infty,0}(\mathbf{A})$ implies the invertibility of all other operators in (i)–(vii).

Let $x \in \mathbb{R}$. Since $H_{x,\infty}(A) = H_{x,\infty}(\mathbf{A})$ by Lemma 6.6.8, the invertibility of A implies that of $H_{x,\infty}(\mathbf{A})$. Further, if $H_{\infty,0}(\mathbf{A})$ is invertible then the sequence

$$\mathbf{B} := (P_\tau((c(-\infty)\chi_- + c(+\infty)\chi_+)W^0(\chi_+)$$
$$+ (d(-\infty)\chi_- + d(+\infty)\chi_+)W^0(\chi_-))P_\tau + Q_\tau)$$

is stable by Lemma 6.6.20. Since $W_{-1}(\mathbf{A}) = W_{-1}(\mathbf{B})$ and $W_1(\mathbf{A}) = W_1(\mathbf{B})$ by Lemmas 6.6.5 and 6.6.6, respectively, then the operators $W_{-1}(\mathbf{A})$ and $W_1(\mathbf{A})$ are invertible.

Similarly, if $t \in \mathbb{R} \setminus \{0\}$, then $H_{\infty,t}(\mathbf{A}) = H_{\infty,t}(\mathbf{B})$ by Lemma 6.6.7, which verifies the invertibility of the operators $H_{\infty,t}(\mathbf{A})$. Finally, the matrix function (vii) is invertible if and only if the point 0 does not belong to the lentiform domains

$$1 + \left(\frac{c(+\infty)}{d(+\infty)} - 1\right)\mathfrak{L}_p \quad \text{and} \quad 1 + \left(\frac{c(-\infty)}{d(-\infty)} - 1\right)\mathfrak{L}_p.$$

That the invertibility of $W_0(\mathbf{A})$ and $H_{\infty,0}(\mathbf{A})$ also implies this condition follows by employing the invertibility criterion for singular integral operators in [74, Section 9.6, Theorem 6.1]. ∎

In the case $p = 2$, part of the conditions of Theorem 6.6.25 can be nicely formulated in geometric terms.

Corollary 6.6.26. *Let $p = 2$. The finite sections method (6.2) applies to the singular integral operator A in (6.36) if and only if the operator A is invertible on $L^2(\mathbb{R})$ and if the point 0 is not contained in the convex hull of the points 1, $\frac{c(-\infty)}{d(-\infty)}$ and $\frac{c(+\infty)}{d(+\infty)}$.*

Indeed, this follows from the previous theorem and the invertibility criterion for singular integral operators since $\mathfrak{L}_2$ is the interval $[0, 1]$.

For a second illustration of Theorem 6.6.24, let A now be the paired operator

$$A = W^0(a)\chi_+ I + W^0(b)\chi_- I \tag{6.37}$$

with $a, b \in PC_p$.

Theorem 6.6.27. *The finite sections method (6.2) applies to the paired operator A in (6.37) if and only if the operator A is invertible on $L^p(\mathbb{R})$, the Wiener-Hopf operators $W(b)$ and $W(\tilde{a})$ are invertible on $L^p(\mathbb{R}^+)$, the operator*

$$P_1\left((a(y^-)W^0(\chi_-) + a(y^+)W^0(\chi_+))\chi_+ I + (b(y^-)W^0(\chi_-) + b(y^+)W^0(\chi_+))\chi_- I\right)P_1$$

is invertible on $L^p([-1,1])$ for every $y \in \mathbb{R}$, and the point 0 does not belong to the lentiform domains

$$a(-\infty) + (a(+\infty) - a(-\infty))\mathfrak{L}_p \quad and \quad b(+\infty) + (b(-\infty) - b(+\infty))\mathfrak{L}_p.$$

Proof. Let again $\mathbf{A} := (P_\tau A P_\tau + (I - P_\tau))_{\tau > 0}$. Theorem 6.6.24 implies that the finite sections method for the operator A is stable if and only if the following operators are invertible:

(i) $W_0(\mathbf{A}) = A$;

(ii) $W_{-1}(\mathbf{A}) = \chi_+ W^0(b)\chi_+ I + \chi_- I$;

(iii) $W_1(\mathbf{A}) = \chi_- W^0(a)\chi_- I + \chi_+ I$;

(iv) $H_{0,\infty}(\mathbf{A}) = \left(a(-\infty)W^0(\chi_-) + a(+\infty)W^0(\chi_+)\right)\chi_+ I$
$\qquad\qquad\qquad + \left(b(-\infty)W^0(\chi_-) + b(+\infty)W^0(\chi_+)\right)\chi_- I$;

(v) $H_{x,\infty}(\mathbf{A}) = a(-\infty)W^0(\chi_-) + a(+\infty)W^0(\chi_+)$ $\quad$ if $x > 0$;
$\quad\, H_{x,\infty}(\mathbf{A}) = b(-\infty)W^0(\chi_-) + b(+\infty)W^0(\chi_+)$ $\quad$ if $x < 0$;

(vi) $H_{\infty,y}(\mathbf{A}) = Q_1 + P_1((a(y^-)W^0(\chi_-) + a(y^+)W^0(\chi_+))\chi_+ I$
$\qquad\qquad + (b(y^-)W^0(\chi_-) + b(y^+)W^0(\chi_+))\chi_- I)P_1$ $\qquad$ for $y \in \mathbb{R}$;

(vii) $(H_{\infty,\infty}(\mathbf{A}))(z) = \begin{bmatrix} a(-\infty)(1-z) + a(+\infty)z & 0 & 0 & 0 \\ 0 & 1 & 0 & 0 \\ 0 & 0 & b(-\infty)z + b(+\infty)(1-z) & 0 \\ 0 & 0 & 0 & 1 \end{bmatrix}$,

for $z \in \mathfrak{L}_p$.

The invertibility of the operators $W_{-1}(\mathbf{A})$ and $W_1(\mathbf{A})$ is equivalent to the invertibility of the Wiener-Hopf operators $W(b)$ and $W(\tilde{a})$, respectively. Thus, the conditions of the theorem are necessary. We show that, conversely, the invertibility of the operator A implies the invertibility of the operators in (iv) and (v). This fact follows immediately from Lemma 6.6.8, where it is shown that $H_{x,\infty}(\mathbf{A}) = H_{x,\infty}(A)$ for every $x \in \mathbb{R}$. $\blacksquare$

Corollary 6.6.28. *Let $p = 2$. The finite sections method (6.2) applies to the paired operator A in (6.37) if and only if the operator A is invertible on $L^2(\mathbb{R})$, the Wiener-Hopf operators $W(b)$ and $W(\tilde{a})$ are invertible on $L^2(\mathbb{R}^+)$ and if, for every $y \in \mathbb{R}$, the point 0 is not contained in the convex hull of the points 1, $\frac{a(y^-)}{a(y^+)}$ and $\frac{b(y^-)}{b(y^+)}$.*

6.7 Finite sections of multidimensional convolution type operators

In this section, we will have a look at different versions of the finite section method for multidimensional convolution type operators in the sense of Section 5.8. The notation is as in that section. In particular, W_a^0 stands for the operator of convolution by the function $a \in L^1(\mathbb{R}^N)$ (see (5.50) for the definition), and we consider all operators on $L^p(\mathbb{R}^N)$ with $1 < p < \infty$.

For simplicity, we will also assume throughout this section that $N = 2$. But the reader will notice that the approach presented in this section will work for $N > 2$ as well. Let us also agree upon the following notion: If K is a cone with vertex at the origin and $x \in \mathbb{R}^2$, then we call the algebraic sum $K + x$ a *cone with vertex at x*. Recall also from Section 5.8 that we do not allow cones to degenerate to a line, but it will be convenient in this section to consider half planes as cones.

6.7.1 Algebras in the background

Let $\mathcal{E}$ be the set of all sequences $(A_n)_{n \in \mathbb{N}}$ of operators in $\mathcal{L}(L^p(\mathbb{R}^2))$ such that $\sup \|A_n\| < \infty$. As before, $\mathcal{E}$ actually forms a Banach algebra and $\mathcal{G}$, the set of all sequences in $\mathcal{E}$ tending in the norm to zero, is a closed two-sided ideal in $\mathcal{E}$. Due to Theorem 6.2.2, the stability of a sequence $(A_n) \in \mathcal{E}$ is equivalent to the invertibility of its coset $(A_n) + \mathcal{G}$ in the quotient algebra $\mathcal{E}/\mathcal{G} =: \mathcal{E}^{\mathcal{G}}$.

The algebra $\mathcal{E}^{\mathcal{G}}$ is in general too large for studying invertibility effectively (e.g., by central localization). So, following the general procedure outlined at the beginning of the chapter, we are going to introduce subalgebras which are more suitable for our purposes. To each function $\varphi \in C(\overline{\mathbb{R}^2})$, we associate a sequence $(\varphi_n I)$, given by the expanded functions $\varphi_n(t) := \varphi(t/n)$. Clearly, this sequence belongs to $\mathcal{E}$, and $\|(\varphi_n I)\|_{\mathcal{E}} = \|\varphi\|_\infty$. We denote by $\mathcal{U}$ the smallest closed subalgebra of $\mathcal{E}$ which contains all sequences $(A_n) \in \mathcal{E}$ such that

$$\lim_{n \to \infty} \|A_n \varphi_n I - \varphi_n A_n\| = 0 \quad \text{for every } \varphi \in C(\overline{\mathbb{R}^2}).$$

Thus, the commutator $(A_n)(\varphi_n I) - (\varphi_n I)(A_n)$ belongs to $\mathscr{G}$ for every $\varphi \in C(\overline{\mathbb{R}^2})$. The following is then easy to check (see Exercise 1.2.14).

Theorem 6.7.1. *The algebra* $\mathscr{U}^{\mathscr{G}} := \mathscr{U}/\mathscr{G}$ *is inverse-closed in* $\mathscr{E}^{\mathscr{G}}$.

Thus, the stability of a sequence $(A_n) \in \mathscr{U}$ is equivalent to the invertibility of the coset $(A_n) + \mathscr{G}$ in the algebra $\mathscr{U}/\mathscr{G} =: \mathscr{U}^{\mathscr{G}}$. By its construction, the latter algebra has a non-trivial center. In particular, the set $\{(\varphi_n I) + \mathscr{G} : \varphi \in C(\overline{\mathbb{R}^2})\}$ is a central C^*-subalgebra of $\mathscr{U}^{\mathscr{G}}$. We shall prove later on that this algebra is isometrically isomorphic to $C(\overline{\mathbb{R}^2})$ in a natural way. Hence, one can localize the algebra $\mathscr{U}^{\mathscr{G}}$ over the space of maximal ideals of $C(\overline{\mathbb{R}^2})$ by Allan's local principle. As previously seen at the beginning of Section 5.8, the maximal ideal space $\overline{\mathbb{R}^2}$ can be thought of as the union of $\mathbb{R}^2$ and an "infinitely distant" sphere (a circle in the case $N = 2$).

To each $x \in \overline{\mathbb{R}^2}$, we associate the maximal ideal $\mathscr{I}_x := \{(\varphi_n I) + \mathscr{G} : \varphi(x) = 0\}$ of $C(\overline{\mathbb{R}^2})$ and the smallest closed ideal $\mathscr{J}_x$ of $\mathscr{U}^{\mathscr{G}}$ which contains $\mathscr{I}_x$. Let $\mathscr{U}_x^{\mathscr{G}} := \mathscr{U}^{\mathscr{G}}/\mathscr{J}_x$ refer to the associated quotient algebra and $\Phi_x : \mathscr{U}^{\mathscr{G}} \to \mathscr{U}^{\mathscr{G}}/\mathscr{J}_x$ to the corresponding canonical homomorphism. The following result characterizing the ideals $\mathscr{J}_x$ is an immediate consequence of Proposition 2.2.4.

Proposition 6.7.2. *Let* $x \in \overline{\mathbb{R}^2}$. *The coset* $(A_n) + \mathscr{G}$ *belongs to* $\mathscr{J}_x$ *if and only if, for every* $\varepsilon > 0$, *there is a* $\varphi \in C(\overline{\mathbb{R}^2})$, *depending on* ε, *with compact support and* $\varphi(x) = 1$ *such that, for n large enough,* $\|A_n \varphi_n I\| < \varepsilon$.

6.7.2 The algebra $\mathscr{F}$ of the finite sections method

The shape of the finite sections will be prescribed by a compact subset Ω of $\mathbb{R}^2$ which contains the origin and has the following properties:

- for every point $x \in \partial\Omega$, there is a cone K_x at x, open neighborhoods U and V of x, and a C^1-diffeomorphism $\rho : U \to V$ such that

$$\rho(x) = x, \quad \rho'(x) = I, \quad \rho(U \cap \Omega) = V \cap K_x;$$

- if $x = 0 \in \partial\Omega$ we require moreover that the associated diffeomorphism ρ is the identity and $\Omega \subset K_0$.

Note that the cone K_x is uniquely defined for every $x \in \partial\Omega$.

For $x \in \partial\Omega$, let K_x^0 be the cone $\{t - x : t \in K_x\}$. The finite sections are then defined by the projection operators $\chi_{n\Omega} I$, where $n\Omega$ denotes the expanded set $\{nt : t \in \Omega\}$, $n \in \mathbb{N}$, and χ_L stands, as usual, for the characteristic function of the set L. The following is almost obvious, and we omit the proof.

Proposition 6.7.3. *The sequence* $(\chi_{n\Omega} I)$ *converges strongly, as* $n \to \infty$,

(i) *to the identity operator* I *if* 0 *is an inner point of* Ω;
(ii) *to* $\chi_{K_0} I$ *if* $0 \in \partial\Omega$, *where* K_0 *is the cone associated with the point* 0.

We will consider the finite sections of operators of the form

$$B := \chi_K A \chi_K I + (1 - \chi_K) I$$

with A in the algebra

$$\mathscr{A}_0 := \text{alg}\{W_a^0, fI : f \in C(\overline{\mathbb{R}^2}), a \in L^1(\mathbb{R}^2)\}$$

with respect to the sequence of projections $(\chi_{n\Omega} I)$. More precisely, we consider the smallest closed subalgebra $\mathscr{F}$ of $\mathscr{E}$ which contains all sequences

- $(\chi_{n\Omega} A \chi_{n\Omega} I + (1 - \chi_{n\Omega}) I)$ with A in $\mathscr{A}_0$,
- (G_n) with $(G_n) \in \mathscr{G}$.

The next result implies that $\mathscr{F}$ is actually a subalgebra of $\mathscr{U}$.

Proposition 6.7.4. *Let $a \in L^1(\mathbb{R}^2)$ and $\varphi \in C(\overline{\mathbb{R}^2})$. Then $\|\varphi_n W_a^0 - W_a^0 \varphi_n I\| \to 0$ as $n \to \infty$.*

Proof. By definition, we have

$$(\varphi_n W_a^0 - W_a^0 \varphi_n I) v = \int_{\mathbb{R}^2} a(x - t) \left(\varphi \left(\frac{x}{n} \right) - \varphi \left(\frac{t}{n} \right) \right) v(t) \, dt,$$

where $v \in L^p(\mathbb{R}^2)$. Functions in $L^1(\mathbb{R}^2)$ can be approximated by functions with compact support, so we assume that the support of the function a is contained in the ball $B_R(0)$ of radius R. If $|x - t| > R$ then $a(x - t) = 0$. For $|x - t| \le R$ we would like to show that, given $\varepsilon > 0$ and n large enough,

$$\left| \varphi \left(\frac{x}{n} \right) - \varphi \left(\frac{t}{n} \right) \right| < \frac{\varepsilon}{\|a\|}.$$

The idea is to use the uniform continuity of the function $\psi = \varphi \circ \overline{\xi}$, where $\overline{\xi}$ is the homeomorphism from $\overline{B_1(0)}$ onto $\overline{\mathbb{R}^2}$ which coincides with ξ on $B_1(0)$ (see Section 5.8), and the inequality

$$|u - s| \le \left| \frac{u}{1 - |u|} - \frac{s}{1 - |s|} \right| \quad \text{for } u, s \in B_1(0), \tag{6.38}$$

which follows immediately from the inequality

$$|x - y| \ge \left| \frac{x}{1 + |x|} - \frac{y}{1 + |y|} \right|$$

derived in (5.53) after the change of variables $s = x/(1 + |x|)$ and $u = y/(1 + |y|)$. From (6.38) we conclude that

$$|u - s| \le |\xi(u) - \xi(s)| \quad \text{for all } u, s \in B_1(0).$$

Setting $u = \xi^{-1}(x/n)$ and $s = \xi^{-1}(t/n)$ we obtain

$$\left| \xi^{-1} \left(\frac{x}{n} \right) - \xi^{-1} \left(\frac{t}{n} \right) \right| \le \left| \frac{x}{n} - \frac{t}{n} \right| \le \frac{R}{n}.$$

The function ψ is uniformly continuous, that is, for each $\varepsilon > 0$ there is a $\delta > 0$ such that

$$y, z \in \overline{B_1(0)} \text{ and } |y - z| < \delta \quad \text{implies} \quad |\psi(y) - \psi(z)| < \frac{\varepsilon}{\|a\|}.$$

Now choose n_0 such that $R/n < \delta$ for $n \ge n_0$. Then

$$\left| \psi \left(\xi^{-1} \left(\frac{x}{n} \right) \right) - \psi \left(\xi^{-1} \left(\frac{t}{n} \right) \right) \right| = \left| \varphi \left(\frac{x}{n} \right) - \varphi \left(\frac{t}{n} \right) \right| < \frac{\varepsilon}{\|a\|}$$

for every $n \ge n_0$. Therefore,

$$\left\| (\varphi_n W_a^0 - W_a^0 \varphi_n I) v \right\| \le \frac{\varepsilon}{\|a\|} \left\| W_a^0 v \right\| \le \varepsilon \|v\|.$$

∎

For the next proposition, recall the definition of the shift operator $V_k : L^p(\mathbb{R}^2) \to L^p(\mathbb{R}^2)$, $(V_k v)(s) := v(s - k)$. Evidently, V_k is a bijective isometry with inverse V_{-k}.

Proposition 6.7.5. *Let $f, \varphi \in C(\overline{\mathbb{R}^2})$, $a \in L^1(\mathbb{R}^2)$, $(G_n) \in \mathcal{G}$ and $x \in \mathbb{R}^2$. Then*

(i) s-$\lim_{n \to \infty} V_{-nx} W_a^0 V_{nx} = W_a^0$;
(ii) s-$\lim_{n \to \infty} V_{-nx} f V_{nx} = f(\theta_\infty) I$ *if $x \ne 0$ where θ satisfies $e^{i\theta} = x/|x|$;*
(iii) s-$\lim_{n \to \infty} V_{-nx} \varphi_n V_{nx} = \varphi(x)$;
(iv) s-$\lim_{n \to \infty} V_{-nx} G_n V_{nx} = 0$;
(v) s-$\lim_{n \to \infty} V_{-nx} \chi_{n\Omega} V_{nx}$ *is $\chi_{K_x^0} I$, I, or 0 depending on $x \in \partial\Omega$, $x \in \text{int } \Omega$ or $x \in \overline{\mathbb{R}^2} \backslash \Omega$;*
(vi) s-$\lim_{n \to \infty} V_{-nx} \chi_{nK_x} V_{nx} = \chi_{K_x^0} I$, *for $x \in \partial\Omega$.*

Proof. Assertions (i) and (iv) are obvious, and (ii) is already proved in Proposition 5.8.5. The proof of (iii) follows easily if one uses

$$V_{-nx} \varphi_n V_{nx} v = \breve{\varphi}_n v \quad \text{with } \breve{\varphi}_n(t) := \varphi(t/n + x).$$

Assertions (v) and (vi) are left to the reader as an exercise.   ∎

Corollary 6.7.6. *The algebra $\mathscr{C} := \{ (\varphi_n I) + \mathcal{G} : \varphi \in C(\overline{\mathbb{R}^2}) \}$ is isometrically isomorphic to $C(\overline{\mathbb{R}^2})$ in the natural way.*

Indeed, this can be proved in the same way as Proposition 5.8.7.

By $\mathscr{F}^{\mathcal{G}}$ we denote the image of $\mathscr{F}$ in $\mathcal{U}^{\mathcal{G}}$ under the canonical homomorphism. Our main task is to study the invertibility of elements of the form $\Phi_x((A_n) + \mathcal{G})$ in $\mathcal{U}_x^{\mathcal{G}}$, with $(A_n) + \mathcal{G} \in \mathscr{F}^{\mathcal{G}}$. In order to avoid complicated notation, as usual, we shall use also Φ_x for the composition $\Phi_x \circ \pi$, where π is the canonical homomorphism from $\mathcal{U}$ onto $\mathcal{U}^{\mathcal{G}}$. Accordingly, we write $\Phi_x(A_n)$ in place of $\Phi_x((A_n) + \mathcal{G})$.

Proposition 6.7.7. *Let $x \in \overline{\mathbb{R}^2}$ and $f \in C(\overline{\mathbb{R}^2})$. Then:*

(i) $\Phi_x(fI) = f(\theta_\infty)\Phi_x(I)$ *for $x \neq 0$;*

(ii) $\Phi_x(\chi_{n\Omega}I) = 0$ *if $x \in \overline{\mathbb{R}^2} \setminus \Omega$ and $\Phi_x(\chi_{n\Omega}I) = I$ if $x \in$ int Ω.*

Proof. Regarding (i), first suppose that $x \in \mathbb{R}^2 \setminus \{0\}$ and $e^{i\theta} = x/|x|$. Recall that $f(\theta_\infty)$ denotes the limit $\lim_{t \to \infty} f(te^{i\theta})$. Since $f \in C(\overline{\mathbb{R}^2})$, given $\varepsilon > 0$ there exists a neighborhood $U_{R,\delta}(\theta_\infty)$ of θ_∞ as defined in (5.49), such that $|f(t) - f(\theta_\infty)| < \varepsilon$ for every $t \in U_{R,\delta}(\theta_\infty)$.

Let $\varphi \in C(\overline{\mathbb{R}^2})$ with $\varphi(x) = 1$ be a function, the support of which is contained in a ball centered on x not including zero and also contained in $U_{R,\delta}(\theta_\infty)$. Then, for n large enough, supp $\varphi_n \subset U_{R,\delta}(\theta_\infty)$, whence

$$\|(f - f(\theta_\infty))\varphi_n I\| < \varepsilon.$$

Thus, from Proposition 6.7.2 it follows that

$$(f - f(\theta_\infty))I + \mathcal{G} \in \mathcal{J}_x.$$

The proof for $x = \theta_\infty \in \overline{\mathbb{R}^2} \setminus \mathbb{R}^2$ is similar. If we choose φ with support contained in $U_{R,\delta}(\theta_\infty)$ and with $\varphi(\theta_\infty) = 1$, then supp $\varphi_n \subset U_{R,\delta}(\theta_\infty)$ for every $n \geq 1$. So, $\|(f - f(\theta_\infty))\varphi_n I\| < \varepsilon$ for every $n \geq 1$. Assertion (i) is verified, and assertion (ii) is immediate from Proposition 6.7.2. $\blacksquare$

For $x \in \partial\Omega$ these considerations are more complex. We start with introducing a Banach algebra $\mathscr{D}_x$ for each point $x \in \partial\Omega$ as follows. For $x \in \partial\Omega$, let $\mathscr{F}_{K_x}$ stand for the set of all bounded sequences (A_n) of operators $A_n : \text{Im } \chi_{nK_x}I \to \text{Im } \chi_{nK_x}I$ and write $\mathscr{G}_{K_x}$ for the set of all sequences in $\mathscr{F}_{K_x}$ which tend to zero in the norm. With respect to pointwise defined operations and the supremum norm, $\mathscr{F}_{K_x}$ becomes a Banach algebra and $\mathscr{G}_{K_x}$ a closed ideal of $\mathscr{F}_{K_x}$. If now $x \in \partial\Omega \setminus \{0\}$, then $\mathscr{D}_x$ is defined as the smallest closed subalgebra of $\mathscr{F}_{K_x}$ which contains the ideal $\mathscr{G}_{K_x}$ and all sequences of the form $(\chi_{nK_x}A\chi_{nK_x}I + (1 - \chi_{nK_x})I)$ with $A \in \text{alg}\{W_a^0, I : a \in L^1(\mathbb{R}^2)\}$. If $x \in \partial\Omega \cap \{0\}$, then let $\mathscr{D}_0$ denote the smallest closed subalgebra of $\mathscr{F}_{K_0}$ which contains the ideal $\mathscr{G}_{K_0}$ and all sequences of the form $(\chi_{nK_0}A\chi_{nK_0}I + (1 - \chi_{nK_0})I)$ with $A \in \text{alg}\{W_a^0, fI : a \in L^1(\mathbb{R}^2), f \in C(\overline{\mathbb{R}^2})\}$.

These algebras will play a crucial role in the stability analysis because, as we shall prove, $\Phi_x(\mathscr{F}^{\mathscr{G}})$ is isometrically isomorphic to $\Phi_x(\mathscr{D}_x)$. For this purpose we have to introduce some intermediate algebras, namely the algebras $\mathscr{U}_x \subset \mathscr{U}$, which will be the object of the next section.

6.7.3 The algebra $\mathscr{U}$ and the subalgebras $\mathscr{U}_x$

Recall that, to each $x \in \partial\Omega$, we associated a C^1-diffeomorphism $\rho : U \to V$ where U and V are neighborhoods of x, and that in the case $0 \in \partial\Omega$ we supposed that $\rho \equiv I$.

Thus, if $x = 0$ and $0 \in \partial\Omega$, then $\chi_{n\Omega}\chi_{nU}I = \chi_{nK_0}\chi_{nU}I$, which implies that $\Phi_x(\chi_{n\Omega}I) = \Phi_x(\chi_{nK_0}I)$ and that the algebras $\Phi_x(\mathscr{F}^{\mathscr{G}})$ and $\Phi_x(\mathscr{D}_x^{\mathscr{G}})$ coincide. So, from now on let x be a fixed point in $\partial\Omega \setminus \{0\}$.

The following operators will help us to describe the local algebras:

$$T_n : L^p(\mathbb{R}^2) \to \operatorname{Im} \chi_{nU}I \subset L^p(\mathbb{R}^2), \quad (T_n v)(t) = \begin{cases} v\left(n\rho\left(\frac{t}{n}\right)\right) & \text{if } t \in nU, \\ 0 & \text{if } t \notin nU; \end{cases}$$

and

$$T_n^{(-1)} : L^p(\mathbb{R}^2) \to \operatorname{Im} \chi_{nV}I \subset L^p(\mathbb{R}^2), \quad \left(T_n^{(-1)}v\right)(t) = \begin{cases} v\left(n\rho^{-1}\left(\frac{t}{n}\right)\right) & \text{if } t \in nV, \\ 0 & \text{if } t \notin nV. \end{cases}$$

Proposition 6.7.8. T_n and $T_n^{(-1)}$ are bounded linear operators satisfying

(i) $\|T_n\| \leq \sup_{t \in V} |J\rho^{-1}(t)|$ and $\|T_n^{(-1)}\| \leq \sup_{t \in U} |J\rho(t)|$, where J stands for the Jacobian;

(ii) $T_n T_n^{(-1)} = \chi_{nU}I$ and $T_n^{(-1)} T_n = \chi_{nV}I$;

(iii) $\chi_{nU} T_n \chi_{nV}I = T_n$ and $\chi_{nV} T_n^{(-1)} \chi_{nU}I = T_n^{(-1)}$.

Assertion (ii) says that T_n and $T_n^{(-1)}$ are locally inverse to each other.

Proof. For (i), let $g \in L^p(\mathbb{R}^2)$. Then

$$\|T_n g\|^p = \int_{nU} |g(n\rho(t/n))|^p dt = \int_{nV} |g(s)|^p |J\rho^{-1}(s/n)|^p ds$$
$$\leq \sup_{s \in V} |J\rho^{-1}(s)|^p \int_{nV} |g(s)|^p ds,$$

whence

$$\|T_n g\| \leq \sup_{s \in V} |J\rho^{-1}(s)| \|g\|_{L^p}.$$

Analogously, one finds the upper bound for the norm of the operator $T_n^{(-1)}$. Assertions (ii) and (iii) are immediate. ∎

Proposition 6.7.9. *The operator*

$$H_\rho : \mathscr{U} \to \mathscr{U}, \quad (A_n) \mapsto \left(T_n^{(-1)} A_n T_n\right)$$

has the following properties:

(i) H_ρ *is a well-defined linear operator;*

(ii) *if* $W \subset U$ *then* $H_\rho(\chi_{nW}I) = \chi_{n\rho(W)}I$;

(iii) *the ideal* $\mathscr{G}$ *is invariant under* H_ρ, *i.e.,* $H_\rho(\mathscr{G}) \subset \mathscr{G}$;

(iv) *if the coset* $A_n + \mathscr{G}$ *belongs to* $\mathscr{J}_x$, *then* $H_\rho(A_n) + \mathscr{G}$ *also belongs to* $\mathscr{J}_x$.

Analogous assertions hold for the operator $H_\rho^{(-1)} : \mathscr{U} \to \mathscr{U}, (A_n) \mapsto \left(T_n A_n T_n^{(-1)}\right)$.

Proof. To prove the first assertion, let $\varphi \in C(\overline{\mathbb{R}^2})$. One has to show that $H_\rho(A_n)$ commutes with $(\varphi_n I)$ modulo a sequence tending in the norm to zero, i.e., that there is a $(G_n) \in \mathscr{G}$ such that

$$T_n^{(-1)} A_n T_n \varphi_n = \varphi_n T_n^{(-1)} A_n T_n + G_n. \tag{6.39}$$

Let $\tilde{\rho}$ and $\widetilde{\rho^{-1}}$ denote the continuous extensions to $\overline{\mathbb{R}^2}$ of ρ and ρ^{-1}, respectively. Observe first, that due to Proposition 6.7.8 (iii),

$$T_n \varphi_n I = (\varphi \circ \tilde{\rho})_n T_n \quad \text{and} \quad T_n^{(-1)} \varphi_n = \left(\varphi \circ \widetilde{\rho^{-1}} \right)_n T_n^{(-1)}.$$

Thus,

$$T_n^{(-1)} A_n T_n \varphi_n I = T_n^{(-1)} A_n (\varphi \circ \tilde{\rho})_n T_n.$$

Since there is a sequence $(F_n) \in \mathscr{G}$ such that $A_n(\varphi \circ \tilde{\rho})_n = (\varphi \circ \tilde{\rho})_n A_n + F_n$, it follows that

$$
\begin{aligned}
T_n^{(-1)} A_n (\varphi \circ \tilde{\rho})_n T_n &= T_n^{(-1)} \left[(\varphi \circ \tilde{\rho})_n A_n + F_n \right] T_n \\
&= \left(\varphi \circ \tilde{\rho} \circ \widetilde{\rho^{-1}} \right)_n T_n^{(-1)} + T_n^{(-1)} F_n T_n \\
&= \varphi_n T_n^{(-1)} + T_n^{(-1)} F_n T_n.
\end{aligned}
$$

Now set $G_n := T_n^{(-1)} F_n T_n$ to obtain the equality (6.39).

Regarding assertion (ii), let g be a function in $L^p(\mathbb{R}^2)$ and $W \subset U$. Then

$$H_\rho (\chi_{nW}) g = T_n^{-1} \chi_{nW} T_n g = T_n^{-1} \chi_{nW} g \left(n\rho \left(\frac{t}{n} \right) \right) = \chi_{n\rho(W)} g.$$

Assertion (iii) is immediate. For Assertion (iv), let $(A_n) + \mathscr{G} \in \mathscr{J}_x$. Thus, given $\varepsilon > 0$, there is a neighborhood $W \subset U$ of x such that for n large enough

$$\|A_n \chi_{nW} I\| < \varepsilon.$$

Since $H_\rho (A_n(1 - \chi_{nU})) = 0$, we have that $H_\rho (A_n) = H_\rho (A_n \chi_{nU})$. By simple computations and (ii) it follows that

$$H_\rho (A_n \chi_{nW} I) = H_\rho (A_n \chi_{nU} \chi_{nW} I) = H_\rho (A_n \chi_{nU}) H_\rho (\chi_{nW} I) = H_\rho (A_n) \chi_{n\rho(W)} I.$$

On the other hand,

$$\|H_\rho (A_n \chi_{nW} I)\| = \|T_n^{(-1)} A_n \chi_{nW} T_n\|.$$

Since $\|T_n\|$ and $\|T_n^{(-1)}\|$ are bounded by $\sup_{t \in V} |J\rho^{-1}(t)|$ and $\sup_{t \in U} |J\rho(t)|$, respectively, we get for n large enough that

$$\|H_\rho (A_n) \chi_{n\rho(W)} I\| < M\varepsilon$$

for some constant M. Since $\rho(W)$ is a neighborhood of x, it follows that the coset $H_\rho(A_n)\chi_{n\rho(W)}I + \mathscr{G}$ belongs to $\mathscr{J}_x$.   ∎

It should be emphasized that the linear operators H_ρ and $H_\rho^{(-1)}$ depend on the point $x \in \partial\Omega \setminus \{0\}$. These operators do not act as homomorphisms, but they generate quotient homomorphisms in the local algebra $\mathscr{U}^{\mathscr{G}}/\mathscr{J}_x$. Indeed, due to Proposition 6.7.9, the map

$$h_\rho : \mathscr{U}^{\mathscr{G}}/\mathscr{J}_x \to \mathscr{U}^{\mathscr{G}}/\mathscr{J}_x, \quad ((A_n)+\mathscr{G})+\mathscr{J}_x \mapsto (H_\rho(A_n)+\mathscr{G})+\mathscr{J}_x \quad (6.40)$$

is well defined. Since $(1 - \chi_{n\mathscr{U}}I)+\mathscr{G} \in \mathscr{J}_x$, Proposition 6.7.8 (ii) leads to

$$\begin{aligned}\left(H_\rho(A_n)H_\rho(B_n)+\mathscr{G}\right) + \mathscr{J}_x &= \left(H_\rho(A_n\chi_{n\mathscr{U}}B_n)+\mathscr{G}\right) + \mathscr{J}_x \\ &= \left(H_\rho(A_nB_n)+\mathscr{G}\right) + \mathscr{J}_x.\end{aligned}$$

Hence, h_ρ is a homomorphism and actually an isomorphism, having

$$h_\rho^{-1} : \mathscr{U}^{\mathscr{G}}/\mathscr{J}_x \to \mathscr{U}^{\mathscr{G}}/\mathscr{J}_x, \quad (A_n+\mathscr{G})+\mathscr{J}_x \mapsto \left(H_\rho^{(-1)}(A_n)+\mathscr{G}\right)+\mathscr{J}_x$$

as its inverse.

Proposition 6.7.10. *Let $x \in \partial\Omega \setminus \{0\}$. Then*

$$h_\rho\left((W_a^0 +\mathscr{G})+ \mathscr{J}_x\right) = \left(W_a^0+\mathscr{G}\right)+ \mathscr{J}_x.$$

Proof. By Proposition 6.7.9 (iv), we need to show that

$$\Phi_x\left(\left(T_n^{(-1)}\chi_{nU}W_a^0\chi_{nU}T_n\chi_{nV}I - \chi_{nV}W_a^0\chi_{nV}I\right) +\mathscr{G}\right) = 0.$$

By Proposition 6.7.2 it is enough to prove that, for every $\varepsilon > 0$ there is an open neighborhood W of x such that $\|R_n\| < \varepsilon$ for all n, where

$$R_n := \chi_{nW}\left(T_n^{(-1)}\chi_{nU}W_a^0\chi_{nU}T_n\chi_{nV} - \chi_{nV}W_a^0\chi_{nV}\right)\chi_{nW}I.$$

Clearly it is sufficient to assume that a is the characteristic function of a rectangle $O = [a_1,b_1] \times [a_2,b_2]$ with $a_1 < b_1, a_2 < b_2$. Let $C := \sup_{s\in V}|(J\rho^{-1})(s)|+1$ and $\varepsilon > 0$ be arbitrarily given. Introduce further

$$H = \max_{1\leq i\leq 2} |b_i - a_i|, \qquad\qquad h = \min_{1\leq i\leq 2}|b_i - a_i|,$$

$$R = \sup_{x\in O} \|x\|, \qquad\qquad \delta = \min\left\{\frac{\varepsilon}{8C(H+h)}, \frac{h}{2}\right\},$$

and choose open, bounded and convex neighborhoods U_1, V_1 of x such that $U_1 \subset U$, $V_1 \subset V$ and

$$\sup_{s \in V_1} |(J\rho^{-1})(s) - 1| < \frac{\varepsilon}{2H^2}, \tag{6.41}$$

$$\sup_{s \in V_1} \|(\rho^{-1})'(s) - I\| < \frac{\delta}{R}, \tag{6.42}$$

$$\sup_{s \in U_1} \|\rho'(s) - I\| < \frac{\delta}{R}. \tag{6.43}$$

Assume further that $W = V_1 \cap \rho(U_1)$. We shall prove that $\|R_n g\| < \varepsilon \|g\|$ for every $g \in L^p(\mathbb{R}^2)$ and all n. By definition, $R_n g$ is equal to

$$\int_{n\rho^{-1}(W)} a\left(n\rho^{-1}(y/n) - s\right) g\left(n\rho(s/n)\right) ds - \int_{nW} a(y-t) g(t) dt$$

if $y \in nW$ and equal to 0 if $y \notin nW$. After substituting $s = n\rho^{-1}(t/n)$, the integral becomes

$$\int_{nW} \left[a\left(n\rho^{-1}(y/n) - n\rho^{-1}(t/n)\right) |J\rho^{-1}(t/n)| - a(y-t) \right] g(t) \, dt,$$

and the kernel function $r_n(y,t)$ of R_n is given by

$$r_n(y,t) := \begin{cases} |(J\rho_n^{-1}(t))| - 1 & \text{if } w_n(y,t) \in O, y-t \in O \\ |(J\rho_n^{-1}(t))| & \text{if } w_n(y,t) \in O, y-t \notin O \\ 1 & \text{if } w_n(y,t) \notin O, y-t \in O \\ 0 & \text{if } w_n(y,t) \notin O, y-t \notin O \end{cases}$$

if $y,t \in nW$ and by $r_n(y,t) := 0$ if y or t does not belong to nW; here $w_n(y,t) := \rho_n^{-1}(y) - \rho_n^{-1}(t)$ and $\rho_n(z) := n\rho(z/n)$. Define

$$O_{+\delta} := \prod_{i=1}^{2} [a_i - \delta, b_i + \delta], \quad O_{-\delta} := \prod_{i=1}^{2} [a_i + \delta, b_i - \delta]$$

and

$$r(y-t) := \begin{cases} \frac{\varepsilon}{2H^2} & \text{if } y-t \in O_{-\delta} \\ C & \text{if } y-t \in O_{+\delta} \setminus O_{-\delta} \\ 0 & \text{if } y-t \notin O_{+\delta}. \end{cases}$$

Our next aim is to compare the functions r_n and r. For this purpose let us mention that $(\rho_n^{-1})'(z) = (\rho^{-1})'(z/n)$ for $z \in nV$. Further, if $y,t \in nW$ and $w_n(y,t) \in O$ then $y - t \in O_{+\delta}$. This can easily be seen as follows: set $y_1 = \rho_n^{-1}(y), t_1 = \rho_n^{-1}(t)$ and consider

$$\|\rho_n^{-1}(y) - \rho_n^{-1}(t) - (y-t)\| = \|y_1 - t_1 - (\rho_n(y_1) - \rho_n(t_1))\|$$

$$= n \left\| \frac{y_1 - t_1}{n} - \left(\rho\left(\frac{y_1}{n}\right) - \rho\left(\frac{t_1}{n}\right) \right) \right\|$$

$$\leq n \sup_{y \in U_1} \|\rho'(y) - I\| \left\| \frac{y_1 - t_1}{n} \right\|$$

$$\leq \frac{\delta}{R} \|y_1 - t\| \leq \frac{\delta}{R} \cdot R = \delta \qquad \text{(by 6.43)}.$$

Analogously, if $y,t \in nW$ and $y-t \in O_{-\delta}$ then (6.42) implies that $\|y-t-w_n(y,t)\| \leq \delta$, and this gives $w_n(y,t) \in O$.

These considerations give, in conjunction with (6.41),

$$|r_n(y,t)| \leq r(y-t)$$

for all $y,t \in \mathbb{R}^2$. Denoting the Lebesgue measure defined on $\mathbb{R}^2$ by μ we have

$$\|R_n\| \leq \|r\|_{L^1} = \frac{\varepsilon}{2H^2}\mu(O_{-\delta}) + C\mu(O_{+\delta} \setminus O_{-\delta})$$

$$< \frac{\varepsilon}{2H^2}H^2 + 4C\delta(H+h)$$

$$\leq \frac{\varepsilon}{2} + \frac{\varepsilon}{2} = \varepsilon,$$

and the proof is finished.   ∎

Let $\mathscr{D}_x^{\mathscr{G}}$ denote the image of $\mathscr{D}_x$ in $\mathscr{U}^{\mathscr{G}}$ under the canonical homomorphism.

Proposition 6.7.11. *The restriction of h_ρ to $\Phi_x(\mathscr{F}^{\mathscr{G}})$ is an isometric isomorphism onto $\Phi_x(\mathscr{D}_x^{\mathscr{G}})$.*

Proof. As already mentioned, $h_\rho : \mathscr{U}^{\mathscr{G}}/\mathscr{J}_x \to \mathscr{U}^{\mathscr{G}}/\mathscr{J}_x$ is an isomorphism. From Proposition 6.7.8 (i) we know that

$$\|h_\rho\| \leq \left(\sup_{s \in V} |J\rho^{-1}(s)| \right) \left(\sup_{s \in U} |J\rho(s)| \right).$$

Given $\varepsilon > 0$, we may assume without loss of generality that U and V are chosen so that $\|h_\rho\| \leq (1+\varepsilon)^2$. Since ε is arbitrary, we have $\|h_\rho\| = 1$. In the same way one proves that $\|h_\rho^{-1}\| = 1$. Hence, h_ρ is an isometry.

Now we show that h_ρ maps the generators of $\Phi_x(\mathscr{F}^{\mathscr{G}})$ onto the generators of $\Phi_x(\mathscr{D}_x^{\mathscr{G}})$, whence the claim follows. From Proposition 6.7.9 (ii) we know that

$$H_\rho(\chi_{n(\Omega \cap U)}I) = \chi_{n(K_x \cap V)}I,$$

and so $h_\rho\left(\Phi_x(\chi_{n\Omega}I)\right) = \Phi_x(\chi_{nK_x}I)$. The coset $\Phi_x(fI)$ is, for $f \in C(\overline{\mathbb{R}^2})$, the same as the constant coset $f(\theta_\infty)\Phi_x(I)$, thus $h_\rho\left(\Phi_x(fI)\right) = \Phi_x(fI)$. These results together with Proposition 6.7.10 give the assertion. ∎

The last proposition shows that we may study $\Phi_x(\mathscr{D}_x^\mathscr{G})$ instead of $\Phi_x(\mathscr{F}^\mathscr{G})$. As previously stated, we are going to introduce algebras $\mathscr{U}_x$ for this purpose. Let $x \in \mathbb{R}^2$, and $\mathscr{U}_x$ be the smallest closed subalgebra of $\mathscr{U}$ containing the sequences (A_n) for which the strong limits

$$\text{s-lim}\, V_{-nx}A_nV_{nx} \quad \text{and} \quad \text{s-lim}\, V_{-nx}A_n^*V_{nx}$$

exist. It follows from the definition of $\mathscr{U}_x$ that the map

$$\mathsf{W}_x : \mathscr{U}_x \to \mathscr{L}\left(L^p(\mathbb{R}^2)\right), \quad \mathbf{A} \mapsto \text{s-lim}\, V_{-nx}A_nV_{nx},$$

with $\mathbf{A} = (A_n)$, is a well-defined homomorphism and that $\|\mathsf{W}_x\| = 1$. Proposition 6.7.5 implies that the algebras $\mathscr{F}$ and $\mathscr{D}_x$ are contained in $\mathscr{U}_x$ for $x \in \partial\Omega$ (for $x = 0$, $\mathscr{U}_x$ equals $\mathscr{U}$). The same is true for the algebra $\mathscr{C}$ defined in Corollary 6.7.6. It is clear that $\mathscr{G}$ is a closed ideal of $\mathscr{U}_x$ and that W_x maps $\mathscr{G}$ onto zero. Thus one can define (and denote by the same symbol) the quotient homomorphism

$$\mathsf{W}_x : \mathscr{U}_x^\mathscr{G} \to \mathscr{L}\left(L^p(\mathbb{R}^2)\right), \quad \mathbf{A} + \mathscr{G} \mapsto \mathsf{W}_x(\mathbf{A}).$$

Proposition 6.7.12. *The ideal $\mathscr{J}_x \subset \mathscr{U}$ is a closed ideal of $\mathscr{U}_x$, and $\mathsf{W}_x(\mathscr{J}_x) = 0$. Moreover, the local ideal $\mathscr{J}_x^x \subset \mathscr{U}_x$, generated by the maximal ideal $\mathscr{I}_x$ of $\mathscr{C}$, coincides with $\mathscr{J}_x$.*

Proof. By Proposition 2.2.5, each coset $\mathbf{A} + \mathscr{G}$ which belongs to $\mathscr{J}_x$ is of the form $(B_n)(\varphi_n) + \mathscr{G}$ with $(B_n) \in \mathscr{U}$ and $(\varphi_n) \in \mathscr{I}_x$. Proposition 6.7.5 (iii) shows that s-lim $V_{-nx}\varphi_nV_{nx} = 0$ if $\varphi(x) = 0$. Hence, s-lim $V_{-nx}B_n\varphi_nV_{nx} = 0$ and $\mathbf{A} \in \mathscr{U}_x$. Now it is clear that $\mathscr{J}_x$ is a closed ideal in $\mathscr{U}_x^\mathscr{G}$ and that $\mathsf{W}_x(\mathscr{J}_x) = 0$. That $\mathscr{J}_x$ indeed equals $\mathscr{J}_x^x$ follows from Proposition 6.7.2, which is obviously true also for the algebra $\mathscr{U}_x^\mathscr{G}$. ∎

From Proposition 6.7.12 and the properties of strong limits we conclude that the map

$$\mathsf{w}_x : \mathscr{U}_x^\mathscr{G} / \mathscr{J}_x \to \mathscr{L}\left(L^p(\mathbb{R}^2)\right), \quad ((A_n) + \mathscr{G}) + \mathscr{J}_x \mapsto \text{s-lim}\, V_{-nx}A_nV_{nx}$$

is a well-defined homomorphism with norm $\|\mathsf{w}_x\| = 1$. Recall that the sequence (I) belongs to $\mathscr{U}_x$ and serves as the unit element. Hence, $((I) + \mathscr{G}) + \mathscr{J}_x$ is the unit element in $\mathscr{U}_x^\mathscr{G} / \mathscr{J}_x$, and w_x takes this element to the identity operator I.

Since $\mathscr{D}_x \subset \mathscr{U}_x$ and $\mathscr{J}_x^x = \mathscr{J}_x$, it follows that $\Phi_x(\mathscr{D}_x^\mathscr{G})$ is isometrically embedded into $\mathscr{U}_x^\mathscr{G} / \mathscr{J}_x$. Define $\mathscr{T}_x$ to be the smallest closed subalgebra of $\mathscr{L}\left(L^p(\mathbb{R}^2)\right)$ containing all operators

$$\chi_{K_x^0}A\chi_{K_x^0}I + (1 - \chi_{K_x^0})I$$

with $A \in \text{alg}\{W_a^0, I\}$. The following observation is crucial: Since $\chi_{nK_x} I = V_{nx} \chi_{K_x^0} V_{-nx}$ for $x \in \partial\Omega \setminus \{0\}$, the equality

$$\chi_{nK_x} A \chi_{nK_x} I + (1 - \chi_{nK_x})I = V_{nx}\left(\chi_{K_x^0} A \chi_{K_x^0} I + (1 - \chi_{K_x^0})I\right) V_{-nx}$$

holds true for all generating elements of $\mathscr{D}_x$ and all $n \in \mathbb{N}$. Hence, the restriction of the mapping W_x to $\mathscr{D}_x$ is continuous, and it maps the generating elements of $\mathscr{D}_x$ onto the generating elements of $\mathscr{T}_x$. Now observe that every element $(A_n) \in \mathscr{D}_x$ can be written as

$$(A_n) = (V_{nx}DV_{-nx}) \quad \text{with} \quad D = W_x(A_n).$$

Conversely, if $D \in \mathscr{T}_x$ then $(V_{nx}DV_{-nx})$ belongs to $\mathscr{D}_x$. It is easy to see that $W_x : \mathscr{D}_x \to \mathscr{T}_x$ is an isometric isomorphism with inverse

$$W_x^{-1} : \mathscr{T}_x \to \mathscr{D}_x, \quad D \mapsto (V_{nx}DV_{-nx}).$$

Moreover, $(A_n) \in \mathscr{D}_x$ is invertible in $\mathscr{U}_x$ if and only if $W_x(A_n)$ is invertible in $\mathscr{L}\left(L^p(\mathbb{R}^2)\right)$. Indeed, if $(A_n) \in \mathscr{D}_x$ is invertible in $\mathscr{U}_x$ then $D := W_x(A_n)$ is invertible in $\mathscr{L}\left(L^p(\mathbb{R}^2)\right)$ since W_x is a homomorphism and $W_x(I) = I$. Conversely, if $D \in \mathscr{T}_x$ is invertible in $\mathscr{L}\left(L^p(\mathbb{R}^2)\right)$ then $(V_{nx}D^{-1}V_{-nx})$ belongs to $\mathscr{U}_x$ and is the inverse to $(A_n) \in \mathscr{D}_x$ in $\mathscr{U}_x$. So the first part of the following proposition is proved.

Proposition 6.7.13. *Let the algebras $\mathscr{U}_x$, $\mathscr{D}_x$ and $\mathscr{T}_x$ be as above. Then*

(i) *the homomorphism $W_x : \mathscr{D}_x \to \mathscr{T}_x$, $(A_n) \mapsto \text{s-lim}\, V_{-nx}A_nV_{nx}$, is an isometric isomorphism. Moreover, (A_n) is invertible in $\mathscr{U}_x$ if and only if $W_x(A_n)$ is invertible in $\mathscr{L}\left(L^p(\mathbb{R}^2)\right)$;*

(ii) *the homomorphism*

$$\mathsf{w}_x : \Phi_x(\mathscr{D}^{\mathscr{G}}) \to \mathscr{T}_x, \quad ((A_n) + \mathscr{G}) + \mathscr{J}_x \mapsto \text{s-lim}\, V_{-nx}A_nV_{nx}$$

is an isometric isomorphism, and an element $a \in \Phi_x(\mathscr{D}_x^{\mathscr{G}})$ is invertible in $\Phi_x(\mathscr{U}_x^{\mathscr{G}})$ if and only if $\mathsf{w}_x(a)$ is invertible in $\mathscr{L}\left(L^p(\mathbb{R}^2)\right)$.

Proof. It remains to prove (ii). The proof is analogous to that of (i). First of all, w_x maps the generators of $\Phi_x(\mathscr{D}^{\mathscr{G}})$ onto the generators of $\mathscr{T}_x$, that is, the image of $\Phi_x(\mathscr{D}^{\mathscr{G}})$ under w_x is dense in $\mathscr{T}_x$. Let π denote the canonical homomorphism from $\mathscr{U}$ onto $\mathscr{U}^{\mathscr{G}}$. Define the homomorphism $\mathsf{w}_x^{-1} : \mathscr{T}_x \to \Phi_x(\mathscr{D}^{\mathscr{G}})$ by $\mathsf{w}_x^{-1} := \Phi_x \circ \pi \circ W_x^{-1}$. Then w_x^{-1} is the inverse to w_x. That w_x is an isometry follows from $\|\mathsf{w}_x\| \leq 1$ and $\|\mathsf{w}_x^{-1}\| \leq 1$. Finally, that $a \in \Phi_x(\mathscr{D}^{\mathscr{G}})$ is invertible in $\mathscr{U}_x^{\mathscr{G}} / \mathscr{J}_x$ if and only if $\mathsf{w}_x(a)$ is invertible in $\mathscr{L}\left(L^p(\mathbb{R}^2)\right)$, follows as the analogous claim in (i). ∎

Note that $\Phi_x(\mathscr{F}^{\mathscr{G}})$ is a subalgebra of $\mathscr{U}_x^{\mathscr{G}} / \mathscr{J}_x$.

Proposition 6.7.14. *Let $x \in \overline{\mathbb{R}^2}$. The local algebra $\Phi_x(\mathscr{F}^{\mathscr{G}})$ is isometrically isomorphic to a Banach algebra $\mathscr{T}_x \subset \mathscr{L}\left(L^p(\mathbb{R}^2)\right)$ of operators, namely to:*

(i) $\mathscr{T}_x = \mathrm{alg}\left\{\chi_{K_x^0}A\chi_{K_x^0}I + (1-\chi_{K_x^0})I : A \in \mathrm{alg}\{W_a^0, I : a \in L^1(\mathbb{R}^2)\}\right\}$ *if* $x \in \partial\Omega \setminus \{0\};$

(ii) $\mathscr{T}_x = \mathrm{alg}\{W_a^0, I : a \in L^1(\mathbb{R}^2)\}$ *if* $x \in (\mathrm{int}\,\Omega)\setminus\{0\}$ *and*
$\mathscr{T}_x = \mathscr{A}_0 := \mathrm{alg}\{W_a^0, fI : a \in L^1(\mathbb{R}^2), f \in C(\overline{\mathbb{R}^2})\}$ *if* $x = 0 \in \mathrm{int}\,\Omega;$

(iii) $\mathscr{T}_x = \mathrm{alg}\left\{\chi_{K_0}A\chi_{K_0}I + (1-\chi_{K_0})I : A \in \mathscr{A}_0\right\}$ *if* $x = 0$ *and* $0 \in \partial\Omega;$

(iv) $\mathscr{T}_x = \mathbb{C}I$ *if* $x \in \overline{\mathbb{R}^2}\setminus\Omega$, *where the isomorphism takes* $(((1-\chi_{n\Omega})I)+\mathscr{G})+\mathscr{J}_x$
to I.

Moreover, in cases (i)–(iii), *the isometry can be given by*

$$\mathsf{w}_x : \Phi_x(\mathscr{F}^{\mathscr{G}}), \quad ((A_n)+\mathscr{G})+\mathscr{J}_x \mapsto \mathrm{s\text{-}lim}\,V_{-nx}A_nV_{nx},$$

and an element $a \in \Phi_x(\mathscr{F}^{\mathscr{G}})$ *is invertible in* $\mathscr{U}_x^{\mathscr{G}}/\mathscr{J}_x$ *if and only if the operator* $\mathsf{w}_x(a)$ *is invertible in* $\mathscr{L}\left(L^p(\mathbb{R}^2)\right)$.

Proof. Assertion (i) is Proposition 6.7.13. For (ii), first let $x \in (\mathrm{int}\,\Omega)\setminus\{0\}$. By Proposition 6.7.7,

$$\Phi_x(fI) = f(\theta_\infty)\Phi_x(I) \quad \text{and} \quad \Phi_x(\chi_{n\Omega}I) = \Phi_x(I).$$

Now consider the homomorphism

$$\mathsf{w}_x : \Phi_x(\mathscr{F}^{\mathscr{G}}) \to \mathscr{L}\left(L^p(\mathbb{R}^2)\right), \quad ((A_n)+\mathscr{G})+\mathscr{J}_x \mapsto \mathrm{s\text{-}lim}\,V_{-nx}A_nV_{nx}.$$

One has

$$\mathsf{w}_x\left(\Phi_x(fI)\right) = f(\theta_\infty)I, \quad \mathsf{w}_x\left(\Phi_x(\chi_{n\Omega}I)\right) = I, \quad \mathsf{w}_x\left(\Phi_x(W_a^0)\right) = W_a^0.$$

Hence, w_x takes $\Phi_x(\mathscr{F}^{\mathscr{G}})$ to $\mathscr{T}_x$ for $x \in (\mathrm{int}\,\Omega)\setminus\{0\}$. Conversely, if $A \in \mathscr{T}_x$, then $(A_n) := (\chi_{n\Omega}A\chi_{n\Omega}I + (1-\chi_{n\Omega})I) \in \mathscr{F}$ and $\mathsf{w}_x(\Phi_x(A_n)) = A$. Therefore, w_x is an isometric isomorphism. If $x = 0 \in \mathrm{int}\,\Omega$ then the proof is analogous, as it is for (iii).

Finally, if $x \in \overline{\mathbb{R}^2}\setminus\Omega$ then, by Proposition 6.7.7, $\Phi_x((1-\chi_{n\Omega})I) = \Phi_x(I)$. Next we prove that $\Phi_x(I) \neq 0$. Let $y \in \mathbb{R}^2\setminus\Omega$ and consider the homomorphism

$$\mathsf{w}_y : \Phi_y(\mathscr{F}^{\mathscr{G}}) \to \mathscr{L}\left(L^p(\mathbb{R}^2)\right), \quad (\mathbf{A}+\mathscr{G})+\mathscr{J}_y \mapsto \mathsf{W}_y(\mathbf{A}).$$

Since $\mathsf{w}_y\left(\Phi_y(\lambda(1-\chi_{n\Omega})I)\right) = \lambda I$ and $\Phi_y(\mathscr{F}^{\mathscr{G}})$ is generated by $\Phi_y(I)$, we get that $\Phi_y(\mathscr{F}^{\mathscr{G}}) \cong \mathbb{C}$. Moreover, this argument also shows that $\Phi_y(\mathscr{F}^{\mathscr{G}})$ is independent of the particular choice of $y \in \mathbb{R}^2\setminus\Omega$. From Theorem 2.2.2 (ii) we infer that the mapping

$$\overline{\mathbb{R}^2} \ni x \mapsto \Phi_x(A_n)$$

is upper semi-continuous. Using that $\mathbb{R}^2$ is dense in $\overline{\mathbb{R}^2}$ we get that $\Phi_z((1-\chi_{n\Omega})I) = \Phi_z(I) \neq 0$ for all $z \in \overline{\mathbb{R}^2}\setminus\mathbb{R}^2$. Hence, $\Phi_x(\mathscr{F}^{\mathscr{G}}) \cong \mathbb{C}$ for all $x \in \overline{\mathbb{R}^2}\setminus\mathbb{R}^2$ and these algebras do not depend on x. The quoted invertibility properties follow as in the proof of Proposition 6.7.13.   ∎

Now we have prepared all the ingredients to prove the main theorem of this section.

Theorem 6.7.15. *A sequence* $\mathbf{A} = (A_n) \in \mathcal{F}$ *is stable if and only if the operator* $W_x(\mathbf{A}) := \text{s-lim}\, V_{-nx} A_n V_{nx}$ *is invertible in* $\mathcal{L}(L^p(\mathbb{R}^2))$ *for each* $x \in \partial\Omega \cup \{0\}$.

Proof. Let $\mathbf{A} \in \mathcal{F}$ be stable. Due to the inverse-closedness of $\mathcal{U}^{\mathcal{G}}$ in $\mathcal{E}^{\mathcal{G}}$, the stability of $\mathbf{A}$ is equivalent to the invertibility of the coset $\mathbf{A} + \mathcal{G}$ in $\mathcal{U}^{\mathcal{G}}$. We shall need also that the algebras $\mathcal{U}_x^{\mathcal{G}}$ are inverse-closed in $\mathcal{E}^{\mathcal{G}}$, which can easily be seen as follows: If (A_n) is stable, then also $(V_{-nx} A_n V_{nx})$ is stable, that is, there is a constant $C > 0$ such that

$$\|V_{-nx} A_n V_{nx} u\| \geq C\|u\|$$

for all $u \in L^p(\mathbb{R}^2)$. Thus, $A_x := \text{s-lim}\, V_{-nx} A_n V_{nx}$ has a closed image and trivial kernel. Repeating these arguments for (A_n^*) we get that A_x is invertible. Now let $(B_n) \in \mathcal{E}$ and $(G_n), (G_n') \in \mathcal{G}$ be sequences such that

$$B_n A_n = I + G_n, \qquad A_n B_n = I + G_n'.$$

Then $V_{-nx} B_n V_{nx} V_{-nx} A_n V_{nx} = I + V_{-nx} G_n V_{nx}$ and

$$V_{-nx} B_n V_{nx} (A_x + (V_{-nx} A_n V_{nx} - A_x)) A_x^{-1} = (I + V_{-nx} G_n V_{nx}) A_x^{-1}.$$

Since $V_{-nx} A_n V_{nx} - A_x$ and $V_{-nx} G_n V_{nx}$ converge strongly to zero, it follows that $V_{-nx} B_n V_{nx}$ converges strongly to A_x^{-1}. Analogously, $(A_x^{-1})^* = \text{s-lim}\, V_{-nx} B_n^* V_{nx}$. Using that $\mathcal{U}^{\mathcal{G}}$ is inverse-closed in $\mathcal{E}^{\mathcal{G}}$, it is thus easily seen that $\mathcal{U}_x^{\mathcal{G}}$ is inverse-closed in $\mathcal{E}^{\mathcal{G}}$, and $(A_n) + \mathcal{G}$ is invertible in $\mathcal{U}_x^{\mathcal{G}}$.

Now from Allan's local principle $(\mathbf{A} + \mathcal{G}) + \mathcal{J}_x$ is invertible in $\mathcal{U}_x^{\mathcal{G}} / \mathcal{J}_x$, which is a subalgebra of $\mathcal{U}^{\mathcal{G}} / \mathcal{J}_x$ by Proposition 6.7.12. Using Proposition 6.7.14 we obtain also that $W_x(\mathbf{A})$ is invertible in $\mathcal{T}_x$ (in particular for $x \in \Omega \cup \{0\}$). The "only if" part is proved.

For the reverse direction suppose that $W_x(\mathbf{A})$ is invertible in $\mathcal{L}(L^p(\mathbb{R}^2))$ for every $x \in \partial\Omega \cup \{0\}$. First, let $x \in \partial\Omega \setminus \{0\}$. By Proposition 6.7.11 we have $h_\rho \Phi_x(\mathcal{F}^{\mathcal{G}}) = \Phi_x(\mathcal{D}_x)$, and Proposition 6.7.13 implies that $\Phi_x(\mathcal{D}_x)$ is isometrically isomorphic to $\mathcal{T}_x$. Moreover, the restriction of w_x to $\Phi_x(\mathcal{F}^{\mathcal{G}})$ is an isometric isomorphism onto $\mathcal{T}_x$ with $w_x(\Phi_x(\mathbf{A})) = W_x(\mathbf{A})$. Proposition 6.7.13 (ii) gives that $\Phi_x(V_{nx} W_x(\mathbf{A}) V_{-nx}) \in \Phi_x(\mathcal{D}_x^{\mathcal{G}})$ is invertible in $\mathcal{U}_x^{\mathcal{G}} / \mathcal{J}_x$ and thus also in $\mathcal{U}^{\mathcal{G}} / \mathcal{J}_x$. Using that $h_\rho \Phi_x(\mathbf{A}) = \Phi_x(V_{nx} W_x(\mathbf{A}) V_{-nx})$ and

$$h_\rho^{-1} : \mathcal{U}^{\mathcal{G}} / \mathcal{J}_x \to \mathcal{U}^{\mathcal{G}} / \mathcal{J}_x$$

is the inverse of h_ρ, we get

$$\Phi_x(\mathbf{A}) = h_\rho^{-1} \Phi_x(V_{nx} W_x(\mathbf{A}) V_{-nx}).$$

Thus, $\Phi_x(\mathbf{A})$ is invertible in $\mathscr{U}^{\mathscr{G}}/\mathscr{J}_x$. If $x = 0$, then the assertion follows from Proposition 6.7.14. Finally we show the invertibility of $\Phi_x(\mathbf{A})$ for all $x \in \overline{\mathbb{R}^2}$.

Let $x \in \operatorname{int} \Omega \setminus \{0\}$. From Corollary 5.8.13 we conclude that $W_x(\mathbf{A})$ is invertible in $\mathscr{L}(L^p(\mathbb{R}^2))$ if $W_0(\mathbf{A})$ is so. We are going to prove that if $W_y(\mathbf{A}) \in \mathscr{T}_y$ is invertible in $\mathscr{L}(L^p(\mathbb{R}^2))$ for at least one $y \in \partial\Omega \setminus \{0\}$, then $\Phi_x(\mathbf{A})$ is invertible for all $x \in \overline{\mathbb{R}^2} \setminus \Omega$. We will use that every sequence $\mathbf{A} \in \mathscr{F}$ can be uniquely represented as

$$(A_n) = (\chi_{n\Omega}B_n\chi_{n\Omega}I + \lambda(1 - \chi_{n\Omega})I), \quad \lambda \in \mathbb{C}$$

and every $A \in \mathscr{T}_y$ as

$$A = \chi_{K_y^0}B\chi_{K_y^0}I + \mu(1 - \chi_{K_y^0})I.$$

Let $\psi : \mathscr{T}_y \to \mathbb{C}$ be the homomorphism that assigns to A the complex number μ. Because

$$\psi(W_y(\mathbf{A})) = \lambda = \Phi_x(\mathbf{A}) \quad \text{for all } x \in \overline{\mathbb{R}^2} \setminus \Omega$$

and $\lambda \neq 0$, we get the invertibility of $\psi_x(\mathbf{A})$. Proposition 6.7.14 now shows that for every $x \in \overline{\mathbb{R}^2}$ the element $(\mathbf{A} + \mathscr{G}) + \mathscr{J}_x$ is invertible in $\mathscr{U}^{\mathscr{G}}/\mathscr{J}_x$. It remains to apply Allan's local principle to obtain that $\mathbf{A} + \mathscr{G}$ is invertible in $\mathscr{U}^{\mathscr{G}}$. ■

We will illustrate this theorem by some examples. Take into account Corollary 5.8.13.

Example 6.7.16. Let $0 \in \partial\Omega$ (then $\Omega \subset K_0$ by assumption), and let A_{ij} be operators in $\mathscr{A}_0 = \operatorname{alg}\{W_a^0, fI : f \in C(\overline{\mathbb{R}^2})\}$. The sequence

$$\sum_{i=1}^{m} \prod_{j=1}^{l} (\chi_{n\Omega}A_{ij}\chi_{n\Omega}I + (1 - \chi_{n\Omega})I)$$

is stable if and only if the following conditions are satisfied:

(i) $\sum_{i=1}^{m} \prod_{j=1}^{l} (\chi_{K_0}A_{ij}\chi_{K_0}I + (1 - \chi_{K_0})I)$ is invertible in $\mathscr{L}(L^p(\mathbb{R}^2))$;

(ii) $\sum_{i=1}^{m} \prod_{j=1}^{l} (\chi_{K_x^0}W_x(A_{ij})\chi_{K_x^0}I + (1 - \chi_{K_x^0})I)$ is invertible in $\mathscr{L}(L^p(\mathbb{R}^2))$ for every $x \in \partial\Omega \setminus \{0\}$.

□

Example 6.7.17. Let Ω be the disk $\overline{B_R(0)}$, and let $A = W_a^0 + fI$ with $a \in L^1(\mathbb{R}^2)$ and $f \in C(\overline{\mathbb{R}^2})$. Then the sequence $\chi_{n\Omega}A\chi_{n\Omega}I + (1 - \chi_{n\Omega})I$ is stable if and only if A is invertible in $\mathscr{L}(L^p(\mathbb{R}^2))$.

□

Example 6.7.18. Let $0 \in \partial\Omega$, $\partial\Omega$ be a smooth curve except at zero and $\Omega \subset K_0$. Consider $A = W_a^0 + fI$ with a and f as in the previous example. Then the sequence $\chi_{n\Omega} A \chi_{n\Omega} I + (1 - \chi_{n\Omega})I$ is stable if and only if

$$\chi_{K_0} A \chi_{K_0} I + (1 - \chi_{K_0})I$$

is invertible in $\mathscr{L}(L^p(\mathbb{R}^2))$.   □

Finally we would like to emphasize that $\mathscr{U}^\mathscr{G}$ and $\mathscr{C}$ constitute a faithful localization pair. Hence, for $\mathbf{A} = (A_n) \in \mathscr{F}$,

$$\|\mathbf{A} + \mathscr{G}\| = \sup_{x \in \partial\Omega \cup \{0\}} \|\mathsf{W}_x(\mathbf{A})\|.$$

If $\mathbf{A}$ is stable, then also

$$\|(\mathbf{A} + \mathscr{G})^{-1}\| = \sup_{x \in \partial\Omega \cup \{0\}} \|\mathsf{W}_x^{-1}(\mathbf{A})\|.$$

6.8 Notes and comments

Since Baxter's pioneering paper [6], the finite sections method for convolution operators has been the subject of numerous investigations by many authors. Gohberg and Feldman established the first systematic and comprehensive theory of projection methods for convolution equations. Their monograph [66] is a basic reference on this topic still. In 1973–4, a new idea appeared in the study of projection methods for one- and multidimensional convolution operators with continuous symbols, namely the use of Banach algebra techniques as pointed out by Kozak in [102]. Essentially, Theorem 6.2.2 is due to him and the main meaning of this theorem is indeed that stability is an invertibility problem.

It is now well understood that Kozak's local theory does not work in the case of discontinuous generating functions. In order to settle this case, new ideas were needed. Section 6.3 presents a general framework which is useful to achieve this aim. It appeared in a particular case in [181] and was then subsequently extended (see for instance [21, 81, 82]). Theorem 6.3.8 is here formulated for the first time; particular cases were used previously.

The material in Section 6.4 is the Wiener-Hopf version of finite sections of Toeplitz operators with continuous generating functions presented, for instance, in [20]. Theorem 6.4.6 is due to Gohberg and Feldman [66].

Theorem 6.5.5 was proved by Elschner [49] using factorization methods which are close to the ideas of Gohberg and Feldman in [66]. Moreover, Elschner studied a variety of approximation methods for Wiener-Hopf and Mellin integral equations with continuous generating functions. His investigations are reflected in [151, Chapter 5] where further historical remarks can be found. Note also the papers [173] and [174] where spline methods are considered in a C^*-algebra context. A general abstract framework for the study of approximation methods for convolution operators with continuous generating functions is in [183]. For piecewise continuous functions of isometries in Hilbert spaces, fulfilling some additional natural conditions, the results are presented in [175].

The results of Section 6.6, as well as the methods of proving them are in the spirit of [165] where the case $p = 2$ was considered. For $p \neq 2$, the results of this section were published in [166]. The proof of Proposition 6.6.21 given above corrects an inaccuracy in the corresponding proof in that paper. Similar methods were also used in [99] to study finite sections of convolution operators generated by slowly oscillating functions.

In Section 6.7 we reproduce and generalize some results from Kozak [102]. We partly follow [118] where the case $p = 2$ is considered, and the PhD thesis of Mascarenhas [117] in regard to some aspects of the problem for $p \neq 2$. The new aspects, in comparison with Kozak's work are the inclusion of the multiplication operators fI, $f \in C(\overline{\mathbb{R}^2})$ into the studied algebra, and the thorough use of limit operator techniques. Note that the paper [119] by Maximenko is also devoted to a particular case of these studies. In particular, he considered ε-pseudospectra and used similar ideas to the ones that appeared for the first time in [164].

Note that in this chapter we study not only single sequences but analyze Banach algebras of sequences. This allows us to get further insight into the structure of these sequences and to get knowledge about the asymptotic behavior of some spectral quantities such as condition numbers, ε-pseudospectra and so on (see for instance [20] and [82]).

References

1. L. Ahlfors. *Complex analysis*. McGraw-Hill, New York, 1978.
2. G.R. Allan. Ideals of vector-valued functions. *Proc. London Math. Soc.*, 18(2):193–216, 1968.
3. W. Arveson. *An invitation to C*-algebras*. Springer-Verlag, New York, 1976.
4. B. Aupetit. *A primer on spectral theory*. Springer-Verlag, New York, 1991.
5. R.E. Avendaño and N. Krupnik. Local principle of computation of the factor-norms of singular integral-operators. *Funkc. Analiz i Priloz.*, 22(2):57–58, 1988. (in Russian; English translation in *Functional Anal. Appl.* 22(2): 130–131, 1988).
6. G. Baxter. A norm inequality for a "finite-section" Wiener-Hopf equation. *Illinois J. Math.*, 7:97–103, 1963.
7. C. Bennett and R. Sharpley. *Interpolation of operators*. Academic Press, London, 1988.
8. B. Blackadar. *Operator algebras. Theory of C*-algebras and von Neumann algebras*. Springer-Verlag, Berlin, Heidelberg, 2006.
9. S. Bochner and R.S. Phillips. Absolutely convergent Fourier expansions for non-commutative normed rings. *Ann. of Math. (2)*, 43:409–418, 1942.
10. F.F. Bonsall and J. Duncan. *Complete normed algebras*. Springer-Verlag, New York, 1973.
11. A. Böttcher. Toeplitz operators with piecewise continuous symbols—a neverending story? *Jahresber. Deutsch. Math.-Verein.*, 97(4):115–129, 1995.
12. A. Böttcher, I. Gohberg, Yu.I. Karlovich, N. Krupnik, S. Roch, B. Silbermann, and I. Spitkovsky. Banach algebras generated by N idempotents and applications. In *Singular integral operators and related topics*, volume 90 of *Oper. Theory Adv. Appl.*, pages 19–54. Birkhäuser, Basel, 1996.
13. A. Böttcher and Yu.I. Karlovich. Toeplitz and singular integral operators on Carleson curves with logarithmic whirl points. *Integral Equations Operator Theory*, 22:127–161, 1995.
14. A. Böttcher and Yu.I. Karlovich. *Carleson curves, Muckenhoupt weights, and Toeplitz operators*. Birkhäuser, Berlin, 1997.
15. A. Böttcher, Yu.I. Karlovich, and V.S. Rabinovich. Emergence, persistence, and disappearance of logarithmic spirals in the spectra of singular integral operators. *Integral Equations Operator Theory*, 25:404–444, 1996.
16. A. Böttcher, Yu.I. Karlovich, and V.S. Rabinovich. Mellin pseudodifferential operators with slowly varying symbols and singular integrals on Carleson curves with Muckenhoupt weights. *Manuscripta Math.*, 95:363–376, 1998.
17. A. Böttcher, Yu.I. Karlovich, and V.S. Rabinovich. Singular integral operators with complex conjugation from the viewpoint of pseudodifferential operators. In *Problems and methods in mathematical physics (Chemnitz, 1999)*, volume 121 of *Oper. Theory Adv. Appl.*, pages 36–59. Birkhäuser, Basel, 2001.
18. A. Böttcher, N. Krupnik, and B. Silbermann. A general look at local principles with special emphasis on the norm computation aspect. *Integral Equations Operator Theory*, 11:455–479, 1988.

S. Roch et al., *Non-commutative Gelfand Theories*, Universitext,
DOI 10.1007/978-0-85729-183-7, © Springer-Verlag London Limited 2011

19. A. Böttcher, S. Roch, B. Silbermann, and I. Spitkovsky. A Gohberg–Krupnik–Sarason symbol calculus for algebras of Toeplitz, Hankel, Cauchy, and Carleman operators. In *Topics in operator theory: Ernst D. Hellinger memorial volume*, volume 48 of *Oper. Theory Adv. Appl.*, pages 189–234. Birkhäuser, Basel, 1990.

20. A. Böttcher and B. Silbermann. *Introduction to large truncated Toeplitz matrices.* Universitext. Springer-Verlag, New York, 1999.

21. A. Böttcher and B. Silbermann. *Analysis of Toeplitz operators.* Springer-Verlag, Berlin, second edition, 2006.

22. A. Böttcher and I. Spitkovsky. Pseudodifferential operators with heavy spectrum. *Integral Equations Operator Theory*, 19:251–269, 1994.

23. A. Böttcher and I. Spitkovsky. A gentle guide to the basics of two projections theory. *Lin. Alg. Appl.*, 432(6):1412–1459, 2010.

24. O. Bratteli and D.W. Robinson. *Operator algebras and quantum statistical mechanics 1: C^*- and W^*-algebras, symmetry groups, decomposition of states.* Springer-Verlag, 2002.

25. L. Castro, R. Duduchava, and F.-O. Speck. Singular integral equations on piecewise smooth curves. In *Toeplitz matrices and singular integral equations (Pobershau, 2001)*, volume 135 of *Oper. Theory Adv. Appl.*, pages 107–144. Birkhäuser, Basel, 2002.

26. J.B. Conway. *A course in functional analysis.* Springer-Verlag, New York, second edition, 1990.

27. H.O. Cordes. *Elliptic pseudo-differential operators – an abstract theory*, volume 756 of *Springer Lecture Notes Math.* Springer, Berlin, 1979.

28. M. Costabel. An inverse for the Gohberg-Krupnik symbol map. *Proc. Royal Soc. Edinburgh*, 87 A:153–165, 1980.

29. M. Costabel. Singular integral operators on curves with corners. *Integral Equations Operator Theory*, 3:323–349, 1980.

30. M. Costabel. Singuläre Integralgleichungen mit Carlemanscher Verschiebung auf Kurven mit Ecken. *Math. Nachr.*, 109:29–37, 1982.

31. M. Cwikel. Real and complex interpolation and extrapolation of compact operators. *Duke Math. J.*, 65(2):333–343, 1992.

32. G. Dales. *Banach algebras and automatic continuity.* Clarendon Press, Oxford, 2000.

33. J. Dauns and K. H. Hofmann. *Representation of rings by sections.* Amer. Math. Soc., Providence, RI, 1968.

34. K.R. Davidson. *C^*-algebras by example.* Amer. Math. Soc., Providence, RI, 1996.

35. C. Davis. Separation of two linear subspaces. *Acta Sci. Math. (Szeged)*, 16:172–187, 1958.

36. V. Didenko and B. Silbermann. *Approximation of additive convolution-like operators: real C^*-algebra approach.* Birkhäuser, Basel, Boston, Berlin, 2008.

37. J. Dieudonné. *Treatise on analysis. Vol. II.* Academic Press, New York, 1976.

38. J. Dixmier. Position relative de deux variétés linéaires fermées dans un espace de Hilbert. *Revue Sci.*, 86:387–399, 1948.

39. J. Dixmier. *C^*-algebras.* North-Holland, Amsterdam, 1977.

40. R.G. Douglas. Local Toeplitz operators. *Proc. London Math. Soc. (3)*, 36(2):243–272, 1978.

41. R.G. Douglas. *Banach algebra techniques in operator theory*, volume 179 of *Graduate Texts in Mathematics.* Springer-Verlag, New York, second edition, 1998.

42. V. Drensky. *Free algebras and PI-algebras.* Springer-Verlag, Singapore, 2000.

43. R. Duduchava. *Integral equations with fixed singularities.* B.G. Teubner Verlaggesellschaft, Leipzig, 1979.

44. R. Duduchava. On algebras generated by convolutions and discontinuous functions. *Integral Equations Operator Theory*, 10(4):505–530, 1987.

45. R. Duduchava and N. Krupnik. On the norm of singular integral operator on curves with cusps. *Integral Equations Operator Theory*, 20(4):377–382, 1994.

46. R. Duduchava, N. Krupnik, and E. Shargorodsky. An algebra of integral operators with fixed singularities in kernels. *Integral Equations Operator Theory*, 33(4):406–425, 1999.

47. R. Duduchava, T. Latsabidze, and A. Saginashvili. Singular integral operators with the complex conjugation on curves with cusps. *Integral Equations Operator Theory*, 22(1):1–36, 1995.

48. T. Ehrhardt, S. Roch, and B. Silbermann. Symbol calculus for singular integrals with operator-valued PQC-coefficients. In *Singular integral operators and related topics*, volume 90 of *Oper. Theory Adv. Appl.*, pages 182–203. Birkhäuser, Basel, 1996.

49. J. Elschner. *On spline approximation for a class of non-compact integral equations.* Akademie der Wissenschaften der DDR Karl-Weierstrass-Institut für Mathematik, Berlin, 1988.

50. I. Feldman and N. Krupnik. On the continuity of the spectrum in certain Banach algebras. *Integral Equations Operator Theory*, 38(3):284–301, 2000.

51. I. Feldman, N. Krupnik, and I. Spitkovsky. Norms of the singular integral operator with Cauchy kernel along certain contours. *Integral Equations Operator Theory*, 24(1):68–80, 1996.

52. G. Fendler, K. Gröchenig, M. Leinert, J. Ludwig, and C. Molitor-Braun. Weighted group algebras on groups of polynomial growth. *Math. Z.*, 245(4):791–821, 2003.

53. A. Figa-Talamanca and G.I. Gaudry. Multipliers of L^p which vanish at infinity. *J. Funct. Anal.*, 7:475–486, 1971.

54. T. Finck and S. Roch. Banach algebras with matrix symbol of bounded order. *Integral Equations Operator Theory*, 18:427–434, 1994.

55. T. Finck, S. Roch, and B. Silbermann. Two projection theorems and symbol calculus for operators with massive local spectra. *Math. Nachr.*, 162:167–185, 1993.

56. T. Finck, S. Roch, and B. Silbermann. Banach algebras generated by two idempotents and one flip. *Math. Nachr.*, 216:73–94, 2000.

57. V. Fock. On certain integral equations in mathematical physics. *Rec. Math. [Mat. Sbornik] N.S.*, 14(56)(1–2):3–50, 1944. (in Russian).

58. J. Galperin and N. Krupnik. On the norms of singular integral operators along certain curves with intersections. *Integral Equations Operator Theory*, 29(1):10–16, 1997.

59. I. Gelfand. Abstrakte funktionen und lineare operatoren. *Rec. Math. [Mat. Sbornik] N.S.*, 4(46)(2):235–284, 1938.

60. I. Gelfand. On normed rings. *C. R. (Doklady) Acad. Sci. URSS (N.S.)*, 23:430–432, 1939.

61. I. Gelfand. To the theory of normed rings. II. On absolutely convergent trigonometrical series and integrals. *C. R. (Doklady) Acad. Sci. URSS (N.S.)*, 25:570–572, 1939.

62. I. Gelfand. Normierte Ringe. *Rec. Math. [Mat. Sbornik] N.S.*, 9(51)(1):3–24, 1941.

63. I. Gelfand and M. Naimark. On the embedding of normed rings into the ring of operators in Hilbert space. *Rec. Math. [Mat. Sbornik] N.S.*, 12(54)(2):197–213, 1943.

64. J. Glimm. A Stone-Weierstrass theorem for C^*-algebras. *Ann. Math.*, 2(72):216–244, 1960.

65. I. Gohberg. On an application of the theory of normed rings to singular integral equations. *Uspekhi Mat. Nauk*, 7:149–156, 1952. (in Russian).

66. I. Gohberg and I.A. Feldman. *Convolution equations and projection methods for their solution.* Amer. Math. Soc., Providence, RI, 1974. First published in Russian, Nauka, Moscow, 1971.

67. I. Gohberg, S. Goldberg, and M.A. Kaashoek. *Classes of Linear Operators, II.* Birkhäuser, Basel, Boston, Berlin, 1993.

68. I. Gohberg and M.G. Krein. Systems of integral equations on the half-line with kernels depending on the difference of the arguments. *Uspehi Mat. Nauk (N.S.)*, 13(2(80)):3–72, 1958. (in Russian; English translation in *Amer. Math. Soc. Transl.* (2), 14, 217–287, 1960).

69. I. Gohberg and N. Krupnik. Algebras of singular integral operators with shifts. *Mat. Issled.*, 8(2(28)):170–175, 1973. (in Russian; English translation in "Convolution equations and singular integral operators", *Oper. Theory Adv. Appl.*, vol. 206 (2010), 213–217).

70. I. Gohberg and N. Krupnik. On a local principle and algebras generated by Toeplitz matrices. *An. Şti. Univ. "Al. I. Cuza" Iaşi Secţ. I a Mat. (N.S.)*, 19:43–71, 1973. (in Russian; English translation in "Convolution equations and singular integral operators", *Oper. Theory Adv. Appl.*, vol. 206 (2010), 157–184).

71. I. Gohberg and N. Krupnik. One-dimensional singular integral operators with shift. *Izv. Akad. Nauk Armjan. SSR Ser. Mat.*, 8(1):3–12, 1973. (in Russian; English translation in "Convolution equations and singular integral operators", *Oper. Theory Adv. Appl.*, vol. 206 (2010), 201–211).

72. I. Gohberg and N. Krupnik. Extension theorems for invertibility symbols in Banach algebras. *Integral Equations Operator Theory*, 15:991–1010, 1992.
73. I. Gohberg and N. Krupnik. *One-dimensional linear singular integral equations (I)*. Birkhäuser, Basel, 1992.
74. I. Gohberg and N. Krupnik. *One-dimensional linear singular integral equations (II)*. Birkhäuser, Basel, 1992.
75. I. Gohberg and N. Krupnik. Extension theorems for Fredholm and invertibility symbols. *Integral Equations Operator Theory*, 16:514–529, 1993.
76. I. Gohberg, N. Krupnik, and I. Spitkovsky. Banach algebras of singular integral operators with piecewise continuous coefficients. General contour and weight. *Integral Equations Operator Theory*, 17:322–337, 1993.
77. L.S. Goldenstein and I. Gohberg. On a multidimensional integral equation on a half-space whose kernel is a function of the difference of the arguments, and on a discrete analogue of this equation. *Soviet Math. Dokl.*, 1:173–176, 1960.
78. D. Goldstein. Inverse closedness of C^*-algebras in Banach algebras. *Integral Equations Operator Theory*, 33:172–174, 1999.
79. I.S. Gradshteyn and I.M. Ryzhik. *Table of integrals, series, and products*. Academic Press, New York, fourth edition, 1965.
80. L. Grafakos. *Modern Fourier analysis*. Springer, New York, second edition, 2009.
81. R. Hagen, S. Roch, and B. Silbermann. *Spectral theory of approximation methods for convolution equations*. Birkhäuser, Basel, 1995.
82. R. Hagen, S. Roch, and B. Silbermann. C^*-algebras and numerical analysis. Marcel Dekker, Inc., New York, 2001.
83. P.R. Halmos. Two subspaces. *Trans. Amer. Math. Soc.*, 144:381–389, 1969.
84. A.Y. Helemskii. *Banach and locally convex algebras*. The Clarendon Press, Oxford, New York, 1993.
85. K.H. Hofmann. Representations of algebras by continuous sections. *Bull. Amer. Math. Soc.*, 78:291–373, 1972.
86. B. Hollenbeck and I.E. Verbitsky. Best constants for the Riesz projection. *J. Funct. Anal.*, 175(2):370–392, 2000.
87. L. Hörmander. Estimates for translation invariant operators in L^p spaces. *Acta Math.*, 104:93–140, 1960.
88. R.A. Horn and C.A. Johnson. *Matrix analysis*. Cambridge University Press, Cambridge, 1990. Corrected reprint of the 1985 original.
89. A. Hulanicki. On the spectrum of convolution operators on groups with polynomial growth. *Invent. Math.*, 17:135–142, 1972.
90. N. Jacobson. The radical and semi-simplicity for arbitrary rings. *American Journal of Mathematics*, 67(2):300–320, 1945.
91. R.V. Kadison and J.R. Ringrose. *Fundamentals of the theory of operator algebras (I)*. Academic Press Inc., New York, 1983. Reprinted in Graduate Studies in Mathematics, 15. American Mathematical Society, Providence, RI, 1997.
92. R.V. Kadison and J.R. Ringrose. *Fundamentals of the theory of operator algebras (II)*. Academic Press Inc., Orlando, FL, 1986. Reprinted in Graduate Studies in Mathematics, 16. American Mathematical Society, Providence, RI, 1997.
93. L.V. Kantorovich and G.P. Akilov. *Functional analysis*. Pergamon Press, 1982.
94. N. Karapetiants and S. Samko. On Fredholm properties of a class of Hankel operators. *Math. Nachr.*, 217:75–103, 2000.
95. N. Karapetiants and S. Samko. *Equations with involutive operators*. Birkhäuser Boston Inc., Boston, MA, 2001.
96. A. Karlovich. Algebras of singular integral operators with PC coefficients in rearrangement-invariant spaces with Muckenhoupt weights. *J. Operator Theory*, 47(2):303–323, 2002.
97. A. Karlovich. Fredholmness of singular integral operators with piecewise continuous coefficients on weighted Banach function spaces. *J. Integral Equations Appl.*, 15(3):263–320, 2003.

98. A. Karlovich. Singular integral operators on variable Lebesgue spaces with radial oscillating weights. In *Operator Algebras, Operator Theory and Applications*, volume 195 of *Oper. Theory Adv. Appl.*, pages 185–212. Birkhäuser, Basel, 2009.

99. A. Karlovich, H. Mascarenhas, and P.A. Santos. Finite section method for a Banach algebra of convolution type operators on Lp(R) with symbols generated by PC and SO. *Integral Equations Operator Theory*, 67(4):559–600, 2010.

100. V. Kokilashvili and S. Samko. Singular operators and Fourier multipliers in weighted Lebesgue spaces with variable index. *Vestnik St. Petersburg Univ. Math.*, 41(2):134–144, 2008.

101. G. Köthe. Die struktur der ringe, deren restklassenring nach dem radikal vollständig reduzibel ist. *Math. Zeit.*, 32(1):161–186, 1930.

102. A.V. Kozak. A local principle in the theory of projection methods. *Dokl. Akad. Nauk SSSR*, 212:1287–1289, 1973. (in Russian; English translation in *Soviet Math. Dokl.* 14:1580–1583, 1974).

103. M.A. Krasnoselskii, P.P. Zabreiko, E.I. Pustylnik, and P.E. Sobolevskii. *Integral operators in spaces of summable functions*. Noordhoff International Publishing, Leiden, 1976.

104. V.G. Kravchenko and G.S. Litvinchuk. *Introduction to the theory of singular integral operators with shift*. Kluwer Academic Publishers Group, Dordrecht, 1994.

105. M.G. Krein. Integral equations on the half-line with a kernel depending on the difference of the arguments. *Uspehi Mat. Nauk*, 13(5(83)):3–120, 1958. (in Russian; English translation in *Amer. Math. Soc. Transl.* (2), 22: 163–288, 1962).

106. M.G. Krein, M.A. Krasnoselski, and D.P. Milman. On the defect numbers of operators in Banach spaces and on some geometric questions. *Trudy Inst. Matem. Akad. Nauk Ukrain. SSR*, 11:97–112, 1948. (in Russian).

107. N. Krupnik. Exact constant in I.B. Simonenko's theorem on an envelope of a family of operators of local type. *Funkc. Analiz i Priloz.*, 20(2):70–71, 1986. (in Russian; English translation in *Functional Anal. Appl.* 20(2): 144–145, 1986).

108. N. Krupnik. *Banach algebras with symbol and singular integral operators*. Birkhäuser, Basel, 1987.

109. N. Krupnik. Minimal number of idempotent generators of matrix algebras over arbitrary field. *Communications in Algebra*, 20:3251–3257, 1992.

110. N. Krupnik and Y. Spigel. Invertibility symbols for a Banach algebra generated by two idempotents and a shift. *Integral Equations Operator Theory*, 17:567–578, 1993.

111. N. Krupnik and Y. Spigel. Critical points of essential norms of singular integral operators in weighted spaces. *Integral Equations Operator Theory*, 33(2):211–220, 1999.

112. N. Krupnik and Y. Spigel. Matrix symbol of order two for some Banach algebras. *Integral Equations Operator Theory*, 40(4):465–480, 2001.

113. N. Krupnik and I. Spitkovsky. On the norms of singular integral operators on contours with intersections. *Complex Anal. Oper. Theory*, 2(4):617–626, 2008.

114. G.S. Litvinchuk. *Boundary value problems and singular integral equations with shift*. Izdat. "Nauka", Moscow, 1977. (in Russian).

115. D. Luminet. A functional calculus for Banach PI-algebras. *Pacific J. Math.*, 125(1):127–160, 1986.

116. A.S. Markus and R. Pöltz. Various measures of non-compactness for linear operators in some Banach spaces. *Seminar Anal. 1987/88 of the Weierstrass-Institut Berlin*, pages 66–84, 1988. (in Russian).

117. H. Mascarenhas. *Convolution type operators on cones and asymptotic spectral theory, PhD Thesis*. TU Chemnitz, Fakultät für Mathematik, 2004.

118. H. Mascarenhas and B. Silbermann. Convolution type operators on cones and their finite sections. *Math. Nachr.*, 278(3):290–311, 2005.

119. E.A. Maximenko. Convolution operators on expanding polyhedra: limits of the norms of inverse operators and pseudospectra. *Sibirsk. Mat. Zh.*, 44(6):1310–1323, 2003. (in Russian; English translation: Siberian Math. J. 44, no. 6: 1027–1038, 2003).

120. S.G. Mikhlin and S. Prössdorf. *Singular integral operators*. Springer-Verlag, Berlin, 1986.

121. G.J. Murphy. *C*-algebras and operator theory*. Academic Press, San Diego, 1990.
122. F.J. Murray and J. von Neumann. On rings of operators. *Ann. of Math. 2nd Ser.*, 37(1):116–229, 1936.
123. F.J. Murray and J. von Neumann. On rings of operators II. *Trans. Amer. Math. Soc.*, 41(2):208–248, 1936.
124. F.J. Murray and J. von Neumann. On rings of operators IV. *Ann. of Math. 2nd Ser.*, 44(4):716–808, 1936.
125. N. Muskhelishvili. *Singular integral equations*. Nauka, Moscow, 1968. (in Russian).
126. T. Nakazi and T. Yamamoto. Norms of some singular integral operators and their inverse operators. *J. Operator Theory*, 40(1):185–207, 1998.
127. J.D. Newburgh. The variation of spectra. *Duke Math J.*, 18:165–176, 1951.
128. N. K. Nikolski. *Operators, functions, and systems: an easy reading. Vol. 1*, volume 92 of *Mathematical Surveys and Monographs*. Amer. Math. Soc., Providence, RI, 2002.
129. B. Noble. *Methods based on the Wiener-Hopf technique for the solution of partial differential equations*, volume 7 of *International series of monographs on pure and applied mathematics*. Pergamon Press, New York, 1958.
130. V. Nyaga. On the symbol of singular integral operators in the case of piecewise Lyapunov contours. *Mat. Issl.*, IX(2):109–125, 1974. (in Russian).
131. V. Nyaga. On the singular integral operator with conjugation on contours with angle points. *Mat. Issl.*, 73:47–50, 1983. (in Russian).
132. V. Ostrovskyi and Y. Samoilenko. *Introduction to the theory of representations of finitely presented *-algebras. I*. Harwood Academic Publishers, Amsterdam, 1999.
133. R.E. Paley and N. Wiener. *Fourier transforms in the complex domain*, volume 19 of *American Mathematical Society colloquium publications*. AMS, New York, 1934.
134. T.W. Palmer. *Banach algebras and the general theory of *-algebras, I*. Cambridge University Press, Cambridge, 1994.
135. T.W. Palmer. *Banach algebras and the general theory of *-algebras, II*. Cambridge University Press, Cambridge, 2001.
136. G.K. Pederson. Measure theory for *C**-algebras, II. *Math. Scand.*, 22:63–74, 1968.
137. G.K. Pederson. *C*-Algebras and their automorphism groups*. Academic Press, London, 1979.
138. V.V. Peller. *Hankel operators and their applications*. Springer Monographs in Mathematics. Springer-Verlag, New York, 2003.
139. B.A. Plamenevskii. *Algebras of pseudodifferential operators*. Kluwer Academic Publishers, 1986.
140. B.A. Plamenevskii and V.N. Senichkin. *C**-algebras of singular integral operators with discontinuous coefficients on a complex contour, I. *Izv. Vyssh. Uchebn. Zaved. Mat.*, (1):25–33, 1984. (in Russian; English translation in *Soviet Math. (Iz. VUZ)* 28(1):28–37, 1984).
141. B.A. Plamenevskii and V.N. Senichkin. *C**-algebras of singular integral operators with discontinuous coefficients on a complex contour, II. *Izv. Vyssh. Uchebn. Zaved. Mat.*, (4):37–64, 1984. (in Russian; English translation in *Soviet Math. (Iz. VUZ)* 28(4):47–58, 1984).
142. R. Pöltz. Operators of local type on spaces of Hölder functions. *Seminar Anal. 1986/87 of the Weierstrass-Institut Berlin*, pages 107–122, 1987. (in Russian).
143. R. Pöltz. Measures of non-compactness for bounded sets and linear operators in some Banach spaces. *Seminar Anal. 1987/88 of the Weierstrass-Institut Berlin*, pages 85–106, 1988. (in Russian).
144. R. Pöltz. Measures of non-compactness of the Cesàro operator in C^n spaces. *Seminar Anal. 1988/89 of the Weierstrass-Institut Berlin*, pages 59–66, 1989. (in Russian).
145. R. Pöltz. Noethericity and local noethericity of operators of local type. *Seminar Anal. 1988/89 of the Weierstrass-Institut Berlin*, pages 67–73, 1989. (in Russian).
146. R. Pöltz. *Various measures of non-compactness and operators of local type in some Banach spaces–Phd Thesis*. Kishinev State University, Kishinev, 1989. (in Russian).
147. S.C. Power. *C**-algebras generated by Hankel operators and Toeplitz operators. *J. Funct. Anal.*, 31:52–68, 1979.

148. S.C. Power. Essential spectra of piecewise continuous Fourier integral operators. *Proc. Royal Ir. Acad.*, 81(1):1–7, 1981.
149. S.C. Power. *Hankel operators on Hilbert space*. Pitman, Boston, London, Melbourne, 1982.
150. S. Prössdorf. *Some classes of singular equations*. North-Holland Publishing Co., Amsterdam, 1978.
151. S. Prössdorf and B. Silbermann. *Numerical analysis for integral and related operator equations*. Birkhäuser-Verlag, Basel, 1991.
152. S. Rabanovich and Y. Samoilenko. On representations of $\mathscr{F}_n$-algebras and invertibility symbols. In *Operator theory and related topics, Vol. II (Odessa, 1997)*, volume 118 of *Oper. Theory Adv. Appl.*, pages 347–357. Birkhäuser, Basel, 2000.
153. V.S. Rabinovich. Algebras of singular integral operators on composed contours with nodes that are logarithmic whirl points. *Izv. Ross. Akad. Nauk, Ser. Mat.*, 60:169–200, 1996. (in Russian; English translation in *Izv. Math.* 60(6): 1261–1292, 1996).
154. V.S. Rabinovich. Mellin pseudodifferential operators techniques in the theory of singular integral operators on some Carleson curves. In *Differential and integral operators (Regensburg, 1995)*, volume 102 of *Oper. Theory Adv. Appl.*, pages 201–218. Birkhäuser, Basel, 1998.
155. V.S. Rabinovich. Criterion for local invertibility of pseudodifferential operators with operator symbols and some applications. In *Proceedings of the St. Petersburg Mathematical Society, Vol. V*, pages 239–259, Amer. Math. Soc., Providence, RI, 1999.
156. V.S. Rabinovich, S. Roch, and B. Silbermann. *Limit operators and their applications in operator theory*. Birkhäuser, Basel, 2004.
157. V.S. Rabinovich and S. Samko. Boundedness and Fredholmness of pseudodifferential operators in variable exponent spaces. *Integral Equations Operator Theory*, 60(4):507–537, 2008.
158. I.M. Rapoport. On a class of singular integral equations. *Dokl. Akad. Nauk SSSR (N.S.)*, 59:1403–1406, 1948. (in Russian).
159. J.P. Razmyslov. Identities with trace in full matrix algebras over a field of characteristic zero. *Izv. Akad. Nauk SSSR Ser. Mat.*, 38, 1974. (in Russian; English translation in *Math. USSR-Izv.* 8 (1974), no. 4: 727–760, 1975).
160. M. Reed and B. Simon. *Methods of modern mathematical physics. Volume 1: Functional analysis*. Academic Press Inc., New York, second edition, 1980.
161. E. Reissner. Note on the problem of distribution of stress in a thin stiffened elastic sheet. *Proc. Natl. Acad. Sci. USA*, 26:300–305, 1940.
162. C. Rickart. *General theory of Banach algebras*. van Nostrand, NJ, Toronto-London-New York, 1960.
163. F. Riesz. *Les systèmes d'equations linéaires à une infinité d'inconnues*. Gauthier-Villars, Paris, 1913.
164. S. Roch. Spectral approximation of Wiener-Hopf operators with almost periodic generating function. *Numer. Funct. Anal. Optim.*, 21(1–2):241–253, 2000.
165. S. Roch, P.A. Santos, and B. Silbermann. Finite section method in some algebras of multiplication and convolution operators and a flip. *Z. Anal. Anwendungen*, 16(3):575–606, 1997.
166. S. Roch, P.A. Santos, and B. Silbermann. A sequence algebra of finite sections, convolution and multiplication operators on $L^p(\mathbb{R})$. *Numer. Funct. Anal. Optim.*, 31(1):45–77, 2010.
167. S. Roch and B. Silbermann. Algebras generated by idempotents and the symbol calculus for singular integral operators. *Integral Equations Operator Theory*, 11:385–419, 1988.
168. S. Roch and B. Silbermann. *Algebras of convolution operators and their image in the Calkin algebra. - Report*, volume 157. R-MATH-05-90 des Karl-Weierstrass-Instituts für Mathematik, Berlin, 1990.
169. S. Roch and B. Silbermann. Representations of non-commutative Banach algebras by continuous functions. *St. Petersburg Math. J.*, 3(4):865–879, 1992.
170. L.H. Rowen. *Polynomial identities in ring theory*, volume 84 of *Pure and Applied Mathematics*. Academic Press Inc., New York, 1980.
171. W. Rudin. *Functional analysis*. McGraw-Hill, New York, 1973. (second edition: MacGraw-Hill, Inc., New York, 1991).
172. N. Samko. Fredholmness of singular integral operators in weighted Morrey spaces. *Proc. A. Razmadze Math. Inst.*, 148:51–68, 2008.

173. P.A. Santos. Spline approximation methods with uniform meshes in algebras of multiplication and convolution operators. *Math. Nachr.*, 232:95–127, 2001.

174. P.A. Santos and B. Silbermann. Galerkin method for Wiener-Hopf operators with piecewise continuous symbol. *Integral Equations Operator Theory*, 38(1):66–80, 2000.

175. P.A. Santos and B. Silbermann. An approximation theory for operators generated by shifts. *Numer. Funct. Anal. Optim.*, 27(3–4):451–484, 2006.

176. R. Schneider. Integral equations with piecewise continuous coefficients in L_p-spaces with weight. *J. Integral Equations*, 9(2):135–152, 1985.

177. H. Schulze and B. Silbermann. One-dimensional singular integral operators on Hölder-Zygmund spaces. *Seminar Anal. 1988/89 of the Weierstrass-Institut Berlin*, pages 129–139, 1989.

178. I.E. Segal. Irreducible representations of operator algebras. *Bull. Amer. Math. Soc.*, 53:73–88, 1947.

179. V.N. Semenyuta and A.V. Khevelev. A local principle for special classes of Banach algebras. *Izv. Severo-Kavkazkogo Nauchn. Zentra Vyssh. Shkoly, Ser. Estestv. Nauk*, 1:15–17, 1977. (in Russian).

180. G.E. Shilov. On decomposition of a commutative normed ring in a direct sum of ideals. *Mat. Sb. (N.S.)*, 32(74(2)):353–364, 1953. (in Russian).

181. B. Silbermann. Lokale Theorie des Reduktionsverfahrens für Toeplitzoperatoren. *Math. Nachr.*, 104:137–146, 1981.

182. B. Silbermann. The C^*-algebra generated by Toeplitz and Hankel operators with piecewise quasicontinuous symbols. *Integral Equations Operator Theory*, 10:730–738, 1987.

183. B. Silbermann. Asymptotic invertibility of continuous functions of one-sided invertible operators. *Facta Univ. Ser. Math. Inform.*, 19:109–121, 2004.

184. G.F. Simmons. *Introduction to topology and modern analysis*. McGraw-Hill, Tokyo, 1963.

185. I.B. Simonenko. A new general method of investigating linear operator equations of singular integral equation type. I. *Izv. Akad. Nauk SSSR Ser. Mat.*, 29:567–586, 1965. (in Russian).

186. I.B. Simonenko. A new general method of investigating linear operator equations of singular integral equation type. II. *Izv. Akad. Nauk SSSR Ser. Mat.*, 29:757–782, 1965. (in Russian).

187. I.B. Simonenko. Convolution type operators in cones. *Mat. Sb. (N.S.)*, 74(116):298–313, 1967. (in Russian; English translation in *Mathematics of the USSR-Sbornik*, 3, 2: 279–293, 1967).

188. I.B. Simonenko. *The local method in the theory of shift-invariant operators and their envelopes*. Rostov. Gos. Univ., Rostov-na-Donu, 2007. (in Russian).

189. I.B. Simonenko and Chin' Ngok Min'. *A local method in the theory of one-dimensional singular integral equations with piecewise continuous coefficients. Noethericity*. Rostov. Gos. Univ., Rostov-na-Donu, 1986. (in Russian).

190. F. Smithies. Singular integral equations. *Proc. London Math. Soc.*, s2-46(1):409–466, 1940.

191. E.M. Stein and G. Weiss. *Introduction to Fourier analysis on Euclidean spaces*. Princeton University Press, Princeton, NJ, 1971. Princeton Mathematical Series, No. 32.

192. M. Takesaki. *Theory of operator algebras I*. Springer-Verlag, 2002.

193. E.C. Titchmarsh. *Introduction to the theory of Fourier integrals*. Oxford University Press, Oxford, 1967. (first edition, 1948).

194. J. Varela. Duality of C^*-algebras. In *Recent advances in the representation theory of rings and C^*-algebras by continuous sections (Sem., Tulane Univ., New Orleans, La., 1973)*, volume 148 of *Mem. Amer. Math. Soc.*, pages 97–108. Amer. Math. Soc., Providence, RI, 1974.

195. N.B. Vasil'ev. C^*-algebras with finite-dimensional irreducible representations. *Uspekhi Matem. Nauk*, 21(1(127)):135–154, 1966. (in Russian; English translation in *Russ. Math. Surv.* 21: 137–155, 1966).

196. N. Vasilevskii and I. Spitkovsky. On the algebra generated by two projections. *Dokl. Akad. Nauk Ukrain. SSR*, 8:10–13, 1981. (in Ukrainian).

197. N.P. Vekua. *Systems of singular integral equations and certain boundary value problems*. Izdat. "Nauka", Moscow, 1970. (in Russian).

198. J. von Neumann. Zur Algebra der Funktionaloperationen und Theorie der normalen Operatoren. *Math. Ann.*, 102(1):370–427, 1929.

199. J. von Neumann. On a certain topology for rings of operators. *Ann. of Math 2nd Ser.*, 37(1):111–115, 1936.

200. J. von Neumann. On rings of operators III. *Ann. Of Math. 2nd Ser.*, 41(1):94–161, 1940.

201. N. Wiener and E. Hopf. Über eine Klasse singulärer Integralgleichungen. *Sitzungsber. Preuss. Akad. Wiss., Phys.-Math. Kl*, pages 696–706, 1931.

202. W. Żelazko. *Banach algebras*. Elsevier Publ. Comp., Amsterdam, London, New York, 1973.

List of Notation

S. Roch et al., *Non-commutative Gelfand Theories*, Universitext,
DOI 10.1007/978-0-85729-183-7, © Springer-Verlag London Limited 2011

Index

S. Roch et al., *Non-commutative Gelfand Theories*, Universitext,
DOI 10.1007/978-0-85729-183-7, © Springer-Verlag London Limited 2011